Advances in Intelligent and Soft Computing 139

Editor-in-Chief: J. Kacprzyk

Advances in Intelligent and Soft Computing

Editor-in-Chief

Prof. Janusz Kacprzyk
Systems Research Institute
Polish Academy of Sciences
ul. Newelska 6
01-447 Warsaw
Poland
E-mail: kacprzyk@ibspan.waw.pl

Further volumes of this series can be found on our homepage: springer.com

Vol. 122. Yinglin Wang and Tianrui Li (Eds.)
Foundations of Intelligent Systems, 2011
ISBN 978-3-642-25663-9

Vol. 123. Yinglin Wang and Tianrui Li (Eds.)
Knowledge Engineering and Management, 2011
ISBN 978-3-642-25660-8

Vol. 124. Yinglin Wang and Tianrui Li (Eds.)
Practical Applications of Intelligent Systems, 2011
ISBN 978-3-642-25657-8

Vol. 125. Tianbiao Zhang (Ed.)
Mechanical Engineering and Technology, 2012
ISBN 978-3-642-27328-5

Vol. 126. Khine Soe Thaung (Ed.)
Advanced Information Technology in Education, 2012
ISBN 978-3-642-25907-4

Vol. 127. Tianbiao Zhang (Ed.)
Instrumentation, Measurement, Circuits and Systems, 2012
ISBN 978-3-642-27333-9

Vol. 128. David Jin and Sally Lin (Eds.)
Advances in Multimedia, Software Engineering and Computing Vol.1, 2011
ISBN 978-3-642-25988-3

Vol. 129. David Jin and Sally Lin (Eds.)
Advances in Multimedia, Software Engineering and Computing Vol.2, 2011
ISBN 978-3-642-25985-2

Vol. 130. Kusum Deep, Atulya Nagar, Millie Pant, and Jagdish Chand Bansal (Eds.)
Proceedings of the International Conference on Soft Computing for Problem Solving (SOCPROS 2011) December 20–22, 2011, 2012
ISBN 978-81-322-0486-2

Vol. 131. Kusum Deep, Atulya Nagar, Millie Pant, and Jagdish Chand Bansal (Eds.)
Proceedings of the International Conference on Soft Computing for Problem Solving (SocProS 2011) December 20–22, 2011, 2012
ISBN 978-81-322-0490-9

Vol. 132. Suresh Chandra Satapathy, P.S. Avadhani, and Ajith Abraham (Eds.)
Proceedings of the International Conference on Information Systems Design and Intelligent Applications 2012 (India 2012) held in Visakhapatnam, India, January 2012, 2012
ISBN 978-3-642-27442-8

Vol. 133. Sabo Sambath and Egui Zhu (Eds.)
Frontiers in Computer Education, 2012
ISBN 978-3-642-27551-7

Vol. 134. Egui Zhu and Sabo Sambath (Eds.)
Information Technology and Agricultural Engineering, 2012
ISBN 978-3-642-27536-4

Vol. 135. Honghua Tan (Ed.)
Knowledge Discovery and Data Mining, 2012
ISBN 978-3-642-27707-8

Vol. 136. Honghua Tan (Ed.)
Technology for Education and Learning, 2012
ISBN 978-3-642-27710-8

Vol. 137. Jia Luo (Ed.)
Affective Computing and Intelligent Interaction, 2012
ISBN 978-3-642-27865-5

Vol. 138. Gary Lee (Ed.)
Advances in Intelligent Systems, 2012
ISBN 978-3-642-27868-6

Vol. 139. Anne Xie and Xiong Huang (Eds.)
Advances in Electrical Engineering and Automation, 2012
ISBN 978-3-642-27950-8

Anne Xie and Xiong Huang (Eds.)

Advances in Electrical Engineering and Automation

Editors
Prof. Dr. Anne Xie
ISER Association
Guangzhou
China

Prof. Dr. Xiong Huang
ISER Association
Guangzhou
China

ISSN 1867-5662 e-ISSN 1867-5670
ISBN 978-3-642-27950-8 e-ISBN 978-3-642-27951-5
DOI 10.1007/978-3-642-27951-5
Springer Heidelberg New York Dordrecht London

Library of Congress Control Number: 2011945165

Printed on acid-free paper

Springer is part of Springer Science+Business Media (www.springer.com)

Preface

EEA2011 is the workshop of EECM2011. In the proceeding, you can learn much more knowledge about Electrical Engineering and Application of researchers all around the world. The main role of the proceeding is to be used as an exchange pillar for researchers who are working in the mentioned field. In order to meet high standard of Springer, the organization committee has made their efforts to do the following things. Firstly, poor quality paper has been refused after reviewing course by anonymous referee experts. Secondly, periodically review meetings have been held around the reviewers about five times for exchanging reviewing suggestions. Finally, the conference organization had several preliminary sessions before the conference. Through efforts of different people and departments, the conference will be successful and fruitful.

During the organization course, we have got help from different people, different departments, different institutions. Here, we would like to show our first sincere thanks to publishers of Springer, AISC series for their kind and enthusiastic help and best support for our conference.

In a word, it is the different team efforts that they make our conference be successful on December 24–25, 2011 Beijing, China. We hope that all of participants can give us good suggestions to improve our working efficiency and service in the future. And we also hope to get your supporting all the way. Next year, In 2012, we look forward to seeing all of you at EEA2012.

October 2011 EEA2011 Committee

Committee

Honor Chairs

Prof. Chen Bin	Beijing Normal University, China
Prof. Hu Chen	Peking University, China
Chunhua Tan	Beijing Normal University, China
Helen Zhang	University of Munich, China

Program Committee Chairs

Xiong Huang	International Science & Education Researcher Association, China
LiDing	International Science & Education Researcher Association, China
Zhihua Xu	International Science & Education Researcher Association, China

Organizing Chair

ZongMing Tu	Beijing Gireida Education Co. Ltd, China
Jijun Wang	Beijing Spon Technology Research Institution, China
Quanxiang	Beijing Prophet Science and Education Research Center, China

Publication Chair

Song Lin	International Science & Education Researcher Association, China
Xionghuang	International Science & Education Researcher Association, China

International Committees

Sally Wang	Beijing Normal University, China
LiLi	Dongguan University of Technology, China
BingXiao	Anhui University, China
Z.L. Wang	Wuhan University, China
Moon Seho	Hoseo University, Korea
Kongel Arearak	Suranaree University of Technology, Thailand
Zhihua Xu	International Science & Education Researcher Association, China

Co-sponsored by

International Science & Education Researcher Association, China
VIP Information Conference Center, China
Beijing Gireda Research Center, China

Reviewers of EEA2011

Chunlin Xie	Wuhan University of Science and Technology, China
LinQi	Hubei University of Technology, China
Xiong Huang	International Science & Education Researcher Association, China
Gangshen	International Science & Education Researcher Association, China
Xiangrong Jiang	Wuhan University of Technology, China
LiHu	Linguistic and Linguidtic Education Association, China
Moon Hyan	Sungkyunkwan University, Korea
Guangwen	South China University of Technology, China
Jack H. Li	George Mason University, USA
Marry Y. Feng	University of Technology Sydney, Australia
Feng Quan	Zhongnan University of Finance and Economics, China
PengDing	Hubei University, China
Songlin	International Science & Education Researcher Association, China
XiaoLie Nan	International Science & Education Researcher Association, China
ZhiYu	International Science & Education Researcher Association, China
XueJin	International Science & Education Researcher Association, China
Zhihua Xu	International Science & Education Researcher Association, China
WuYang	International Science & Education Researcher Association, China
QinXiao	International Science & Education Researcher Association, China
Weifeng Guo	International Science & Education Researcher Association, China
Li Hu	Wuhan University of Science and Technology, China,
ZhongYan	Wuhan University of Science and Technology, China
Haiquan Huang	Hubei University of Technology,China
Xiao Bing	Wuhan University, China
Brown Wu	Sun Yat-Sen University, China

Contents

Research of Multiple Internal Model Control Method Based on Fuzzy Neural Network

Gaohua Chen, Jinggang Zhang, Zhicheng Zhao, and Zheng Li

School of Electronic Information Engineering,
Taiyuan University of Science & Technology,
Taiyuan, China, P.R.
1978cgh@163.com

Abstract. For the key problems of conventional internal model control in the application, a highly robust multiple internal model controller is designed in the paper, it is a controller which suitable for large delay and parameter variation. The on-line intelligent tuning method based on fuzzy neural network for the multiple internal model controller was proposed. The theory analysis and simulation results show that the proposed method is simple, parameter tuning easy, and can make the system has a good target tracking performance, disturbance rejection properties and robustness.

Keywords: fuzzy neural network, internal model control, parameter tuning, robustness.

1 Introduction

Internal model control (IMC) is a very practical control method, its main feature is that structure is simple, design is simple, online adjusting few parameters and easy to adjust, the control effect is particularly significant to the control of the large time delay system and the improvement of robustness and disturbance rejection properties [1-2].The conventional IMC there is still some problems in the applications, such as IMC often occurs for the instability when the object is time-varying, IMC is ineffective when the model mismatch is too large or nonlinear serious[3].

In order to give full play to the superiority of IMC, so that it has more adaptability, a highly robust multiple IMC controller is designed in the paper, it is a controller which suitable for large delay and parameter variation. To avoid the tradeoff between robustness and dynamic property when adjusting the controller parameters, an on-line intelligent tuning method based on fuzzy neural network for the multiple IMC controller was proposed. Using the scheme, the parameter of the controller is online regulated according to the system error and its change, the system has a better target tracking performance, disturbance rejection properties and robustness.

A. Xie & X. Huang (Eds.): Advances in Electrical Engineering and Automation, AISC 139, pp. 1–7.
springerlink.com

2 Multiple Internal Model Control Method Based on Fuzzy Neural Network

2.1 Design of Multiple IMC Controller

IMC there is still some problems in the applications, in order to give full play to the superiority of IMC, so that it has more adaptability, a highly robust multiple IMC controller is designed in the paper, it is a controller which suitable for large delay and parameter variation. Multiple IMC structure is shown in Fig.1.

The structure consists of the controller and the multiple internal model compensator, where *Go(s)* is the controlled object, $\hat{G}_0(s)$ is the mathematical model of the object, *P(s)* is the compensation module.

The equivalent structure of Fig. 1 is shown in Fig.2.

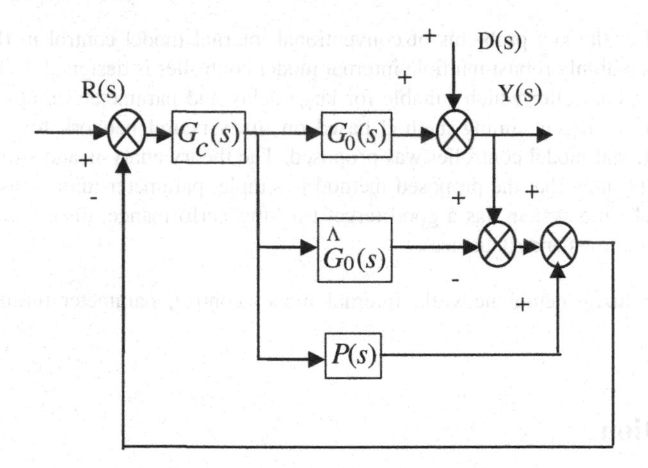

Fig. 1. Multiple internal model control structure

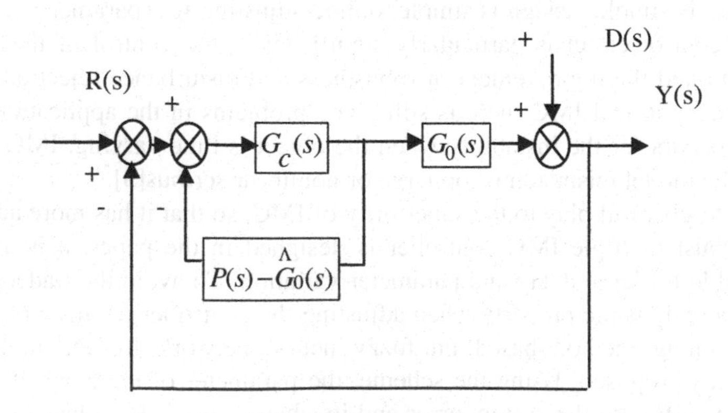

Fig. 2. The equivalent structure of multiple internal model control

The system transfer function as follows:

$$\frac{Y(s)}{R(s)} = \frac{G_c(s)G_0(s)}{1+G_c(s)P(s)+G_c(s)[G_0(s)-\hat{G}_0(s)]} \tag{1}$$

$$\frac{Y(s)}{D(s)} = \frac{1+G_c(s)[P(s)-\hat{G}_0(s)]}{1+G_c(s)P(s)+G_c(s)[G_0(s)-\hat{G}_0(s)]} \tag{2}$$

When the guarantee $\hat{G}_0(s) \approx G_0(s)$, in order to meet the conditions of the system without static error, in addition to $G_c(s)$ with integral functions, yet to meet:

$$\lim_{t\to T_s} L^{-1}[\frac{Y(s)}{R(S)}R(s)] = 1 \qquad \lim_{s\to 0}\frac{G_c(s)G_0(s)}{1+G_c(s)P(s)}R(s) = 1 \tag{3}$$

$$\lim_{t\to T_s} L^{-1}[\frac{Y(s)}{D(s)}D(s)] = 1 \qquad \lim_{s\to 0}\frac{G_c(s)\hat{G}_0(s)}{1+G_c(s)P(s)}D(s) = 1 \tag{4}$$

From the above analysis, it can be seen that the static gain of $\hat{G}_0(s)$, *P(s)* and *Go(s)* should be the same.

According to IMC design principle, $G_c(s)$ is designed as follows:

$$G_C(s) = F(s)\hat{G}_{0-}^{-1}(s) \qquad F(s) = 1/(\alpha ns+1) \tag{5}$$

Where $G_{0-}(s)$ contains minimum phase part of model, *F(s)* is low-pass filter, in general take a first-order filter, then *n=1*, in which α for filter parameter.

2.2 Intelligent Tuning of Controller Parameter

Control parameter α directly affects the stability and robustness of closed-loop system, but the election of the appropriate parameter to meet the performance requirements is very difficult.

Combine the fuzzy control and the neural network, and using the learning function of neural network to learn membership functions, fuzzy rules and other parameters, and can dynamically adjust the weights of fuzzy neural network to speed up the convergence rate of the entire neural network [4-5]. A fuzzy neural network system is designed in this paper, and input of network is deviation *e* and the changes of deviation *de*, in order to make the system also has a good target tracking performance and disturbance rejection properties, filter parameter α is on-line tuned through the output of the network. Fuzzy neural network structure is shown in Fig.3.

The role of the input layer is directly transmitting input values of variable x to the next layer of neurons. Input of network is deviation *e* and the changes of deviation *de*.

$$x = (x_1, x_2)^T = (e, de)^T \tag{6}$$

$$O_i^1 = x_i \quad i = 1,2 \tag{7}$$

Where O_i^1 denote the output of neurons *i* of the first layer.

The role function of fuzzy layer is Gaussian function, its role is fuzzy processing input variables. Fuzzy output value as follows:

$$O_{ij}^2 = e^{\frac{(x_i - c_{ij})^2}{\sigma_{ij}^2}} \quad (i = 1,2;\ j = 1,2,...m) \tag{8}$$

Where c_{ij} is the center of Gaussian function, σ_{ij} is the width of Gaussian function.

Each node of rules layer expresses a fuzzy rule. The output of each node can be expressed as follows:

$$O_k^3 = O_{1i}^2 \times O_{2j}^2 \quad (i = 1,2,...m \quad j = 1,2,...m \quad k = 1,2,...m \times m) \tag{9}$$

Where O_k^3 denote the output of neurons *k* of the third layer.

The output layer is expressed as follows:

$$y = \sum_{k=1}^{m \times m} w_k \times O_k^3 \quad (k = 1,2...m \times m) \tag{10}$$

Where w_k is weight value of the clear processing.

c_{ij}, σ_{ij} and w_k are corrected by using gradient descent method, learning algorithm is as follows:

Take the error function as follows:

$$E = \frac{1}{2}(y_d - y)^2 \tag{11}$$

Where y_d is the desired output, y is the actual output.

$$c_{ij}(k+1) = c_{ij}(k) - \eta \frac{\partial E}{\partial c_{ij}} \tag{12}$$

$$\sigma_{ij}(k+1) = \sigma_{ij}(k) - \eta \frac{\partial E}{\partial \sigma_{ij}} \tag{13}$$

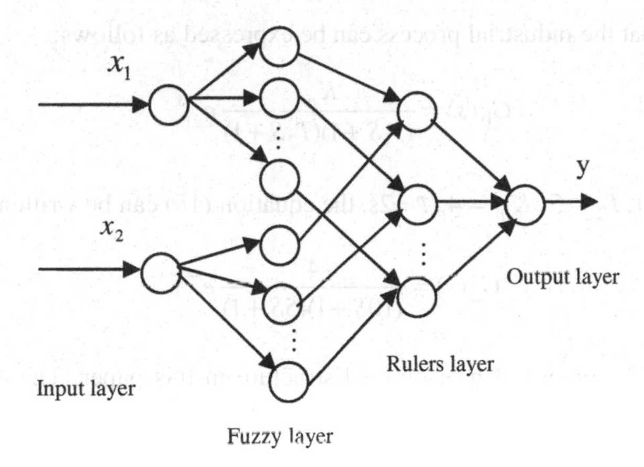

Fig. 3. Structure of fuzzy neural network

$$w_k(k+1) = w_k(k) - \eta \frac{\partial E}{\partial w_k} \tag{14}$$

Where η is learning efficiency.

Control structure of multiple IMC based on fuzzy neural network is shown in Fig.4.

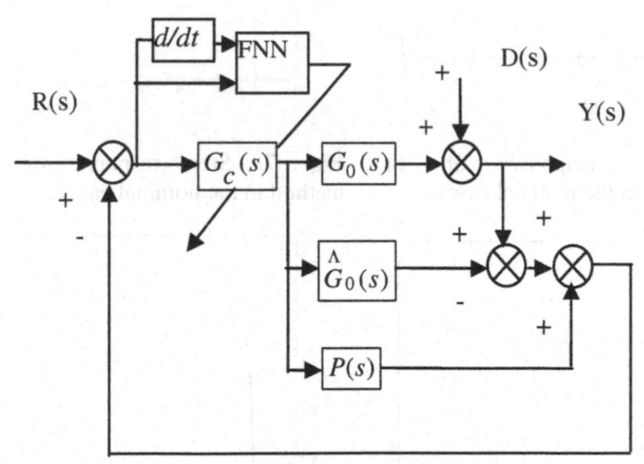

Fig. 4. Structure of multiple IMC based on fuzzy neural network

3 Experimental Simulation and Analysis

To demonstrate the effectiveness of the proposed method, the conventional IMC scheme and the proposed control scheme are used to the simulation research for an industrial process control system respectively.

Suppose that the industrial process can be expressed as follows:

$$G_0(s) = \frac{K_m}{(T_1S+1)(T_2S+1)} e^{-\tau s} \tag{15}$$

Where $T_1 = 10$, $T_2 = 5$, $K_m = 4$, τ =2s, the equation (15) can be written as follows:

$$G_0(s) = \frac{4}{(10S+1)(5S+1)} e^{-2s} \tag{16}$$

P(s) is taken the model of near-optimal structure in this paper. The equation as follows:

$$P(s) = \frac{4}{(10s+1)\ (5s+1)} \tag{17}$$

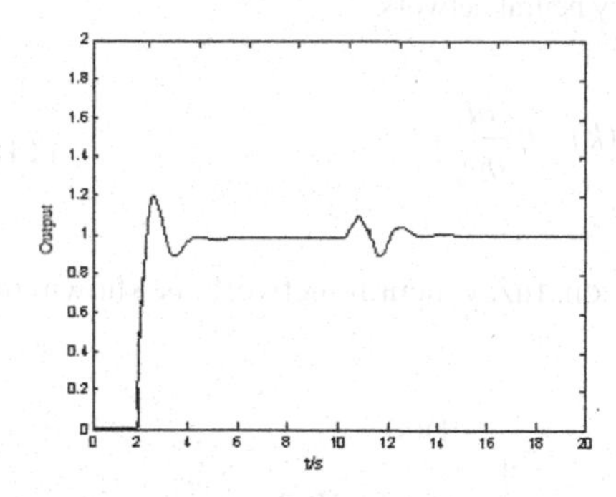

Fig. 6. Step response of the conventional IMC in the nominal cases

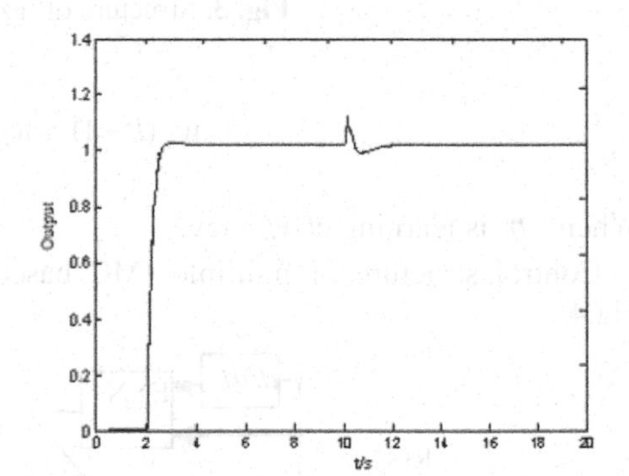

Fig. 7. Step response of the paper method in the nominal cases

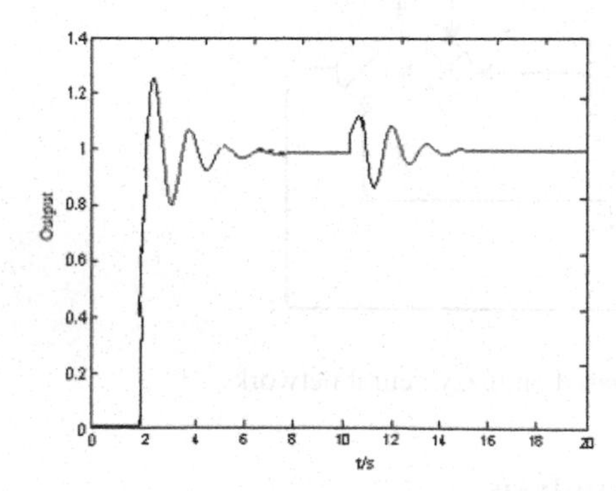

Fig. 8. Step response of the conventional IMC with 15%mismatches model in T_1 and T_2

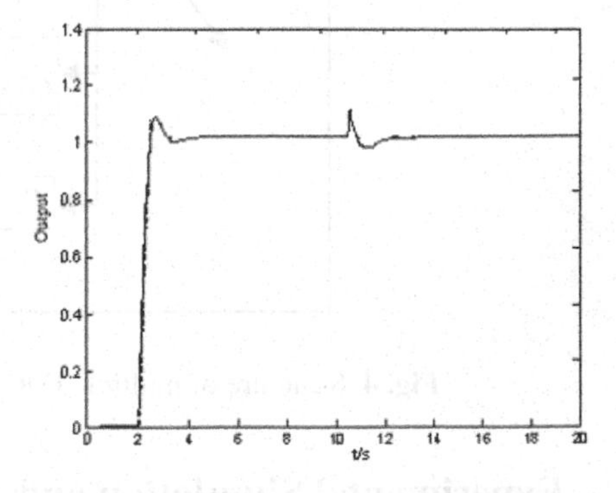

Fig. 9. Step response of the paper method with 15%mismatches model in T_1 and T_2

If target input of the system for $r(t) = \varepsilon(t)$, and adding disturbance input *d(t)* When *t=10s*, the amplitude of *d(t)* is taken as 0.5. When $G_0(s) = \hat{G}_0(s)$, the step responses of the conventional IMC is shown in Fig.6, the step responses of the paper method is shown in Fig.7, it is clearly seen that the proposed method to respond faster, small overshoot, disturbance rejection properties is better. when T_1 and T_2 increase 15% respectively, the step responses of the conventional IMC is shown in Fig.8, the step responses of the paper method is shown in Fig.9, the simulation results show that the proposed method can make the system has a good target tracking performance and robustness when the model mismatch.

4 Conclusion

For the shortage of conventional IMC, a highly robust multiple internal model controller is designed in the paper. And using fuzzy neural network, a controller parameters on-line intelligent tuning method was proposed. Compared with the conventional IMC, the proposed method in this paper to control algorithm is simple, parameter easy to adjust, small overshoot, disturbance rejection properties is better, and has a good target tracking performance and robustness when the model mismatch. The proposed method in this paper has good prospects and promotion of value.

Acknowledgement. This work is partially supported by university technology development project of Shanxi province, China (20091024) and UIT of Taiyuan University of Science & Technology (XJ2010046).

References

1. Zhao, Y.: A survey of development of internal model control. Information and Control 29(6), 526–531 (2000)
2. Shao, X., Zhang, J., Zhao, Z., Chen, Z.: Study of DC speed regulating system of double loop circuit. Journal of Electrical & Electrical Education 30(1), 75–78 (2008)
3. Arputha, J., Selvi, V., Radhakrishnan, T., Sundaram, S.: Model based IMC controller for processes with dead time. Instrumentation Science & Technology 34(4), 463–474 (2006)
4. Aoyama, A., Doyle III, F.J.: A fuzzy neural- network approach for nonlinear process control. Engineering Applications of Artificial Intelligence 8(5), 483–498 (1995)
5. Cui, C., Qu, W., Lv, D., et al.: Unknown Environment Based on Fuzzy Neural Network of Robot Path Planning. Transactions of Beijing Institute of Technology 29(8), 686–689 (2009)

An Improved Algorithm for Negative Inventory Balances in Inventory Cost Accounting under ERP Environment

LinKun Li[1], KaiChao Yu[1,*], and YiMing Li[2]

[1] Faculty of Mechanical and Electronic Engineering,
Kunming University of Science and Technology, Kunming 650093, China
[2] Kunming YUNNEI Power CO., LTD, Kunming 650024, China
glimmer_lin@live.cn

Abstract. In the manufacturing industry, negative inventory balances is an especially sticky subject for each enterprise. If enterprise does not allow negative inventory balances, that will affect transfer speed of logistics and information, but allowing negative inventory balances will exist inventory inaccurate risk. This paper mainly analyzed in the inaccurate problems caused by calculating inventory costs based on allowing negative inventory balances exists, and how to improved algorithm to solve them.

Keywords: Negative Inventory Balances, Cost Accounting, Moving Weighted Average Method, ERP.

1 Introduction

Inventory of the enterprise is a very important asset, which has large proportion in the current assets total amount of most enterprises. For inventory is directly bound up with purchase, consumption and sales, the quantity and value confirmation (i.e. inventory valuation) are crucially important to inventory management [1].The main methods of stock cost accounting in various occupations are moving weighted average method, FIFO and specific identification method. Different methods of inventory valuation lead to different influence of enterprise's financial position and operating results [2].

In the manufacturing sector, the common method of cost accounting is moving weighted average method. In the ERP or MIS, it won't appear error when storage department account cost by moving weighted average method. Since all inventory affairs data input need artificial type and adjustment, the negative inventory balances, caused by advance material issue notes or postpone material receipts notes in practice, would bring some problems as inaccurate stock price, inaccurate cost accounting in certain period.

2 Conventional Method of Inventory Cost Accounting

According to cost accounting problems caused by negative inventory balances in warehouse management, ERP or MIS has a solution. Taking the Oracle ERP as an

* Corresponding author.

A. Xie & X. Huang (Eds.): Advances in Electrical Engineering and Automation, AISC 139, pp. 9–16.
springerlink.com

example, the option of negative inventory balances in Oracle ERP determines whether inventory transactions can drive the inventory balances of an item negative [3]. For transactions that decrement inventory from a locator, such as issues and transfers, verify the result of the transaction does not drive the on-hand (packed or unpacked) or available quantity below zero. If negative inventory balances are allowed and the new on-hand quantity is negative or zero after the time of the receipt, inventory splits the transaction into two parts-- quantity from negative to zero and quantity from zero to positive--if necessary [4]. Upon a larger scale, this feature insures that the accounting transactions reported to the general ledger and inventory value reports balance at all times. For short-term transaction processing, deviation might appears in inventory value. From the microscopic view, inventory transaction processing influenced by negative inventory balances is analyzed below.

3 Inventory Cost Accounting Problems Caused by Negative Inventory Balances

Based on moving weighted average algorithm for calculation method of stock price, the new on-hand cost equals to the inventory average cost—weighted average cost of current average cost and receipt cost, after a material received.

New average cost = (current cost + receipt cost) / (on-hand quantity + receipt quantity)

The new average unit cost of the item is not affected by issue cost but unit cost fluctuation of receipt transaction. With exist of negative balances, average cost might have many unusual changes. Some examples are provided for revealing the impact of negative inventory balances below.

First, it is assumed that the average cost of the item is 50 and quantity is 100, so cost of inventory is 5000. Then the issue transaction is created, the issue quantity is 200, current quantity is -100 and inventory cost is -5000, so average cost is 50. Secondly, according to the traditional calculating formula above, it will show the variability of average cost caused by negative balances according to different receipt transactions below in brief.

3.1 Unit Cost of Receipt Transaction Equals to Current Average Cost

Transactions 1

Date	Receipt Quantity	Receipt Unit Cost	Issue Quantity	Average Cost	On-hand Quantity	Inventory Cost
2011-12-1				50	100	5000
2011-12-2			200	50	-100	-5000
2011-12-3	200	50		**50**	100	5000

When receipt unit cost (50) in 2011-12-3 equals to current (2011-12-2) average cost (50), the new (2011-12-3) average cost will not change whatever receipt quantity.

3.2 Unit Cost of Receipt Transaction Is Less Than Current Average Cost

Transactions 2

Date	Receipt Quantity	Receipt Unit Cost	Issue Quantity	Average Cost	On-hand Quantity	Inventory Cost
2011-12-1				50	100	5000
2011-12-2			200	50	-100	-5000
2011-12-3	200	40		**30**	100	3000

When unit cost (40) of receipt transaction is less than current average cost (50), new average cost (30) is decreased comparatively and less than receipt unit cost or current average cost.

3.3 Unit Cost of Receipt Transaction Is Less Than Current Average Cost and Receipt Quantity Equals to Absolute Value of Current On-hand Quantity

Transactions 3

Date	Receipt Quantity	Receipt Unit Cost	Issue Quantity	Average Cost	On-hand Quantity	Inventory Cost
2011-12-1				50	100	5000
2011-12-2			200	50	-100	-5000
2011-12-3	100	40		***(50)**	**0**	**-1000**

After receipt quantity writes off current on-hand quantity, because of the change of receipt unit cost, the new on-hand quantity is zero rather than new inventory cost, which means that there aren't materials in storage, but the inventory cost is exit (-1000).

3.4 Receipt Quantity Is More Than Receipt Quantity of Each Tree Transactions above and Unit Cost of Receipt Transaction Is Less than Current Average Cost

Transactions 4

Date	Receipt Quantity	Receipt Unit Cost	Issue Quantity	Average Cost	On-hand Quantity	Inventory Cost
2011-12-1				50	100	5000
2011-12-2			200	50	-100	-5000
2011-12-3	300	40		**35**	200	7000

Compare this with Transactions 1, if receipt unit cost does not change and receipt quantity goes up, the average cost will be increased accordingly.

3.5 Reversing the Order of Receipt Transaction and Issue Transaction of Transactions 4

Transactions 5

Date	Receipt Quantity	Receipt Unit Cost	Issue Quantity	Average Cost	On-hand Quantity	Inventory Cost
2011-12-1				50	100	5000
2011-12-2	300	40		42.5	400	1700
2011-12-3			200	**42.5**	200	8500

Compare this with Transaction 4, just only reversing the order of transactions by first receipt transaction; the last average cost becomes 42.5 which are different from 35 of Transaction 4.

3.6 Receipt Quantity Is Less Than Receipt Quantity of Each Five Transactions above and Unit Cost of Receipt Transaction Goes Up

Transactions 6

Date	Receipt Quantity	Receipt Unit Cost	Issue Quantity	Average Cost	On-hand Quantity	Inventory Cost
2011-12-1				50	100	5000
2011-12-2			200	50	-100	-5000
2011-12-3	120	60		**110**	20	2200

Compare this with five transactions above, with receipt unit cost rises 10, the last average cost rises to 110 which is above double original average cost (50) and nearly two times of receipt unit cost (60).

3.7 Receipt Quantity Is Less Than Receipt Quantity of Each First Five Transactions above and Unit Cost of Receipt Transaction Falls

Transactions 7

Date	Receipt Quantity	Receipt Unit Cost	Issue Quantity	Average Cost	On-hand Quantity	Inventory Cost
2011-12-1				50	100	5000
2011-12-2			200	50	-100	-5000
2011-12-3	120	35		**-40**	20	-800

Compare this with fist five transactions above, with receipt unit cost falls 15, the last average cost is negative balance, which means that issue transaction later will issue cost at same time.

3.8 Enlarge the Effects of Receipt Transaction of Two Transactions above

Transactions 8

Date	Receipt Quantity	Receipt Unit Cost	Issue Quantity	Average Cost	On-hand Quantity	Inventory Cost
2011-12-1				50	100	5000
2011-12-2			200	50	-100	-5000
2011-12-3	101	60		**1060**	1	1060

Transactions 9

Date	Receipt Quantity	Receipt Unit Cost	Issue Quantity	Average Cost	On-hand Quantity	Inventory Cost
2011-12-1				50	100	5000
2011-12-2			200	50	-100	-5000
2011-12-3	101	35		**-1465**	1	-1465

Compare these with two transactions above, receipt quantity down to nearly absolute value of current on-hand quantity (-100), the average costs appear greater deviation.

4 The Reason of Deviation of Average Cost

Through the analysis of transactions above, with the influence of negative inventory balances, because of different receipt quantity and unit cost, last average cost in nine transactions appears obvious deviation with receipt cost and current cost in various degrees.

In effect, most negative inventory balances problems are caused by failures in management. In some urgent shipment case, warehouse keeper has marked the out-bound order, but godown entry hasn't done by purchasing department or manufacture department. In other words, material issue notes have done before its receipt notes caused negative inventory balances and average cost variance.

From the perspective of transaction cost, the cost variance between issue cost and receipt cost for negative balance parts is allocated to actual receipt item (last on-hand balance), which balances inventory cost. The less last on-hand quantity, the larger cost allocated to it. According to Transaction 9 above, this cost variance is almost forty times higher than receipt unit cost. Actually, the cost variance can be hundreds times or more. But for a long time, because of continuously receipt material, it keeps average costs and physical inventory counts in normal level. However, if negative inventory balances still exist in ERP, the short term average cost will also fluctuate with receipt quantity, which cause that managers can't control the recent inventory cost accurately.

5 Improved Algorithms for Negative Inventory Balances

In order to address the impact of short term average cost fluctuation caused by negative balances, it is possibly for managers to carry out more management measures in general affairs to avoid negative balances. Strengthen cooperation and communication with various departments; it will keep logistics and information flow accuracy and timeliness and reduce the inventory cost distortion [5]. But it also badly influences the convenience of inventory transactions. Moreover, the negative inventory balances inevitably appear in early intervention in ERP or MIS.

For solving the problem with inventory cost inaccuracy caused by negative balances, the original moving weighted average method is improved to reflect the average cost deviation and modify the average cost while negative balance exist. By introducing the deviation below, it is easy to separate the inaccurate cost from inventory cost. The deviation can also be used as basis of issue cost for cost regulation.

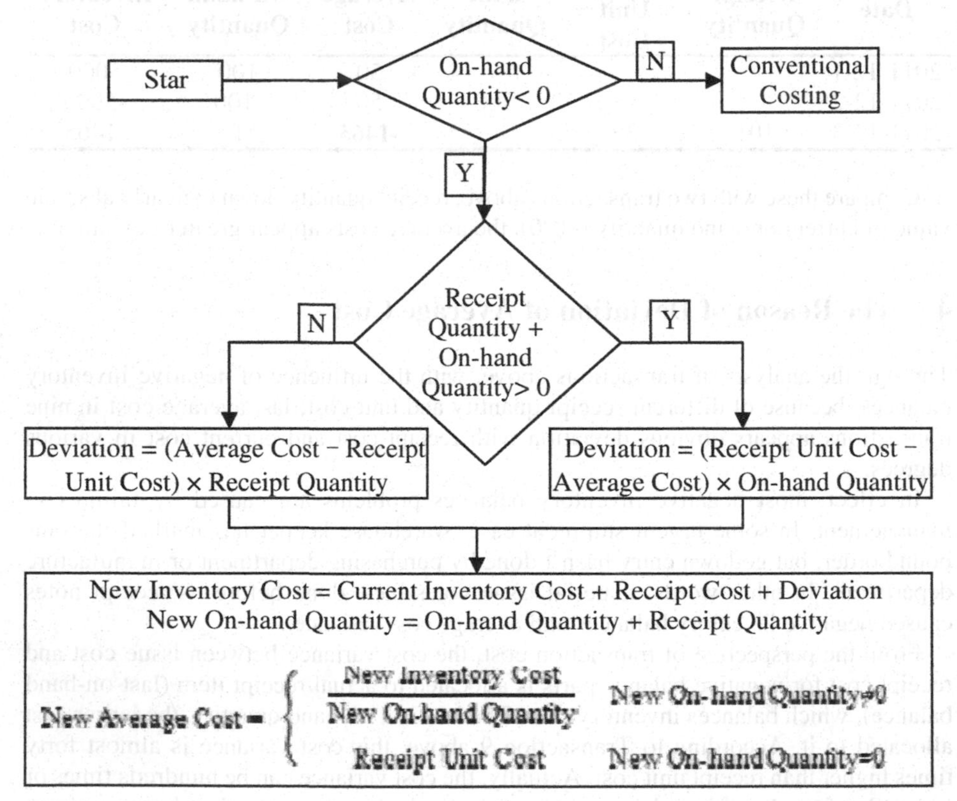

Finally, Transactions 8 and 9 above are verified by using this improved algorithm and results follow as:

Improved Transaction 8

Date	Receipt Quantity	Receipt Unit Cost	Issue Quantity	Average Cost	On-hand Quantity	Inventory Cost
2011-12-1				50	100	5000
2011-12-2			200	50	-100	-5000
2011-12-3	101	60		**60**	1	60

Deviation: -1000

Improved Transaction 9

Date	Receipt Quantity	Receipt Unit Cost	Issue Quantity	Average Cost	On-hand Quantity	Inventory Cost
2011-12-1				50	100	5000
2011-12-2			200	50	-100	-5000
2011-12-3	101	35		**35**	1	35

Deviation: 1500

This improved algorithm mainly solves the inventory average cost inaccuracy problems caused by negative inventory balances. Average cost can be adjusted for normal inventory average cost by deviation value. In the ERP, if the receipt quantity completely offset the negative balance and inventory also has a surplus, the calculation result of new average cost equal to receipt unit cost. Otherwise, if receipt quantity cannot offset the negative balance, the potential inaccuracy still exists. But it is unusual in practice. Because the negative balances caused by advance material issue item notes or other reasons made by actually demand—rather than the mistake of warehouse keepers—are wrote off by receipt notes at one time. Even though the negative balances cannot be wrote off at one time, the average cost will be regulated by later receipt notes though this algorithm until on-hand quantity is positive.

The Deviation is not virtual value in computational process. When Deviation is positive, the receipt unit cost is less than current average cost and the early issue material caused negative balance was delivered as more cost than actual inventory cost. Therefore, the practical significance of Deviation is the cost the storage owe receiving department. The Deviation finally need come to its destination, because it is outstanding account. There are some steps that can be taken in order to handle this Deviation—charged off by financial department from purchasing department reports, taken into finished cost when receiving department accounting cost, or marked and reported to approver by warehouse keeper, which cannot influence regular inventory affairs and convenience of transactions.

References

1. Nie, Q.: The Influence of Inventory Cost Accounting on Inventory Cost. Guide to Business 1, 125–126 (2010)
2. Su, X.: The Influence of Inventory Cost Accounting Method Change on Business Accounting. Commercial Accounting 13, 25–26 (2006)
3. Oracle Inventory User's Guide Part No. A83507-10, (DB/OL) (February 2011), http://download.oracle.com/docs/cd/B25516_18/current/acrobat/115invug.zip
4. Negative Inventory Balances (DB/OL) (February 2011), http://download.oracle.com/docs/cd/A60725_05/html/comnls/us/cst/recalc08.htm
5. Li, R.: Recognition of "Moving Weighted Average Method" for Stock Accounting. Sci. / tech. Information Development & Economy 2 (2004)

Improved Simulated Annealing Algorithm Used for Job Shop Scheduling Problems

Shao-zhong Song[1,2], Jia-jun Ren[2], and Jia-xu Fan[3]

[1] Department of Information Engineering, Jilin Business and Technology College, Changchun 130062, China
[2] College of Mechanical Science and Engineering, Jilin University, Changchun, 130022, China
[3] College of Computer Science and Technology, Jilin University, Changchun, 130012, China
shshry@163.com, 75737398@qq.com, jiaxu_fun@hotmail.com

Abstract. In dealing with shop scheduling, by optimizing solver to improve the simulated annealing algorithm, it can be solved quickly and effectively. The stability of the mean completion time and the output of optimal completion time were compared with other simulated annealing algorithm.

Keywords: simulated annealing algorithm, job shop scheduling, optimization solution.

1 Introduction

Since 1954, Johnson published his first paper on the production scheduling ,the scheduling problem of manufacturing systems have gone through a single machine scheduling, multi-machine scheduling, flow shop scheduling, job shop scheduling, flexible scheduling manufacturing systems from simple to complex development process [6]. Currently there are many different types of scheduling model, generally can be divided into two categories: deterministic models and stochastic models. In each category, the scheduling model is further subdivided into single model, multi-machine models, flow shop and job shop models etc [5]. Scheduling problem is a combinatorial optimization problem, so scheduling solution to the problem to a large extent dependent on the scheduling algorithm. This paper attempts to use the improved simulated annealing algorithm for job shop scheduling problem.

2 Principle and Realization

Because the job shop scheduling problems are strongly NP hard problems [2], the classic simulated annealing algorithm does not guarantee high search efficiency and always converge to global optimal solution.Based on the certain requirements to ensure optimal quality of characteristics of the issue and improve the efficiency of simulated annealing search.

Sign Convention: Scurr-The current state of the annealing process, Sbest-The optimal state of Annealing process , fcurr-the current objective function value of

A. Xie & X. Huang (Eds.): Advances in Electrical Engineering and Automation, AISC 139, pp. 17–25.
springerlink.com

annealing process, p-control parameters of annealing process ,fbest-the optimal objective function value of annealing process, Si-The current state of the sampling process, Sj-Neighborhood state of Sj, Sopt -optimal state of sampling process fopt-optimal objective function value of sampling process L-Markov chain length, q-Sampling process control parameters.

2.1 Improved Annealing Process

Step 1: For a given initial temperature t_0, randomly generated initial state S_0 ; order $S_{curr}=S_{best}=S_0$ calculated f_{curr} and to make $f_{best}=f_{curr}$; order p=0, i=0.

Step 2: Order $t=t_i$ use t, S_{best}, S_{curr}call the following improvements in the sampling process, and record the optimal state of the sampling process and the optimal objective function value so that $S_{curr}=S_{opt}$, $f_{curr}=f_{opt}$.

Step 3: Judge $f_{best}<f_{curr}$ is set up, and if so, make p=p+1 ; otherwise $S_{best}=S_{curr}$, $f_{best}=f_{curr}$, set p=0.

Step 4: Return temperature $t_{i+1}=at_i$, $0<a<1$ make i=i+1 .

Step 5: Judge $p>a$ or $t_i<t_e$ is set up, if so, output S_{best}, $f_{best}$and end ; otherwise turn to step 2. Improve the annealing process flow shown in Figure 1

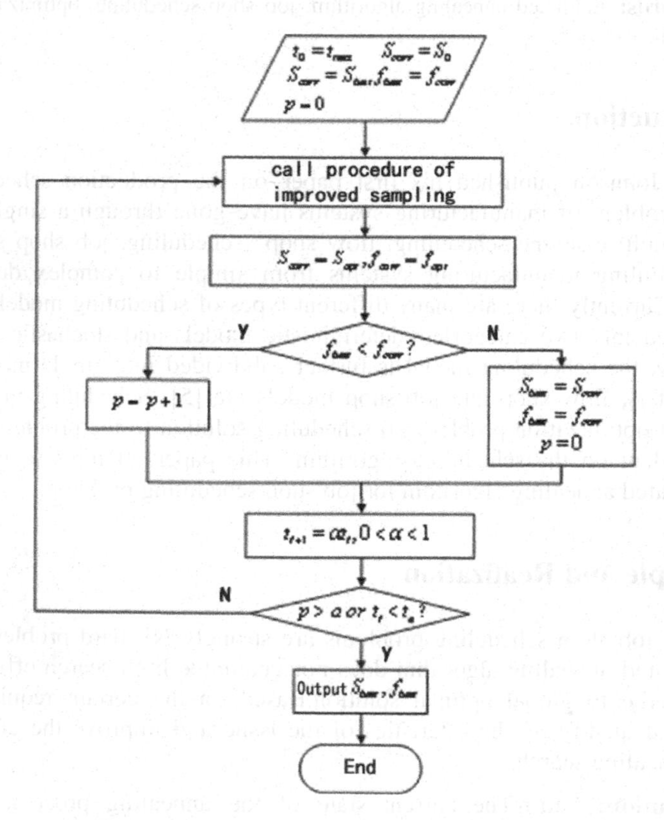

Fig. 1. Flow chart of improved annealing process

2.2 Improved Sampling Process

Step 1: Read from the current state of annealing process $S_i=S_{curr}$,$S_{opt}=S_{best}$, $f_{opt}=f_{curr}$, make q=0 ,k=0.

Step 2: Build neighbourhood state S_j, and calculate $\Delta f_{ij}=f(S_j)-f(S_i)$;judge $\Delta f_{ij}\le 0$ is set up, if so, turn to step 3; otherwise turn to step 4.

Step 3: Order $S_i=S_j$ to judge $f_{opt}<f(S_j)$ is set up, and if so , q=q+1, turn to step 5; otherwise make $S_{opt}=S_j$, $f_{opt}=f(S_j)$, q=0 , turn to step 5.

Step 4: Judge $\exp\left(-\frac{\Delta f_{ij}}{t_i}\right) > random(0,1)$, If so $S_i=S_j$ otherwise turn to step 5.

Step 5: Judge q>b or k>L is set up, if so return S_{opt} , f_{opt} ,turn into the annealing process; otherwise, turn to step 2.

Improved sampling process flow shown in Figure 2.Here annealing control parameter a and sampling the control parameters b were taken a=2n,b=5n n as the number of parts.

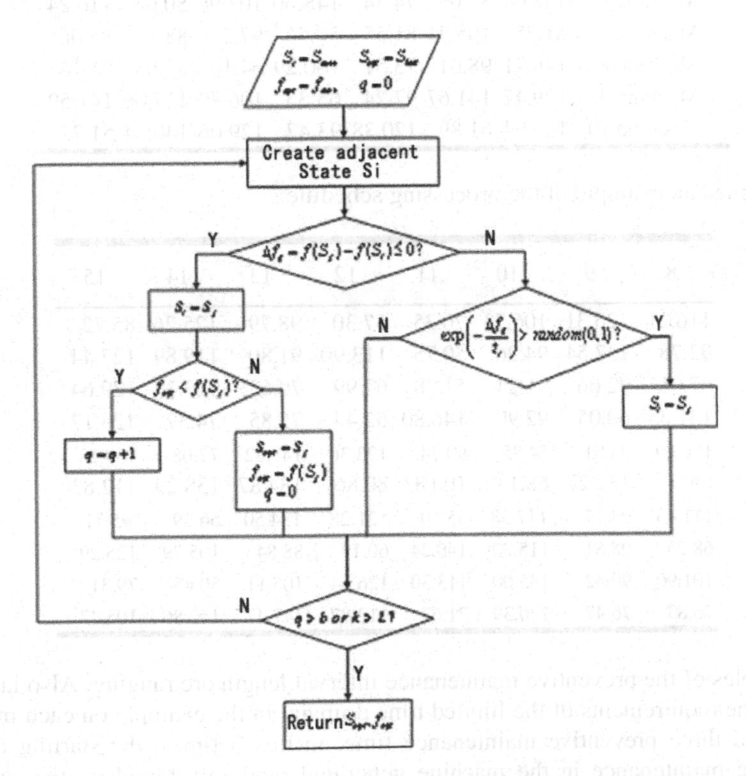

Fig. 2. Flow chart of the improved sampling process

3 Numerical Example

Now consider the availability of flow shop scheduling problem example lack appropriate benchmark to test the performance of the algorithm. Therefore, [1] gives the scheduling of data generation method, randomly generated example data needed data. An example here is designed as a 15 piece in the 10 machine scheduling problems; the work piece in thc machine processing time is shown in Table 1.

Table 1. An example of the processing schedule

Workpiece / Machine	1	2	3	4	5	6	7
Machine 1	77.08	101.91	106.94	109.53	125.13	89.65	102.55
Machine 2	67.34	78.07	59.19	61.89	71.66	120.75	147.00
Machine 3	64.36	96.66	144.33	136.78	140.51	86.62	64.02
Machine 4	136.14	60.97	121.55	71.07	125.89	123.53	74.95
Machine 5	124.18	61.68	86.15	51.77	59.03	93.30	99.31
Machine 6	148.04	87.05	74.74	148.60	103.96	80.64	116.24
Machine 7	51.75	105.71	81.55	56.56	97.21	88.11	85.06
Machine 8	146.41	98.01	73.51	100.29	64.10	83.64	96.14
Machine 9	129.47	141.67	97.24	65.35	106.77	117.06	141.59
Machine 10	130.94	61.89	120.38	93.47	129.06	146.84	51.72

Continued an example of the processing schedule.

8	9	10	11	12	13	14	15
116.21	126.31	106.73	70.35	57.30	98.79	125.26	85.72
92.78	137.54	94.26	50.75	113.90	91.80	119.89	127.44
67.01	72.66	91.24	51.78	67.99	70.63	123.38	129.64
131.35	60.05	92.90	146.80	62.44	75.85	74.37	123.17
130.09	58.81	54.85	99.74	123.36	148.82	77.08	138.28
65.63	132.27	68.13	100.83	80.86	134.87	138.29	112.83
111.43	93.17	117.58	55.10	121.28	134.50	66.99	95.71
68.25	98.81	115.30	140.24	60.19	88.84	105.79	125.29
101.90	99.42	135.00	113.30	126.64	105.11	50.45	79.31
76.83	76.47	120.39	71.67	112.87	147.47	142.86	105.47

Examples of the preventive maintenance interval length are ranging. Also take into account the requirements of the limited time domain, in the example on each machine considered three preventive maintenance time, in this 3 times, the starting time of preventive maintenance in the machine generated randomly based on the selection process in time domain, The length of time for each preventive maintenance is equal to average processing time of work piece in the machine (sample obtained by the normal distribution) as shown in Table 2.

Table 2. Examples of non-time table

Unavailable time Machine	Period 1	Period 2	Period 3
Machine 1	[395.46,501.48]	[955.31,1054.86]	[1288.48,1387.58]
Machine 2	[338.15,436.47]	[836.22,931.64]	[1095,1189.42]
Machine 3	[442.14,538.96]	[964.20,1060.22]	[1015,1110.66]
Machine 4	[136.14,227.24]	[845.45,941.28]	[1283.50,1387.43]
Machine 5	[323.78,418.05]	[476.11,570.48]	[1268.17,1368.18]
Machine 6	[235.09,340.41]	[957.16,1063.29]	[1025.29,1131.27]
Machine 7	[239.01,324.93]	[677.38,764.91]	[1199.00,1292.55]
Machine 8	[146.41,245.99]	[944.47,1038.43]	[1233.73,1340.17]
Machine 9	[433.73,545.70]	[657.56,764.09]	[1480.51,1588.92]
Machine 10	[130.94,238.26]	[535.74,650.40]	[1340.00,1445.78]

Because of the large solution space of the example, solution, need to make some amendments to the cooling schedule, the specific parameters are as shown in Table 3. Scheduler runs a scheduling result shown in Figure 3- Figure 4.

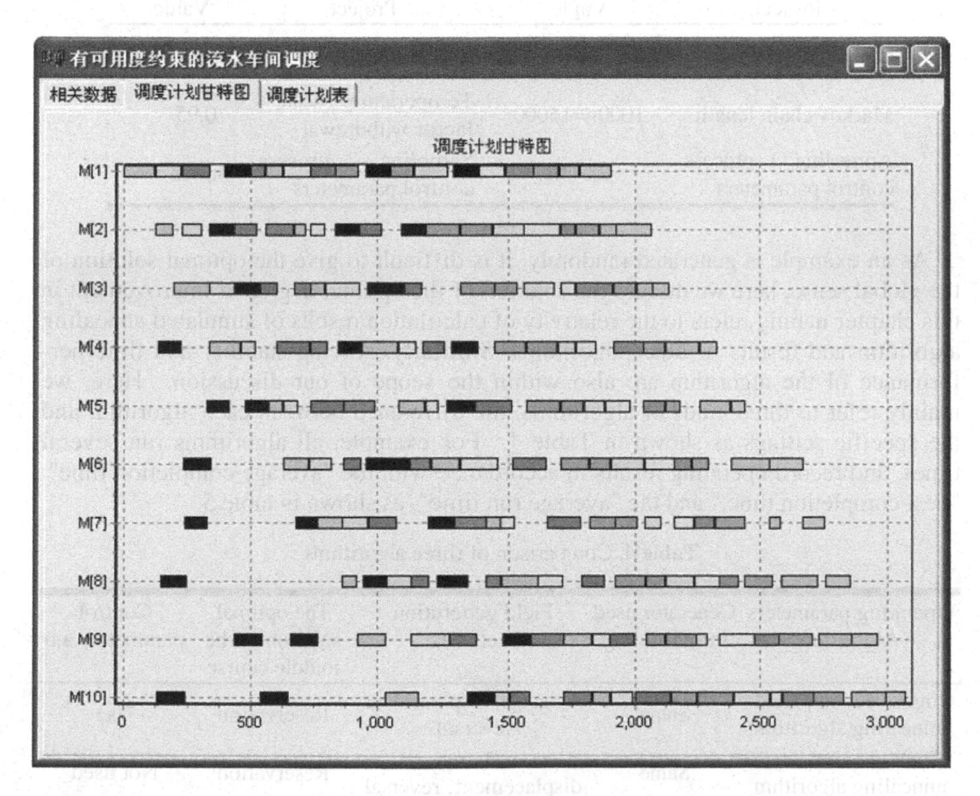

Fig. 3. Machine scheduling schedule (example)

有可用度约束的流水车间调度

相关数据 | 调度计划甘特图 | 调度计划表

最优完成时间为：3070.69970703125

Machine[1]	5	2	8	12	11	4	13	15	6	14
Machine[2]	5	2	8	12	11	4	13	15	6	14
Machine[3]	5	2	8	12	11	4	13	15	6	14
Machine[4]	5	2	8	12	11	4	13	15	6	14
Machine[5]	5	2	8	12	11	4	13	15	6	14
Machine[6]	5	2	8	12	11	4	13	15	6	14
Machine[7]	5	2	8	12	11	4	13	15	6	14
Machine[8]	5	2	8	12	11	4	13	15	6	14
Machine[9]	5	2	8	12	11	4	13	15	6	14
Machine[10	5	2	8	12	11	4	13	15	6	14

Fig. 4. Gantt chart scheduling plan (examples)

Table 3. An example of the cooling schedule

Project	Value	Project	Value
Onset temperature	Initial acceptance probability of 0.98	End temperature	0.000001
Markov chain length	1000n=15000	Temperature scale factor withdrawal	0.95
Annealing process control parameters	2n=30	Sampling process control parameters	5n=75

As an example is generated randomly, it is difficult to give the optimal solution of the global sense, here we discuss the solution of the optimal degree of improvement in this chapter mainly refers to the relativity of calculation results of simulated annealing algorithm and results of other algorithms. Similarly, solving stability and time performance of the algorithm are also within the scope of our discussion. Here, we mainly refer to three kinds of algorithms, the difference between each algorithm and the specific settings as shown in Table 4. For example, all algorithms run several times, and record operating results in accordance with the "average completion time", "best completion time," and the "average run time", as shown in table 5.

Table 4. Comparison of three algorithms

Operating parameters Algorithm	Generator used in scheduling	Field generation methods	The optimal solution of the middle course	Control parameters a,b
Improved simulated annealing algorithm	Same	exchange, displacement, reversal	Reservation	Use
Standard simulated annealing algorithm	Same	exchange, displacement, reversal	Reservation	Not used
Classical simulated annealing algorithm	Same	exchange	Not retained	Not used

Table 5. Comparison of results from three algorithms

Algorithm	Run times	Mean completion time	Optimal completion time	Average run time (seconds)
Improved simulated annealing algorithm	25	3068.17	3049.71	0.17
Standard simulated annealing algorithm	5	2889.70	2888.00	4620.00
Classical simulated annealing algorithm	5	3101.43	2889.18	4325.15

4 Discussion

From the output point of view of optimal degree results, the standard simulated annealing algorithm were ahead of the other two methods in the mean completion time and the optimal completion time, mainly due to the sampling step length of 15,000 and retained optimal value occurs in the sampling process. Although the classical simulated annealing algorithm also uses a sample length of 15,000, but it doesn't reserve the "so far optimal value" in the sampling process, moreover, because adjacent field generated by a single two exchange operators, leads to the algorithm a lower searching performance.

Judging from the stability of output, difference of the average completion time and optimal completion time in improved simulated annealing algorithm is 18.46, while the standard simulated annealing algorithm is only 1.70, almost negligible. Relative to the first two, difference of average completion time and optimal completion time of classic simulated annealing algorithm is 212.25. We think it is the mechanism of "keep the optimal value so far" that makes the first two algorithms results more stable than the third.

Judging from the time efficiency of the algorithm, run time of the improved simulated annealing algorithm is only 0.17 seconds, while the standard simulated annealing algorithm needed about 1 hour and 30 minutes, although the classical simulated annealing algorithm simplifies the field generating operator and also shorten the computing process, the actual computing time has not significantly reduced however. Overall, the improved simulated annealing algorithm is faster than the latter two almost 30,000 times. Mainly due to improved simulated annealing algorithm does not determined by the termination of the temperature and the number of Markov chain length uniquely, the annealing process control parameters And the sampling process control parameters have great influence on the efficiency of the algorithm, which will be further discussed later.

To compare three types of indicators above, the classic simulated annealing algorithm is the worst, in the following discussion will no longer be involved.While the former two, two indicators in the three indicators indicates the performance of simulated annealing algorithm is inferior to the standard simulated annealing algorithm, the improved simulated annealing algorithm does not seem to be more desirable than the standard simulated annealing algorithm. However we should point out that the standard simulated annealing algorithm consumes nearly 30,000 times longer only to optimize results of the 5.82% increase. It can be inferred from the calculation flow of improved annealing algorithm introduced above that the key to improve the performance of simulated annealing algorithm is the value of annealing process control parameters. With the comparison to standard simulated annealing algorithm, the improved simulated annealing algorithm requires further consideration of relationship between calculation results optimal degree and run-time efficiency. Therefore, we conducted a series of tests based on improved simulated annealing algorithm. Operating parameters and operating results as shown in Table 6.

Table 6. Different parameters operating results of improved simulated annealing algorithm

Test number	Annealing process control parameters	Sampling process control parameters	Run times	Mean completion time	Optimal completion time	Average run time (seconds)
1	5n=75	10n=150	25	3048.97	3031.24	0.31
2	10n=150	20n=300	25	3030.81	3022.65	0.51
3	20n=300	40n=600	25	3013.79	3005.88	1.41
4	40n=600	80n=1200	25	3002.12	2991.31	6.13
5	80n=1200	160n=2400	25	2994.26	2987.41	21.05

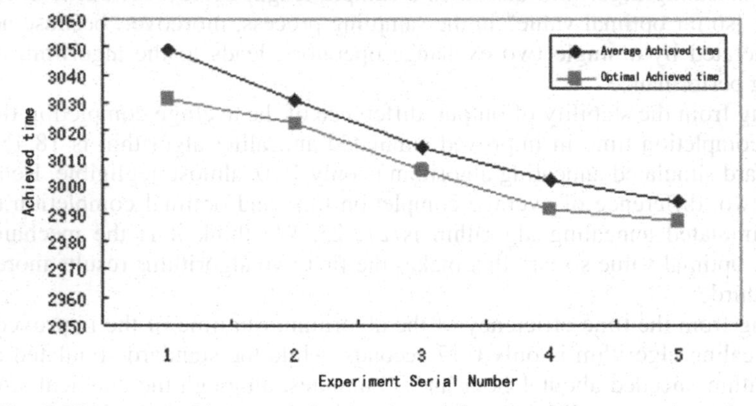

Fig. 5. The output of improved simulated annealing algorithm comparison chart

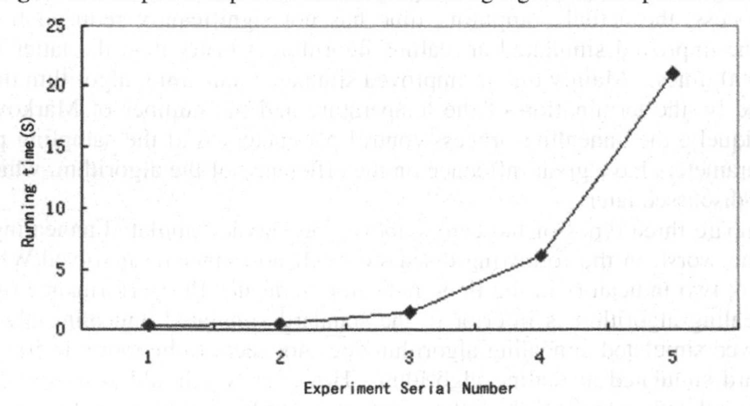

Fig. 6. Algorithm running time comparison chart

It can be drawn from the operating results of Table 6, annealing process control parameters and the sampling process control parameters improve the performance of simulated annealing is more evident. With increases, the mean and the optimal completion time of algorithm outputs were gradually reduced, that optimal degree of algorithm results increase; at the same time, the time is rapidly rising when each time algorithm runs, as is shown in Figure5,6. When the annealing process control parameters is more than 600, the sampling process control parameters is more than 1,200, the improvement of the results of previous tests is not so obvious; when the annealing process control parameters more than 1,200, the sampling process control parameters is more than 2,400 after measured the output of the algorithm is stably around 3,000. Compared with the standard simulated annealing algorithm to improve the simulated annealing algorithm for optimal results with a difference of 3.43% is still, fortunately, improved simulated annealing algorithm is still leading in the run time, almost faster than the standard simulated annealing out of 200 times.

Acknowledgement. The authors gratefully acknowledge the funding Jilin province Education Science "Eleventh Five Years" Scheme Item (2008275,2009322, 2010222,2010224).

References

1. Demirkol, E., Mehta, S., Uzsoy, R.: Benchmarks for shop scheduling problems. European Journal of Operational Research 109, 137–141 (1998)
2. Ruiz, R., Garcia-Diaz, J.C., Maroto, C.: Considering scheduling and preventive maintenance in the flow-shop sequencing problem. Computers & Operations Research 34, 3314–3330 (2007)
3. Xiao, S., Lu, H., Fan, A., Song, H.: Applied Research of Simulated Annealing Algorithm on Solving Combinatorial Optimization Problems. Journal of Sichuan University of Science & Engineering (Natural Science Edition) 23, 116–118 (2010)
4. Tao, Q., Zou, Q.: Simulated annealing algorithm for two dimensional placement problems. Journal of Hunan University of Arts and Science (Science and Technology) 21, 11–13 (2009)
5. Pinedo, M., Zhang, Z. (translation).: Scheduling theory, algorithms and systems, 2nd edn. Tsinghua University Press (2007)
6. Wu, Q., Qiao, F., Li, L., Wang, Z.: Semiconductor manufacturing systems scheduling. Electronic Industry Press (2006)

Convex Geometry on Partially Ordered Sets

YaoLong Li

College of Mathematics and Information Science,
Weinan Teachers University,
714099, Weinan, Shaanxi, China
liyaolong188@yahoo.com.cn

Abstract. The definition the closure operator for poset covex geometry are presented, the closure axioms of convex geometry on partially ordered sets is given, and a characterestic for the convex geometry on partially ordered sets is investigated. Finally, we study some properties of the convex geometry on partially ordered sets.

Keywords: poset convex geometry, rank function, feasible set, rank axioms of poset convex geometry.

1 Introduction

The theory of matroids was introduced by Whitney as a combined abstraction of linear independence and the cycle structure of graphs. This theory has been used in many fields such as Lattice Theory, Combinatorial Optimization and Greedy Algorithm. The latter was then shown by what called meet-distributive lattice-a concept equivalent to antimatroids. Different notions of convexity in groughs, directed groughs and ordered sets were linked to convex geometry.

The theory of poset matroids is another extension of the classial matroids, which was developed by replacing the underlying sets of amatroids by a partially ordered set[2,3]. Consequently the notion the subsets of the underlying sets is replaced by that of the filters. Since antimatroid is With the same ideal as this, we presented the theory the poset convex geometry.

In this paper, we mainly investigate some properties of the rank function and study the rank axioms for poset convex geometry.

Let $(P,\leq)$ be a finite partially ordered set. For any given subset A of P. we define

$$\mathrm{Max}\,(A)=\{x\in A: x \text{ is maximal in } A\},$$

$$\mathrm{Min}\,(A)=\{x\in A: \text{x is minimal in } A\},$$

For every x and y in a poset P, $x\leq y$, the interval $[x,y]$ is defined to be the set

$$[x,y]=\{z\in P: x\leq z\leq y\}.$$

A. Xie & X. Huang (Eds.): Advances in Electrical Engineering and Automation, AISC 139, pp. 27–31.
springerlink.com

If $|[x, y]| = 2$, then we call x is covered by y, denoted $x \prec y$. A filter A of P is a subset of P such that for every $x, y \in P$, if $x \geq y$ and $y \in A$, then $x \in A$, we let $F(P)$ denote the family of all filters of P.

Let P be a finite set, $\Im$ a family of a filter of P satisfying:

(I1) $\varnothing, P \in \Im$;

(I2) $X, Y \in \Im$ implies $X \cap Y \in \Im$;

The family $\Im$ gives rise to the following operator:

$$\sigma(A) = \{X : A \subseteq X, X \in \Im\}.$$

It is straightforward to check that has the properties of a closure operator of convex geometry. Moreover, it has the following **anti-exchange property:**

(P) If $y, z \notin \sigma(X)$ and $z \in \sigma(X \cup y)$, then $y \notin \sigma(X \cup z)$.

This property (P) is a combinatorial abstraction of a property of the usual convex closure in Euclidean spaces. Namely, for two points x and y not in the convex hull of the set X, if z is in the convex hull of $X \cup y$, then y is outside the convex hull of $X \cup z$.

2 Convex Geometry on Partially Ordered Sets

Theorem 2.1. Let P be a finite set, $\Im$ a family of a filter of P satisfying:

(I1) $\varnothing, P \in \Im$;

(I2) $X, Y \in \Im$ implies $X \cap Y \in \Im$;

(I3) If $X \in \Im \setminus P$, then there exists $x \in$ Max $(P - X)$) such that $X \cup x \in \Im$.

Then we call $(P, \Im)$ a poset convex geometry of P, and the member of $\Im$ feasible sets.

If P is a trivially ordered set, the preceding definition yields the classial notion of convex geometry [1].

In the following, we give a characterization of poset convex geometry.

Preposition 2.2. Let P be a finite set and $\Im$ a family of a filter of P. Then $(P, \Im)$ a poset convex geometry of P if and only if

(I1)' $\varnothing, P \in \Im$;

(I2)' For every non-empty set $X \in \Im$, there exists $x \in$ Min (X) such that $X - x \in \Im$;

(I3)' For every $X, Y \in \Im \setminus P$ with $|X| = |Y| + 1$, there exists $x \in$ Max $(X - Y)$ such that $Y \cup x \in \Im$;

Proof: Suppose that $(P,\Im)$ a poset antimatroids of P (P is a finite set). By (I1), $\varnothing, P \in \Im$, then the application of (I2) implies the existence of an $x \in \mathrm{Min}(X)$, such that $X - x \in \Im$. Since (I3)' is equivalent to $\Im$ being under intersection, so we have that (I3)' hold.

Conversely, suppose that $(P,\Im)$ satisfies (I1)'-(I3)', and $X, Y \in \Im \setminus P$ is non-empty in P. We will show that $(P,\Im)$ a poset antimatroid. By (I1)', there exists $x_1 \in \mathrm{Min}(X \cap Y)$ such that $X - x_1 \in \Im$. Define $X_{k-1} = X - x_1$, since $X_{k-1} \in \Im$, by (I1)', there exists $x_2 \in$ Min (X) such that $X - (x_1 \cup x_2) \in \Im$. With the repeated using this mothed, we have that $X \cap Y \in \Im$. By (I3)' we can easily see that $\Im$ is closed under unions. Thus (I3) holds.

In the following, we will give two examples of poset convex geometry [1].

Example 2.3. (*poset shelling*) Let $(P,\leq)$ be a poset and let $\Im = \{X \subseteq E \mid x \in X, y < x \Rightarrow y \in X\}$ (namely, $\Im$ is the family of ordered ideal of $(P,\leq)$) consist of all filters of the poset. Then $(P,\Im)$ is a poset convex geometry on poset $(P,\leq)$. We say this kind of convex geometry is a **poset shelling** on $(P,\leq)$

Example 2.4. (*tree shelling*) Let X be the vertex set of a (graph- theoretic) tree Y, and define $\Im = \{A \subseteq X :$ the subgraph induced by X is connected$\}$. Then we can see that $\Im$ is a covex geometry on X, and we say this kind of covex geometry is a **tree shelling.**

Example 2.5. (*graph search*) Let $G = (V,E)$ be a connected graph with rooted $r \in V$, and define $\Im = \{X \subseteq V \setminus r :$ the subset induced by $V \setminus X$ is connected$\}$. Then we can see that $\Im$ is a covex geometry on $V \setminus r$, and we say that this kind of covex geometry is a **graph search.**

3 Closure Operator for Convex Geometry on Poset

In this section, we define the rank function r of a subset in a poset covex geometry $(P,\Im)$ as

$$r(X) = \mathrm{Max}\{|A| : A \subset X, A \in \Im\}$$

and the closure operator of X is that $\sigma(X) = \{x \in P \mid r(X \cup x) = r(X)\}$.

Then we study the closure axioms of poset covex geometry and give some properties of poset covex geometry.

Theorem 3.1. (*Closure axioms for poset convex geometry*) A mapping $\sigma : F(P) \rightarrow F(P)$ is the closure operator of a poset covex geometry if and only if for all $X, Y \in F(P)$, the following conditions hold:

(CL1) $X \subseteq \sigma(X)$;

(CL2) $X \subseteq Y \subseteq \sigma(X)$ implies $\sigma(X) = \sigma(Y)$;

(CL3) If $X \subseteq Y$ and $X, Y \in F(P)$, thenMax $(\sigma(X) - X)$ $\subseteq$ Max $(\sigma(Y) - Y)$;

Proof: Assume that $\sigma : F(P) \rightarrow F(P)$ is the closure operator of a poset covex geometry $(P, \Im)$, (CL1) and (CL2) is obvious true. To see (CL3), let $X \subseteq Y \subseteq P$ and $x \in$ Max $(P - \sigma(X))$, then there exists a feasible set T of X , with $T \cup x \in \Im$. If $x \notin$ Min (X) then we have that $r(Y \cup x) \geq |T \cup x| \geq r(Y)$, then $x \in$ Min $\sigma(X)$. Therefore, Max $(\sigma(X) - X) \subseteq$ Max $(\sigma(Y) - Y)$.

Conversely, suppose that $\sigma : F(P) \rightarrow F(P)$ satisfies properties (CL1), (CL2) and (CL3). In the following, we show that it is the rank function associated with a poset covex geometry on the poset P.

Define $\Im$ = $\{X : x \notin \sigma(Max(X - x)), \forall x \in Min(X)\}$, we shall show that $\Im$ satisfies (I1)-(I3).

Obviously, $\varnothing \in \Im$ and $P \in \Im$. In the following, we will show that $\Im$ satisfies (I2). Assume that $X, Y \in \Im$. If $X \cap Y \notin \Im$, by the definition of $\Im$, there exists $x \in Min(X)$ such that $x \in \sigma(Max((X \cap Y) - x))$. By (CL3), and we have that $x \in \sigma(Max((X \cap Y) - x)) \subseteq \sigma(Max(X - x))$. Contradicting that $X \in \Im$. In the following, we shall show (I3). If $X \in \Im \setminus P$, since P is a finite set, by (CL2) we can obtain $r(Y) \leq r(X \cup x) < |Y|$. Obviously this contracts the assumption that $Y \in \Im$ and (I3) hold.

Since (CL2) and (CL3) hold, by $X \subseteq Y \subseteq \sigma(X)$ implies $\sigma(X) = \sigma(Y)$, which means that $X \cup x \in \Im$. Then (I3) holds.

4 Conclusion

In this paper, the closure operator and closure axioms for poset covex geometry are presented, this theory has been used in many fields such as Combinatorial Optimization and Greedy Algorithm. The closure axioms for poset antimatroids will contribute to the theory of poset antimatroids and theory of ordered sets.

Acknowledgments. This work was supported by the Natural Science Foundation of Shaanxi Province (2011JM1010), the Scientific Research Program Founded by Shaanxi Provincial Education Department(Program No. 11JK0480).

References

1. Korte, B., Lovsz, L., Schrader, R.: Greedoids. Springer, Berlin (1991)
2. Barnabei, M., Nicoletti, G., Pessoli, L.: Symmetric property for poset matroids. Adv. Math. 102, 230–239 (1993)
3. Barnabei, M., Nicoletti, G., Pessoli, L.: Matroids on partially ordered sets. Adv. Appl. Math. 21, 78–112 (1998)
4. Li, S., Feng, Y.: Global rank axioms for poset matroids. Acta Mathematic Sinica, English Series 20(3), 507–514 (2004)
5. Gratzer, G.: Gerenal Lattice Theory. Birkhauser -Verlag, Berlin (1998)
6. Schroder, B.: Ordered Sets. Birkhauser, Boston (2003)
7. White, N.: Matroid Applications. Cambridge University Press, Cambridge (1992)
8. White, N.: Theory of Matroids. Cambridge University Press, Cambridge (1992)
9. Koshevoy, G.A.: Choice function and abstract convex geometries. Mathematical Social Science 38, 35–44 (1999)
10. Kashiwabara, K., Nakamura, M., Okamoto, Y.: The affine representation theorem for abstract covex geometries 30, 129–144 (2005)
11. Adaricheva, K.V., Gorbunov, V.A., Tumanov, V.I.: Join-semi distributive lattices and convex geometries. Advances in Mathematics 173, 1–49 (2003)
12. Danilov, V., Koshevoy, G.: Mathematics of Plott choice functions. Mathematical Social Science 49, 245–272 (2005)

A Multiple Relay Selection Scheme in Cognitive Relay Network

Jiali Xu, Haixia Zhang, and Dongfeng Yuan

School of Information Science and Engineering, Shandong University, 250100 Jinan, China
{xujiali,haixia.zhang,dfyuan}@sdu.edu.cn

Abstract. This paper studies multiple relay selection scheme in cognitive relay network to maximize the signal-to-noise ratio (SNR) of the secondary receiver. We first formulate the relay selection problem as a 0-1 integral programming problem, then propose an iterative relay selection scheme, which is suboptimal but simple. Numerical results indicate that the proposed iterative scheme can always achieve the near-optimal SNR performance.

Keywords: Cognitive relay network, multiple relay selection, maximization SNR, iterative scheme.

1 Introduction

Cognitive radio (CR) [1,2] has been proposed to significantly improve the spectrum efficiency by exploiting the available spectrum holes or sharing the spectrum resource. Based on the nature of broadcast and multiple access of the wireless transmission, cooperative relay technique [3,4] can yield spatial diversity to alleviate the multipath fading. Motivated by the above two techniques, cognitive relay network (CRN) has been an interesting area in recent years [5,6]. In CRN, spectrum efficiency and network capacity can be dramatically improved by allowing the secondary relay nodes help other's transmission task.

A key challenge in CRN is how to effectively select the cooperative relay nodes. There exist some literatures on relay selection and power allocation[7]-[11]. In [7], several relay selection schemes were proposed and the corresponding diversity orders were investigated. In [8], based on the combinatorial optimization theory, an optimal relay selection method was addressed. However, these studies do not consider the CR environment. In [9], it was assumed that the relay node cooperated only when the primary user in the corresponding network cluster was absent. In [10], the optimal single relay selection and power allocation were studied, but without considering the multiple relay selection. In our previous work [11], based on greedy scheme and best single scheme, a hybrid scheme was proposed but it still could not guarantee to get the optimal performance. The optimal multiple relay selection scheme is so far still left open.

The contributions of this paper are as follows. First, we generalize the multiple relay selection problem in CRN as a 0-1 integral programming problem,

A. Xie & X. Huang (Eds.): Advances in Electrical Engineering and Automation, AISC 139, pp. 33–39.
springerlink.com

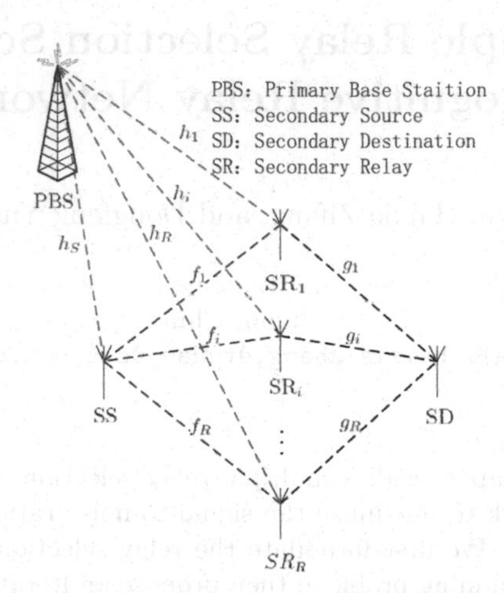

Fig. 1. System model for multiple secondary relay nodes

in which all the secondary nodes work in underlay mode, i.e., sharing the spectrum with primary users. Second, an iterative scheme is proposed to resolve this problem, the computational complexity as well as the amount of feedback bits are analyzed. Third, along with the proposed schemes in [11], the performances of the schemes are compared and analyzed.

The rest of this paper is organized as follows. In Section 2, we describe the system model. In Section 3 we first formulate the multiple relay selection in CRN and then describe the iterative scheme. Section 4 shows and analyzes numerical results. Section 5 concludes the whole paper.

2 System Model

The system model is depicted in Fig. 1. The CRN considered in this paper is composed of a primary base station (PBS) and a cognitive radio system. The whole cognitive radio system consists of a secondary source (SS), a secondary destination (SD) and R secondary relay (SR) nodes, which can help the communication between SS and SD.

Assume there is no direct link between SS and SD and each SR node works in two-hop AF mode and the ith relay node is denoted by $\mathrm{SR}_i, i = 1, ..., R$. It is worth to point out that our analysis can be easily extended to the case, in which there exists direct link. Denote the channel gain from SS to SR_i with f_i and the channel gain from SR_i to SD with g_i. Denote h_s and h_i as the channel gain from SS and SR_i to PBS, respectively. Further assume that each relay knows the CSI of himself, i.e., f_i and g_i, and SD node knows all the CSI, i.e., $f_1, ..., f_R$ and $g_1, ..., g_R$ and $h_1, ..., h_R$. The transmit power of SS and SR_i denoted by P_s and P_i, respectively.

Each transmission period is consisted of two time slots. In the first time slot, SS broadcasts information $\sqrt{P_s}x$, in which x is selected randomly from a BPSK codebook. Denote the received signal at SR$_i$ with

$$y_i = f_i\sqrt{P_s}x + n_i. \quad (1)$$

Then in the second half time slot, SR$_i$ first scales y_i with the following factor

$$G_i = \frac{\alpha_i\sqrt{P_i}}{\sqrt{1+|f_i|^2P_s}}, \quad (2)$$

then forwards the scaled signal

$$x_i = y_iG_i. \quad (3)$$

In (2), the coefficient α_i denotes the power percentage that SR$_i$ uses to transmit, thus we have $\alpha_i = 1$ if SR$_i$ is selected as the relay, or $\alpha_i = 0$ otherwise. For ease of exposition, we assume that all SR nodes transmit simultaneously and share the same spectrum with PBS in the second time slot. SD combines received signals from SR$_1, \cdots, SR_N$ as follows

$$y = \sum_{i=1}^{R} g_ix_j + n_d = \sqrt{P_s}\sum_{i=1}^{R}\frac{\alpha_i|f_ig_i|\sqrt{P_i}}{\sqrt{1+|f_i|^2P_s}}x + \sum_{i=1}^{R}\frac{\alpha_i|g_i|\sqrt{P_i}}{\sqrt{1+|f_i|^2P_s}}n_i + n_d, \quad (4)$$

where n_i and n_d are the i.i.d. additive white Gaussian noise received at SR$_i$ and SD, respectively, all with zero-mean and unit-variance. Considering that the above secondary system is within the transmission range of the PBS, the interference power constraint (IPC) of PBS must be exactly guaranteed at each phase of transmission.

3 Problem Statement and Proposed Scheme

3.1 Problem Statement

In our problem, what we want is to maximize the signal to interference plus noise ratio (SINR) perceived at SD node provided that the IPC of the PBS is guaranteed. For simplicity, we assume that PBS works at a constant power. Thus, the interference power from PBS to SD is also constant, and maximizing the SINR of SD is equivalent to maximizing the corresponding SNR. The generalized SNR at the SD writes

$$SNR_{general} = SNR_{\{1,\cdots,R\}} = \frac{P_s\left(\sum_{i=1}^{R}\frac{\alpha_i|f_ig_i|\sqrt{P_i}}{\sqrt{1+|f_i|^2P_s}}\right)^2}{1+\sum_{i=1}^{R}\frac{\alpha_i^2|g_i|^2P_i}{1+|f_i|^2P_s}}. \quad (5)$$

Let $\boldsymbol{\alpha} = \{\alpha_1, ..., \alpha_N\}$ denotes the relay selection vector. The multiple relay selection problem can be formulated as

$$\boldsymbol{\alpha}^* = \arg\max_{\boldsymbol{\alpha}} SNR_{general}, \quad (6)$$

$$\text{s.t.} \quad \alpha_i \in \{0, 1\}, \quad (7)$$

$$P_s|h_s|^2 \leq I_p, \quad (8)$$

$$\sum_{i=1}^{R} I_i = \sum_{i=1}^{R} \alpha_i^2 P_i |h_i|^2 \leq I_p, \quad (9)$$

where I_p and I_i denote the IPC of PBS and the interference from SR_i to PBS, respectively. Constraints in (8) and (9) are imposed by the IPC of PBS at the two transmission slots, respectively. There will be no transmission established if any one of constraints in (8) and (9) is not guaranteed.

3.2 The Proposed Iterative Scheme

The proposed iterative scheme is based on the recursion. Let $S_0 = \emptyset$ and $T_0 = 0$. Assume that there are i SR nodes that have been selected as cooperative relay nodes. Denote the set of these nodes with S_i, and let T_i represents the summation of interference to PBS resulted from these selected SR nodes. Then the $(i+1)$th cooperative SR node can be selected based on the following criterion,

$$S_{i+1} = S_i \cup j_i, \quad (10)$$

$$T_{i+1} = T_i + I_{j_i}, \quad (11)$$

where

$$j_i = \arg \max_{j \in \{1,...,R\} \backslash S_i} SNR_{S_i \cup \{j\}}, \quad (12)$$

$$\text{s.t.} \quad T_{i+1} \leq I_p, \quad (13)$$

$$S_{i+1} \geq S_i. \quad (14)$$

Thus, the set of selected SR nodes writes

$$S_{iterative} = \arg \max_{S_i, 1 \leq i \leq R} SNR_{S_i}. \quad (15)$$

In other words, for each step, under the constraint of the IPC, we add one SR node to previous set. This SR node is the one in $\{1, \ldots, R\} \backslash S_i$ that results in the best and constructive receive SNR. The recursion stops until the sum of the interference caused by cooperative relays exceeds the IPC, or the new added relay plays a deconstructive role in SNR, i.e., $SNR_{S_i \cup \{j\}} < SNR_{S_i}$. Finally we choose the set with the largest received SNR as our multiple relay selection decision. In an ideal case that all the relay node can be selected, the number of computations in the ith step is $R - i + 1$. Thus the total complexity is

$$\sum_{i=1}^{R}(R-i+1)=\frac{R(R+1)}{2}\sim O(R^2). \tag{16}$$

After the multiple relay selection procedure is finished, the SD broadcasts the results to all the relay nodes via the dedicated channel. This scheme needs R bits to distinguish whether SR_i node is selected or not, so the amount of feedback bits of iterative scheme is R. If node SR_i is selected, it will cooperate with its full power, otherwise it will keep silent during this transmission period.

4 Numerical Results

In this section, some simulation results are presented to evaluate the performance of the proposed scheme. System setup is as follows.

We adopt following simplified path loss model [12]

$$h=\sqrt{K\left(\frac{d_{ij}}{d_{ref}}\right)^{-\rho}}, \tag{17}$$

where d_{ij} represents the distance from node i to node j, d_{ref} indicates the reference distance, K is a unitless constant depending on the antenna characteristics and the average channel attenuation, and ρ stands for the path loss exponent. In the experiments, following assumptions are made: the default unit of distance is meter; SS and SD are located at (0, 0) and (60, 0), respectively; PBS is set at (30, 40); R relay nodes are randomly located within a rectangular area described by (-10, -10), (-10, 10), (70, -10), (70, 10); $\rho = 3$ and $K = 0.0265$; the reference distance $d_{conf} = 100$; SUs have the same transmit power, i.e., $P_s = P_1 = ... = P_N$.

In [11], we proposed some other schemes to resolve the multiple relay selection in CRN: conventional greedy scheme in knapsack problem, the best single scheme which selects the node with the best SNR performance, exhaustive search scheme and the hybrid scheme which combines the best single scheme and greedy scheme.

In Fig. 2, we compare the performance of proposed iterative scheme and above mentioned four schemes, in the case of $I_p = 20$ dB. We can see that the iterative and the hybrid scheme can get the same performance as the exhaustive scheme, while the greedy scheme can outperform the best single scheme in most cases. When only one node is selected, in other words, when the transmit power $P \geq 26$ dB, all the schemes get the same results.

In Fig. 3, we consider the case of $I_p = 30$ dB, $R = 15$ relay nodes are randomly allocated in the rectangular area. Results show that the iterative scheme outperforms all other schemes and can always achieve the near optimal performance.

The analytical and confirming simulation results show that the proposed iterative scheme can achieve better tradeoff between the performance and complexity than other schemes. It can achieve near-optimal performance with acceptable complexity.

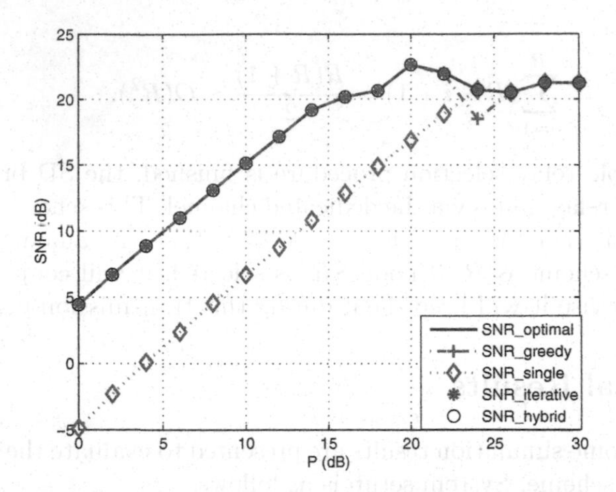

Fig. 2. SNR performance of five schemes, exhaustive, best single, greedy, hybrid, and iterative. Five fixed relay nodes are employed, which are located at (30, 15), (30, 10), (30, 0), (30, -5), (30, -20), respectively.

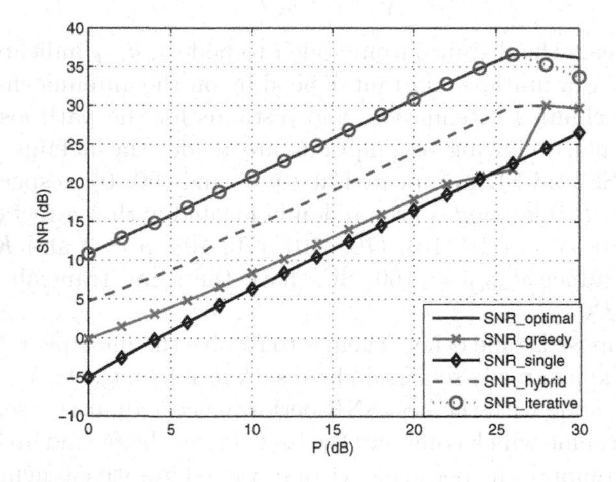

Fig. 3. SNR performance in the case of $Ip = 30$ dB. R=15 relay nodes are randomly allocated in the rectangular area.

5 Conclusion

In this paper, we work on the multiple relay selection problem in cognitive relay network. We first formulate the problem as a 0-1 integral programming problem, then we propose a novel iterative scheme to resolve it. Numerical results indicate that near optimal performance can be achieved by the proposed scheme. Complexity of the proposed scheme is also analyzed, it is shown that iterative scheme is of low computational and communication complexity.

Acknowledgments. This work is supported by NSFC (No. 60832008, No. 61071122) and Independent Innovation Foundation of Shandong University.

References

1. Mitola, J., et al.: Cognitive Radio: Making Software Radios More Personal. IEEE Pers. Commun., 13–18 (August 1999)
2. Wang, B., Liu, K.: Advances in Cognitive Radio Networks: A Survey. IEEE Journal of Selected Topics in Signal Processing 5, 5–23 (2011)
3. Nosratinia, A., Hunter, T.E., Hedayat, A.: Cooperative Communication in Wireless Networks. IEEE Commun. Magazine 42, 74–80 (2004)
4. Zhang, X., Wang, J., Qian, Z.: Advances in cooperative wireless networking: Part I [Guest Editorial]. IEEE Commun. Magazine 49, 54–55 (2011)
5. Letaief, K., Zhang, W.: Cooperative Communications for Cognitive Radio Networks. Proceedings of the IEEE 97, 878–893 (2009)
6. Jia, J., Zhang, J., Zhang, Q.: Cooperative Relay to Improve Diversity in Cognitive Radio Networks. IEEE Commun. Magazine 47, 111–117 (2009)
7. Jing, Y., Jafarkhani, H.: Single and Multiple Relay Selection Schemes and their Achievable Diversity Orders. IEEE Trans. on Wireless Commun. 8(3), 1414–1423 (2009)
8. Michalopoulos, D.S., Karagiannidis, G.K., Tsiftsis, T.A., Mallik, R.K.: An optimized user selection method for cooperative diversity systems. In: Proc. IEEE Globecom 2006, San Francisco, CA (November-December 2006)
9. Lee, K., Yener, A.: Outage performance of cognitive wireless relay networks. In: Proc. IEEE Globecom 2006 (November 2006)
10. Sun, C., Letaief, B.: Uer Cooperation in Heterogeneous Cognitive Radio Networks with Interference Reduction. In: Proc. of IEEE ICC 2008 (May 2008)
11. Xu, J., Zhang, H., Yuan, D., Jin, Q., Wang, C.: Novel Multiple Relay Selection Schemes in Two-Hop Cognitive Relay Networks. In: Proc. of CMC 2011 (April 2011)
12. Goldsmith, A.: Wireless Communication. Cambridge University Press (2005)

Flexible Index System of Human Resources Considering the Organizational Flexibility Factor

Yuping Chu, Shucai Li, and Yuran Jin

School of Business Administration, University of Science and Technology, Liaoning, Anshan, 114501, China
jinyuran@163.com

Abstract. The organizational flexibility, such as the diversity of the enterprise's organizational structure, directly affects the efficiency and effect of human resources management. However, the traditional human resource flexible index system in this point has obviously been ignored. Therefore, the paper introduced the organizational flexibility index to the flexible index system of human resources management. Analytic Hierarchy Process was adopted to determine the weight of indexes. The new index system can reflect the characteristics of various flexible index factors effectively and provide a more reasonable and efficient analysis and improvement method.

Keywords: Human resources, Flexible management, Flexible index, Analytic Hierarchy Process.

1 Introduction

In the management activities, human are subjects as well as objects.Flexible management is a human-oriented and humanized management, it bases on the study of the human psychology and behavior rules. It doesn't adopt mandatory ways but produce a potential persuasion in people's mind and then makes the organization plan become a self-consciousness of the individual action [1]. It is mainly by "Humanity" for mark and lays emphasize on equality and respect, rational and emotional, create and intuition, as well as the jumping and the change, sensitive and flexibility, the initiative to meet the change, the use of change[2].The basic characteristics of the flexible management mainly manifest the inner driver, durability and guidance[3]. Now, many scholars begin to study the the new human resources management form in the practice, and assert their own views in various ways. Zhao ShuMing, one of famous human resources experts in our country, has mentioned about the flexible human resources management content[4], like the flexible work system, flexible working hours, flexible working place, elastic welfare system and so on; XuChunYan has analysized the organization knowledge and human resources flexible management system[5]. She thinks that many kinds of organization must develop towards knowledge and it requires the reconstruction of the human resource management system; Jiang Xiaolan thinks flexible human resources management should be: flexible human resources management must have a feasible method to ensure its effective operation in the mechanism[6], mainly by democratic management and internal communication

A. Xie & X. Huang (Eds.): Advances in Electrical Engineering and Automation, AISC 139, pp. 41–46.
springerlink.com

as well as personalized management; According to Blyton's view, Xie Lin and Du Gang divide human resources flexible index into four forms: function flexible, quantity flexible, function of time and salary flexible. Among them, function flexible is the most important flexible dimension, including the working integration ability, the vertical integration ability and the skills of the staff [7]. Compared to the above research, this study has many advantages. It makes a more detailed explanation and provides certain basis for the study of human resources flexibility. But it also has some disadvantages, mainly shows in that flexible index selection attaches too much importance to the index of the external performance, and lack of implicit flexible research and the composition of the dimension of flexible slightly single, etc.

In view of the above questions, this article adopts analytic hierarchy process (AHP) to have a further analysis of human resources flexible index. AHP is a multi-objective decision analysis method which combines with qualitative analysis and quantitative analysis [8]. Analytic hierarchy process (AHP) is to determine the index so that on one hand it can reflect the characteristics of flexible management index, on the other hand, it can manifest the requirements of the qualitative and quantitative.

2 The Framework Analysis of Human Resources Flexible Index System

The functions, quantity, time and the salary, and other aspects of the flexibility are very important to an enterprise. In addition, because a series of basic activities are in the organization, whether the enterprise can make timely recognition and processing to the external information will directly affect the other human resource flexibility being improved. Therefore, based on the previous research, this article innovatively introduces the organization flexible index in human resources flexible index system and constructs the following the deferred class time Fig. 1:

The meaning of each index as following:

Functions flexibility: The integrate ability is the mobility and creative ability of the employees when they are in work, that is the ability of enlarging work content; As for the vertical integration ability, it means to make the working content more colorful in the process of operation organization, that is to advance the staff's ability.

The organizational flexibility: The labor intensity flexible reflects the enterprise labor intensity; The diversity of the changing structure has influences on the staff's future; The consistency of the process helps to improve the product quality and enterprise reputation; The post organization of the flexibility is to change the traditional design of rigid post mode and provide a vast development space.

Salary flexibility: Performance and skills by flexible formulated and the change of elastic wages. Salary is flexible with the changes in the external environment such as rising prices, the life level and the performance of the company, the final salary levels; Dividend and year-end bonus is the share of the profit. Namely, with reference to the enterprise operation situation form the year-end bonus; fixed salary and changeable salary is said to change by many factors. The changeable salary in large proportion will change fast.

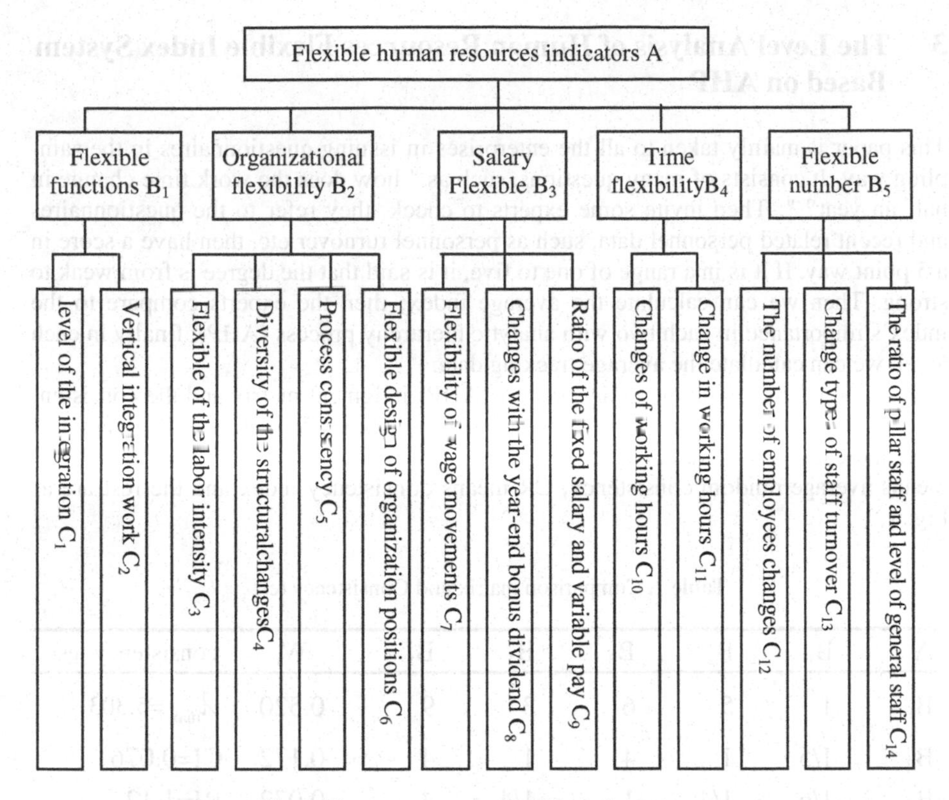

Fig. 1. Human resources flexible index system

Time flexibility: The work time variation degree of flexibility should base in total fixed working time and change according to the work's requirements or the employee's demand; The working hour's variation degree is determined by the changed job. And the number of work hours of time variation degree influences the flexible working time of the enterprise.

Quantity flexibility: The employee's quantity fluctuation range changes with human's needs and produces the number of employee's change; The change of Employee turnover's kinds change according to the nature of the work and the change of mission therefore change the type of human resources, adding new jobs or reduce some staffs from other department or draft internal personnel and let external personnel in; The proportion of backbone employees and ordinary employees shows the proportion of key personnel and non-critical personnel. Quantitative variation of amplitude, type changes fast, high percentage of ordinary staff. As a result, the human resources of enterprises become more flexible in Numbers [9].

3 The Level Analysis of Human Resources Flexible Index System Based on AHP

This paper is mainly taken to all the enterprises in issuing questionnaires in the sampling way. It consists of many questions, such as " how does the work time change in half an year? "; Then invite some experts to check, they refer to the questionnaires and recent related personnel data, such as personnel turnover etc, then have a score in a 5 point way. If it is in a range of one to five, it is said that the degree is from weak to strong. Then we can calculate the average index; then the experts compare to the index's importance in each two with analytic hierarchy process (AHP); finally in each level, we can calculate the average tracking data.

Based on the analytic hierarchy process (AHP) judgment matrix and the consistency test, the results are as the following Table 1 to Table 6, Wi refers to the weight of each indicator, λmax refers to the biggest characteristic root judgment matrix, RI means average random consistency, CR means consistency index and the rest are as Fig. 1.

Table 1. Comparison matrix and Consistency test

A	B_1	B_2	B_3	B_4	B_5	W_i	consistency test
B_1	1	5	6	3	9	0.520	λ_{max} =5.303
B_2	1/5	1	4	1	4	0.172	CI=0.076
B_3	1/6	1/4	1	1/4	4	0.072	RI=1.12
B_4	1/3	1	4	1	5	0.199	CR=0.068<0.1
B_5	1/9	1/4	1/4	1/5	1	0.037	

Table 2. Comparison matrix of the flexibility functions and Consistency test

B_1	C_1	C_2	W_1	consistency test
C_1	1	8	0.889	λ_{max} =2 , RI=0
C_2	1/8	1	0.111	CI=0, CR=0<0.1

Table 3. Comparison matrix of the organizational flexibility and Consistency test

B_2	C_3	C_4	C_5	C_6	W_2	consistency test
C_3	1	2	3	1	0.376	λ_{max} =4.112
C_4	1/2	1	1	1	0.202	CI=0.043
C_5	1/3	1	1	1	0.182	RI=0.9
C_6	1	1	1	1	0.240	CR=0.043<0.1

Table 4. Comparison matrix of the flexible wages and Consistency test

B_3	C_7	C_8	C_9	W_2	consistency test
C_7	1	5	6	0.717	$\lambda_{max}=3.094$
C_8	1/5	1	3	0.195	CI=0.047, RI=0.58
C_9	1/6	1/3	1	0.088	$CR=0.081<0.1$

Table 5. Comparison matrix of the time flexible and Consistency test

B_4	C_{10}	C_{11}	W_3	consistency test
C_{10}	1	5	0.833	$\lambda_{max}=2$, RI=0
C_{11}	1/5	1	0.167	CI=0， $CR=0<0.1$

Table 6. Comparison matrix of the number flexible and Consistency test

B_5	C_{12}	C_{13}	C_{14}	W_2	consistency test
C_{12}	1	3	2	0.550	$\lambda_{max}=3.018$
C_{13}	1/3	1	1	0.210	CI=0.009 ,RI=0.58
C_{14}	1/2	1	1	0.240	$CR=0.016<0.1$

4 Conclusion

By using AHP to the research of flexible index, this paper draws conclusion as follows: (1) by the calculation of table 1, we can see that in these factors of flexible management of human resources, the order to important degree is function flexible>salary flexible>organizational flexibility>time flexible>quantity flexible. From here, we can easily see the importance of organizational flexibility in an enterprise. Every enterprise should strengthen the organizational flexibility; (2) From table 2: We can see, the level of the integrate ability to work>the vertical integration ability to work; (3) From table 3: we can see, labor intensity flexible>the position of flexible design organization>the diversity of structural change>process consistency; diversity; (4)From table 4, we can see, the changes of the salary's raising>the rate of fixed salary and changeable wage>variation degree of dividends and year-end bonus; (5)From table 5: we can see, the variations of work stage> the work hours variations(6) From table 6, quantity flexibility>the rate of ordinary employees and the key employees>the variation of employee's flow. Of course, we should not neglect the importance of them just because of the low index. But we can relatively control it a little loosing.

Faced with the new economic wave, flexible management will become the new trend of development of human resources in 21st century. We should remember the flexible management's "soft" should be regularly, otherwise you will get backfire effect. Based on the study of the flexible system process, because my knowledge and time are limited, the study of flexible index remains a long way to go. And this also urged me to do more. For example, I can collect more information and data on the enterprise human resources flexible literature, pay more attention to the flexible index system, build more perfect system, and put forward more practical and more perfect system construction mode and management scheme. Only in this way, can we improve the overall competitiveness of the enterprise.

References

1. Golden, W., Powell, P.: Towards a Definition of Flexibility: In Search of the HolyGrail. Omega 28, 373–384 (2000)
2. Xie, L., Du, G.: Human resources the flexible management the evaluation index system and analysis model. Journal of Xi'an University of Electronic Science and Technology (social science edition) 2, 103–107 (2006) (in Chinese)
3. Handy, C.: Inside Organizations-21 Itleas for Managers. Penguin Books (1990)
4. Chen, R.: Some studies of flexible. Management Engineering 3, 36–38 (1998) (in Chinese)
5. David: Examining the Human Resource Architecture: The Relationships Among Human Capital, Employment and Human Resource Configurations. Journal of Management 28 (2002)
6. Jiang, X.: Enterprise flexible management analysis, pp. 56–59. Qingdao Ocean University (2002) (in Chinese)
7. Atkinson, J.: Manpower Strategies for Flexible Organizations. Personal Management 8, 28–31 (1984)
8. Vickery Supply Chain Flexibility An Empirical Study. The Journal of Supply Chain Management (1999)
9. Feng, Z.: To flexible management, vol. 13. China Social Science Press, Beijing (2003) (in Chinese)

Research on an Integrated Model of Information Resources Based on Ontology in the Power System*

BaoYi Wang, Wei Zhou, and ShaoMin Zhang

School of Control and Computer Engineering,
North China Electric Power University, Baoding, China
{wangbaoyi,zhangshaomin}@126.com, wnvjungle@yahoo.cn

Abstract. With the development of the electric power informatization,the problem of data integration is becoming more and more serious. Due to lack of overall planning and design, it is difficult to make system integration for the characteristics such as numerous information islands, heterogeneous database, and the low information integration level. In view of the above questions this paper has proposed a power system integrated research method based on ontology. This paper discusses the mapping between the Local Ontology and the data sources. Firstly we discuss how to convert from the relational database schema to OWL ontology, then give a corresponding conversion algorithm, finally the algorithm is proved to be feasible by an application example.

Keywords: information integration, data schema, ontology.

1 Introduction

The data of the power system is mainly made up of five parts, such as the business process data, the equipment data, the action records data, the time sampling data and the relevant information data. These data is distributed in several heterogeneous information systems which are maybe in different places. The difference of these systems is not only about the platform, the geographical position, the business logic and storage, but also about the professional naming rule of different fields. But as the electric power system standardization requires more and more highly, enterprise application may need to visit several heterogeneous systems at the same time.

In view of the above questions this paper has proposed a power system integrated research method based on ontology. The mapping between the Local Ontology and the data sources is discussed, and we give an algorithm about how to convert from the ER Schema to the Local OWL.

---

* This work is supported by Hebei higher education science research plan funding subject of research on an integrated model of information resources based on ontology in the power system.

2 The Information Resources Integration Model

2.1 System Model

This paper makes a study of the ontology construction and ontology mapping. According to the structure of the mixed ontology, we propose an information integrated system model based on ontology. The model of the system architecture is shown in figure 1.

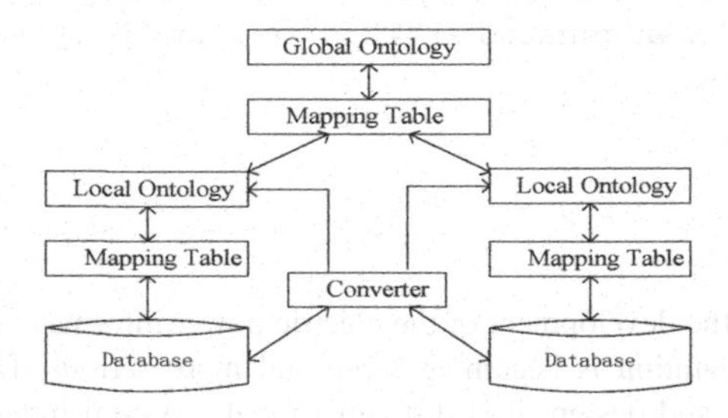

Fig.1. An Integrated Model of Information Resources Based on Ontology

2.2 The Syntax and Semantics of the ER Schema and the OWL Ontology

In this section we give the syntax and semantic of the ER Schema and the OWL ontology; and then propose an algorithm.

Definition 1: For the limited set X and Y, we call a function about a subset of X to Y a tuple T, which maps $x_i \in X$ to $y_i \in Y$ (i=l, 2, ..., k). T is represented as $[x_1{:}y_1,\ldots,x_k{:}y_k]$.So we can use $T[x_i]$ to represent y_i.

Definition 2: an ER Schema is a five tuple $S=(L_s,isa_s,att_s,rel_s,card_s)$, footnote s means set, among them:(1) $L_s= E_s \cup A_s \cup R_s \cup U_s \cup D_s$ is an organized alphabet, which has no intersection with each other, it contains E_s a set of entity symbols, As a set of attribute symbols, R_s a set of relationship symbols, U_s a set of role symbols, D_s a set of domain symbols. (2) $isa_s \subseteq E_S \times E_S$ is the injective and acyclic binary relation on E_s, which is used to represent the ISA relationship between two entities.(3) att_s is a function which maps each entity symbol of E_s to a A_s marking D_s.(4) rel_s is a function which maps each relationship symbol of R_s to a U_s marking E_s. (5) $card_s$ is a function from $E_s \times R_s \times U_s$ to $N_0 \times (N_1 \cup \{\infty\})$(N_0 means the nonnegative integer, N_1 means the positive integer, ∞ means infinite), and meets the following conditions: for each role $U \in R_S$, and $rel_s(R)= [\ldots,U{:}E,\ldots]$,we define $card_s(E,R,U)=(mincard_s(E,R,U), maxcard_s(E,R,U))$, The first component $mincard_s$ (E, R,U) means the minimum times which the entity E involves the relationship R through the role U for(0 means no constraint), the second component $maxcard_s(E,R,U)$ means the corresponding maximum times(∞ means no constraint).

Definition 3: An OWL DL Ontology is a two-tuples O=(ID_0,$axiom_0$),among them:(1) ID_0 = $CID_0 \cup DRID_0 \cup OPID_0 \cup DPID_0$ is a limited identifier set, and has no intersection: a class identifier set CID_0 includes two predefined classes identifier owl:Thing and owl:Nothing besides identifiers defined by users, a data range identifier set $DRID_0$, each data range identifier has a predefined XML Schema data type identifier, a object property identifier set $OPID_0$, a datatype property identifier set $DPID_0$.(2) $axiom_0$ is a limited axiom sequence, which contains class axiom and attribute axiom.

2.3 The Algorithm Converting from ER Schema to OWL DL Ontology

There is the corresponding relationship between ER Schema and OWL DL Ontology. Based on the above formal presentation, the following algorithm gives the format of how to convert ER Schema to OWL DL Ontology.

Algorithm: R2O(L_s,isa_s,att_s,rel_s,$card_s$,O)
Input: ER schema S=(L_s, isa_s, att_s, rel_s, $card_s$)
Output: OWL DL ontology: $O = \phi(S)$, $ID_0 = \phi(L_S)$,
$axiom_0 = \phi(isa_S, att_S, rel_S, card_S)$

Step 1: LettertoIdentifier(L_s)// realize the transformation from the alphabet to //the identifier set.

```
{
    foreach(object obj in Ls)
    {
    if(obj∈Es)
        ϕ( obj )∈ CID0 // create a class identifier ϕ (E) ∈ CID0 , the
                        //identifier has the same name of E.
    if(obj∈Es)
        ϕ( obj )∈ CID0 // create a class identifier ϕ (R) ∈ CID0 , the
                        //identifier has the same name of R.
    if(obj∈As)
        ϕ( obj )∈ DPID0 // create a class identifier ϕ (A) ∈ DPID0 ,
                        //the identifier has the same name of A.
    if(obj∈Ds)
        ϕ( obj )∈ DRID0 // create a class identifier ϕ (D) ∈ DRID0 ,
                        //the identifier has the same name of D.
    if(obj∈Us)
        ϕ( obj ) ∈ OPID0 //Create a class identifier
        V = inv_ϕ( obj )∈ OPID0 // ϕ (U) ∈ OPID0 and
                        //V = inv_ϕ( U )∈ OPID0 the
                        // identifier has the same name of U.
    }
}
```

Setp 2: ElementoOWL(isa_s,att_s, rel_s,$card_s$)

{

foreach(E_1, $E_2 \in E_s$ && E_1 isa_s E_2,)//if E_1, $E_2 \in E_s$ and E_1 isa_s E_2,

// create a class axiom

subClassof $(\phi(E_1)\phi(E_2))$ (1)

foreach($E \in E_S$&&$att_s(E)=[A_1:D_1,\ldots,A_h:D_h]$)

{

CreatClassaxiom1(E,att_s(E)) (2)

for(i=0;i<h;i++)

CreatePropertyaxiom1(A_i,D_i,E) (3)

}

foreach($R \in R_s$ &&$rel_s(R)=[U_1:E_1,\ldots,U_h:E_h]$)

{

CreatClassaxiom2(R, rel_s(R)) (4)

for(i=0;i<k;i++)

{

CreatePropertyaxiom2(R,V_i, U_i,E_i) (5)

CreatClassaxiom3(R,V_i,E_i) (6)

}

}

foreach($U \in U_S$,&&$Rel_S(R)=[\ldots,U:E,\ldots]$)

{

CreatePropertyaxiom3(U,R,E) (7)

if($m=mincard_S(E,R,U)\neq\infty$)

CreatClassaxiom4(E,m) (8)

if($n=maxcard_S(E,R,U)\neq\infty$)

CreatClassaxiom5(E,n) (9)

}

foreach($X,Y \in E_S \cup R_S$&&$X \neq Y$&&$X \in R_S$)

DisjointClasses$(\phi(X)\ \phi(Y))$ (10)

}

Among them:

CreatClassaxiom1(E, att_s(E))

{

$$Class(\phi(E)\ partial$$

$$restriction\ (\phi(A_1)\ allValuesFrom(\phi(D_1))\ cardinality(1))$$

$$...restriction\ (\phi(A_h)\ allValuesFrom(\phi(D_h))\ cardinality(1))$$

}

CreatePropertyaxiom1 (A_i, D_i,E)

{

$$DataTypeProperty(\phi(A_i)\ domain(\phi(E))$$

$$range\ (\phi(D_i)))$$

}

CreatClassaxiom2(R, rel_s(R))

{

Class($\phi(R)$ *partial*
restriction ($\phi(U_1)$ *allValuesFrom*($\phi(U_1)$) *cardinality*(1))
...restriction ($\phi(U_k)$ *allValuesFrom*($\phi(D_k)$) *cardinality*(1))
}
CreatePropertyaxiom2(R, U_i, E_i)
{
Object Property(V_i domain ($\phi(E_i)$) range (ϕ(R))
inverseof($\phi(U_i)$)
}
CreatClassaxiom3(R,V_i,E_i)
{
Class($\phi(E_i)$ *partial*
restriction (V_i *all*ValuesFrom($\phi(R)$)))
}
CreatePropertyaxiom3 (U,R,E)
{
Object Property(ϕ(U) domain
($\phi(R)$) range (ϕ(E)))
}
CreatClassaxiom4(E,m)
{
Class($\phi(E)$ *partial*
restriction (V minCardinality(m)))
}
CreatClassaxiom5(E,n)
{
Class($\phi(E)$ *partial*
restriction (V maxCardinality(n)))
}

3 Application Example

In order to verify the effectiveness of the algorithm, there is an example about a personnel management system in the power system.

Table team means the information of team; table workarea means the information of work area; Table user means user. Among them, team and workarea, user and team, user and workarea have the ISA relationship.

team(team_name(varchar),team_infro(clob),team_sort(number),team_workarea(varchar));workarea(workarea_name(varchar),workarea_infro(clob));user(user_id(number),user_name(varchar),user_pwd(char),user_teamname(varchar),user_workarea(varchar)).

According to the above definition of ER schema, its alphabet L is as follows: Entity symbols set: E={team,workarea,user}; Attribute symbols set: A={team_name,team_infro,team_sort,workarea_name,workarea_infro,user_id,user_name,user_pwd}; Role symbols set: U={user_id,team_name,workarea_name}; Domain symbols set: D={varchar,clob,number,char}; Relationship symbols set: R={user, team}.

First realize the conversion from the alphabet of the ER schema to the ontology identifier set ID= LettertoIdentifier(L). The ontology identifier is as follows:

The class identifier set: CID={ team,workarea,user }; The data type identifier set: DPID={ team_name,team_infro,team_sort,workarea_name,workarea_infro,user_id, user_name,user_pwd }; The data domain identifier set: DRID={xsd:string,xsd: integer}; The object attribute identifier set: OPID={user_id,inv_user_id,team_name, inv_team_name,workarea_name,inv_workarea_name}.

Then realize the conversion from the child elements of the ER Schema to the ontology axioms=ElementoOWL(isa_s,att_s,rel_s,$card_s$).The number of the ontology axioms is as follows:(1) Three SubClassof axioms;(2) Three Class axioms;(3) Eight DataTypeProperty axioms;(4) Ten Object Property axioms.

The axioms output by the algorithm are the same as these sorted out manually, which means the algorithm is reasonable and can construct ontology automatically.

4 Conclusion

That how the Local Ontology is built is mainly discussed in this paper. Starting from the ER Schema, this paper discusses how to convert from the relational database schema to OWL ontology, and gives the algorithm about how to convert from the ER Schema to OWL Ontology. Compared with the similar method, this method can construct ontology automatically, and overcome the heavy manual work.

References

1. McGuinness, D., Harmelen, F.: OWL Web Ontology Language Overview (February 10, 2004), `http://www.w3.org/TR/2004/REC-owl-features-20040210/.W3CRecommendation`
2. William, S., Austin, T.: Ontologies. IEEE Intelligent Sytstems, 18–19 (January/February 1999)
3. W3C Working Draft. Web Services Architecture (August 2003)
4. W3C Working Draft. Web Services Glossary (August 2003)
5. Xu, X.-Z.: Research and Implementation of Ontology-based Heterogeneous Data Integration System. Master's Thesis. Xidian University (2007) (in Chinese)
6. Yuan, G.-J.: Analysis on heterogeneous database integration system. Laboratory Science (4), 105–107 (2009) (in Chinese)
7. Cui, X.-J., Wang, J.-C.: Research and Realization of Integrated Environment for Data Warehouse. Computer Application Research (12), 178–184 (2006) (in Chinese)
8. Yang, Q.: Research on Information Integration for Teaching Management Based on Ontology. Master's Thesis. Shandong University (2007) (in Chinese)

Research of the Distributed Time-Slot Assignment of WSN for Meter Reading System

BaoYi Wang, Xiangning Jiang, and ShaoMin Zhang*

School of Control and Computer Engineering,
North China Electric Power University, Baoding, China
{wangbaoyi,wlsiamca,zhangshaomin}@126.com

Abstract. Wireless sensor network for automatic meter reading system has fixed topology, change of the network is small and the energy consumption is relatively insensitive to other WSN, according to these features, most of the traditional MAC protocol does not apply to this network. Take the characteristics of residential automatic meter reading system into consideration and based on the strong edge coloring of graph theory, this paper presents a distributed slot allocation algorithm. The algorithm achieves to assign different time slots for nodes within two-hop in a distributed way, which solves the "hidden nodes" problem of wireless sensor network, assures the successful communication of no conflict and makes the protocol possible to be applied in the network later. Simulation shows that the algorithm can finish the slot assignment quickly and accurately.

Keywords: WSN, MAC protocols, TDMA, timeslot allocation.

1 Introduction

Wireless sensor network (WSN) is a two-way wireless communication technology with the characteristics of close distance between nodes, low complexity, low power consumption, low data rate and low cost [1-2]. The WSN for Automatic meter reading system takes meters as leaf nodes has characteristics of fixed topology, small change and high tolerance of the message delay [3]. Therefore, we can assign fixed time-slots for each node in WSN in the network establishment period. In this way, nodes can communicate with each other without conflict.

Researchers have done a lot of works in the time-slot assignment period[4-6]. Taking the special topology of residential meter reading system into consideration, we present a new distributed slot assignment algorithm in this paper based on the strong edge coloring theory. The simulation shows the algorithm can finish the time-slot assignment within a short period in a distributed way, and assure nodes within two-hop are assigned different slot (color).

* This work is supported by Hebei higher education science research plan funding subject of research of information integration based on IEC61970 and the security techniques of smart grid. (Z2010290).

A. Xie & X. Huang (Eds.): Advances in Electrical Engineering and Automation, AISC 139, pp. 53–58.
springerlink.com

2 Topology of Meter Reading System

As Figure 1 shows, the network consists of a base station, data collection terminals(DCT) and meters, each data collection terminal manages N meters. (In this paper, we only consider the part from the meters to the base station). Figure 1 shows the topology of the meter reading system. Devices all communicate wirelessly. Clustering structure of this WSN naturally formed by the geographic features. A DCT and its meters generates a sub-cluster. DCT is the cluster head which has strong information processing and data transfer ability while other meters' processing ability is weaker. When the data the meter collected reaches a certain threshold, the meter passes the data to DCT within one hop. When data reaches, the DCT packs the data with the corresponding meter information and passes it to the base station in a multi-hop way.

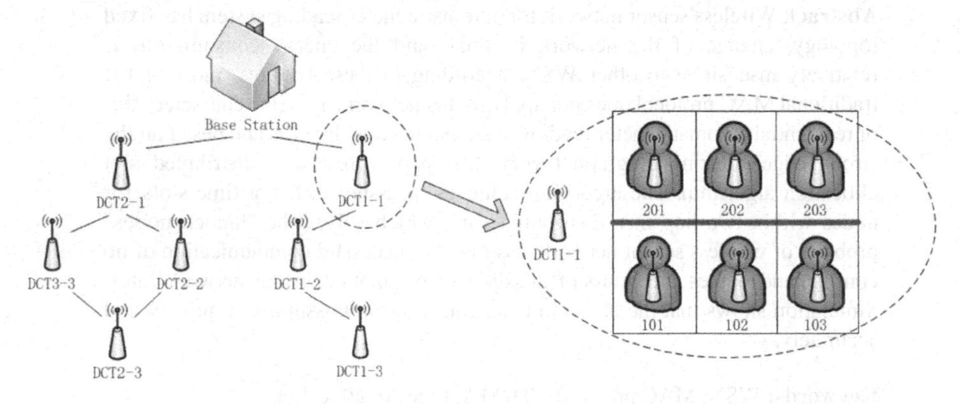

Fig. 1. WSN topology of the residential automatic meter reading system

3 Time-Slot Assignment Algorithm

3.1 Background

Let undirected graph $G=(V,E)$ represents the network where V is the set of nodes and E is the set of edges. Edge $e=(u,v)$ exists only when u and v are all belong to V and are able to communicate with each other. As the WSN(if not specified, the WSN below are all refer to the network for the meter reading system) is clustered, the vertex v are on behalf of a whole cluster include cluster head and other nodes.

As a wireless sensor node broadcasts to other nodes, conflict can happen among all the nodes within a two-hop distance, the time-slot assigned to v should be different from its neighbors within two-hop.

3.2 Time-Slot Assignment Algorithm

A node has six states: UNSET, UNSET_1, REQUEST, GRANT, SET and DONE. As the algorithm runs by cluster, if not specified, the "node" in the article below represents just the cluster head in the network. Figure 2 shows the node transition of the time-slot assignment algorithm.

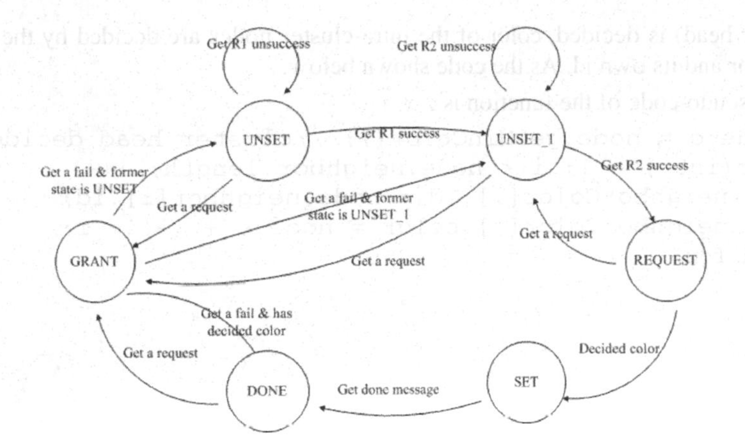

Fig. 2. Node transition of the time-slot assignment algorithm

Assuming that each node has already known the information of its one hop neighbors, and also has its default color between [1,4].

Distributed time-slot (color) assignment algorithm runs as follows. Initially all the nodes are in UNSET state. While node A is in UNSET, A generates a random number R_1 between [0,1], if R_1 is less than $p_1 = 1/3$, means A generated a successful number and then A change itself into UNSET_1 state. Otherwise, A remains in UNSET, waiting for another T_A, T_A is set to $4d_A$ and d_A is a rough estimate of the maximum one-way message delay.

After entering UNSET_1, A generates a new random R_2, if R_2 is less than $p_2 = 1/k$, A gets the permission to communicate with its neighbors. let C_j be the number of neighbors of node n within two-hop which color have not decided, k is the maximum value of the set C_j. Otherwise, A remains in UNSET_1, waiting for another T_A, and then re-generates R_2. Through information exchange, node A can then determine its own color.

If A generates R_2 successfully, A has the right to communicate. It then changes its state into REQUEST and broadcasts *request*$_A$ to all its neighbors.

When B, a neighbor of A, receives *request*$_A$ from A, and if B is in UNSET, UNSET_1, DONE, then B switches to GRANT and sends *grant*$_B$ to A. *grant*$_B$ carries the color information of B and its neighbors (which is carried in *release*).

If B receives *request*$_A$ from A and it's in REQUEST or GRANT, B sends *reject*$_B$ to A. If A receives a *reject* from any nodes, A broadcasts *fail*$_A$ to its one-hop neighbors and change its status back to UNSET_1. If B Gets a *fail*$_A$ from node A, and A is the node which sent request to B changing its status into GRANT, then B returns to the previous UNSET or UNSET_1 state if its color has not yet determined, otherwise, B will return to the DONE state.

Having received the *grant* messages sent from all the neighbors, A can decide its own color which should be different from its two-hop neighbors. Colors of intra-cluster nodes should be fixed along with the cluster head. As Figure 3 shows, Intra-cluster nodes are listed in A, with their id as offset value. Once color of the

A(cluster head) is decided, color of the intra-cluster nodes are decided by the cluster head color and its own id. As the code shown below.

The pseudo-code of the function is :

```
node.c = node.getMinColor(); //cluster head decided
for(int i = 1; i < node.neighbor.length; i++)
    neighborColor[i].id = node.neighbor[i].id;
    neighborColor[i].color = node.c + '-' + i;
end for;
```

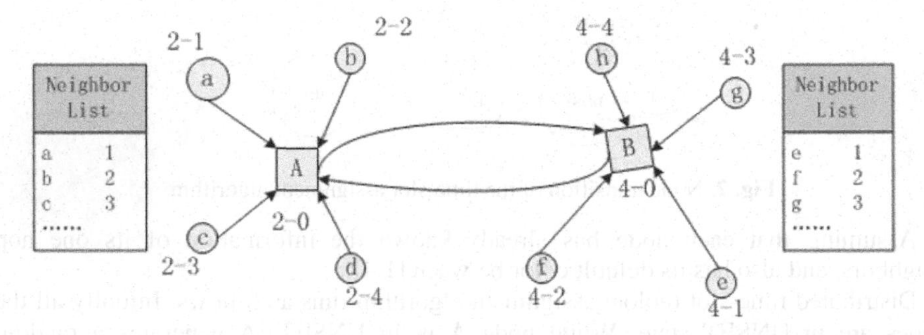

Fig. 3. Intra-cluster color assignment

Function description

getMinColor(): If the node's the default color is included in two-hop neighbor set, get a minimum integer that is not included in two-hop neighbor's color set and return the number.

When color of each node of the entire cluster is decided, cluster head A change its status into SET and broadcast set_A which contains its own color to its neighbors.

Receiving set_A , neighbors(cluster heads) of A will update the relevant information while the ordinary nodes , which are in the same cluster with A, will record their own color and return $done_i$ to node A, A enters DONE.

4 Analysis of Correctness

After algorithm completes, the color of each node is different from the neighbors within two-hop.

Proof: To prove that algorithm to each node is assigned the correct color, it is easy to prove that any two nodes within the two-hop does not select the same color.

Reasons are: (1) a node must get a *grant* message form its one-hop neighbors;

(2) Any node within two-hop distance has at least one same neighbor node;

(3) Each node sends at least one *grant* message at each round. The *grant* contains the color information of its own and its one-hop neighbors. This feature ensures that each node choose the smallest possible color which does not conflict with neighbors within two-hop distance.

5 Simulation

Suppose that there are three buildings in a residential, each building has 20 floors. According to the network topology shown in section two, there are 30 clusters in the network under the assumption that each cluster has a cluster head and 6 normal nodes. Suppose the transmission distance of cluster head is 10m, bandwidth is 2Mbps. In our experiment, we also only consider the cluster heads in the network.

Figure 4 shows the color assignment results of 10 independent tests under the assumption above. As network topology is fixed, It can be seen that during the test, the number fluctuate slightly.

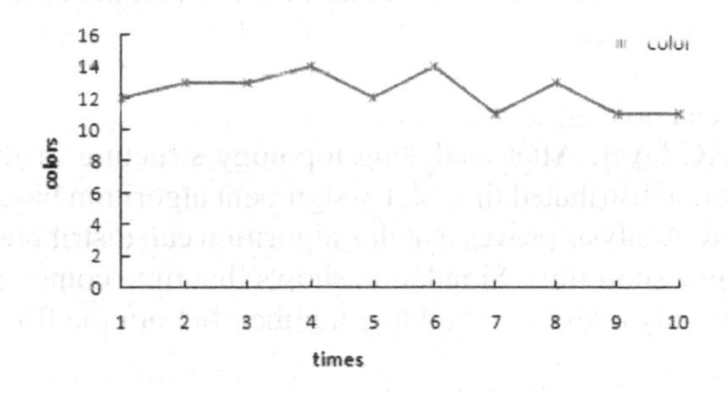

Fig. 4. Color assignment results

Figure 5 shows the relationship between algorithm running time and network scale. As algorithm is specially designed for residential users, cluster head increases as the floor and building number increase. From this figure, algorithm operation time rises obviously as the building number increases, while the building floor does not seen to have some certain influence.

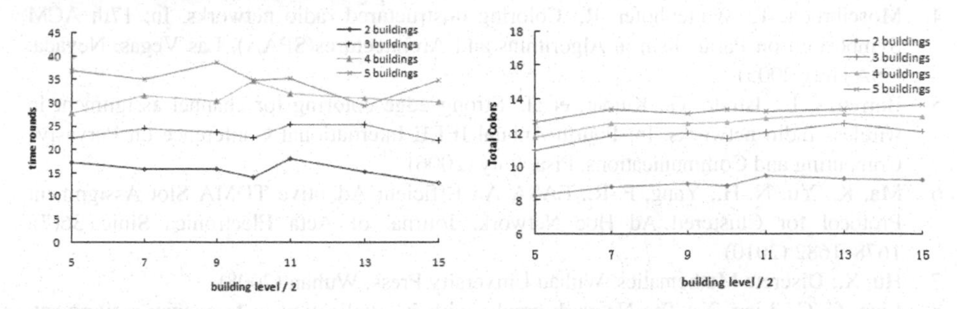

Fig. 5. Algorithm running time and network scale

Fig. 6. Total colors and network scale

Figure 6 shows the relationship between total colors and network scale. It can be seen from the figure that, also, the total colors tend to be bigger along with the increase of building number, and has little thing to do with the number of floors. There' a great difference between only 2 buildings and the other conditions. Considering the geographic characteristics, the estimate of the maximum node degree of the network with 2 buildings is 5, and 8 while the network with building number exceeds 2. From which we can draw the conclusion that in this algorithm, the node degree influence the total colors a lot, as the increase of the node degree, the total count of colors will have an obvious increase.

6 Conclusion

Wireless sensor network used for meter reading system has much difference with traditional wireless sensor network, which takes meter as leaf nodes, has characteristics of fixed topology, small change and high tolerance of the message delay. Therefore, in the initial stage of network establishment, assign fixed time slot for each node, thus, in MAC layer, nodes can do conflict-free communication after applying a simple TDMA protocol in the MAC layer. After analyzing topology structure of the meter reading system, in this paper, a distributed time slot assignment algorithm based on strong edge coloring is proposed. Analysis proves that this algorithm can distributedly allocate time slot for each node in a short time. Simulation shows that time complexity and the total number of color are only relevant to building number, but not the floor.

References

1. Yu, H., Li, O., Zhang, X.: Wireless sensor network theory, technology and implementation. National Defence Industry Press, Beijing (2008)
2. Karl, H., Willig, A.: Protocols and Architectures for Wireless Sensor Networks. Publishing House of Electronics Industry, Beijing (2007)
3. Khalifa, T., Naik, K., Nayak: A Survey of Communication Protocols for Automatic Meter Reading Applications. IEEE Communications Surveys & Tutorials 13(2), 168–182 (2011)
4. Moscibroda, T., Wattenhofer, R.: Coloring unstructured radio networks. In: 17th ACM Symposium on Parallelism in Algorithms and Architectures(SPAA), Las Vegas, Nevada, USA (July 2005)
5. Barrett, C.L., Istrate, G., Kumar, et al.: Strong edge coloring for channel assignment in wireless radio networks. In: Fourth Annual IEEE International Conference on Pervasive Computing and Communications, Pisa, Italy (2006)
6. Ma, K., Yu, N.-H., Yang, F.-R.: EASA An Efficient Adaptive TDMA Slot Assignment Protocol for Clustered Ad Hoc Network. Journal of Acta Electronica Sinica 36(7), 1678–1682 (2010)
7. Hu, X.: Discrete Mathematics. Wuhan University Press, Wuhan (2008)
8. Lian, G.-C., Lian, X.: The Network graphs with its application in frequency assignment. Journal of Jinling Institute of Technology 25(1), 1–4 (2009)

The Drivers Analysis of Strategic Cost Based on AHP

Liwen Niu, Ming Chen, Jun Jiang, and Qi Zhou

School of Economics and Management, Hebei University of Engineering, Handan, China
{7428780,wichenming}@163.com
kittycatcat@126.com, zhoyqi@yahoo.com

Abstract. With the continuous development of knowledge economy, a profound influence have had on enterprises management as well as economic operation by new ideas, new concepts, Information technology advances, emerging new industries, changing customer demand, the pace of the international trend of the times, making an increasingly competitive market. Enterprises are to survive and develop, we must take a strategic level, cultivate long-term competitive advantage to adapt to the changing competitive environment. In this environment, strategic cost management to replace the traditional cost management for modern enterprises to provide a more scientific guidance. This paper used AHP made a brief analysis of cost drivers of strategic cost management . Provide recommendations for various industries when they carry out the strategic cost management.

Keywords: strategic cost management, strategic cost applications, AHP.

With the continuous development of modern market economy and international competition, companies are facing more complex environment, uncertainty also increases. In this context, enterprises want to survive and develop, must consider the issue by standing on a strategic level, not only to focus on immediate development, but also to develop their own long-term competitive advantage.

1 The Concept of Strategic Cost Management

With the intensification of Enterprise global competition, enterprises continue to make every effort to reduce costs in order to Snatch market, to enhance their corporate profits. However, the cost reduction is often accompanied with a decline of product quality and other issues, which will reduce the competitiveness of enterprises. This is not conducive to long-term development. Therefore, enterprises should enhance their competitiveness, not only through this cost as the core of traditional methods, but also take into account the company's strategic development. Strategic cost management get out of the traditional cost management ideas, combined the cost management and business strategy together.

A. Xie & X. Huang (Eds.): Advances in Electrical Engineering and Automation, AISC 139, pp. 59–65.
springerlink.com

2 The Objectives of Strategic Cost Management

On the one hand objectives of strategic cost management should be subordinated to the strategic management objectives, on the other should comply the functional characteristics of cost-management system itself ,but the ultimate goal is to implement the sustainability of the reduce the cost. Although at different times to achieve and maintain a competitive business advantage by taking the cost of management objectives, methods and means are different, but the cost of managing corporate strategic positioning of the main objectives is the continued pursuit of cost reduction.

3 Strategic Cost Drivers

3.1 Definition, Characteristics and Classification of the Strategic Cost Drivers

Cost driver is the cause of the cost of the product. these reasons constitute the cost of the decisive factor. The so-called strategic cost driver is the strategic factors of Company's products cost. It has the following characteristics:

① Closely linked with business strategy, such as the scale of the business, integration and so on.
② Their impact on product cost more long-term, more durable, more far-reaching.
③ Compared with the operating cost drivers, the formation and dynamics of these changes are more difficult.

Strategic cost can be divided into structural cost drivers and performance cost drivers. Because of these cost drivers often not considered in the costing, it is often overlooked by traditional cost management.

3.2 Structure Cost Drivers

The structure cost drivers refers to the basic economic structures, such as long-term corporate investment-related cost drivers. Their formation often takes a long time; and often difficult to change once established; At the same time, these factors tend to occur before the start of production, it must be careful in spending prior to full assessment and analysis. Therefore, the choice of the structural cost drivers can be deciding the cost trend. Structural cost drivers are:

① Economies of scale: the so-called economies of scale is larger in value chain activities, the activity or activities to improve cost efficiency can be assessed on a larger scale due to the volume of business leaving the unit costs.
② Degree of integration: Horizontal integration associated with economies of scale, and degree of integration refers to the degree of vertical integration. Integration is responsible for their own business to business area more widely and more directly in the enterprise business flow directly to the ends extends to sales, parts and components within the system and to provide raw materials and so on.

③ Learning and overflow: Enterprise value chain activities can improve operational efficiency through the learning process, so that Through the study of the factors that can reduce costs.

3.3 Geographical Location

Business location can affect the cost in several ways. Main features:

① Wages and tax rates in different countries, different cities differences affect the company's wage costs and tax expenses.

② Business environment in which the ease of transportation and infrastructure available to the situation will affect the production and operating costs.

③ Enterprise in different climate, culture, concepts such as cultural environment, not only affects the demand for the product, but also affected the business concepts and methods.

④ Location may largely determine the inflow of talent. In has a superior living environment, a good cultural atmosphere and high standard of living cities tend to attract more business people

⑤ Geographical location for operating costs has important implications. Relative energy and raw material suppliers that affect the location of an important factor in the purchase cost. The relative location of the buyer will affect the marketing costs and marketing costs such as freight.

3.4 Performance cost drivers

Performance cost drivers refers to the business operating procedures. It sets up after the structural cost driver decided. And these cost drivers are mostly non-quantifiable, the impact of other costs vary by company.

① The use of production capacity: Mode of production capacity, mainly affect the cost level through the use of fixed costs. The fixed costs associated with the range of increase in production is not changed, when the company's production capacity utilization, production rise, the unit share of fixed costs is relatively small, which led companies to reduce unit costs.

② The relationship between enterprise and customer, supplier.This association can be divided into two categories: one is the internal link; the other is the enterprise with suppliers (upstream), the customer (downstream).

③ Total quality management: The different from traditional quality management is that total quality management emphasized that the scope of quality management, quality control of the whole process. Total quality management philosophy is based on the quality of the least cost, best product quality. Therefore, total quality management improvement can always reduce costs, is an important cost driver, can provide significant opportunities to reduce costs.

The using of AHP in the Strategic Cost

4.1 To Establish Hierarchical Model

In order to study better cost driver, we investigate the use of experts to study the law. First, segment the cost driver. First level: strategic cost A; second floor: implementation of cost driver B1, Geographical location B2, the structural cost driver B3; third layer: economies of scale C1, integration of C2, learning and overflow C3, wages and interest rates differences in C4, transportation convenience C5, climate and culture C6, personnel quality C7, operating costs C8, the use of production capacity model C9, business and the relationship between customers and suppliers C10, total quality management C11.

4.2 The Construction of Pairwise Comparison Matrix to Determine

The first step of the hierarchical model based on pairwise comparison of each factor to compare the outcomes of the series to determine the matrix.

Judgment matrix to include:

1. (B1, B2 B3) importance inter-comparison to A
2. (C1 C2 C3 C4) importance inter-comparison to B1
3. (C5 C6 C7 C8) importance inter-comparison to B2
4. (C9 C10 C11 C12 C13 C14) importance inter-comparison to B3

At inter-comparison more medium, the advertent problem is to ensure to judge the consistency of matrix. (For example judgment A>B, B>C, but A<the C is a kind of judgment inconformity). Table below is AHP inter comparison mark degree.

Table 1. AHP inter-comparison scale

Opposite importance	Define	Exegesis
1	Equal importance	Two factor is, the j is also important
3	A little bit importance	Experience judges factor i more inching than factor j importance
5	Very importance	Experience judges factor i more important than factor j
7	Is obvious importance	Feel keenly factor i to compare factor j importance, and have already been practiced a confirmation
9	Is absolute importance	Strongly feel the factor i comparing factor j to have an absolute predominance position
2,4,6,8	Two close together middle status between the judgments	Mean to need to take to compromise a value between two judgment
Count down	a_{ij} =1/aji	Factor i and factor j relatively gain of judgment value

4.3 Applies an Eigen Value to Solve a Technique to Seeking Maximum Eigen Value and Eigenvector of Judging the Matrix, Then Carries on a Consistency Examination

Judge the R.I of matrix among them. Be worth the following form:

Table 2. The R.I. value of judgment matrix

order	1	2	3	4	5	6	7	8	9
R.I.	0.00	0.00	0.58	0.90	1.12	1.24	1.32	1.41	1.45

Only C.R. <0.01 just think to judgment matrix to have satisfied consistency

Connect down and then carry on applied the Eigen value solve a technique to seeking maximum Eigen value of judging the matrix value and eigenvector and carry on a consistency examination. The knot is in case.

Table 3. Judgment matrix and weight with the code of Strategic cost

A	B1	B2	B3	weight	max-Eigen value
B1	1	5	3	0.652	
B2	1/5	1	1/2	0.125	3.0037
B3	1/3	2	1	0.222	

The form 3 gets C.R=0.032<0.01

Judgment matrix to have satisfied consistency through examination

Table 4. Judgment matrix and weight with the code of implementation of the type

B1	C1	C2	C3	weight	max-Eigen value
C1	1	7	3	0.677	
C2	1/7	1	1/3	0.091	3.007
C3	1/3	3	1	0.231	

The form 4 gets C.R=0.006<0.01

Judgment matrix to have satisfied consistency through examination

Table 5. Judgment matrix and weight with the code of geographical location

B2	C4	C5	C6	C7	C8	weight	max-Eigen value
C4	1	4	3	7	1	0.367	
C5	1/4	1	1	2	1/4	0.095	
C6	1/3	1	1	2	1/3	0.118	5.0126
C7	1/7	1/2	1/2	1	1/6	0.056	
C8	1	4	3	6	1	0.364	

The form 5 gets C.R=0.0028<0.01

Judgment matrix to have satisfied consistency through examination

Table 6. Judgment matrix and weight with the code of performance cost drivers

B3	C9	C10	C11	weight	max-Eigen value
C9	1	2	1/2	0.286	
C10	1/2	1	1/3	0.167	3.0092
C11	2	3	1	0.545	

The form 6 gets C.R=0.0079<0.01

Judgment matrix to have satisfied consistency through examination

Table 7. Influence on synthesis weight by the bottom hierarchical factor

	B1	B2	B3	C Hierarchical total	Sequence
	0.652	0.125	0.222	row preface result	
C1	0.677			0.441	1
C2	0.091			0.059	5
C3	0.231			0.151	2
C4		0.367		0.046	6
C5		0.095		0.012	10
C6		0.118		0.015	9
C7		0.056		0.007	11
C8		0.364		0.046	7
C9			0.286	0.063	4
C10			0.167	0.037	8
C11			0.545	0.121	3

Derived from the table, strategic cost of various factors, the impact on the cost of corporate strategy is different, C1,C3,C11 weight greater than 0.1, the cost impact of corporate strategy is relatively large, the business strategy focused on cost control should be considered; C9,C2,C4,C8 Weights 0.05 or so, the firm's strategy of cost control is also an important influence. In addition, factors in the second layer, B1 strategic cost impact on the cost of corporate strategy is relatively large, is a major factor. In the strategy of cost control, should distinguish between primary and secondary, to consider and develop appropriate methods of cost control measures.

In short, strategic cost management is the traditional cost management positive change to adapt to environment, is the combination of cost management and strategic management, is the development trend of the times of cost management. The combination of cost management and strategic management as a strategic cost management, Not only enhance the influence of cost management but also is the cost management depth in the new social competitive environment, From a practical level, this research will help to enhance the effectiveness of China's enterprise management strategy, and will help solve the current implementation of the strategic reorganization of state-owned enterprises have faced the strategic cost management issues.

References

1. Chen, K.: The management of the strategic cost of enterprises study. Chinese public finance economic publisher, Peking (2001)
2. Xia, K.: The cost management of strategic and its mode and method. Foreign Economy and Management (2) (2000)
3. Yuan, T.: The system contents of strategic cost accountancies of. The Finance Attends Meeting to Account to Lead to Publish (March 2001)
4. Le, Y.: Cost management of. strategics and enterprise competitive advantage. Reply Dan university publisher, Shanghai (2006)

K-Harmonic Means Data Clustering with PSO Algorithm

Fangyan Nie, Tianyi Tu, Meisen Pan, Qiusheng Rong, and Huican Zhou

Institute of Graphics & Image Processing Technology,
Hunan University of Arts and Science,
Changde, 415000, P.R. China
niefyan@gmail.com

Abstract. Clustering is a useful tool to explore data structures and have been employed in many disciplines. One of the most used techniques for clustering is based on *K*-means such that the data is partitioned into *K* clusters. Although *K*-means algorithm is easy to implement and works fast in most situations, it suffers from several drawbacks due to its choice of initializations and convergence to local optima. The *K*-harmonic means clustering solves the problem of initialization, but for the convergence to local optima, the *K*-harmonic means is hopeless. In this paper, a new method is proposed to solve the problem of convergence to local optima, namely particle swarm optimization *K*-harmonic means clustering (PSOKHM) algorithm. The experiment results on the three well known datasets show the effectiveness of the PSOKHM clustering algorithm.

Keywords: Clustering, *K*-harmonic means, Particle swarm optimization.

1 Introduction

Cluster analysis is used for clustering a data set into groups of similar individuals. *K*-means (KM) algorithm is one of the most popular and widely used clustering techniques due to its easiness of implement and efficiency, with linear time complexity. However, the KM algorithm suffers from several drawbacks. The objective function of the KM is not convex and hence it may contain many local minima. Consequently, in the process of minimizing the objective function, there exists a possibility of getting stuck at local minima, as well as at local maxima and saddle points. The outcome of the KM algorithm, therefore, heavily depends on the initial choice of the cluster centers [1-2]. *K*-harmonic means (KHM) algorithm is a new clustering approach presented by B. Zhang et al [3-5]. It is insensitivity to the initialization of the centers. Usually, KHM algorithm does do better job on optimizing performance than KM. While, like the KM algorithm, KHM algorithm does not ensure that it converges to a solution which is global optimum also. Particle swarm optimization (PSO) [6] is a population-based and classical global search algorithm. It searches automatically for the optimum solution in the search space, and is not trapped into local optimal solutions easily. In this paper, we present a new clustering method based on KHM algorithm with PSO, named as PSOKHM, to overcome the shortage of the KHM algorithm.

A. Xie & X. Huang (Eds.): Advances in Electrical Engineering and Automation, AISC 139, pp. 67–73.
springerlink.com

2 *K*-Harmonic Means Clustering Algorithm

KHM algorithm minimizes the harmonic average from all points in N to all centers in K, where N denotes the number of data point in data set, and K denotes the number of classes. The harmonic average of K numbers $\{a_1,\ldots,a_k\}$ is defined as

$$HA(\{a_1,\cdots,a_K\}) = K \Big/ \sum_{i=1}^{K} \frac{1}{a_i} . \tag{1}$$

The harmonic average is small if *one* of the numbers in $\{a_1,\ldots,a_k\}$ is small. Therefore, *HA()* behaves more like the *MIN()* function than an averaging function. When we assign the patterns to the clusters we use minimum of the distances in KM algorithm and harmonic average in KHM algorithm

$$\min\{\| x-c \|^2 | c \in C\} \rightarrow HA\{\| x-c \|^2 | c \in C\} . \tag{2}$$

where $\|\cdot\|^2$ denotes the L^2-distance. The performance function for the KM then is

$$Perf_{KM}\left(\{x_i\}_{i=1}^{N}, \{m_k\}_{k=1}^{K}\right) = \sum_{k=1}^{K} \sum_{x \in S_k} \| x - m_k \|^2 = \sum_{i=1}^{N} MIN\{\| x_i - m_k \|^2 | k = 1,\cdots,K\} . \tag{3}$$

Replacing *MIN()* by *HA()*, the performance function of KHM is gained

$$Perf_{KHM}(X,M) = \sum_{i=1}^{N} HA\{\| x_i - m_k \|^2 | k = 1,\cdots,K\} = \sum_{i=1}^{N} \frac{K}{\sum_{k=1}^{K} \frac{1}{\| x_i - m_k \|^2}} . \tag{4}$$

In [5], the authors used *pth* power of the L^2-distance, it was found that KHM works better with values of $p>2$, then gain the performance function of KHM_p

$$Perf_{KHM_p}(X,M) = \sum_{i=1}^{N} HA\{\| x_i - m_k \|^p | k = 1,\cdots,K\} = \sum_{i=1}^{N} \frac{K}{\sum_{k=1}^{K} \frac{1}{\| x_i - m_k \|^p}} . \tag{5}$$

To derive an algorithm for minimizing the KHM_p performance function, we can take partial derivatives of the KHM_p performance function (5) with respect to the center positions m_k, $k=1,\ldots,K$ and set them to zero

$$\frac{\partial Perf_{KHM_p}(X,M)}{\partial m_k} = -K \sum_{i=1}^{N} \left(p(x_i - m_k) \Big/ d_{ik}^{p+2} \left(\sum_{l=1}^{K} \frac{1}{d_{il}^p} \right)^2 \right) = \vec{0} . \tag{6}$$

where $d_{i,l}=\|x_i-m_l\|$. Solving m_k from the last set of equations, its recursive formula can be gained

$$m_k = \sum_{i=1}^{N} \frac{1}{d_{i,k}^{p+2} \left(\sum_{l=1}^{K} \frac{1}{d_{il}^p} \right)^2} x_i \Bigg/ \sum_{i=1}^{N} \frac{1}{d_{i,k}^{p+2} \left(\sum_{l=1}^{K} \frac{1}{d_{il}^p} \right)^2} . \tag{7}$$

Starting with a set of initial positions of the centers, KHM$_p$ calculates $d_{i,l}=\|x_i-m_l\|$, and then the new positions of the centers from (7). The recursion is continued until the performance value stabilizes. For KM, each data point belongs to the closest center 100% in every iteration, but the KHM is not. KHM has a "built-in" dynamic weighting function, which boosts the data that are not close to any center by giving them a higher weight in the next iteration. This like the soft version of KM - Fuzzy *c*-means algorithm (FCM), The KHM's membership function and dynamic weighting function of each iteration are showed as follows

$$u_{ik} = \frac{1}{d_{ik}^{p+2}} \Big/ \sum_{l=1}^{K} \frac{1}{d_{il}^{p+2}}, \quad i=1,\cdots,N; k=1,\cdots,K \quad . \tag{8}$$

$$w(x_i) = \sum_{l=1}^{K} \frac{1}{\|x_i - m_l\|^{p+2}} \Big/ \left(\sum_{l=1}^{K} \frac{1}{\|x_i - m_l\|^{p}} \right)^2, \quad i=1,\cdots,N \quad . \tag{9}$$

3 Particle Swarm Optimization Algorithm

PSO algorithm is a parallel evolutionary computation technique based on the social behavior metaphor [6]. The procedure of PSO is illustrated as follows.

i. Initialization: Randomly generate a population of the potential solutions, called "particles", and each particle is assigned a randomized velocity.

ii. Evaluate the fitness $f(x_i^t)$ of each particle according to object function.

iii. Compare the personal best of each particle to its current fitness, and set $pbest_i^t$ to the better performance, i.e.

$$pbest_i^t = \begin{cases} x_i^t & if \;\; f(x_i^t) > f(pbest_i^{t-1}) \\ pbest_i^{t-1} & if \;\; f(x_i^t) \le f(pbest_i^{t-1}) \end{cases} . \tag{10}$$

iv. Set the global best $gbest^t$ to the position of the particle with the best fitness within the swarm, i.e.

$$gbest^t \in \{pbest_1^t, \cdots, pbest_m^t\} \mid f(gbest^t) = \max\{f(pbest_1^t), \cdots, f(pbest_m^t)\} \quad . \tag{11}$$

v. Velocity and position update: The particles then "fly" through hyperspace while updating their own velocity, which is accomplished by considering its own past flight and those of its companions. The particle's velocity and position are dynamically updated by the following equations:

$$v_i^{t+1} = wv_i^t + c_1 r_1^t (pbest_i^t - x_i^t) + c_2 r_2^t (gbest^t - x_i^t) \quad . \tag{12}$$

$$x_i^{t+1} = v_i^{t+1} + x_i^t \quad . \tag{13}$$

where c_1, c_2 are learning factors, w is an inertia weight and rand is a uniformly generated random number from the range[0,1], which is produced every time for each iteration, $r_1^t, r_2^t \sim U(0,1)$ are random numbers between zero and one; x_i^t is the current position of the particle i, v_i^t is the current velocity of the particle i, $pbest_i^t$ is the personal best position of the particle i, $gbest^t$ is the global best position of the current generation t. Clerc and Kennedy [7] has pointed out that the use of a constriction factor is needed to ensure the convergence of the algorithm by replacing Eq. (12) with the following:

$$v_i^{t+1} = K(v_i^t + c_1 r_1^t (pbest_i^t - x_i^t) + c_2 r_2^t (gbest^t - x_i^t)) \quad . \tag{14}$$

and

$$K = 2 \Big/ \left| 2 - \varphi - \sqrt{\varphi^2 - 4\varphi} \right| \tag{15}$$

where $\varphi = c_1 + c_2$ and $\varphi > 4$. Typically, φ is set to 4.1 and K is thus 0.7298 .

vi. Let $t=t+1$. Go to step **ii** and repeat until meets the stop criteria.

4 Hybrid PSO and KHM Clustering Algorithm

In the context of clustering, a single particle represents the M cluster centroid vectors. That is, each particle x_i is constructed as follows:

$$x_i = (m_{i1}, \cdots, m_{ij}, \cdots, m_{iM}) \tag{16}$$

where m_{ij} refers to the jth cluster centroid vector of the ith particle in cluster C_{ij}. Therefore, a swarm represents a number of candidate clusterings for the current data vectors. The fitness of particles is measured as the Eq. (5).

1. **Initialization.** Generate a population of size P.
Repeat
2. **Evaluation.** Evaluate the fitness of each particle according to Eq. (5).
3. **KHM Method.** Apply KHM algorithm to the global best particle and replace the global best particle with the update.
4. **PSO Method.** Apply PSO algorithm for updating P particles.
   Selection. From the population select the global best particle and the local best particles according to Eqs. (10) and (11).
   Velocity Update. Apply velocity update to the P particles according to Eqs. (14) and (13).
Until a termination condition is met.

Fig. 1. The PSOKHM algorithm

The KHM algorithm tends to converge faster than the PSO, but usually led to a less accurate clustering because it will trap a local optimum solution if the problem contains many local minima. This section shows that the performance of the PSO algorithm can further be improved by embedding the KHM algorithm. The main idea of the proposed algorithms is to use PSO to generate non-local moves for the cluster centers and to select the suitable best solution. The KHM data clustering with PSO algorithm is show in Fig. 1.

5 Experimental Results and Comparative Performances

This section compares the results of the KM, KHM and PSOKHM on three classification problems. The clusterings quality is measured according to the following three criteria:

- the quantization error, it is defined as follows

$$J = \sum_{i=1}^{M} \sum_{\forall x \in C_j} \frac{\| x - m_j \|}{|C_j|} \Big/ M \quad (17)$$

where $|C_j|$ is the number of data vectors belonging to cluster C_j.

- the intra-cluster distances, i.e. the distance between data vectors within a cluster, where the objective is to minimize the intra-cluster distances;
- the inter-cluster distances, i.e. the distance between the centroids of the clusters, where the objective is to maximize the distance between clusters.

For all the results reported, averages over 50 simulations are given. All algorithms are run for 1000 function evaluations, and the PSOKHM algorithms used 20 particles. For PSOKHM, c_1=2.1, c_2=2, then K=0.7298.

The well known classification problems used for the purpose of this paper are

- **Iris plants database:** Iris dataset is a standard benchmark in the pattern recognition literature. The dataset contains 150 instances each composed of four measurements of an iris flower. There are three types of flowers represented by 50 instances each.
- **Wine:** This is a classification problem with "well behaved" class structures. There are 13 inputs, 3 classes and 178 data vectors.
- **Breast cancer:** The Wisconsin breast cancer database contains 30 relevant inputs and 2 classes, 569 instances. The objective is to classify each data vector into benign or malignant tumors.

Table 1 shows the results obtained from the three clustering algorithms for the problems above. The values reported are averages over 50 simulations with mean and standard deviations to indicate the range of values to which the algorithms converge. First, if we consider the quantization error, the PSOKHM had the smallest average quantization error for all the problems. When considering inter- and intra-cluster distances, the former ensures larger separation between the different clusters, while the latter ensures compact clusters with deviation from the cluster centroids. With reference to these criteria, the PSOKHM approaches succeeded most in finding clusters with larger separation than the KM and KHM algorithms.

Fig. 2 illustrates the convergence behavior of the KHM and PSOKHM algorithm for the first problem. The KHM algorithm exhibited a faster, but premature convergence to a larger fitness value, while the PSOKHM algorithm had slower convergence, but to lower fitness value. As indicated in Fig. 2, the KHM algorithm converged from a higher fitness value to a lower fitness value with a very faster speed, while the convergent speed of the PSOKHM algorithm is a slower and smoothness from a lower fitness value.

Table 1. Experiments of K-means, KHM and PSOKHM

Problem	Algorithm		Quantization Error		Intra-cluster Distance		Inter-cluster Distance	
			Mean	Std.	Mean	Std.	Mean	Std.
Iris	K-means		0.6925	0.1372	36.1283	21.5113	3.7774	0.1812
	KHM	p=0.5	0.7245	0.1055	40.9597	26.7610	3.6516	0.6146
		p=1.5	0.6512	6.3e-17	16.3782	7.2e-15	3.9383	3.8e-15
		p=2.0	0.6494	0.0000	16.5971	1.9e-14	3.9401	2.1e-15
		p=2.5	0.6490	5.6e-16	16.5303	1.0e-14	3.9379	4.0e-15
		p=3.5	0.6474	0.0000	16.9357	1.4e-14	3.9434	5.4e-15
	PSOKHM	p=0.5	0.6858	0.0876	32.0785	21.5173	3.7523	0.2899
		p=1.5	0.6512	6.7e-17	16.3782	7.6e-15	3.9383	3.9e-15
		p=2.0	0.6494	0.0000	16.5971	2.0e-14	3.9401	2.0e-15
		p=2.5	0.6490	5.6e-16	16.5303	1.3e-14	3.9379	4.0e-15
		p=3.5	0.6474	0.0000	16.9357	1.5e-14	3.9434	5.4e-15
Wine	K-means		143.556	297.737	1.62e06	8.43e05	8.02e04	5.37e03
	KHM	p=0.5	101.123	3.8196	2.09e06	1.67e06	7.76e04	1.02e04
		p=1.5	97.7925	4.3e-14	7.18e05	4.39e-10	8.55e04	7.97e-11
		p=2.0	98.6952	6.8e-14	7.21e05	4.98e-10	8.54e04	7.34e-11
		p=2.5	100.723	1.0e-13	7.39e05	6.87e-10	8.52e04	5.02e-11
		p=3.5	106.716	2.4926	1.25e06	9.82e05	8.29e04	4.81e03
	PSOKHM	p=0.5	96.5786	0.3780	8.93e05	9.09e04	8.32e04	1.02e03
		p=1.5	97.7925	4.3e-14	7.18e05	4.27e-10	8.55e04	7.91e-11
		p=2.0	98.6952	6.7 e-14	7.21e05	5.09e-10	8.54e04	7.34e-11
		p=2.5	100.723	1.0e-13	7.39e05	6.87e-10	8.52e04	4.93e-11
		p=3.5	107.149	0.3296	1.05e06	2.53e04	8.39e04	60.7068
Breast Cancer	K-means		427.178	55.6972	1.03e08	4.36e07	2.24e05	5.91e04
	KHM	p=0.5	337.975	2.6920	3.78e07	7.49e06	2.79e05	2.66e04
		p=1.5	334.390	2.3e-13	3.08e07	9.98e-09	3.13e05	2.94e-10
		p=2.0	335.765	5.7e-14	3.18e07	3.00e-08	3.14e05	5.88e-11
		p=2.5	343.667	4.6e-13	3.73e07	1.08e-08	3.12e05	2.94e-10
		p=3.5	364.325	3.4e-13	5.16e07	3.76e-08	2.97e05	2.94e-10
	PSOKHM	p=0.5	336.085	3.7243	4.91e07	1.73e07	2.52e05	4.18e04
		p=1.5	334.390	2.3e-13	3.08e07	1.15e-08	3.13e05	2.94e-10
		p=2.0	335.765	5.7e-14	3.18e07	3.05e-08	3.14e05	5.88e-11
		p=2.5	343.667	4.6e-13	3.73e07	1.04e-08	3.12e05	2.94e-10
		p=3.5	361.853	0.5047	4.99e07	3.35e05	2.99e05	419.3463

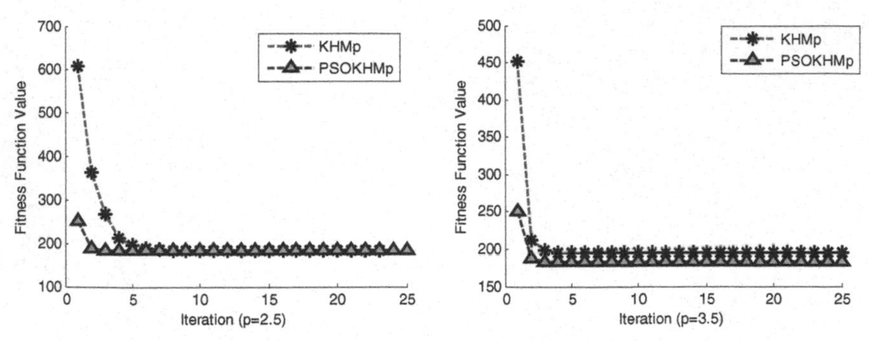

Fig. 2. Algorithm convergence for Iris clustering problem when p=2.5 and p=3.5

6 Conclusions

This paper investigated the application of the KHM with PSO algorithm to cluster data vectors. Three algorithms were tested, namely the KM, the KHM and PSOKHM algorithm. Compared with KM and KHM algorithm, the PSOKHM showed better convergence to lower quantization error, and in general, larger inter-cluster distances and smaller intra-cluster distances, while the KM algorithm may get stuck at a local optimum, depending on the choice of the initial cluster centers. In addition, the PSOKHM converge at a better global optimum solution compared with KHM.

Acknowledgments. This work is partially supported by the Scientific Research Project of Institutions of Higher Education of Hunan Province, China (Grant No. 11C0913), the Young Core Instructor Foundation of Hunan Provincial Institutions of Higher Education, China, the Scientific Research Foundation for Doctor of Hunan University of Arts and Science, the Foundation of 11th Five-year Plan for Key Construction Academic Subject (Optics) of Hunan Province, China and the Scientific Research Foundation of Hunan University of Arts and Science (Grant No. JJQD06025).

References

1. Xu, R., Wunsch, D.I.I.: Survey of clustering algorithms. IEEE Trans. Neural Networks 16(3), 645–678 (2005)
2. Yi-Tung, K., Erwie, Z., I-Wei, K.: A hybridized approach to data clustering. Expert Systems with Applications 34, 1754–1762 (2008)
3. Zhang, B., Hsu, M., Dayal, U.: K-harmonic means - A data clustering algorithm. Technical Report HPL-1999-124. Hewlett-Packard Laboratories (1999)
4. Zhang, B., Hsu, M., Dayal, U.: *K*-Harmonic Means - A Spatial Clustering Algorithm with Boosting. In: Roddick, J., Hornsby, K.S. (eds.) TSDM 2000. LNCS (LNAI), vol. 2007, p. 31. Springer, Heidelberg (2001)
5. Zhang, B.: Generalized K-harmonic means – dynamic weighting of data in unsupervised learning. In: Proc. of the 1st SIAM International Conference on Data Mining, IL, Chicago, pp. 1–13 (2001)
6. Kennedy, J., Eberhart, R.C.: Particle swarm optimization. In: Proc. of the IEEE International Joint Conference on Neural Network, pp. 1942–1948 (1995)
7. Clerc, M., Kennedy, J.: The particle swarm – Explosion, Stability, and Convergence in a Multidimensional Complex Space. IEEE Trans. on Evolutionary Computation 6(1), 58–73 (2002)

Mobile WEKA as Data Mining Tool on Android

Pengfei Liu[1], Yanhua Chen[1], Wulei Tang[2], and Qiang Yue[3]

[1] College of Science, South China Agricultural University, Guangzhou, 510642, China
[2] SoftPark of Guangdong, Guangzhou, 510663, China
[3] Guangdong Nortel Telecommunications Co. Ltd, 510665, China
pfliu@scau.edu.cn, {chen_yanhua123,atota}@126.com,
stevenyue@gdnt.com.cn

Abstract. Mobile data mining is an exciting research area that aims at finding interesting patterns from datasets on mobile platform. Limited to the computing power and operating system of traditional mobile devices, mobile data mining lacks attention before. Nowadays mobile devices have a stronger and stronger computation power also the advanced operating system supporting the demand of data mining anywhere and anytime. This paper presents and implements a Java based framework to extend data mining tool Weka to mobile platform. It provides a friendly graphic user interface and simplifies the classification, clustering and associate rule mining functions on android platforms. As an example of usage, we test the model on some datasets and illustrate the feasibility of the proposed approach. A Java implementation of the model demonstrated in this article is available from mobileWeka project website. http://mobileweka.googlecode.com/files/MobileWeka.zip.

Keywords: Mobile, Data mining, Machine learning.

1 Introduction

Mobile computing increasingly present in recent years with stronger and stronger computation power however being ignored, but with right computing models it could be used to solve practical problems such as data mining. At present we focus on the Android platform running on Linux kernel due to its advantages such as open source and free, that make it the best choice for mobile devices.

Data mining plays an important role in many research areas and is mostly done on computers in the past. Supposing that one day every mobile device can easily connect to other devices including data generating devices and can exchange data freely via Bluetooth or WIFI, and then we can process data just by mobile devices. This idea of data mining anywhere and anytime is very amazing but not crazy.

Most of the popular data mining models are designed to work on computers and they often do not pay attention to mobile computing environments. In support of mobile data mining, we proposed an extensible model, which does not need to send local data to other workstations or servers and receive results from them thereby not causing extra communication cost.

The goal of this work is to provide a user-friendly model to simplify data mining on mobile devices, based on an extensible design for enabling adding new data

A. Xie & X. Huang (Eds.): Advances in Electrical Engineering and Automation, AISC 139, pp. 75–80.
springerlink.com

mining algorithms. These algorithms can be of any kind, including classical ones such as Bayes, Neural Network and K-Means, as the cases we will present.

2 State of the Art

Mobile computing enables the processing with a more convenient way than traditional desktop processor architectures. At present there are many interesting applications in mobile devices. Many of them are task-specific and for research purpose, such as measuring generalization of visuomotor perturbations in wrist movements using mobile phones [1]; Shih G *et al.* proposed the possibility of imaging informatics on Android or IPhone platforms [2].

Data mining on mobile platform is developing in recent years. Several data mining tools on mobile platform have been proposed. For example, The AnalyticDroid from Togaware is Rattle for Android essentially [3], that is an experimental application for controlling analytics from a mobile device using an R Server.

One client programs on mobile devices can invoke remote execution of data mining tasks and show analysis results locally [4]. A distributed and mobile algorithm for global association rule mining was proved to work well for large problem sizes [5].

A model performs minor data analysis and summary early before the source data arrives to the data mining machine. It aims to reducing the amount of further processing required in order to perform data mining [6]. In addition, there is another Java-based project, being the main reference Weka-for-Android (https://github.com/rjmarsan/Weka-for-Android), designed to be a port of weka 3 to the Android platform; the project is to get GUI components removed from Weka project to make it work on Android.

On traditional desktop computers, Weka is very popular and is a landmark in data mining and machine learning [7]. It is widely admitted within academia users as a useful tool for data mining research, especially for bioinformatics. However, it lacks the mobile platform version and is not applied to the mobile data mining domain, which is worth devoting time.

We think it is reasonable now to localize mobile data mining with a user friendly graphic user interface while mobile devices have a stronger and stronger computation power; also there are a lot of minor or medium size datasets, and the mobile processors can tackle them without taking too much time.

In this way, the mobile data mining would be a very attractive technique to increase the flexibility of data mining, especially for bioinformatics researchers.

3 Framework

This section dives in the functionality and design of the proposed framework, called mobileWeka. It is designed to easily launch data mining on mobile device, and be capable of seamlessly employ all of the algorithms embedded in Weka.

MobileWeka should support standard data mining tasks, e.g. clustering, classification and association rule mining. The main issue to solve in developing course is to abstract basic control flow of data mining, and providing users with graphic interface that having some key features: simple and extensible.

The framework composes of four layers. The first one is Android OS; the second one is Android Java SDK that depends on the first one; the third one is Weka which is in charges of bottom data mining algorithms. The last one is the presentation layer, which provides a friendly user interface and control logic. Fig. 1 presents an illustration of the proposed system architecture.

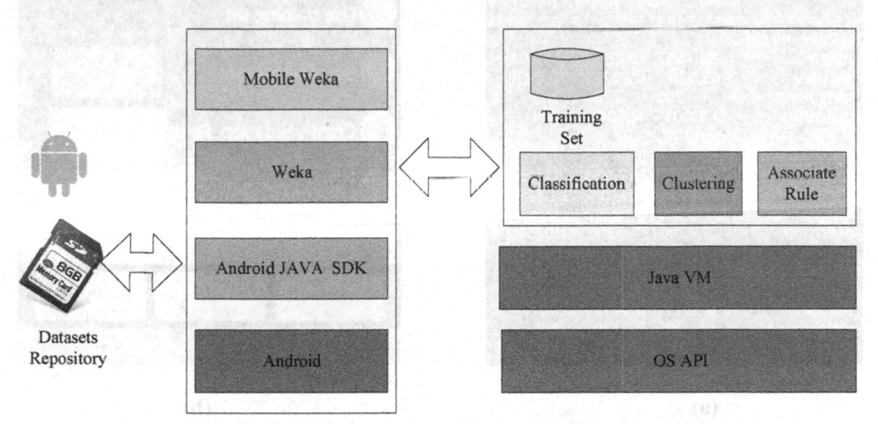

Fig. 1. Architecture of the proposed mobile data mining model

MobileWeka is written in Java since it is supported by Android SDK also platform independent. The developing environment is Eclipse IDE for Java Developers, with Java SDK 1.6 and android SDK 1.5. What we mainly to do list as following.

(i). Write XML configuration files according to the design of GUI.
(ii). Fulfill the file chooser.
(iii). Import Weka package to mobileWeka.
(iv). Enroll Weka's embedded algorithms to mobileWeka.
(v). Design suitable processing logic of data mining.

4 Result

As we have expressed, mobileWeka is the mobile version of Weka and the main differences between mobileWeka and Weka are the running efficiency and user interface. So this section mainly presents efficiency test results between both of them.

To fully test mobileWeka, some datasets contained in WEKA are introduced. The experimental mobile device is the HTC Dream mobile phone (also known as TMobile G1), which was the first phone to the market using the Android OS. The processor of G1 is ARM-based MSM7201A with 528 Mhz speed and a 192MB RAM memory; while computer's processor is Intel® Core™2 Duo T6670 with 2.20 GHz.

During test procedure, we take some screenshots of mobileWeka. Figure 2 shows the classifier model option setting, algorithms selection, model parameters setting and result evaluation. More screenshots such as clustering and association rule mining can refer to supplementary data.

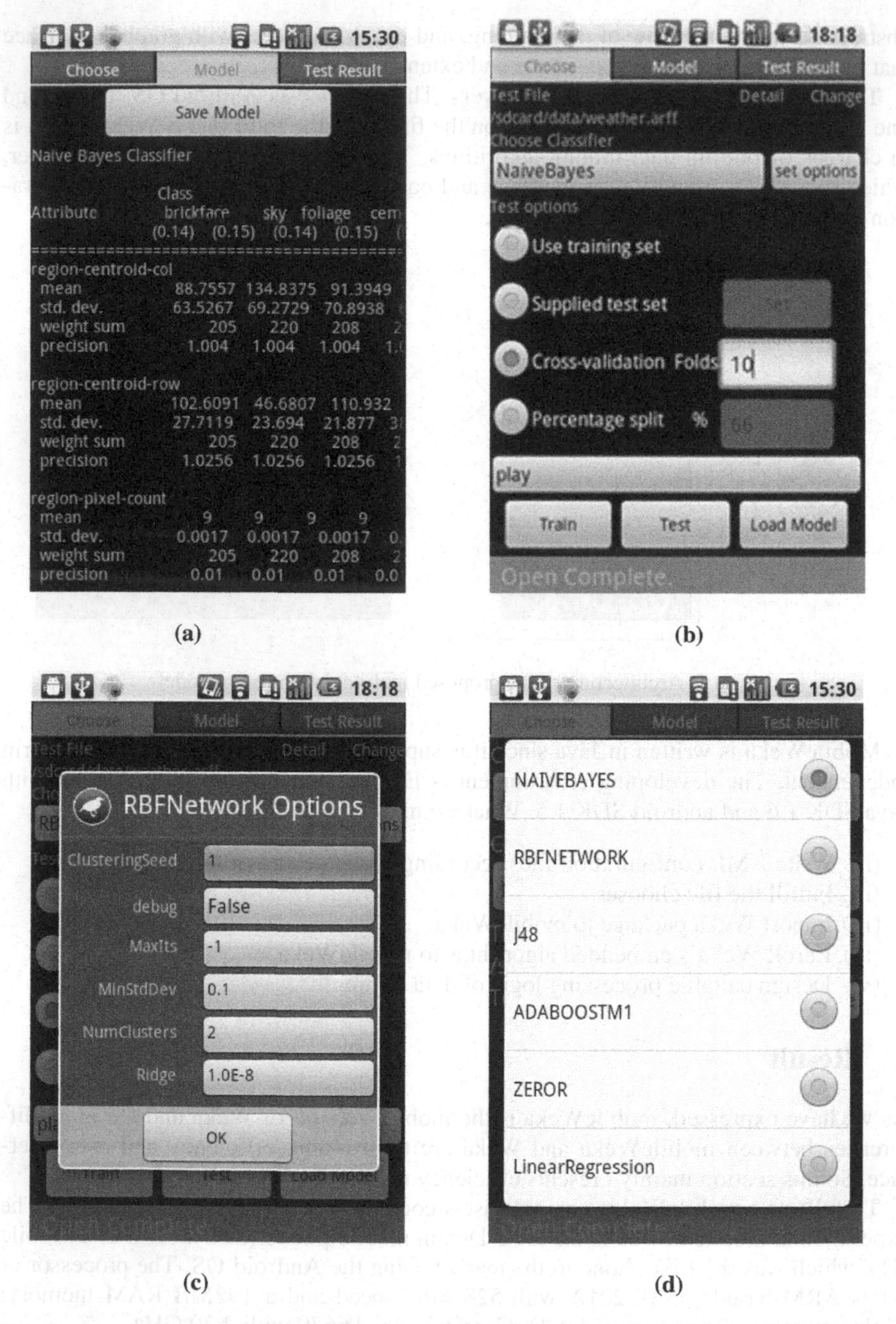

Fig. 2. (a): Classifier models evaluation result. 2 (b): Classifier model option setting. 2(c): Classifier model parameters setting. 2(d): Classifier algorithm selection.

The test datasets of bank-data, vote and segment-test have the sizes of 37k bytes, 40k bytes and 108k bytes respectively. During the classification tests, we choose the 66% percentage split method as default standard. Every experiment runs on G1 (Mobile-Weka) and computer (Weka) both. The clustering results are shown in table 1 while the results of classification (sum of training time and test time, in seconds) of are shown in table 2 and the results of association rule mining are shown in table 3. More test results can refer to the supplementary data.

It can be seen that computing time of MobileWeka varied from about 0.01 second to hundreds of seconds, depending on the mining algorithms and the sizes of the data. Some test results have the computing time of 0, because the processing time on computer is too short to record.

The results indicates that mobileWeka uses more computing time than desktop computer but still falls within normal range in most cases. As the sizes of datasets increase, mobileWeka's processing time grows dramatically.

Without doubts the computing models on mobile platforms have lower performance than in computers. Nevertheless, they have great room of improvement with the rapid development of mobile technology.

Table 1. Result of clustering

Algorithm	Dataset	MobileWeka	Weka
SimpleKMeans	bank-data	8.8	0
EM	bank-data	unfinished	49
FarthestFirst	bank-data	1.23	0
DBScan	bank-data	410	1

Table 2. Result of classification

Algorithm	DataSet	MobileWeka	Weka
NAIVEBAYES	vote	1.36	0.01
RBFNETWORK	vote	16.41	0.07
J48	vote	1.89	0.02
ADABOOSTM1	vote	6.02	0.02
ZEROR	vote	0.28	0
NAIVEBAYES	segment-test	34.76	0.03
RBFNETWORK	segment-test	1300.45	6.29
J48	segment-test	60.2	0.11
ADABOOSTM1	segment-test	11.49	0.02
ZEROR	segment-test	1.56	0.01

Table 3. Result of association rule mining

Algorithm	Dataset	MobileWeka	Weka
Apriori	vote	22.76	0
FilteredAssociator	vote	15.8	0
FPGrowth	vote	13.6	0

5 Conclusion

We have presented a mobile data mining model named mobileWeka and demonstrated the feasibility of data mining on mobile devices. MobileWeka is easy to use also works well when datasets are small size or medium size; On the other hand, it does sacrifice execution time while working on large datasets due to low computing capability.

Mobile data mining is not being used as often as would be expectable, taking into account that mobile devices' capabilities. However, this situation will change with the rapid development of mobile hardware in short years, and then mobile data mining will enjoy widespread acceptance in both academia and business.

As future developments of the proposed model we are studying how to make the model more useful and high efficiency, also completing other functions such as data editing and graph presentation.

Furthermore, MobileWeka can be transplanted to any other mobile platform that supporting Java environment. We plan to release this work as an open source project in future.

Acknowledgements. We would like to express our sincere thanks to all the people who have devoted their time to Weka project. Without their effort, it would be impossible for us to finish our study and this paper.

Funding: Supported by Scientific Research Foundation of South China Agricultural University (4900-k11045) and Natural Science Foundation of Guangdong Province, China (S2011040004387).

References

1. Fernandes, H.L., Albert, M.V., Kording, K.P.: Measuring Generalization of Visuomotor Perturbations in Wrist Movements Using Mobile Phones. Plos ONE 6(5), e20290 (2011)
2. Shih, G., Lakhani, P., Nagy, P.: Is Android or iPhone the Platform for Innovation in Imaging Informatics. Journal of Digital Imaging 23(1), 2–7 (2010)
3. Williams, G.: Rattle: A Data Mining GUI for R. The R Journal 1, 45–55 (2009)
4. Talia, D., Trunfio, P.: Mobile Data Mining on Small Devices through Web Services. In: Mobile Intelligence, John Wiley & Sons, Inc., NJ (2010)
5. Wang, F., Helian, N., Guo, Y., Jin, H.: A distributed and mobile data mining system. Parallel and Distributed Computing, Applications and Technologies 27-29, 916–918 (2003)
6. Goh, J.Y., Taniar, D.: An Efficient Mobile Data Mining Model. In: Cao, J., Yang, L.T., Guo, M., Lau, F. (eds.) ISPA 2004. LNCS, vol. 3358, pp. 54–58. Springer, Heidelberg (2004)
7. Frank, E., Hall, M., Trigg, L., Holmes, G., Witten, I.H.: Data mining in bioinformatics using Weka. Bioinformatics 20(15), 2479–2481 (2004)

Multi-relay Cooperative Performance Based on LT Fountain Coding

ChunKai Chen[1,2] and Hong Xie[1]

[1] Harbin Engineering University, 150001, Harbin, China
[2] Heilongjiang Institute of Science and Technology, 150027, Harbin, China
{Chenckoffice,xiehonghrbedu}@163.com

Abstract. We investigate a cooperative communications scheme with N relays, where both the transmissions from the source to the relays and from the relays to the destination use fountain codes. Fountain code is more suitable cooperative communication. In order to achieve high-efficiency cooperative communication, when CSI is unknown to the sender, a multi-relay cooperative model is proposed, and LT code apply in it, furthermore, we discussed two transmission modes, error bit rate and the probability density. Simulation results show that: comparing to direct transmission, the scheme have lower bit error rate, as well as a faster decoding speed.

Keywords: cooperative communication, fountain codes, LT codes, amplify-and-forward, diversity order.

1 Introduction

Cooperative diversity can solve a difficult problem that some antennas are installed in the mobile terminal; it will promote the practicability of MIMO technology. Cooperative diversity can effectively use space resources to improve communication system performance, such as system power consumption, BER performance, outage probability, coverage, etc. However, cooperative diversity is also facing a major problem that base station must know the number of collaborative terminal information to correctly decode. Coding cooperation is a product combined by channel coding technology ideas and cooperative diversity. As it has spatial diversity and time diversity advantage, and it can get diversity and coding gain. At the same time, a large number of classical channel coding techniques can be used, different schemes of the channel encoding and decoding become focus in research of coding cooperation recently.

In recent years, Luby proposed a new encoding scheme, that is, fountain codes. It has been continuously extended to the physical layer and applying [1-5]. When the channel erasure probability is high, it has a large amount of data feedback in transmission. On the other hand, in some broadcast or multicast situation, when any receiver does not receive sender information correctly, source terminal is required to send the information again. When the number of receiver terminals is more, the channel is poor, the resend information will be greatly increased, resulting in transmission efficiency decreased, when transmission terminal does not know channel information, fountain codes can adjust code rate according to the channel state adaptively, firstly,

A. Xie & X. Huang (Eds.): Advances in Electrical Engineering and Automation, AISC 139, pp. 81–86.
springerlink.com

transmission terminal encode to transmission information using the fountain encoding scheme, as long as the receiver receives enough encoded information, the original information can be translated [6-7]. This nature of code rate independence makes fountain code become coding scheme of cooperation communication than other fixed code rate. Jeff Casture proposed fountain coding cooperative scheme in the single-relay cooperative system [8]. This paper looks at the structure based on digital fountain codes and it is suitable for relay channel under effective communications encoding and decoding scheme, we expect that fountain codes excellent performance can bring more coding gain in cooperative communication system, and discussed fountain cooperation in two ways and direct transmission performance.

2 System and Channel Model

Cooperative diversity idea derived from Cover and El Gamal's research on the relay channel, the main purpose of relaying is to help the main channel transmission, however, it does not distinguish between primary and secondary in cooperative communication, the user is both a information source and relaying. In order to obtain diversity gain, the cooperative model is considered as downlink mode in this paper, base stations communicate with a single-user with the help of some relay nodes, to simplify the analysis, and we consider only one Level relay case.

Channel is erasure channel among users, it is assumed that erasure probability is same form the base station to any user node, defined as p_e^D , erasure probability distribution of the base station to the relay node and relay node to the user terminal is defined as p_e^R and p_e^{RD} separately, it is transmitted according to two ways.

2.1 The First Transmission Way

Its block diagram is shown in Fig. 1, after the source transmit information to the fountain encoder, it will continue to transmit information to all relays and the destination nodes, if one of relays decoded successfully, the data is encoded with fountain scheme, then the relay forwards to the destination node directly, this method has the advantage of its simplicity, since only one node (whether successful decoding nodes or base stations) can transmit at any time, so there is no mutual interference among nodes. As there is smaller erasure probability from base station to relay nodes, or relay to user nodes relative to link of base station to user, so this method is superior to the performance of direct transmission.

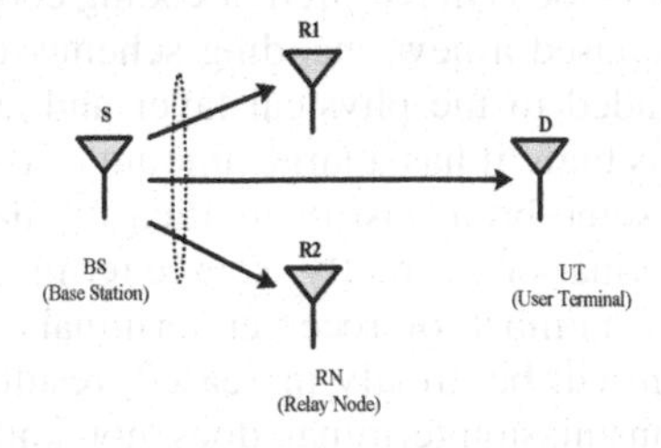

Fig.1. Cooperative relay transmission model (method 1,2)

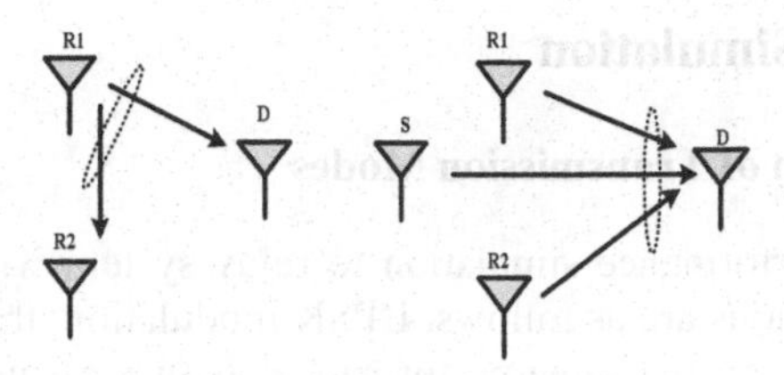

Fig. 2. Model of cooperative relay receiver

2.2 The Second Transmission Way

The base station transmits information, all relay nodes are involved in the cooperative transmission, after it is decoded correctly then forwarded to the users, the user terminals also receives the information decode correctly from the base station and all the relay nodes, information received by the user continue to accumulate, thus the fountain decoding time is shortened. However, there is mutual interference problem among relay nodes, if it is not handled properly this will result in a decline in overall performance, [9] proposed for each transmitter is assigned a unique spreading codes, RAKE receiver process information from the sending terminal at the receiving terminal, this method ensures its orthogonality, but the spreading code reduces the bandwidth efficiency, in other words, the bandwidth efficiency must be increased to ensure the continued cycle of the symbol, which will reduce the system throughput. [10] Proposed the amplitude modulation scheme, this scheme is belong to erasure multiple access channel, and it does not decline throughput and bandwidth efficiency. In order to all relay nodes receive a packet correctly, the sending node needs to send the packet, the average number $\bar{N}$ can be expressed as

$$\bar{N} = E[N] = \sum_{n=1}^{\infty} n \cdot P\{N=n\} = (1-P_{ek})^L + \sum_{n=2}^{\infty}\left\{n \cdot \left[\left(1-P_{ek}^{n}\right)^L - \left(1-P_{ek}^{n-1}\right)^L\right]\right\} \quad (1)$$

Where P_{ek} is the packet error probability, when fountain transmitted mode is used, assuming the source node implement fountain coding by the original packet of M as a group, any relay node needs to receive M packets from the source node sends the encoded information correctly, so original M packets can be encoded. We assume that all relay nodes had encoded the original information correctly, the source node sends the number of packets is N_M, the conditions of $N_M = n$ include the following: at least one node in L relay nodes can encode the original information after this node send nth data packets , but other nodes encode original information only source node transmit $M \sim n-1$ packet.

3 Analysis and Simulation

3.1 BER Comparison of Transmission Modes

We had carried out performance simulation to relay system with fountain transmission; Simulation conditions are as follows, BPSK modulation, the channel is Rayleigh fading channels. There are the same performances between source - the relay nodes and relay - target node. To compare the performance of the fountain relay Transmit, we had carried out simulation to common relay which use the same error correction coding and modulation mode. Number of transmitting bits is K = 10000, LT code encoding is selected.

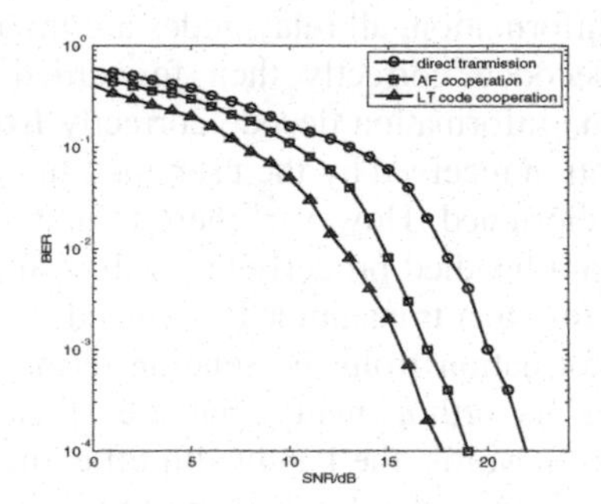

Fig. 3. Comparison of three transmission error bit rate

As illustrated in Fig. 3, relative to direct transmission, relay is involved in system; AF transmission will increase the size of signal to noise ratio certainly, cooperative bit error rate is smaller than direct transmission bit error rate. The amplify-and-forward cooperation based on fountain codes (AF mode) has higher accuracy than the other mode; this will improve system performance greatly.

3.2 Strategy Comparison of PDF

In order to compare performance with direct transmission (no relay nodes are involved in system), we have analyzed the performance based on fountain cooperative system performance, the numbers of relay is defined as U=20, erasure probability is $P_e^D = 0.3$, the numbers of data packet is defined as M = 100. We use the probability density function (PDF) as performance parameter, and describe transmission strategy 1, 2 of cooperative transmission performance.

When there is a valid node, as channel conditions from the base station to relay and relay to the user node are better than condition from the base station to the user node, so, the total performance can be improved. We assume that the erasure probability is $P_e^R = P_e^{RD} = 0.1$. There are two possible scenarios. First, the target node can decode successfully before the relay node, the probability of this event is $\sum_{j \geq N} f^{ER}\left(j, P_e^R\right)$, where j means the j th data packet from source node, in this case, successful decoding probability of the destination node is $f^{ER}\left(N, P_e^D\right)$. Second, the occurrence probability ($j < N$) that the relay node can decode successfully before the target relay node is $f^{ER}\left(j, P_e^R\right)$, in this case, the target node receives data, which include information packets of base station and relay nods. Define the following two auxiliary equations:

$$g_{y,P_e}(x) = C_x^y (1-P_e)^x P_e^{y-x} \qquad x \in \{0:y\} \tag{2}$$

$$h_{y,P_e}(x) = C_{x-1}^{y-1} (1-P_e)^x P_e^{y-x} \qquad x \in \{1:y\} \tag{3}$$

While relay nodes can decode after transmitting the j th data packet, the probability that the s th data packet is received is $g_{j,P_e^D}(s)$, for the remaining $N-j$ packets, the probability received of t th packet is $h_{N-j,P_e^{RD}}(t)$. Taking into account U users, so the total PDF is

$$f^{RI}(N) = \sum_{j \ge N} f^{ER}\left(j, P_e^R\right) f^{Dir}(N) + \sum_{j=M}^{N-1} f^{ER}\left(j, P_e^R\right) \sum_{i=1}^{U} C_i^U \left[A(j,N)\right]^i \left[B(j,N)\right]^{U-i} \tag{4}$$

For different number of users, we use the formula (4) and get the PDF curve shown in curve b in Fig.4. The first 1 transmission mode PDF curve is more concentrated than the direct transmission. When the base station and relay nodes transmitted information to the user nods simultaneously, the same formula (4) is used. However the auxiliary formula is replaced with following formula:

$$A'(j,N) = \sum_{s=0}^{j} \sum_{q=1}^{N-j} g_{j,P_e^D}(s) h''_{N-j}(q) f(s+q) \tag{5}$$

$$h''_N(q) = \frac{1}{C_2}\left[h'_N(q) - \left(h'_{N-1}(q) P_e^D P_e^{RD}\right)\right] \tag{6}$$

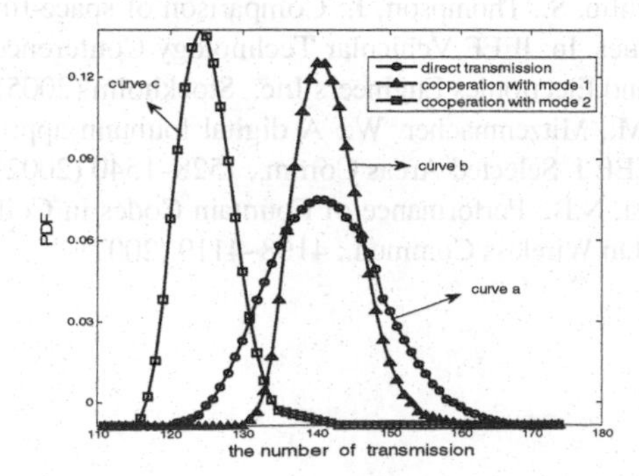

Fig. 4. Comparison of PDF in several ways

Where C_2 is the normalization constant, the PDF curve for the transmission mode 2 is shown in curve c in Fig.4. Transmission mode 2 is better than the other two cases.

4 Conclusions

In this paper, we research cooperative communication systems based on Fountain codes. At first, encoding and decoding processes of LT codes in fountain code have been analyzed in detail. Taking full advantage of Fountain code, such as fighting against fast time-varying fading, and cooperative diversity gain, in order to accelerating decode speed, two methods were proposed. Finally, we simulate and test performance of the entire cooperative communication system, simulation results show that cooperative scheme based on fountain code not only receive efficiency tend to practical applications, and it can improve overall system reliability effectively. In addition, the fountain code encoding and decoding algorithms, the degree of design, real-time and other aspects need further research.

References

1. Luby, M.: LT codes. In: Annu. Symp. Found. Comput. Sci. Proc., pp. 271–280. Institute of Electrical and Electronics Engineers Computer Society, Vancouver (2002)
2. Shokrollahi, A.: Raptor codes. IEEE Transactions on Information Theory, 2551–2567 (2006)
3. MacKay, D.J.C.: Fountain codes. IEE Proceedings: Communications, 1062–1068 (2005)
4. Etesami, O., Shokrollahi, A.: Raptor codes on binary memory symmetric channels. IEEE Trans. Inf. Theory, 2033–2051 (2006)
5. Castura, J., Mao, Y.: Rateless coding over fading channels. IEEE Commun. Lett., 46–48 (2006)
6. Gummadi, R., Sreenivas, R.S.: Relaying a Fountain Code Across Multiple Nodes. In: IEEE Inf. Theory Workshop, ITW, pp. 149–153. Inst. of Elec. and Elec. Eng. Computer Society, Porto (2008)
7. Sendonaris, A., Erkip, E., Aazhang, B.: User Cooperation Diversity- Part I: System Description. IEEE Trans. on Commun., 1927–1948 (2003)
8. Jardine, A., McLaughlin, S., Thompson, J.: Comparison of space-time cooperative diversity relaying techniques. In: IEEE Vehicular Technology Conference, pp. 2374–2378. Institute of Electrical and Electronics Engineers Inc., Stockholm (2005)
9. Byers, J.W., Luby, M., Mitzenmacher, W.: A digital fountain approach to asynchronous reliable multicast. IEEE J. Selected Areas Comm., 1528–1540 (2002)
10. Molisch, A.F., Mehta, N.B.: Performance of Fountain Codes in Collaborative Relay Networks. IEEE Trans. On Wireless Commun., 4108–4119 (2007)

The Wireless Communication System Based on NRF905

GuangWei Liu[1,2] and LuHong Mao[1]

[1] School of Electronic and Information Engineering, Tianjin University,
[2] Nankai University, Binhai College,
Tianjin , China
nkbhlgw@126.com

Abstract. In view of the tiring wiring and the high cost of traditional short-range Wireline communication system, the design concept and the implementation of wireless communication system based on NRF905 RF chip has been introduced in this paper. In the design of wireless communication system, it is paramount to ensure the reliable data communication between the main system and the subsystem. The structure of this system is simple and the data transmission is reliable, which can be widely applied in the field of short-range wireless communication.

Keywords: NRF905, Microcontroller, wireless communication, DS18B20.

1 Introduction

It is required to real-timely measure the temperature in some fields such as the grain depot temperature control system and the frig temperature control system[1]. The traditional detection system needs to carry on the wiring work and the constraints of the site environment will bring some trouble to the wiring work. At the same time, the cost is high and the maintenance is difficult. The NRF905 is a wireless communication chip provided by Nordic company, which has employed the GFSK modulation and interiorly integrated the Enhanced Shock Burs protocol of Nordic. The communication speed is up to 2Mbps and the standard SPI has been also provided so as to facilitate the interface with the microcontroller. It will be easy to establish a temperature measurement and control wireless communication system combining with the DS18B20 temperature sensor.

2 System Structure

The temperature measurement and control wireless communication system is consisted of three parts: the slave system, the host system and the management terminal. The system structure can be shown in Fig. 1.

The slave systems are distributed in various places of application areas and it is required to collect the temperature information of each area. The slave systems have composed the wireless data communication network which takes the host system as the center through the self-organized method. And the collected temperature information has been also wirelessly transmitted to the host system by slave systems.

A. Xie & X. Huang (Eds.): Advances in Electrical Engineering and Automation, AISC 139, pp. 87–91.
springerlink.com

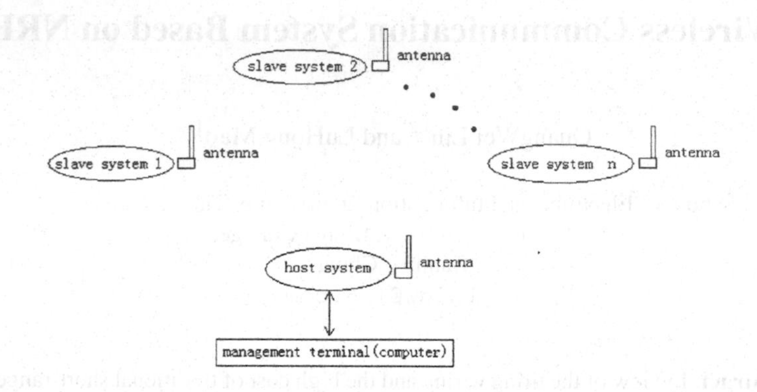

Fig. 1. The structure of the temperature measurement and control wireless communication system

The host system has managed the all connected slave systems and received the temperature information from slave system[2]. Then the all collected temperature information of slave systems will be reported to the management terminal. The management terminal will record and display the obtained temperature data. In view of the different applications, it will analyze them so as to make the corresponding control treatment.

3 The Design and the Implementation of Slave System

The slave system has realized the collection and the processing of temperature information as well as the wireless transmission. The slave system has employed the STC89C52 Microcontroller to control the NRF905 RF chip and the DS18B20 temperature sensor in order to realize the functions of slave system. The hardware structure can be shown in Fig.2.

The NRF905 is the highly-integrated single-chip wireless transceiver solution promoted by Nordic company, which has the auto-response and the auto-retransmission capabilities. The speed can be up to 2Mps and there are 125 selectable work channels. The switching time of channels is short and it can be used for the frequency hopping. The output power, the channel selection and the protocol setting can be set through the SPI port. The Enhanced Shock Burst mode can simultaneously control the response and retransmission functions without increasing the workload of microcontroller. The NRF905 also has the function that the same channel can receive the six-way data in different routes[3]. The use of FDMA technology can realize the data collection of 750 spots, which makes it become the best choice for the implementation of the hardware of temperature measurement and control wireless communication system.

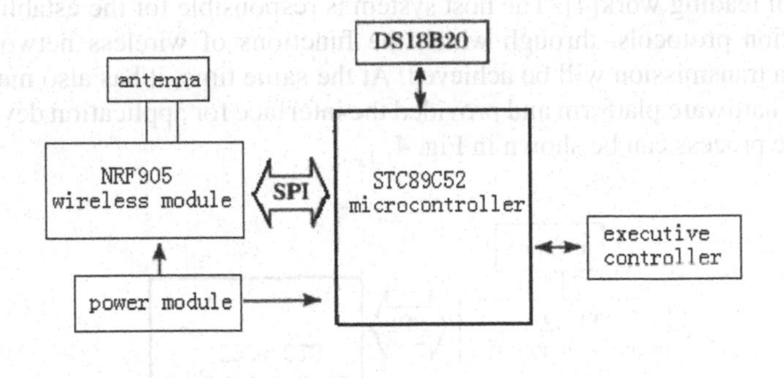

Fig. 2. The hardware structure of slave system

The DS18B20 is an intelligent digital temperature sensor introduced by American Dallas Semiconductor Company. It can directly read the measured temperature. According to the actual requirements, the reading ways of the number with digits between nine and twelve can be achieved through the programming. It can measure the temperature from -55°C to +125°C and have the ±0.5°C accuracy under the temperature from -10°C to +85°C. The use of DS18B20 can save the system resources and make the system structure much simpler and more reliable.

The microcontroller of slave system and host system needs the 5V power supply and the NRF905 wireless module requires the 3.3V power supply. It is necessary to use AMS1117-5 and AMS1117-3.3 to transform the voltage and the stable effect is good.

4 The Host System and the Management Terminal

The host system is responsible to establish the wireless network and receive the temperature information from slave system as well as send the information to the computer of management terminal through the serial port. The design of the NRF905 wireless module and the power module in host system is the same as the one of slave system. The computer is considered as the management terminal. The management terminal will receive the information reported by host system through the serial port and then save them. Meanwhile, it can also send and receive the real-time information received by software supervision through the serial part. According to the obtained information, it is required to make the corresponding control treatment for the different applications. The hardware structure of host system can be shown in Fig. 3.

5 The Software Process

After the starting of slave system, it is necessary to firstly initialize the devices and the registers as well as look for the host system and the slave system to establish a connection. Then the information of each temperature sensor in nodes will be read in turn and the temperature data will be sent to the host system. After reading the information of all sensors, the sensors will enter into the timed dormancy, which are waiting for the next round of reading work[4]. The host system is responsible for the establishment of communication protocols, through which the functions of wireless networking and wireless data transmission will be achieved. At the same time, it has also managed the resources of hardware platform and provided the interface for application development. The software process can be shown in Fig. 4.

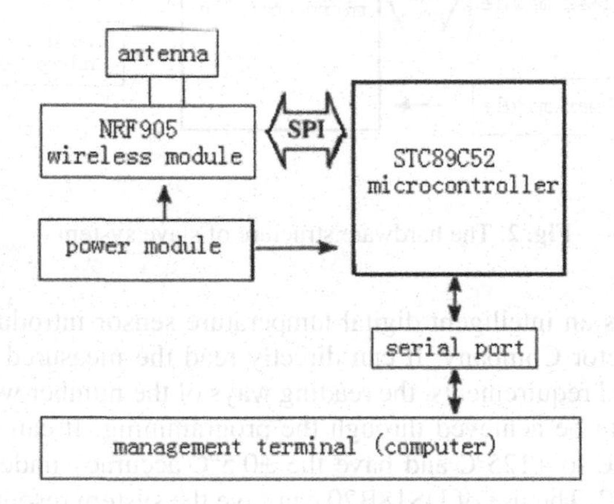

Fig. 3. The hardware structure of host system

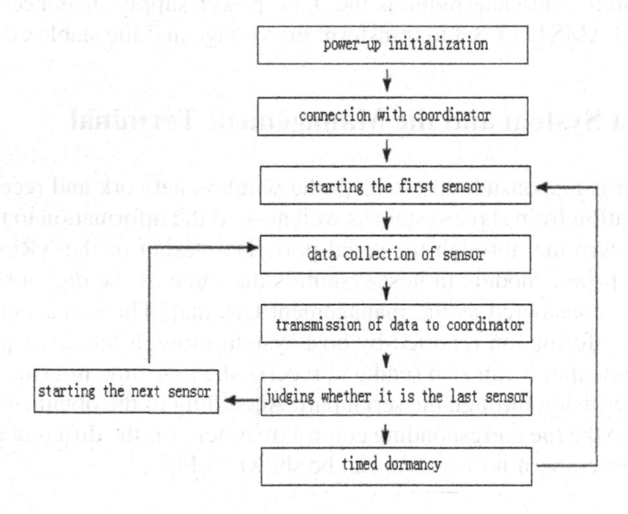

Fig. 4. The software process

6 The Implementation of System

The system has realized the wireless communicaiton function between a host system and five slave systems. The five slave systems have been distributed into a larege warehouse and the distance between the slave system and the host system is about 25m. Through the software supervision of mangement terminal, the real-time temperature information will be collected and the temperature control informaition can be also sent, which has realized the temperature measurement and control wireless communication between the host system and the slave system. The management terminal software can be shown in Fig.5.

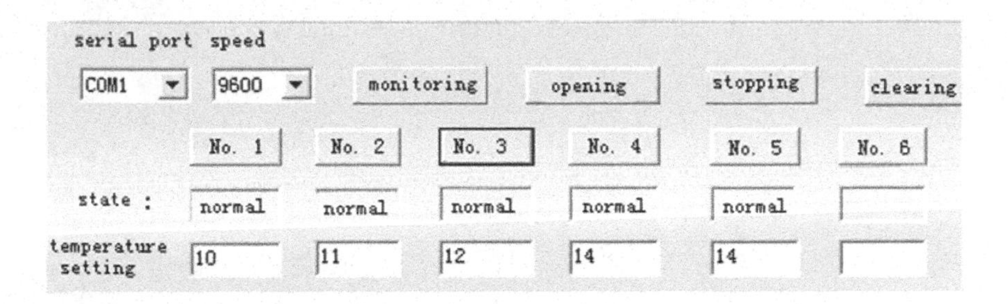

Fig. 5. The monitoring interface of management terminal

7 Conclusion

The structure of the temperature measurement and control wireless communication system proposed in this paper is simple and it can be easily realized as well as the performance is stable, which can be widely applied in the places where the wiring work is inconvenient. The system can be expanded or converted into the humidity sensor, the gas sensor, the smoke sensor and the infrared sensor, etc, and the wireless communication system can be easily realized.

References

1. Wu, J., Yuan, S.: Design and implementation of a general node for wireless sensor networks. Chinese Journal of Scientific Instrument (9), 32–36 (2006)
2. Zhao, H., Zhao, X.: Principle and Application of DS18B20 Intellect Thermometer. Modern Electronics Technique (4), 32–34 (2003)
3. Xie, W., Yang, J.: Principle and Application of Microcontroller and the C51 Program, pp. 35–38. Tsinghua University Press, Beijing (2006)
4. Li, W., Duan, C.: Introduction and Application of Short Distance Wireless Data Communication, pp. 210–227. Beijing Aerospace University Press, Beijing (2006)

Knowledge Sharing Behaviors in Knowledge Management System

Chih-Chung Chen and Chiung-En Huang

Aletheia University (Tainan campus),
70-11, Beishiliao, Madou Dist., Tainan City, 72147, Taiwan (R.O.C.)
jason556@mail.au.edu.tw, a3126747@gmail.com

Abstract. This study mainly aims to explore the knowledge sharing behaviors in the knowledge management system among firefighters. The study is theoretically based on the planned behavior theory and the technology acceptance model. Also, by reviewing literatures with some exogenous variables like trust and organizational citizenship behavior added. There are totally 250 questionnaires delivered with valid receipt rated at 85.6%. Finally, the hypothesis testing is operated through regression analysis. It is found the added exogenous variables also identically affect the knowledge sharing behaviors among firefighters.

Keywords: Knowledge Sharing, Planned Behavior Theory, Organizational Citizenship Behavior.

1 Introduction

Alavi and Leidner[1]contend knowledge sharing means one of the factors highly critical to whether the knowledge management system is successful. However, the widely seen problems existing in firms internally mean there are considerable obstacles for peers' knowledge sharing. Most employees are very unwilling to share their own most knowledge assets with others[2]. Therefore, personal knowledge cannot be transferred unavailable for the best applications and layouts. Furthermore, employees retreat from using the organizational knowledge management system platforms available for employees.

In the past, regarding the interpretation for the applications of information technology systems, the planned behavior theory and the technology acceptance model are both functioned with excellent interpretation capabilities to predict the behavioral intentions of using systems. About the difference between them both, the technology acceptance model focuses on general information technology users to explore a simplified model of theory[3]. Although the technology acceptance model can interpret more variables, it is still unavailable to cover another model with empirical background. Additionally, although the technology acceptance model provides broadly applicable structures for information systems, yet the planned behavior theory can express more concrete messages[4]. Furthermore, Chau and Hu [5] also propose an integrated model with both the technology acceptance model and the planned behavior theory. They contend these two theories can compatibly co-exist with mutually

A. Xie & X. Huang (Eds.): Advances in Electrical Engineering and Automation, AISC 139, pp. 93–98.
springerlink.com

complementary functions to achieve the goal with the interpretation capabilities for expanding individual model.

In the researches on the technology acceptance model[6], researchers clearly indicate trust directly affect the behavioral intentions of users. With further exploration in the dimension of perceived usefulness, especially under the Internet environment, due to the influence caused by the factor of trust, users expect from these transaction behaviors on Internet, they can get the feelings of perceived usefulness. On the other hand, in terms of perceived ease of use, it is also positively correlated. Trust can make consumers have a good impression on sellers and therefore, a seamless exchange relationship is constructed mutually[7].

Through the planned behavior theory, it is available to understand the factors to trigger knowledge sharing behaviors. However, there are probably some other factors affecting knowledge sharing behaviors internal to organizations. Morrison [8] reveals organizational citizenship behaviors belong to an important part for member activities internal to organizations. Employees are willing to engage in some behaviors out of their own tasks contributing to organizations. Therefore, an employee with high organizational citizenship behaviors, he is certainly more willing to share knowledge when organizations expect employees to exchange knowledge. Therefore, this research adopts both TPB and TAM models as the basis. It is also added with trust and organizational citizenship behaviors to explore the factors to affect knowledge sharing behaviors.

2 Literatures Review and Hypothesis

2.1 Trust

McAllister[9] suppose trust is cognitive-based and influence-based. After members trust organizations, they start to share their own opinions and feelings, and also invest personal sentiments. Differently put, the stronger trust the users have, the higher perceived ease of use and perceived usefulness they will be. Therefore, the hypotheses are proposed as below.

H1: Higher trust of users to cast on knowledge management system platforms triggers higher perceived usefulness.

H2: Higher trust of users to cast on knowledge management system platforms triggers higher perceived ease of use.

2.2 Combine the Technology Acceptance Model and the Planned Behavior Theory(C-TAM-TPB)

Taylor and Todd [10] integrate both the technology acceptance model and the planned behavior theory with two variables, namely subjective norms and perceived behavior control added into the technology acceptance model. The combined TAM and TPB, namely the C-TAM-TPB, is proposed. Based on the research results made by Taylor and Todd[11], the C-TAM-TPB model with both the technology acceptance model and the planned behavior theory combined together show very high fit to users' behaviors in using technologies.

Regarding the technology acceptance model, it has been verified by numerous researches with effective prediction on behavioral intentions of users [3,7]. About the planned behavior theory, it is also proven to have predicting capabilities by numerous researches[11,12]. In the C-TAM-TPB with both the technology acceptance model and the planned behavior theory combined [11], perceived ease of use positively affect perceived usefulness. Both perceived ease of use and perceived usefulness positively affect the attitude. The attitude, subjective norms, perceived behavior control positively affect behavioral intentions. Behavioral intentions positively affect using behaviors. Therefore, the hypotheses are proposed as below.

H3: Users' higher perceived ease of use on knowledge platforms triggers higher extents on perceived usefulness.

H4: Users' higher perceived usefulness on knowledge platforms triggers higher extents on the attitude.

H5: Users' higher perceived ease of use on knowledge platforms triggers higher extents on the attitude.

H6: Users' higher "perceived behavior control" on knowledge platforms triggers higher extents on "behavioral intentions ".

H7: Users' higher "attitude" on knowledge platforms triggers higher extents on "behavioral intentions ".

H8: Users' higher "subjective norm" on knowledge platforms triggers higher extents on "behavioral intention".

H9: Users' higher "behavioral intentions" on knowledge platforms triggers higher extents on "actual behaviors".

2.3 Organization Citizenship Behavior

As Morrison[8]indicates, organizational citizenship behaviors are an important part belonging to member activities internal to organizations. Employees engage in some behaviors outside tasks contributing to their own organizations. Bateman and Organ[13] suppose the concept basis of organizational citizenship behaviors is the positive sentiments of the social exchange theory. According to the perspective of social exchange theory, when behaviors get some benefits or goodness from the social exchange relationships, the reciprocation is done in a self-controllable method. Organizational citizenship behaviors are probably the reciprocation of employees. If employees believe they can be rewarded with fair remuneration, employees probably feedback the informal role behaviors expected by their own organizations. Therefore, when organizational members can form good knowledge atmosphere and employees can also spontaneously conduct knowledge sharing behaviors as reciprocation, knowledge sharing behaviors become a kind of organizational citizenship behaviors. Therefore, for employees vested with high organizational citizenship behaviors, when organizations expect all employees can conduct knowledge sharing behaviors mutually, this employee is probably higher willing for knowledge sharing. Through employees' sharing the knowledge about their own organizations, senior employees aggressively exchange their own experience and knowledge allowing junior employees to know the situations of organizational management or culture. Therefore, the hypothesis is proposed as below

H10: Higher employees' organizational citizenship behaviors trigger higher extents on "actual behaviors".

3 Research Method

3.1 Measurement

The reference like actual behaviors, behavioral intentions, attitudes, subjective norms, perceived behavior control, perceived usefulness and perceived ease of use are based on the questionnaires[5,11]with the relevant questions separately 3, 5, 4, 3, 4, 4 and 4 items. Trust refer to Gefen et al.[6], totally 6 questions. Organization Citizenship Behavior refer to Organ[14]and Podsakoff & Mackenzie[15], totally 22 questions.

The aforementioned measurement are conducted with the Likert5-point scales with respondents' answers based on their own agreement extents (1: totally disagree; 5: totally agree).

Demographic Information: includes genders, marriages, ages, academic levels, positions, seniorities, computer using experience and Internet touching experience, totally 9 items.

3.2 Sample

The experimental subjects are selected from the firefighters in the southern Taiwan. There are 250 questionnaires delivered with 214 valid replies received, rated at 85.6%.

4 Research Result

The regression analysis is conducted to verify the research hypotheses. Based on the results of regression analysis, it can be found the standardized regression coefficients for "trust" to "perceived usefulness" is 0.685(p-value is 0.000). There is significance shown and it means the Hypothesis 1 of "trust" really affecting "perceived usefulness" is established. The standardized regression coefficient for "trust" to "perceived ease of use" is 0.642 (p-value is 0.000). There is significance shown and it means the Hypothesis 2 of "trust" really affecting "perceived ease of use" is established. The standardized regression coefficient for "perceived ease of use" to "perceived usefulness" is 0.784(p-value is 0.000.). There is significance shown and it means the Hypothesis 3 of "perceived ease of use" really affecting "perceived usefulness" is established. The coefficients for both "perceived usefulness" and "perceived ease of use" to "attitude" are separately 0.811 and 0.817(two p-value are 0.000.). There is significance shown and it means both "perceived usefulness" and "perceived ease of use" really affect "attitude". Namely, both the Hypothesis 4 and 5 are established. The standardized regression coefficients for "perceived behavior control", "attitude" and "subjective norm" to "behavioral intentions" are separately 0.792, 0.814 and 0.714 (three p-value are all 0.000.). There is significance shown and it means "perceived behavior control", "attitude" and "subjective norm" really affect "behavioral intention". Hypothesis 6, 7 and 8 are established. The standardized regression coefficient for "behavioral intentions" to "actual behavior" is 0.589(p-value is 0.000.). There is significance shown. It means "behavioral intentions" really affects" actual behavior" and the Hypothesis 9 is established. The standardized regression coefficient for "organizational citizenship behavior" to "actual behavior" is 0.594 (p-value is 0.000).There is significance shown. It means "organizational citizenship behavior" really affects "actual behavior" and the Hypothesis 10 is established.

5 Conclusion

Based on the regression analysis results, it is found all hypotheses are established with detailed descriptions for every hypothesis as below. "Trust" really affects users' "perceived ease of use" and "perceived usefulness" on knowledge management system platforms consistent with the results in this research[9]. Users' "perceived ease of use" to knowledge management system platforms really affects "perceived usefulness". Both "perceived ease of use" and "perceived usefulness" really affects users' attitude. The results in this research is consistent with those[3,7,13]. Users' "perceived behavior control", "attitude" and "subjective norm" affect users' "behavioral intention". Users' "behavioral intention" affects their "actual behavior". The results in this research are consistent with those[12].

On the whole, as it is found this research, TPB and TAM can be used to explain the knowledge sharing behaviors for firefighters to use knowledge management system platforms. The applicability to this realm of both TPB and TAM is also verified. Additionally, it is also found from this research, trust and organizational citizenship behaviors are both important variables to affect knowledge sharing behaviors.

In the systems of public sectors, the design of knowledge sharing platform systems probably cannot be as conveniently operable as the systems designed by private firms themselves. Also, the attitude of government authorities is less organized and disciplined than that of private firms. Under such a culture system, the performance for the operation of knowledge sharing platforms cannot be positively shown. The employee turnover rates in government organizations are low and the competence is weak. In terms of knowledge sharing, employees show lower willing. When compared with private firms, it can be clearly seen the civil servants show less aggressive attitude toward the behaviors of knowledge sharing. Therefore, for government agencies to establish knowledge sharing platforms, it is supposed to integrate the ideas and organizational culture of internal members helpful for creating the knowledge sharing space and enhancing the employees' willing for knowledge sharing.

References

1. Alavi, M., Leidner, D.E.: Review: Knowledge Management and Knowledge Management System: Conceptual Foundations and Research Issue. MIS Quart. 25(1), 107–136 (2001)
2. Senge, P.: Sharing Knowledge: The Leader's Role Is Key To a Learning Culture. Exec. Excell. 4(11), 17–18 (1997)
3. Mathieson, K.: Predicting User Intention: Comparing the Technology Acceptance Model with the Theory of Planned Behavior. Inf. Syst. Res. 2(3), 173–191 (1991)
4. Alrafi, A.: Technology Acceptance Model: Critical Analysis with Reference to Managerial Use of Information and Communication Technology. In: Innovation North Research Conference, Leeds Metropolitan University, UK, pp. 18–20 (July 2005)
5. Chau, P.Y.K., Hu, P.J.H.: Investigating Healthcare Professionals' Decisions to Accept Telemedicine: Technology An Empirical Test of Competing Theories. Inf. Manag. 39(4), 297–311 (2002)
6. Gefen, D., Karahanna, E., Straub, D.: Trust and TAM in Online Shopping: An Integrated Model. MIS Quart. 27(1), 51–90 (2003)
7. Wu, I.L., Chen, J.L.: An Extension of Trust and TAM Model with TPB in the Initial Adoption of On-Line Tax An Empirical Study. Int. J. Hum. Comput. Stud. 62(6), 784–808 (2005)
8. Morrison, E.W.: Role Definitions and Organizational Citizenship Behavior: The Importance of the Employee's Perspective. Acad. Manag. J. 37(6), 1543–1567 (1994)
9. McAllister, D.J.: Affect- and Cognition-Based Trust as Foundations for Interpersonal Cooperation in Organizations. Acad. Manag. J. 38(1), 24–59 (1995)
10. Taylor, S., Todd, P.: Understanding Information Technology Usage: A Test of Competing Models. Inf. Syst. Res. 6(2), 144–176 (1995)
11. Planned Behavior: A Study of Consumer Adoption Intention. Int. J. Res. Mark. 12(2), 137–155 (1995)
12. Ryu, S., Ho, S.H., Han, I.: Knowledge Sharing Behavior of Physicians in Hospitals. Expert. Sys. Appl. 25(1), 113–122 (2003)
13. Benamati, J., Rajkumar, T.M.: The Application Development Outsourcing Decision: An Application of The Technology Acceptance Model. J. Comput. Inf. Syst. 42(4), 35–44 (2002)
14. Organ, D.W.: Organizational Citizenship Behavior: It's Construct Clean-Up Time. Hum. Perform. 10(2), 85–97 (1997)
15. Posdakoff, P.M., Mackenzie, S.B.: Organizational Citizenship Behaviors and Sales Unit Effectiveness. J. Mark. Res. 31(3), 351–363 (1994)

An Effective Inventory Management Control Strategy

Zhu-lin Li, Fen Liu, and Kelin Qiao

Institute of Computer Science, Yan'an University,
Yan'an City, Shaanxi Province, China
ydlizhulin@163.com

Abstract. An inventory strategy for spare parts of equipment maintenance is proposed. Its main content includes defining for inventory rate η, extracting mathematic model of $\eta(t)$, and getting the result that the distribution rule of $\eta(t)$ accords with Weibull distribution. Further more, this paper defines task and task time section H(t), and regards them as constraint to get some strategies. To control the climbing edge of model curve to ascend rapidly, we try to shorten the time of ordering spare parts, equal important, to control the falling edge to descend rapidly, try to reduce the time of carrying out task. The experiment shows the strategies are effective. It can provide science foundation for spare parts inventory, and make inventory rate to reach minimum format.

Keywords: maintenance spare parts, storage control, inventory rate, weibull distribution.

1 Introduction

The spare part is indemnificatory material for the equipment maintenance and contingency processing, and is important factor for ensuring equipment good status. In enterprise, the maintenance plan is made, and order and store amount are need decision-making in order to ensure the equipment running well, especially to the costly spare part. With the rapid development of the science technology, the equipments have become more and more complex, and the automatization degree has become higher and higher. To an enterprise, if the equipment stopping, the loss is very big, even catastrophic. Hence, to ensure production continuity, the equipment can be maintained and replaced in time, and enough spare parts are need stored. However, with more and more kinds and amount for spare parts, higher and higher cost for purchasing and storing spare parts. In fact, the high cost for storing spare parts has impropriated plentiful flow fund, and prevent from the enterprise developing. Hence, the theory technology of inventory control for equipment maintenance spare parts is one of the hotspots and difficulties in inventory management field [1,2]. It involves history data analysis, model extraction of demand task and reliability analysis for equipment and etc [3-5]. Their constraint conditions are complex and dynamic, and the process of storage is unstructured [6,7]. The research for spare parts inventory management includes the classification [8-10], the demand forecast [11-13], the inventory control method, and the inventory control strategy and model[14-16] etc. The classification of the spare parts has been many methods [8,9,17,18], for example,

A. Xie & X. Huang (Eds.): Advances in Electrical Engineering and Automation, AISC 139, pp. 99–105.
springerlink.com

maintainable and non-maintainable pare parts, urgent and non-urgent pare parts, crucial and non-crucial pare parts etc. In China, the main method is ABC classification method. Based on this ABC classification method, create inventory control strategy and model, and it can guide the demand forecast analysis. The control inventory need solve three problem, they are inventory check period, order number and order time.

In this paper, to create some control inventory strategies which making inventory rateη reach to minimum in the time section H(*t*), the authors analyzed lots of history data of maintenance spare parts and found the distribute rules for the working spare parts. Owing to analysis and study to the information database of maintenance spare parts, we define the inventory rate, extract its model, and educe some control strategies. All these works are based on actual inventory data for long time. Hence, they have important value to spare parts management.

2 Math Model and Inventory Rule

We will discuss the math model, inventory rate and inventory amount in a certain time section.

2.1 Definition of Inventory Rate η and Consumption δ

The definitions of inventory rate $\eta(t)$ and consumption $\delta(t)$ are following.

Definition 2.1. The inventory rate $\eta(t)$ is

$$\eta(t)=\frac{x(t)}{D}\times 100\% \qquad t\in[a,b], a,b\in(-\infty,+\infty) \tag{1}$$

with D order amount for a sort of spare parts, $x(t)$ correlative inventory variant.

If we name $\delta(t)$ as consumption, then $\delta(t)$ is

$$\delta(t)=1-\eta(t) \tag{2}$$

It is obvious $\eta(t)$, $\delta(t)\in[0,1]$.

2.2 Math Models of η and δ

The following is the models for $\eta(t)$ and $\delta(t)$.

A. Data structure of information source. To realize digitization and informationizion management method, we have been built Database Management System, MEDB. A sample of relation mode $R(a_1,a_2,\cdots,a_n)$ is showed in Table 1.

Table 1. The inventory information for maintenance spare parts

No.	In-depot No.	In-depot date	In-depot reason	In-depot amount	Factory	Quality	State	Classification
000001	040001	2004.04.02	spare parts	6	211FAC	eligibility	good	*A*
000002	050003	2005.03.02	spare parts	4	207FAC	eligibility	good	*A*
000003	060120	2006.12.01	spare parts	10	214FAC	eligibility	good	*B*
000004	070002	2007.02.11	spare parts	3	205FAC	eligibility	good	*C*
000005	040007	2004.03.02	spare parts	7	211FAC	eligibility	good	*C*
000006	050011	2005.11.23	spare parts	9	207FAC	eligibility	good	*B*

Using the language of database, we will do three kinds of operations to sort the data of spare parts.

$\underset{a_9\in A}{S}\ MEDB=F_A$, $\underset{a_9\in B}{S}\ MEDB=F_B$, $\underset{a_9\in C}{S}\ MEDB=F_C$,in which F_A, F_B, F_C denote the file of three kinds of spare part A, B and C respectively. S denotes the selection operation, $a_i \in A$ denotes selection condition. The same to the expression $a_i \in B$ and $a_i \in C$.

B. Math Model. If we select spare part A as example, the result will show in Fig.1(a). According by the statistic result, enough random data of $\eta(t)$ appears a certain disciplinarian change. We found it accords with Weibull distribution by means of curve fitting method[8,9]. In the same way, the fitting curves of $\eta(t)$ for B, and C are showed in Figure 1(b).

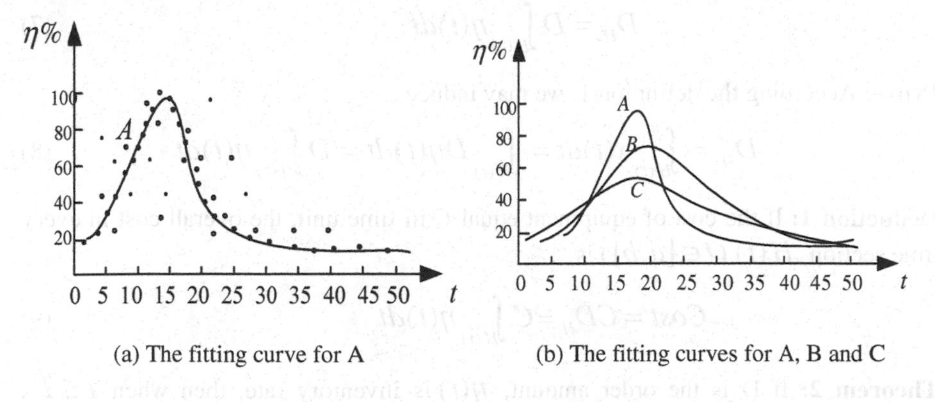

(a) The fitting curve for A (b) The fitting curves for A, B and C

Fig. 1. The fitting curve of $\eta(t)$ for A, B and C

The following equation is the computing expression for $\eta(t)$

$$\eta(t) = \frac{m}{t_0}(t-\gamma)^{m-1} e^{-\frac{(t-r)^m}{t_0}} \tag{3}$$

Where, m denotes shape parameter, t_0 denotes scale parameter, γ denotes starting location parameter.

For A sort curve, m=2,r=0, t0=1, so model expression changes into formula (4)

$$\eta(t) = 2te^{-t^2}, -\infty < t < +\infty \tag{4}$$

Similarly, For B and C sort curves, model expressions change into formula (5) and formula (6) respectively.

$$\eta(t) = 2te^{-t^2/2}, -\infty < t < +\infty \tag{5}$$

$$\eta(t) = 2te^{-t^2/4}, -\infty < t < +\infty \tag{6}$$

The physical meanings of all parameters of $\eta(t)$ are following:

The sign γ is referred to starting time for order; the high of $\eta(t)$ denotes whether order is full or not by planning. If it is full order, then $\eta(t)$ can reach to 100%; if not, then $\eta(t)$ is less than 100%. t_0 is the width of the curve, it denotes the length of inventory time. Comparing three curves in Fig.1, we can educe the results: the spare parts of A is most costly, its order time and storage time are all short, order is full nearly; the spare parts of C is cheaper and less important than A, its order time and storage time are all longer than A, and order is not full. The meaning of spare parts C is between A and C.

Theorem 1: If D is the order amount, $x(t)$ is correlative inventory variant, η(t), $t \in [a,b]$, then the inventory amount D_H in the time sections $H(t)$ is

$$D_H = D\int_{H(t)} \eta(t)dt \tag{7}$$

Prove: According the definition 1, we may induce

$$D_H = \int_{H(t)} x(t)dt = \int_{H(t)} D\eta(t)dt = D\int_{H(t)} \eta(t)dt \tag{8}$$

Deduction 1: If the cost of equipment equal C in time unit, the overall cost in every time section $H(t)$ ($t \in [a,b]$) is

$$Cost = CD_H = C\int_{H(t)} \eta(t)dt \tag{9}$$

Theorem 2: If D is the order amount, $\eta(t)$ is inventory rate, then when $t \leq T$, inventory logistics capacity is

$$\int_{-\infty}^{T} D\eta(t)dt = D\int_{-\infty}^{T} \eta(t)dt = D\int_{-\infty}^{T} \frac{m}{t_0}(t-r)^{m-1} \cdot e^{-(t-r)^{m}/t_0} dt \tag{10}$$

Theorem 3: $\eta(t)$ is inventory rate, then distribution function $F(x)$ can reflect the size of inventory logistics capacity, $F(x)$ is

$$F(x) = \int_{-\infty}^{x} \eta(t)dt \tag{11}$$

Especially, $F(+\infty) = 1$ denotes logistics capacity is over.

The prove for theorem2 and theorem3 can concluded by some property of probability density function of $\eta(t)$ Weillbull distribution curve. The prove process omitting.

3 Inventory Control Strategy

We can conclude some inventory control strategy for maintenance spare parts by above discussion:

- Try to shorten in-depot period of spare parts. The strategy can be explained by inventory model, that is, the more precipitous of climbing section of Weibull curve, the less of in-depot time section of spare parts.
- Try to shorten the performing time of task in order to reduce the storage time for spare parts. The shorter of the performing time, the less of the parameter t_0 for model curve.
- Try not to order full goods in not high risk condition. This can control the peak of the model curve. Generally, the peak is about 80%.

To reduce the storage time, the spare parts are needed to send early after entering storeroom. This viewpoint can be explained by the falling edge of model curve, the rapider of the edge, the better of the model.

4 Example Analysis

To account for the theory and the control strategies, spare part A as example to analyze all process from in-depot to task finishing.

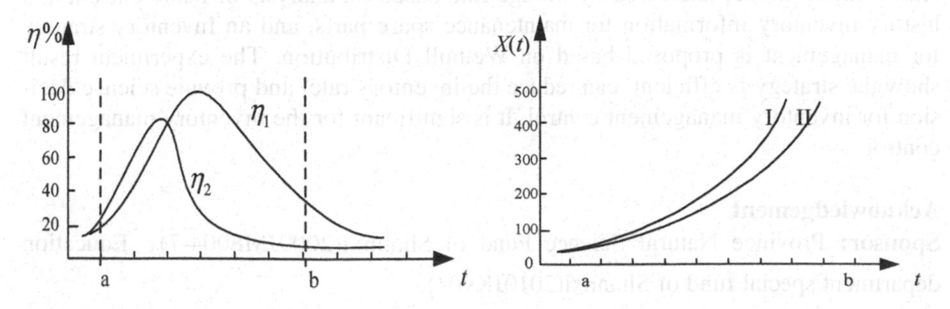

Fig. 2. The compare curves of original model and control model

Fig. 3. The curve I for $x(t)$ before adopted control strategy and curve II after adopted

Denoting: *In* Fig.3, *I is the curve of* $x(t)$ *before adopted control strategy, and II is the curve after adopted strategy*

Table 2. The statistic data of the curve in Fig.2

$\eta(t)\%$	5	10	17	19	25	30	35	40	45
η_1	18.32	42.61	80	99.73	80.28	47.54	37.60	20.9	17.52
η_2	21.67	60.73	65.87	30.18	17.83	8.36	7.21	5.97	5.13

With the time flowing the inventory rate has changed, some discrete data are showed in table 2. In table 2, η_2 denotes the data for have used inventory control strategy, and η_1 denotes the data for not used. From Fig.2 and table 2, we can see obviously that the inventory time of η_2 is shorter than η_1. In Fig.3, the curve I and II of the inventory amount $x(t)$ correspond to the curve η_1 and η_2 in Fig.2 respectively.

If D_H is the inventory amount in time section $[a,b]$, we will compute it by following equation

$$D_H = D\int_{H(t)} \eta(t)dt \approx D\sum_{i=1}^{n} \eta(t_i)\Delta t_i, \qquad t_1 = a, t_n = b \tag{12}$$

If $\Delta t_i = 1$, then the equation(7) will become the equation (13)

$$D_H = D\int_{H(t)} \eta(t)dt \approx D\sum_{i=1}^{n} \eta(t_i) \tag{13}$$

5 Conclusions

An inventory model is created by storage rate based on analysis of many current and history inventory information for maintenance spare parts, and an Inventory strategy for management is proposed based on Weibull Distribution. The experiment result show the strategy is efficient, can reduce the inventory rate, and provide science decision for inventory management control. It is significant for the inventory management control.

Acknowledgement

Sponsor: Province Natural Science Fund of Shannxi(2009JM8004-7), Education department special fund of Shannxi(2010JK904).

References

1. Syntetos, A.A., Boylan, J.E., Disney, S.M.: Inventory forecasting and planning: a 50-year review. Journal of the Operational Research Society 60(special issue), 149–160 (2009)
2. Lo, C.-Y.: Advance of Dynamic Production-Inventory Strategy for Multiple Policites Using Genetic Algorithm. Information Technology Journal 7(4), 647–653 (2008)
3. Li, G.-Q., Wang, H.-B., Wang, M.-C., Hu, N.-L.: A Management Information System of Spare Parts Based on Workflow for Mining Enterprises. Gold Science Technology 17(2), 63–67 (2009)
4. Bayindir, Z.P., Birbil, S.I., Frenk, J.B.G.: A deterministic inventory production model with general inventory cost rate function and piecewise linear concave production costs. European Journal of Operational Research 197(1), 114–123 (2007)
5. Panda, D., Kar, S., Maity, K., Maiti, M.: A single period inventory model with imperfect production and stochastic demand under chance and imprecise constraints. European Journal of Operational Research 188(1), 121–139 (2008)
6. Guo, Y.-J.: Reliable Engineering Theory. Publishing House of Tsinghua University (2002)
7. Xiong, Z.-P., Huang, J.-L.: The Inventory Management. Publishing House of Machine Industry (2007)
8. Cheng, X.-C., Zhang, Z.-M.: The Equipment System Engineering. Publishing House of National Defence Industry, Beijing (2005)
9. Allen, S.G., D'esopo, D.A.: An ordering policy for repairable stock items. Operations Research 16(3), 482–489 (1968)
10. Cohen, M.A., Kleindorfer, P.R., Lee, H., Pyke, D.: Multi-item service constrained (s, S) policy for spare parts logistics system. Naval Research Logistics 39, 561–577 (1992)
11. Braglia, M., Grassi, A., Montanari, R.: Multi-attribute classification method for spare parts inventory management. Journal of Quality in Maintenance Engineering 10(1), 55–65 (2004)
12. Gupta, V., Rao, T.: On the M/G/1 machine interference model with spares. European Journal of Operational Research 89, 164–171 (1996)
13. Rajashree, K.K., Pakkala, T.P.M.: A Bayesian approach to a dynamic inventory model under an unknown demand distribution. Computers & Operations Research 29, 403–422 (2002)
14. Aronis, K.-P., Magou, I., Dekker, R., Tagaras, G.: Inventory control of spare parts using a Bayesian approach: a case study. European Journal of Operational Research 154, 730–739 (2004)
15. Ghobbar, A.A., Friend, C.H.: Sources of intermittent demand for aircraft spare parts within airline operations. Journal of Air Transport Management 8, 221–231 (2002)
16. Aronis, K.-P., Magou, I., Dekker, R., Tagaras, G.: Inventory control of spare parts using a Bayesian approach: a case study. European Journal of Operational Research 154, 730–739 (2004)
17. Wagner, S.M., Lindemann, E.: A case study-based analysis of spare parts management in the engineering industry. Production Planning & Control 19(4), 397–407 (2008)
18. Mann Jr., L.: Toward. A systematic maintenance program. The Journal of Industrial Engineering 17(9), 461–473 (1966)
19. Aronis, K.-P., Magou, I., Dekker, R., Tagaras, G.: Inventory control of spare parts using a Bayesian approach: a case study. European Journal of Operational Research 154, 730–739 (2004)

Construction of Characteristic Major in Electronic and Information Engineering

Xiao Chen[1,2]

[1] School of Electronic and Information Engineering, Nanjing University of Information Science and Technology, Ningliu Road 219,
210044 Nanjing, China
[2] Jiangsu Key Laboratory of Meteorological observation and Information Processing
chenxiao@nuist.edu.cn

Abstract. China universities and colleges characteristic specialty construction is an important method to optimize the professional structure, improve the quality of personnel training, and professional characteristics. Based on construction of national characteristic specialty in Electronic and Information Engineering of Nanjing University of Information Science and Technology, this paper introduced its specialty training objectives and positioning, and proposed building program. We proposed specific reforms and innovative measures from eight aspects, i.e., the conciseness of the professional characteristics, teachers team construction pattern, personnel training model, deepening of curriculum development, enrich the content of practice teaching, improvement of the scientific research, consummating of teaching management system, and internationalization education. This is suited to the requirements of the new century training model, forming significant professional characteristics.

Keywords: Electronic and information engineering, Specialty construction, Characteristic major.

1 Introduction

Different levels of education and training at different levels of talent, is the country's economic development needs. Different types of schools, training of personnel of different characteristics, are the community's demands and the inherent laws of development of higher education. Formally implemented in 2007, "the quality of undergraduate teaching and teaching reform project", is an important measure of China higher education [1]. Specialty Construction is an important part of the "quality engineering". It is comprehensive system engineering, encompassing all aspects of personnel training; the goal is to comprehensively improve the quality of undergraduate training. Construction is to optimize the characteristics of professional higher education and professional structures to improve the quality of personnel training, establish an important measure of professional characteristics. Specialty Construction is currently an important carrier of training colleges and universities innovation, deepen teaching reform, improvement the quality of education.

A. Xie & X. Huang (Eds.): Advances in Electrical Engineering and Automation, AISC 139, pp. 107–112.
springerlink.com

The electronic information industry in China national economy is strategically pillar industry, it is to promote the social employment, pull move economic growth, adjust industrial structure, transforming the mode of development and safeguarding national security plays an important role. Early in the last century 80's, along with the computer technology, communications technology as a symbol of the rapid development of information technology, electronic information science and technology is in the position in national economy is more and more outstanding. From the 1990s, the Ministry of Education revised the "University undergraduate catalog" 2 times, and in the late 1990s established the Electronic and Information Engineering. The targeting of professional training is to develop electronic technology and information systems with the basic knowledge of wide caliber, engage in various types of electronic equipment and information systems research, design, manufacture, application and development of advanced engineering and technical personnel.

The electronic information engineering major of Nanjing University of Information Science and Technology, as national characteristic specialty building points, firmly grasp the development trend of the subject under the new situation with innovation of characteristic specialty construction connotation and practice, working closely with the state, the local economic and social development, reform the personnel training plan, strengthen practice teaching, optimize the curriculum system, strengthening the teachers team and the construction of teaching materials. We adapt to the new century requires efforts to build a training model.

2 Background

Characteristic refers to things by the performance of the unique color, style, etc. With the popularization of higher education, whether a school's characteristics, is related to the school's survival and development. It is more important whether there is a group obvious advantage of professional features different from other institutions [2]. Specialty refers to institutions of higher learning in teaching reform and professional construction process, the education idea, training objectives, training mode, training quality and has distinctive characteristics, students of certain aspects of the quality and ability of the students is better than that of other colleges and universities, and has been widely recognized, have a higher reputation [3]. The electronic information engineering of Nanjing University of Information Science and Technology is the Jiangsu province professional characteristics and national characteristic specialty construction. At all levels of leadership and the strong support of the relevant units, staff members rely on the collective wisdom and strength, make concerted efforts, reform and practice, after years of construction and teaching practice, the initial formation of professional training scheme, teaching staff, perfect excellent teaching experiment equipment, complete specifications advanced teaching management system, teaching research and training fruitful, science research and subject construction effect is distinct of the good momentum of development, has formed the remarkable professional characteristics and advantages.

3 Orientation and Aim of Professional Characteristics

A professional characteristic is a professional in the development process of the formation of relatively stable is recognized by the society unique good development mode or a school characteristic. Professional characteristics in long-term running process of accumulation, is the professional development of long period of practice are highly generalized and sum up scientifically. Professional orientation is a professional construction and development starting point. Only a clear positioning can do a characteristic. Professional characteristics of the location and characteristics of running a school, professional development history and resource advantage are consistent [4].

According to "Chinese ordinary university undergraduate specialty" published by the Ministry of education in 2007, the Electronic Information Engineering opened is as much as 568 colleges and universities. In the professional field of the top colleges and universities, through the electronic information engineering specialty and the school development and traditional advantage combination, they formed distinctive professional characteristics. For example, Beijing University of Posts and Telecommunications, relying on the rapid development of the cause of the civil communication, Beihang University, mainly focus on the aerospace industry of information processing and information transmission.

As the first Chinese Meteorological college, Nanjing University of Information Science and Technology (formerly Nanjing Institute of meteorology) was born to meet meteorological service demand. Founded in 1960, is was identified as the national key university in 1978, renamed in 2004 the Nanjing University of Information Science and Technology, China Meteorological Administration and the Jiangsu Provincial People's government to build a national key university, Jiangsu province key university. It in the spring tide of reforming and opening, closely around the Chinese meteorological service development needs, and to seek economic construction, social development of the combination of points, to personnel training, scientific research and social service, had taken a paragraph of period of sweat drenched but shine in glory history [5].

In order to construct the electronic information engineering specialty, we carry out four in one of the reform and development of ideas "the discipline construction, specialty construction, curriculum construction and laboratory construction", in-depth understanding of professional disciplines, on the foundation of meteorological industry and local economic development closely, rely on the school of electronic and information engineering, Jiangsu Province Meteorological Sensor Network Technology Engineering Center, Jiangsu meteorological detection and signal processing laboratory. The goal is meteorological electronic and information processing as the bright characteristic, the leading domestic and international impact of national characteristics specialty, relying on meteorological industry, local economy, combined with modern electronic technology and new and high technology in meteorological information acquisition and processing, meteorological observation and weather measurement equipment in the a large number of applications.

4 Content of Characteristic Specialty Construction

According to my school undergraduate three teaching guide - study, employment and abroad, the concrete construction plan are as follows.

First, Refining Professional Characteristics. Organization and to employ domestic scholars, experts and teachers in the school for further study and analysis of the professional development direction. Further refining our electronic information engineering school characteristics, so that the professional meteorological electronic and meteorological information processing features more distinct. The specialty is building to a leading domestic, international have a certain impact on the national characteristic specialty.

Second, Teachers and Teaching Team. According to the professional development direction and development requirements, and the professional subject construction needs, combined with the characteristics of the major, according to the professional, the discipline of short-term goals and long-term planning, we formulate detailed introduction of talent and talent training scheme, improve the introduction of talent and training qualified personnel in the use of norms and requirements, give full play to the introduction of talent the enthusiasm and initiative, for the introduction of talents and create a harmonious working environment and democratic, scientific and academic atmosphere. Improve the professional influence at home and abroad, to form a professor, associate professor, senior engineer as the leading team of professional teachers in the school. the existing professional direction is clear, theory and practice teaching, scientific research ability, high level of academic structure and a reasonable age structure, educational background and professional title, at home and abroad, have a certain visibility of teachers.

Third, talent training mode. According to the professional training requirements and characteristics, we establish the professional guidance committee consisting of domestic scholars, experts, teachers in the school, meteorological industry experts. According to the school rules, the committee working system guides and promotes the professional construction. Prominent in weather electronics and meteorological information processing for the characteristics of the application type talents training mode, accurate positioning of our electronic information engineering specialty training goal training program, reform, obtained in recent years teaching research and reform achievement is reflected in the new training program, formulate scientific and rational electronic information engineering specialty training plan.

Forth, Course Reform. The curriculum includes public basic course, specialized basic and specialized courses and advanced professional and characteristic classes, curriculum planning on thick foundation, practical spirit, strengthening basic public and professional course teaching, set up the experiment teaching, production practice, course design, practice, electronic process practice, metalworking practice, graduation design, practice of professional practice teaching system. According to the professional personnel training needs tremendous change and development, the profession in recent years to improve the professional course teaching system in a very important position. It is mainly reflected in the frontier of knowledge, scientific thinking, scientific method, scientific research and development, the introduction of international advanced technique of teaching content and methods of introducing, bilingual teaching, curriculum and graduation practice link enhancement. Strengthen teaching reform research, optimize the curriculum system and teaching content, to high-quality curriculum and top-quality teaching materials construction as a leader, to build a high level of professional curriculum and professional teaching. In order to reform project, high-quality goods curriculum, excellent teaching material construction as the leading organization of teaching team, teaching effect, teaching quality and teaching experience of selecting teaching team leader. Taking the teaching team to study and optimize the construction of curriculum system and teaching content reform.

Fifth, practice teaching. To organize the school and outside experts, scholars, business and senior engineer of the National Meteorological Bureau atmospheric detection center, information center and cooperation with experts, combined with the professional characteristics, research, improvement and construction of the professional experiment, practice teaching system. This system should attach importance to students' innovation consciousness, innovation spirit, practice ability, independent analysis of issues and the ability to solve problems. Adhere to the experiment teaching method and experiment teaching content reform, experimental means innovation. Adhere to the student-oriented teaching concept. Adhere to the knowledge training and ability by equal. The study of theory and practice to promote each others advanced the idea of experimental teaching. Focus on the cultivation of students' comprehensive use of the knowledge, the ability to solve practical problems and the ability of innovation research.

Sixth, scientific research. Strengthening scientific research, motivate and encourage teachers to actively carry out scientific research with teaching, create a harmonious environment for scientific research and scientific research atmosphere, focus on the cultivation of academic leaders, give full play to the role of young scientific research enthusiasm and initiative, to promote scientific research and teaching, scientific research led to teaching, so that the two coordinated development.

Seventh, teaching management. Advancing with the times, people-oriented, to further improve and perfect the system of teaching management, teaching management. Make the teaching order health, stable, orderly, perfect teaching quality monitoring system. At the same time, introducing, absorbing domestic and foreign advanced management experience of teaching, the teaching management level has been improved. On the basis of this, strict teaching management. Ensure that the rich teaching experience of teachers in undergraduate teaching first-line, and professors are undergraduate students in basic courses teaching, reward outstanding teachers, and actively create conditions for the cultivation of young teachers.

Finally, internationalization. Research and analysis of the mode of international cooperation, international cooperation of road running, expand domestic and foreign universities, research institutes, large enterprises and atmospheric sounding meteorological department cooperation, formed the internationalization of the teaching, scientific research and talent training mode. Employ well-known domestic and foreign scholars enrich teaching team, established outside the professional Steering Committee of the building. With foreign university to establish teacher-student exchange system, realize the professional undergraduate and postgraduate training joint, and accepted part of foreign students and teachers came to our school to learn. At the same time, to establish a survey of graduate mechanism, to establish online feedback system, complete the external teachers and the teachers' scientific research and teaching complementary study, completed a survey of graduate feedback system construction.

5 Conclusion

Major in Electronic and Information Engineering is the society need strong universality, adaptability specialty, student employment situation is valued. We will continue, with the characteristic specialty construction as an opportunity, to deepen the reform of education and teaching, to promote professional and subject construction to enhance the overall level, efforts to explore the electronic information engineering specialty construction of new ideas, new modes and new mechanism, cogent raise electronic information engineering undergraduate personnel training quality.

Acknowledgments. This work is supported by the Teaching Research and Reform Program of Nanjing University of Information Science and Technology, China (No. 11JY041).

References

1. Ministry of Finance on the implementation of undergraduate teaching quality and teaching reform project of opinions (1) (2007)
2. Yang, Z., Zhu, G., Lu, J.: Metallurgical Engineering Specialty construction and practice. Education and Occupation 6, 93–97 (2009)
3. Ministry of Education: On the strengthening of "Quality Engineering" undergraduate Specialty Construction of guidance. Teach High Division Letter 28(28) (2008)
4. Zhang, X.: Orientation and cultivation of specialty characteristics. Meitan Higher Education 27, 19–23 (2009)
5. Nanjing University of Information Science and Technology, http://www.nuist.edu.cn

An Improved APIT Location Algorithm for Wireless Sensor Networks

WenHua Cheng[1], Jia Li[1], and Huaizhong Li[1,2]

[1] College of Physics and Electronic Information Engineering, Wenzhou University, Zhejiang China, 325035

[2] School of Computer and Security Science, Edith Cowan University, Australia, 6050

Chengwenhua861112@163.com

Abstract. In order to improve localization accuracy and coverage of the APIT technique, we proposed an improved localization algorithm which further employed the RSSI range method with the APIT technique. Namely, we performed an extra RSSI value comparison in carrying out Point-In-Triangulation test to reduce the Out-To-In errors. Furthermore, we used intersection of two anchor circles to solve the problem when the number of nearby anchor nodes was insufficient for localization. Simulation results show that using the proposed algorithm, localization accuracy can be increased by about 11% when the density of nodes is relatively low. Meanwhile, the proposed algorithm can achieve better coverage comparing with the classic APIT algorithm.

Keywords: wireless sensor networks, APIT localization, received signal strength indicator, Out-To-In error, algorithm.

1 Introduction

Wireless sensor networks(WSNs) is new generation of sensor networks. It's wildly applied with the rapid development of sensor technique, embedded calculate technique, modern networks and wireless communication technique.

One of critical methods of WSNs is the localization technique. Recently, localization technique has been the hot spot of WSNs investigation and people have done plenty of research on many localization algorithms. As one of commonly used localization algorithm, APIT algorithm puts emphasis on the research of increasing the localization coverage and preserving the excellent localization accuracy. On the basis of original APIT algorithm, [1] leads into the effective coefficient of anchor nodes to unknown nodes, and solve the problem of APIT algorithm my fail when nodes rarefaction. [2] uses the counter to decrease the wrong-judgement, meanwhile, and increase the localization coverage by updating located nodes to anchor nodes thus heighten the localization density . [3] proposes a improved APIT algorithm by making use of the relations between the received signal strength of the nodes and their distance. This method reduces the error caused by Point-In-Triangulation Test (PIT Test), and also increase the density of nodes.

A. Xie & X. Huang (Eds.): Advances in Electrical Engineering and Automation, AISC 139, pp. 113–119.
springerlink.com

This paper proposes an improved localization algorithm which further employs the RSSI range method with the APIT technique. In order to make sure the accuracy of APIT judgment, we add to the Comparison-RSSI-value when the nodes are having PIT Test. So, it can decrease the Out-To-In error and increase localization accuracy, Besides, when the number of anchor nodes is 2, we locate unknow node through using the intersection of two circles. This will increase the localization coverage.

2 APIT Localization Algorithm

2.1 The Idea of APIT Algorithm

The idea[4] of APIT algorithm is as follows: unknown nodes monitor the anchor nodes all around, and then get the position information of anchor nodes and energy information of received signal energy. Assume the number of monitoring anchor nodes is N, APIT chooses 3 nodes among the N anchor nodes every time, and then tests whether unknown nodes is in the triangle composed by 3 choosed anchor nodes or not.After one by one testing, at last, we calculate the overlapping area in all of the triangles, and the centroid of this area is the position of unknown nodes.

2.2 Point-In-Triangulation Test

The theoretical basis[4] of APIT algorithm is perfect Point-In-Triangulation Test. The theory of PIT test is as follows: assume that it exsist one direction, and along this direction, the unknown nodes M will away from or close to,at the same time, the three endpoints A B C of the triangle. So we can say M locates outside of the ΔABC,otherwise ,M is in the ΔABC.

2.3 In-To-Out Error and Out-To-In Error

The advantages of APIT algorithm are obvious. Due to the node density and other special reasons, APIT will make some mistakes on localization by making use of the information of neighbor nodes to judge whether unknown nodes is in the triangle composed by anchor nodes or not.

If unknown node M closes to one side of triangle,M will fay away from three endpoints A B C at the same time. Neighbor node 4 lies outside of triangle and it is far away from three endpoints A B C compared to M. According to the definition of APIT, we make the wrong judgement that the node M lies outside of the ΔABC, but in fact, M lies in the ΔABC. In [4],This mistake calls In-To-Out error. Generally, this situation happens when M closes to one side of ΔABC also called boundary effect.

We can see that unknown node M and it's neighbor nodes both lie outside of the triangle that composed by anchor nodes ABC, and neighbor nodes 2 and 3 close to the anchor nodes A and B respectively. Therefore, one thing will happen when unknown node M received the information of anchor node from the neighbor node: neighbor node 2 gets the stronger signal from the anchor node A than the other nodes, neighbor nodes 3 also gets the stronger signal from the anchor node B than the other nodes. So according to the APIT rules, unknown node M should be judged inside the ΔABC,but the fact is in the opposite.In [4], This situation that node should be judged outside the triangle but be wrong judged inside the ΔABC is called Out-To-In error.

3 Improvement of APIT Algorithm

3.1 Shortage of APIT Algorithm

In the WSNs that stochasticly arranged, sometime, the number of anchor nodes may be less than 3 monitored by some unknown nodes. In this situation, it can't process with APIT Test. In original APIT algorithm,it marks these unknown nodes to the undetermined nodes when the number of anchor nodes is 3.

Sometimes, because of localization of unknown nodes, we still can't judge the localization of the unknown nodes accuratly though the number of neighbor anchor nodes that monitored by unknown nodes is more than 3, particularly, existing the circumstance of In-To-Out error and Out-To-In error mentioned above.

Fig. 1 is the APIT test result given by [4]. We can figuire that Out-To-In error is more than In-To-Out of APIT test. Meanwhile, node density has a great influence on Out-To-In error. The node density is bigger,the less on Out-To-In error.

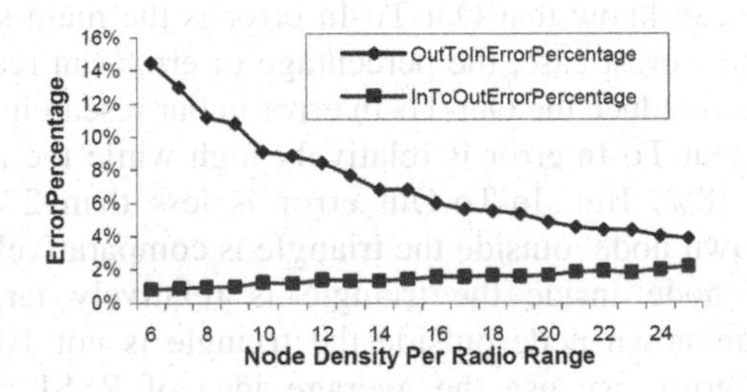

Fig. 1. The relation between node density and error

According to the analysis above, the improvement of APIT is mainly in the following two aspects: realizing positioning of unknown nodes when the number of neighbor anchor nodes is less than 3, and minimizing two types of errors. We will elaborate these two improvements below.

3.2 The Number of Anchor Nodes Is 2

We can't use APIT when the number of anchor nodes is less than 3. So in order to locate these kinds of unknown nodes, we employe the RSSI range technology further. The hardware requirement on RSSI is not high, and we can realize the localization of nodes without any fringe device.

In order to imitate the relation between propagation distance and signal strength, we need to build RSSI Path-Loss-Model[5]. In this paper,the signal model we took as follows:

$$P_R(d)[dBm] = P_T[dBm] - PL(d_0)[dBm] - 10n\log(d/d_0) \ . \tag{1}$$

Here, P_R is the signal's power of received point, P_T is the signal's power of sending point, and $PL(d_0)$ is the signal's energy loss when the distance is d_0, n is the constant of the path loss.

Suppose the signal strength that unknown nodes received is P, then put it into (1) above and we can calculate the result that the distance is d'. In fact, because of different elements that cause signal attenuation,so it makes $P \le P_R$ .Put it into (1), we can get the result $d' \ge d$ .Once the number of neighbor anchor nodes is 2, we can draw two circles respectively by taking anchor nodes as the centers, and d' as the radius. Through analysis, the unknown nodes must fall on the intersection of two circles, and we can know the approximate position of the nodes. In this paper, we take the centroid of intersection as the location of the unknown nodes, shown in Fig. 2, and O' is the estimated location of the unknown node O.

3.3 Reduce Out-To-In Error

According to [4], we can know that Out-To-In error is the main source of the APIT algorithm error. In the worest case, the percentage of error can reach to 15%. So we need to focus on how to reduce the Out-To-In error in our research.

Fig. 1 shows that Out-To-In error is relatively high while the percentage of node density is less than 18%. But, In-To-Out error is less than 2%. This means the judgement that unknown node outside the triangle is comparatively accurate, but the error that unknown node inside the triangle is relatively large. Certainly, the judgement that the unknown node outside the triangle is not 100% correct. So in order to reduce the error, we use the average idea of RSSI value, depicting as follows[6]:

Suppose there are N unknown nodes outside the ΔABC, through the APIT test (Fig. 3). We need to get the sum of RSSI value from all the nodes to the three endpoints A B C:

$$R = \sum R_{iA} + \sum R_{iB} + \sum R_{iC} (i = 1, \cdots, n) \ . \tag{2}$$

Here, R_{iA} means the RSSI value from node i to the anchor node A. Then, get the average value of R, marking it with $\overline{R}$.

For comparing conveniently, we draw the reference node O' which RSSI value is $\overline{R}$ (Fig. 3). We assume that node O is inside the triangle after APIT test and the sum RSSI value from O to three endpoints is R', if $R' < \overline{R}$ we consider that Out-To-In error takes place[7].

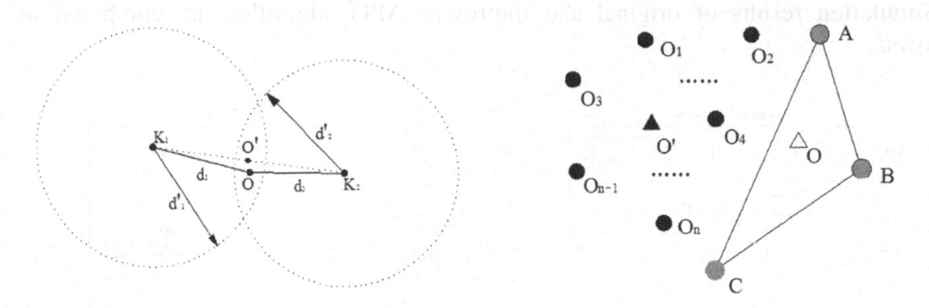

Fig. 2. The number of anchor nodes is 2

Fig. 3. Reduce Out-To-In error

By the above method, we can avoid some wrong judgements effectively. Simulation result show when the node density is relatively lower, we can increase localization accuracy by plus this step.

3.4 New Algorithm Description

(1)After networks deployment, all anchor nodes broadcast messages,unknown nodes receive and exchange them with each other(nodes localization, signal strength etc.);
(2)Suppose the number of anchor node received is i, if $i \geq 3$ then proceed APIT test. Otherwise , turn to (5);
(3)By APIT test, we judge the location relation between node and triangle. If node lies outside the triangle, then calculate the average value of RSSI value from all the nodes outside the triangle to the three endpoints, and marks it with $\overline{R}$. If node lies inside the triangle, then calculate the sum of RSSI value from node to three endpoints,and marks with R'. If $R' < \overline{R}$, node lies outside the triangle otherwise node should be inside the triangle.
(4)Procee APIT localization;
(5)If $i = 2$,according to the description of 3.1 to deal with, and if $i < 2$,mark this unknown node with undetermined node.

4 Simulation of Improved Algorithm

In this paper, we used the MATLAB R2009b software. The result of each simulation we got is after running algorithm 50 times and then take their average value. Simulation parameters are as follows:

(1)Nodes are deployed stochasticly in a square which side length is 50m.

(2)Communication radius of all nodes(contain anchor nodes) is 10m.

(3)Calculate relations of neighbor nodes.

(4)Total number of nodes has two case,100 and 250.The number of anchor nodes in the ratio of the total nodes is vary.We observe the variation of localization accuracy and coverage under the new algorithm in order to findout the appropriate application environment for this algorithm .

Simulation results of original and improved APIT algorithm are compared as follows:

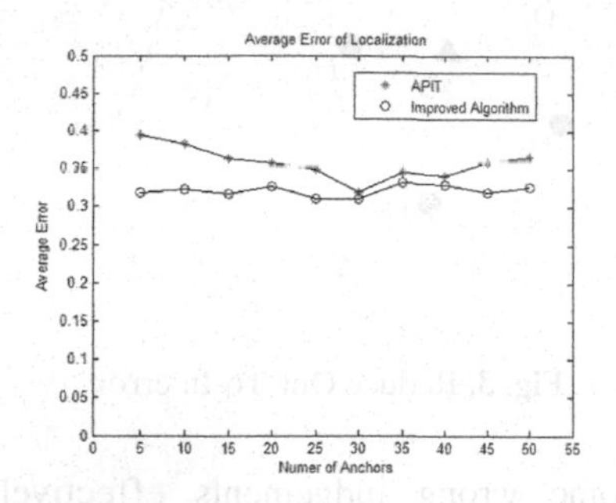

Fig. 4. Localization error of 100 nodes

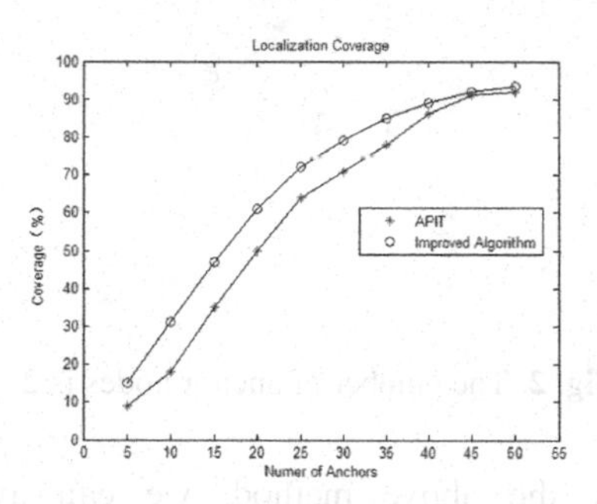

Fig. 5. Localization coverage of 100 nodes

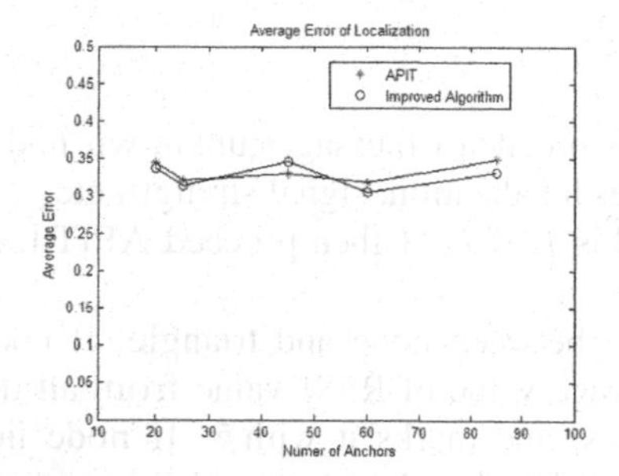

Fig. 6. Localization error of 250 nodes

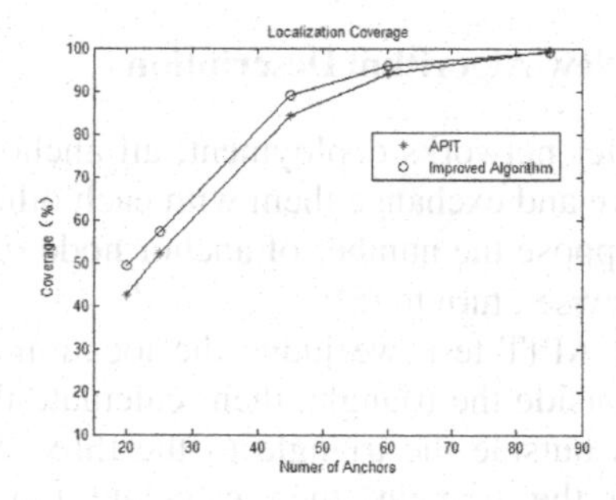

Fig. 7. Localization coverage of 250 nodes

From the simulation, we know new APIT algorithm localization accuracy has increased. Especially, when the number of nodes is 100, accuracy has increased by 11%(Fig. 4). But when the number of node is 250, the accuracy has increased not so high relatively(Fig. 6). Besides, for coverage issue of APIT, when the number of neighbor anchor nodes is 2, we locate unknow node through using the intersection of two circles. So we can see improved algorithm has also some improvements from the simulation(Fig. 5,Fig. 7).

5 Conclusion

In this paper, we investigated the APIT technology in WSNs. When node density is not so high, it would cause much Out-To-In error with using APIT technology. Therefore, we take Comparing-RSSI-value method when we will proceed APIT test to make sure whether nodes inside the triangle or not.

Overall, when the number of nodes is not large, new algorithm have more advantages than traditional APIT algorithm on localization accuracy and coverage. From the idea of this paper, we can see new algorithm is based on the knowned nodes outside the triangle, resulting that the judgements of outside the triangle is not entirely correct. So, how to judge the nodes outside the triangle correctly is the critical issue of our next study.

References

1. Han, B., Xu, C.-B., Yuan, H., et al.: An improved APIT localization algorithm for wireless sensor networks. Computer Engineering and Applications 44(4), 122–124 (2008) (in Chinese)
2. Zhao, J., Pei, Q.-Q., Xu, Z.-Q.: Approximate Point-In-Triangulation teat in WSNs. Computer Engineering 33(5), 109–111 (2007) (in Chinese)
3. Feng, X.-F., Cao, M. L., Sun, C.: Hybrid positioning algorithms for WSNs based on APIT. Microelectronics and Computer 26(6), 58–61 (2009) (in Chinese)
4. He, T., Huang, C., Blum, B.M., et al.: Range- free localization schemes in large scale sensor networks. In: Proceedings of the 9th Annual International Conference on Mobile Computing and Networking, MOBICOM 2003, pp. 81–95. ACM Press, New York (2003)
5. Zhang, J.-W., Zhang, L., Ying, Y.: Research on RSSI location based on ZigBee. Journal of Sensor Technology 22(2), 285–288 (2009) (in Chinese)
6. He, T., Stankovic, J.A., Lu, C., Abdelzaher, T.F.: SPEED:A Stateless Protocol for Real-Time Communication in Sensor Networks. In: Proceedings of IEEE ICDCS 2003, Providence, RI (May 2003)
7. Doherty, L., Ghaoui, L.E., Pister, K.S.J.: Convex Position Estimation in Wireless Sensor Networks. In: Proceedings of the IEEE INFOCOM 2001, Anchorage, AK (April 2001)

To Design the University Online Teaching System Based on the Features of the Cultural Psychology of College Students

Gao Caiyun[1], Qiu Zhiwei[2], and Zhao Dongmei[3]

[1] Foreign Language College, Hebei University of Economics and Business Shijiazhuang, China
[2] Fnformation Technology College, Hebei University of Economics and Business, Shijiazhuang, China
[3] Public Administration College, Hebei University of Economics and Business, Shijiazhuang, China
caiyungao80@126.com

Abstract. Thanks to the rapid development of information technology, online teaching has been popularized in modern education. College students are the main user groups of online teaching system, and the quality of online teaching system's design will have a direct impact on their cultural psychology. Based on the characteristics of college students' cultural psychology, this paper will design the online teaching system, which aims at effectively improving the efficiency of online teaching, reducing the adverse influence on the students' cultural psychology and further promoting the development of online teaching.

Keywords: College Students, Cultural Psychology, Internet Environment.

1 Introduction

This instruction file for Word users (there is a separate instruction file for LaTeX users) may be used as a template. Kindly send the final and checked Word and PDF files of your paper to the Contact Volume Editor. This is usually one of the organizers of the conference. You should make sure that the Word and the PDF files are identical and correct and that only one version of your paper is sent. It is not possible to update files at a later stage. Please note that we do not need the printed paper.

Since the 90s of last century, the Internet technology has been popularized and developed rapidly .Computer network technology began to penetrate into the economy, education and social life in all areas, and have profound impacts and changes in people's work and lifestyle. So online teaching came into being and becme a necessary product of technological development. Online teaching is an important part of modern education that can help people to easily achieve the purpose of acquiring knowledge, and meet their needs of updating knowledge.

Online teaching has the advantages that traditional teaching model can not match, for example, asynchronous discussion and counseling can be achieved; one can set the text, sound, image, animation on the whole, and all these advantages can greatly increase the student's interests and improve the quality of teaching. While as the

A. Xie & X. Huang (Eds.): Advances in Electrical Engineering and Automation, AISC 139, pp. 121–126.
springerlink.com

learners of online teaching can break through the limits of time and space to control and interfere with obtaining and treating information, this open, independent learning there have been some inevitable problems, for example, improper using of the network will decrease the level of college students' cultural psychology and cause physical and mental diseases. Therefore, this paper studies on different psychological characteristics of students caused by psychological are using the online teaching system, especially focusing on depression and anxiety, the two psychological health factors, to continuously optimize the design of online teaching system; to strive to improve college students' psychological health; to create a people-centered teaching model; to achieve the goal of network supporting teaching and management; to promote the development of online teaching.

2 Theoretical Bases

Education is the most important way to solve the problem of college students' cultural psychology. Modern education is not only to foster the person that is full of wisdom, but to nurture people of cultural psychology [1]. As an important tool for modern college students' education, the online teaching should be given full play of the effectiveness of education.

2.1 The Definition of Online Teaching

Based on the computer network environment, online teaching is a teaching process achieved through the database, network configuration, educational software and instructional design of the external environment. It is a collection of the teaching content and suspensor network environment for teaching, which are organized according to certain teaching objectives and teaching strategies [2]. Online teaching has many distinctive features:

•The teaching model of asynchronous discussion and counseling can be achieved, that is, teaching progress can be conducted when students and teachers are not face to face.

•The autonomy of students is increased. As the traditional "classroom" concept is changed, students can break through the limits of time and space to accept the guidance of different teachers.

•Teaching resources become rich, for teachers can make the teaching contents expanded by combining the knowledge of books with the Internet's rich resources in accordance with the teaching requests.

•Online teaching is so interactive that any user can be both a recipient and a sender of information, thus facilitating the two-way communication of teaching and learning in the teaching process.

2.2 The Development of Online Teaching

With the rapid development of science and technology, online teaching has been vigorously developed in China. In 1994, the Ministry of Education has put the "China Education and Research Network Demonstration Project" into effect, marking the beginning of online teaching in China. Along with the application of network in China, online teaching has been developed rapidly:

■In 1997, the first online university was built by Hunan University and Hunan Telecom.

■In 2001, the Ministry of Education, continued to expand the scope of the pilot universities to 45.

■ In 2003, the Ministry of Education issued the "Ministry of Education's notice on starting the work of teaching quality and teaching reformation engineering's excellent course construction" to put out by modern network technology

■In 2007, the Ministry of Education, continued to expand the pilot areas to 68

The gradual emergence of various new online universities, virtual universities, online education bases marked the flourishing of online teaching and provided a new direction for China's educational reform.

2.3 Cultural Psychology of College Students

In China, Internet has been used in various industries. According to the latest CNNIC statistics, up to December 31, 2002, the total number of online computers in China has reached 20.83 million, and the total number of Internet users was 5910 million. While according to CNNIC statistics on October 31, 1997, the number of online computers in China was 299,000 units, and the number of Internet users was 62 million. Nearly five years, Internet users in China increased by nearly 100 times [3]. In 2001, Bowker surveyed the occupations of the chat rooms users and discovered that most of them were students [4]. This showed that the computer network technology has gradually become an important factor impacting people's study and work .As college students are the main users of the network, their cultural psychology are directly affected by the network.

3 Theoretical Models

Nowadays the popular network systems only to emphasize the design of teaching and learning environment, while ignore the effects of psychological health of the learners on the learning. Therefore, it is necessary to apply the theory of college students' cultural psychology into the design of the online teaching system. Based on the traditional design of online teaching system, this study applied the psychological theory to the design of online teaching system, as the theoretical model shown in Figure 1. After registration, online teaching system will make an evaluation on the user's cultural psychology, and then transcribe them to the different application areas. The study mainly measured the characteristics of college students with depression and anxiety. According to the different psychological characteristics of the college students, this system can automatically assign the students to different learning subsystems, and add the right elements guiding students to the access of Internet Moreover, lectures on cultural psychology and related information will be carried out in all aspects of online teaching system... At the same time, the system administrators make different settings according to different students. This theoretical model aims at optimizing the online teaching system to improve the cultural psychology of college students and to improve the efficiency of online teaching.

4 Research Methods

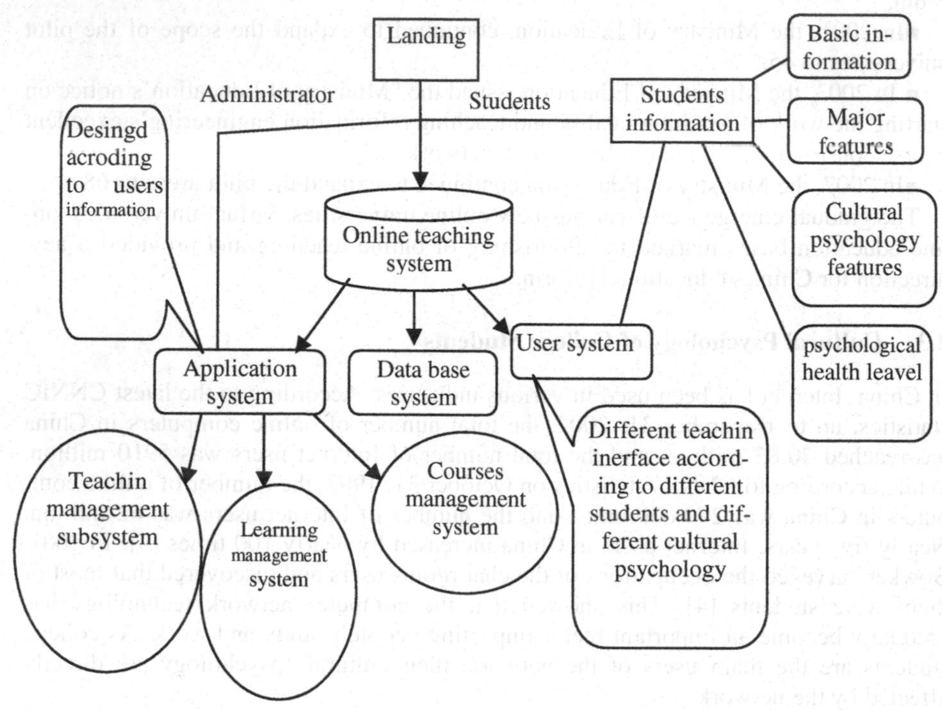

4.1 Subjects

By Stratified random sampling method, 500 subjects were selected from the full-time college students who used online teaching in the colleges of Hebei, 470 were valid files.

4.2 Research Tool

(1) Demographic Information Survey

The subject's gender, grade, school, major, students' source, whether being the only child in family etc.

(2) Beck Depression Inventory (Beck Depression Inventory, BDI)

(3) State - Trait Anxiety Inventory (State-Trait Anxiety Inventory, SATI)

4.3 Procedure

First, distribute the questionnaires to the subjects. This study used the method of Group Test and was supervised by professional teachers of psychology as the major test. Before the survey, the main test tried to present the meaning and purpose of the survey, emphasized on the items that should be paid on attention, tried to dispel the students' doubts, and encouraged the subjects to answer the question seriously, then read the guide words. The students began answering questionnaires, and then collected the questionnaires. Analyzed the collected questionnaires' data by SPSS16.0.

5 Results

5.1 Analysis on Depression and Anxiety of College Students

In this study, the scores of each dimension of SATI Anxiety Scale and the BDI depression scale are analyzed by multivariate variance, and other dimensions of anxiety and depression scores are also made, including major ,gender, source and whether the only child .Check Table 1 of the results:

Table 1. Anxiety and Depression in College Students

	SATI	BDI
Liberal arts	43.00 ± 8.93	10.61 ± 10.11
Science	43.83 ± 9.35	10.86 ± 10.21
Male	44.28 ± 9.37	1.275 ± 1,121
Female	42.56 ± 8.85	8.76 ± 8.57
t	2.047 *	4.322 ***
City	42.31 ± 9.65	10.40 ± 10.30
Countryside	44.07 ± 8.77	10.93 ± 10.07
t	-1.983 *	-0.549
Only-child	41.53 ± 9.54	10.38 ± 10.08
Non-only-child	44.03 ± 8.93	10.85 ± 10.18
t	-2.497 *	-0.437

It can be seen from the test results that college students SATI anxiety scores have significant differences in gender, source, and whether being the only child in family. On anxiety, boys score higher than girls, and rural students score higher than students from cities, so as the non-only-child is to the only-child. However, the scores on the depression merely show distinctive differences in gender.

6 Summary

The findings shows that the students' cultural psychology are different because of their different grade ,gender and sources. Psychological health of university students, to some extent, affects their academic study and life, and their personality development, therefore the application of theory of cultural psychology teaching system applied to the design of the network is necessary. Psychological theory will be applied to the design in the theoretical model. Then after registration, the users will be given a psychology evaluation, and then the results will be transcribed to different application areas. According to the different psychological characteristics of students, they will be automatically assigned to different learning subsystem in the system which introduces and adds proper elements to guide students to the access of Internet. Moreover, lectures on cultural psychology and related information will be carried out in all aspects of online teaching system. Meanwhile, the system administrator will make different settings for different students. The application of this theoretical model will effectively improve the cultural psychology of students and their learning efficiency, and promote the development of online teaching.

Acknowledgements. This Scientific research achievement is supported by Hebei funds of Social Science Project（The Item Number:HB11JY016）and Hebei Association of Social Sciences （The Item Number:201104009）.

References

1. Lin, Z.C., Fang, W.Z.: Cultural psychology. Economic Science Press (2010)
2. Wu, B., Liu, J.: Teaching courses on Internet use and discussion. Luoyang Technology College 15(2), 65–66 (2005)
3. According to the China Internet Network China Internet Network Information Center, CNNIC 11th China Internet Network Development Statistic Report 2003 / 1 statistics (2003)
4. Bowker, R.M., Atkinson, P.J., et al.: Effect of contact stress in bones of the distal interphalangeal joint on microscopic changes in articular cartilage and ligaments. Am. J. Vet. Res. 62(3), 414–424 (2001)

The Application of Internet Technology in Translation

Li Qingjun[1], Li Lina[2], and Gao Caiyun[3]

[1] College of Foreign Languages, Hebei University of Economics and Business, Shijiazhuang, China
[2] College of Foreign Languages, Shijiazhuang University of Economics, Shijiazhuang, China
[3] College of Foreign Languages, Hebei University of Economics and Business, Shijiazhuang, China
Qingjunli@126.com

Abstract. The Internet technology provides great help for the translation to a certain extent, the translator can solve many difficulties in the translation practice with the help of the network information resources, including resource sharing and network translation tools, sometimes they might even find corresponding version directly and examine the correctness of translation through the machine translation, thus making the translation more efficient and accurate.

Keywords: internet-assisted translation, network information resources, machine translation.

1 Introduction

The increasing globalization , the deepening of China's opening up, the frequent exchange and cooperation with foreign countries and the growing demand of information localization have brought the unprecedented market opportunity for China's translation service providers and promoted the industrialization of translation, and the informationization of translation is an important way to realize it, which means accomplishing the modernization of translation by using computer, the auxiliary translation software, the Internet, various of retrieval tools, speech input tools etc.

2 The Introduction of the Internet Tools into Translation

Generally, many people regard that translation ability involves bilingual and bicultural ability and some translation skills, most researchers home and abroad regard the translation ability as a kind of language transformation and communication, but some researchers mentioned the ability of application of assistant translation tools. In 1972 Holmes pointed out that the applied translation research should include translator training, the auxiliary translation and translation criticism in his translation research frame (Holmes, 2006:88). Although the cognition foundation of the specialized translation comes from the bilingualism or foreign language user's cognition skill, the ability involved discourse transformation also contains some knowledge structures that do not belong to the bilingual ability. In the process of translation, the translator

A. Xie & X. Huang (Eds.): Advances in Electrical Engineering and Automation, AISC 139, pp. 127–132.
springerlink.com

will use various interdisciplinary knowledge and skills to complete a task, such as a specific subject knowledge, world knowledge, research skills, and cognitive potential, as well as strategies for tackling problems(Jacket & Massey, 2005: 48). "Thus, translation not only involves the communicative abilities of the language itself, but also the translator's ability to use tools to complete the translation task in the translation process" [1].

3 The Online Search Engines and Dictionaries

Translation is a complex activity which involves knowledge of many fields, no matter what type of translation belongs to, there are two phases: comprehension and presentation. Usually the difficulties of translation are not caused by the incomprehensible words, but due to the lack of background knowledge, especially when translating unfamiliar materials, the translator must know the basic background knowledge, but the translator's personal knowledge and information are limited, so translators need to have a large number of reference materials, such as encyclopedia, dictionaries, and so on. But it's very difficult for the translators to own these traditional references due to the objective conditions. Now, the network provides a huge shared data bank for the translator, which includes encyclopedia, the dictionaries and some special resources. All of these provide many conveniences to the resolution of translation difficulties. With the network technology, the translator can find related information to get a general understanding of the related subject through search engines, online encyclopedia, electronic dictionary, online terminology and online newspapers and magazines, thus the translation quality can be improved by reducing understanding errors. As time passes, the translator will use these knowledge in the later translation practice, thus the translator's ability to understand will be expanded and enriched.

Along with the technical and social development, the new-created things make the language renewed greatly. "It's obvious that the renewal of traditional dictionary can not catch the speed of the language development. Moreover the practical translation often deals with the different specialty, ranging from the popular information, computer, biology to fire prevention, marine transportation and so on. Without doubt, in order to deal with so great variety of translation materials, achieving support from the Internet through the search function is the only way to get the information promptly and accurately" [2].

A. The advantages and characteristics of the network retrieval

With the rapid development of Internet and the swift growth of network information, the network search engine has obtained more and more favor of scholars, researchers and translator's .The Internet is a free giant corpus to the translators. "It is very convenient for translators to use search engine to search related knowledge, the idiomatic translation of words and expressions, thus enhancing the translation efficiency and quality" [3]. It should say that the inquiry way of Internet search is a basic skill to the E - time translators.

.The network inquiry is time-saving and effort -saving

Take Google as an example. If you want to search the English expression of UNESCO, as long as you input the appropriate key word, Google can find 2,260,000 items that conformed to the inquiry result in 0.05 seconds. Obviously the internet search engine is very efficient. But if you want to obtain the information with the traditional inquiry way, you should read a huge collection of reference books, and sometimes your work is fruitless. Thus many problems which meet in the translation can be solved easily and rapidly through the network search.

.The accuracy and error correction function of network inquiry

Many English words have established translation, for example, the translation of proper names such as the unit name, the place name, the personal name and so on belong to this category. "A random translation of a material with many unfamiliar words will lead reader to misunderstanding, but the help of internet will improve the accuracy of translation" [4].

B. Several network translation inquiry tools and characteristics

.Internet search engines

"The search engine is a system with many reorganized internet information and provides the inquiry service for the user" [5]. The frequently used English search engines are the following:

http: //www.google.cn it is the most widespread specialized search engine with the most users. It is characterized by its high accuracy and speed.

http: //www.yahoo.com It is the one of the most famous catalog indexes and search engines and provides the category, the website and the full text search function, it has a rich collection of website and a high accurate retrieval result.

http: //www.altavista.com It is widely regarded as one of best search engines. It can distinguish the big and small letters and the proper names and provides bidirectional online translation between English and other several country languages.

http: //www.excite.com It is one of most popular Internet search services. It can not only search the key words, but also infer the related content automatically.

.On-line English dictionaries

"The online dictionaries are important internet-based on-line reference tools for the users to inquire real-time glossary service. They are characterized by their convenient retrieval, large vocabulary, and prompt renewal and so on" [6].

The frequently used online dictionaries:

http: //www.iciba.com

http: //www.dreye.com.cn

http: //www.godict.com

http: //dict.cn

http: //www.chinafanyi.com

4 Machine Translation and Problems

The application of Internet-based translation tools will make the translation more efficient, especially for scientific and technical translation. Specifically, the main translation tools include a machine translation system and machine translation software. The translator can get the version of the text through machine translation and translation software when he inputs the text directly into the computer, but it is poorly readable, so the translation memory technology is always used in the process of translation to improve translation efficiency and accuracy. The user can establish memory storehouse with the existing original text and the translation by using the translation memory technology. In the translation process, the translation memory system can give the reference translation through searching the same or similar translation resources in the translation memory storehouse, thus the redundant work is avoided. There are many difficult problems of the natural language processing branches in machine translation, it's hard to get breakthrough before solving some basic problems because of the ambiguous meanings in morphology analysis, the part-of-speech tagging, the syntax analysis and the semantic analysis. The main problems in the machine translation exist in the following aspects:

A. Polysemy. Polysemy exists in both Chinese and Western languages. In machine translation process, sometimes at first you need to determine the specific meanings of a word, as well as the correspondence in the target language, but the relationship between any two different languages is not one-to-one correspondence, because the words can derive extended meaning in its own evolution in order to meet the need of the presentation, therefore polysemy appeared. The solution to polysemy is the word sense disambiguation, also called the semantic tagger, which means giving every word in the text its semantic encoding according to its respective context. At present, polysemy typography study is still in its beginning stage; mainly include artificial intelligence methods, dictionary-based methods and corpus-based approaches.

B. Word Concurrency. Theoretically it refers to the syntax distribution of one word with two or even more kinds of parts of speech, these words are called the concurrent word. Therefore, although the part of speech of a word in a sentence is definite, because it belongs to a certain context, sometimes the part-of-speech tagging cannot guarantee the correctness completely. The so-called part-of-speech tagging determines grammat-icalategory of each word, thus determines its part of speech. Part-of-speech tagging automatically is the automatic elimination of ambiguous meanings of word concurrency.

C. Ambiguous meanings of phrase structure. Different adhesive relations of nouns, prepositions and so on can result different reasonable meanings. Distinguishing the structure ambiguity of the machine translation is a widespread problem, because it involves a deep analysis and comprehension of natural language. At present, we can not do machine translation on the basis of the comprehension of natural language, what we can do now is just to study it as a kind of applied technology. Using the strategy of ambiguity to ambiguity is only an expedient way before the system has the capacity to identify structural ambiguity.

D. The quality of translation cannot be guaranteed. The difficulties in distinguishment of semantic ambiguity and choice of semantic meaning degrade the quality of translation, which was caused mainly by the lag behind of the theory research on the development basis of machine translation. At present, the entire translation system was estabilished mainly on the syntactic level, the limited semantic which information system used is used for the syntactic level, the target language relies solely on simple syntactic structure conversion from the source language, but this is not enough. In addition, the language rules which the system used are still based on traditional grammar, they can not reflect the complex characteristics of natural language, the quality of the translation has not made substantial progress, a lot of problems existed 50 years ago has not resolved now. The existing translation progress relied more on the computer-resource development, but not on the translation method. Since the 1990s, the computer and network technology developed a lot, therefore the range of available technologies are further expanded, which includes full-text search technology, voice input, network information gathering and the query engine, and so on.

In short, the difficulties in machine translation mainly include two difficulties. One of the difficulties is the difference between the natural language and computer language, there are always exceptions in grammar ,which makes it impossible for the computer to store; the other difficulty is the word ambiguity. A sound translation needs the combination of translator's English ability and computer skills with machine translation, which can be seen in the following figure:

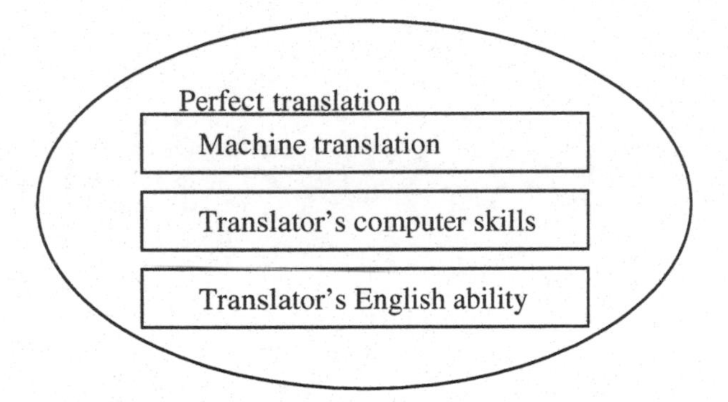

5 Conclusion

"A handy tool makes a handy man." It is not enough for us to face the challenging translation task merely by depending upon individual knowledge and information. In the internet informationalized era, the translator must be good at using Internet to assist the translation practice for the high efficiency and high quality of translation, thus"getting twice the result with half the effort" [7]. However, although the internet technology can improve the efficiency of the translation greatly, it is only a highly effective tool; the internet tools are useless without solid basic language skills, the instruction of correct translation theories and methods, so the translator can not ignore the training of translation basic skills when applying the internet to enhance translation quality and efficiency.

Acknowledgements. This research is sponsored by Hebei Association of Social Science（The Item Number : 201003093).

References

1. Ge, J., Fan, X.: The internet-assisted translation ability. Shang Hai Journal of Translators, 62–65 (January 2008)
2. Zhang, J.: Language polymorphism and internet-assisted translation. China Journal of Translators for Science and Technology, 32–35 (Feburary 2003)
3. Wang, J.: The application of internet sources in translation. China Journal of Translators for Science and Technology, 36–40 (Feburary 2007)
4. Tan, Y., Gao, Y.: Internet and translation studies in new era. Chinese Translation, 58–60 (Feburary 1999)
5. Wang, Y.: Search engine and translation. China Journal of Translators for Science and Technology, 28–30 (January 2005)
6. Zhao, C.: The helper of dictionary compilation — network search engine. Lexicographical Studies, 106–109 (March 2005)
7. Wan, Z.: The internet-assisted translation. Shang Hai Journal of Translators, 77–80 (March 2008)

A Fuzzy Set Approach to Group Decision Making with Ordinal Interval Information on Alternatives

Quan Zhang[1], Xiao-dong Su[2], and Wei Chen[2]

[1] School of Information Engineering
[2] School of Fundamental Education,
ShenYang University of Technology, 110870, Shenyang, China
isqzhang@sut.edu.cn

Abstract. This paper focuses on the group decision making problems with preference information on alternatives in form of ordinal intervals. A fuzzy set approach is proposed, where the ordinal interval information on alternatives is matched into intervals in [0,1], and further transformed into the fuzzy set over basic linguistic evaluation set. 2-tuple linguistic values are then obtained for ranking the alternatives. An example is given for illustrating the proposed approach.

Keywords: Group decision making, Preference orderings, Interval, Linguistic evaluation, Fuzzy set, Basic linguistic evaluation set.

1 Introduction

Group decision making (i.e., GDM) is characterized by a number of experts who express their preference information (e.g., on alternatives) based on their own ideas, attitudes, motivations and knowledge, in order to find the best alternative(s) from a feasible set and achieve a common consensus[1]-[4]. Usually, experts have to express their preferences in the forms of numeric values [1]-[2], linguistic values[3]-[4], preference orderings or intervals[1][2][5][6].

In [1], preference orderings, utility value vectors and fuzzy preference relation are adopted in GDM with fuzzy preference relation as the uniform format. In [2], three formats, i.e., the fuzzy preference relation, the fuzzy linguistic relation and the interval-valued preference relation, are employed to handle the uncertain GDM problems. In the information uniformity process, numerical values in [0, 1], linguistic terms and interval values, are all transformed into the fuzzy sets on BLTS (i.e., *basic linguistic term set*). In the aggregation process, the unified preference relations with fuzzy sets over the BLTS, are combined into a collective one. In the preference exploitation process, the collective preference relation with fuzzy sets over the BLTS, is transformed into 2-tuple linguistic values to obtain the alternative selection.

Due to the problem complexity and uncertainty of decision situation, preference orderings or intervals are easier for experts use to express their opinions [5]-[6]. However, research on the GDM with ordinal interval information on alternatives is not so common [5]-[6]. In [5], an integer program model is employed to solve GDM

A. Xie & X. Huang (Eds.): Advances in Electrical Engineering and Automation, AISC 139, pp. 133–138.
springerlink.com

problems with ordinal interval information on alternatives. But it is more complicated with the large number of experts and alternatives, which results in larger computation operation and difficulty in finding the solution. In [6], a TOPSIS method is proposed for the GDM problems with ordinal interval information on alternatives. But it is not so precise by defining the distance functions.

In this paper proposes a new fuzzy set approach which will overcome the shortcomings of the current research on GDM problems with ordinal interval information on alternatives.

2 Problem Descriptions and Theory Foundation

2.1 Problem Descriptions

The following notations are employed in this study for describing the GDM problems with ordinal interval information on alternatives：

$S=\{S_1, S_2, \dots S_m\}$: is the set of alternatives, $m \geq 2$.

$E=\{E_1, E_2, \dots E_k\}$: is the set of experts, $K \geq 2$.

$W=(w_1, w_2, \dots, w_K)$: is the weight vector, where w_j 为is the weight of expert E_j, $0 \leq w_j \leq 1$,且 $\sum_{j=1}^{n} w_j = 1$.

In this study, the experts express their opinions on alternatives in the form of ordinal interval.

Definition 1. $O = (o(1), \dots, o(m))$ is called the preference orderings on alternative set S, if $o(i)$ is the ranking position of alternative S_i, $i = 1, \dots, m$.

Definition 2. $ran = [ran_l, ran_u]$ is called the ordinal interval on alternative set S, if $1 \leq ran_l \leq ran_u \leq m$, and both ran_l and ran_u are positive integers.

Specially, if $ran_l = ran_u$, then the ordinal interval degenerates into usual preference orderings. Without loss of generality, the smaller the number of ran_l (or ran_u), the higher (better) the position it represents, for example, if $[ran_l, ran_u]=[1, 2]$,then it show that some alternative is ranked on the first or the second position, or between the two positions.

2.2 Concepts of Linguistic Evaluations

Linguistic Terms

In complex or uncertain decision environment, linguistic terms can be used to express decision makers' subjective opinions or judgments more precisely.

Definition 3. A linguistic term $\tilde{T}$ on real number set is defined as a triangular fuzzy number (denoted as (u, α, β)), if its membership function $\mu_{\tilde{T}}$ ($R^+ \rightarrow [0, 1]$) is defined as,

$$\mu_{\tilde{T}}(x) = \begin{cases} \dfrac{x-\alpha}{u-\alpha}, & x \in [\alpha, u], \\ \dfrac{x-\beta}{u-\beta}, & x \in [u, \beta], \\ 0, & \text{others} \end{cases} \tag{1}$$

where, $\alpha \leq u \leq \beta$, and u is the model value, α and β stand for the lower value and the upper value of linguistic term $\tilde{T}$ respectively.

Linguistic Term Set

Suppose $TERMSET=\{t_0, t_1, \ldots, t_g\}$ is a linguistic term set for evaluating the alternatives. *TERMSET* is defined as the ordering set, which is composed of a number of linguistic terms with odd number (i.e., $g+1$ is odd number). For example, consider a set of seven terms, i.e., *TERMSET* ={t_0="none", t_1="very poor", t_2="poor", t_3="fair", t_4="good", t_5="very good", t_6="perfect"}. The following properties of *TERMSET* are assumed[3] : (1) The *TERMSET* is ordered: $t_i \geq t_j$, if $i \geq j$. The symbol "$\geq$" denotes "better or equal"; (2) There is the negation operator "*Neg*": Neg(t_i)=t_j, such that $j=g-i$, where $g+1$ is the number of elements in *TERMSET*, and the largest term in *TERMSET* is t_g; (3) There is the max operator and the min operator: Max{t_i, t_j}=t_i and Min{t_i, t_j}=t_j if $t_i \geq t_j$.

Basic Linguistic Evaluation Set

Denote $TERMSET^B$={$term_0^B$, $term_1^B$, …, $term_g^B$} as the basic linguistic evaluation set, which is of the properties of the linguistic term set as discussed above. In this study, the ordinal interval information on alternatives is matched into intervals in [0,1], and then transformed into fuzzy set over the basic linguistic evaluation set $TERMSET^B$.

The membership functions (including the model value μ_i, the lower value α_i and the upper value β_i) of triangular fuzzy numbers for the elements in the basic linguistic evaluation set (i.e., $TERMSET^B$={$term_0^B$, $term_1^B$, …, $term_g^B$}), $\gamma_i = (\mu_i, \alpha_i, \beta_i)$ ($i = 0,1,\ldots,g$), are defined as follows,

$$\gamma_i = \begin{cases} \alpha_0 = 0 & \\ \mu_i = \dfrac{i}{g-1} & 0 \leq i \leq g-1 \\ \alpha_i = \dfrac{i-1}{g-1} & 1 \leq i \leq g-1 \\ \beta_i = \dfrac{i+1}{g-1} & 0 \leq i \leq g-2 \\ \beta_{g-1} = 1 & \end{cases} \tag{2}$$

Where g+1 is the number of elements in the basic linguistic evaluation set $TERMSET^B$.

3 The Proposed Approach

3.1 Transform Preference Ordering into Intervals

Set up the corresponding intervals for m ranking positions respectively, $inter_i = [\frac{m-i}{m}, \frac{m+1-i}{m}]$, $1 \le i \le m$, for example, the first position corresponds to $[\frac{m-1}{m}, 1]$.

Set up the membership functions for the corresponding intervals for m ranking positions respectively,

$$y(i) = \begin{cases} 0 & others \\ 1 & [\frac{m-i}{m}, \frac{m+1-i}{m}] \end{cases}, \quad 1 \le i \le m \tag{3}$$

Thus, preference ordering or ordinal interval information on alternatives is transformed into the sub intervals of [0, 1].

3.2 Transform the Ordinal Interval Information into the Fuzzy Set over the Basic Linguistic Evaluation Set

Based on the discussions in 3.1, the following method can be used to transform the ordinal interval information into the fuzzy set over the basic linguistic evaluation set $TERMSET^B$[3].

$$\tau(inter_i) = \{(term_k^B, \delta_k^i) \mid k = 0, 1, \ldots, g\} \tag{4}$$

Where, $\delta_k^i = \max_x \min\{y(i), \gamma_k\}$, and $y(i)$ is the membership functions of interval $inter_i$, γ_k is the membership functions of triangular fuzzy number for linguistic term $term_k^B$ in the basic linguistic evaluation set $TERMSET^B$.

3.3 Aggregate Experts' Opinions

Given the ordinal interval information on alternative S_i from expert E_j, ran_i^j, $i = 1,\ldots,m$, $j = 1,\ldots,K$,the method proposed in 3.1 can be used to transform ran_i^j into a sub interval in [0, 1], then, the method proposed in 3.2 can be sued to transform the obtained sub intervals in [0, 1] into the fuzzy set over the basic linguistic evaluation set $TERMSET^B$, denoted as FS_i^j,

$$FS_i^j = \{(term_0^B, o_0^{ij}), (term_1^B, o_1^{ij}), \ldots, (term_g^B, o_g^{ij})\}, \; i = 1,\ldots,m, \; j = 1,\ldots,K \tag{5}$$

If the experts are given weights information w_j (w_j is the weight of expert E, $0 \le w_j \le 1$, and $\sum_{j=1}^{n} w_j = 1$), the simple weighted sum method can be used to calculate the overall values $Overall_i$ of alternative S_i,

$$Overall_i = \{(term_0^B, o_0^i), (term_1^B, o_1^i), \ldots, (term_g^B, o_g^i)\}, \quad i = 1, \ldots, m \tag{6}$$

Where,

$$o_l^i = \sum_{j=1}^{K} w_j o_l^{ij}, \qquad i = 1, \ldots, m, \; l = 0, \ldots, g \tag{7}$$

3.4 Rank the Alternatives

It is obvious that the overall values $Overall_i$ of alternative S_i obtained in 3.3 is still the fuzzy set over the basic linguistic evaluation set $TERMSET^B$, i.e., $\{(term_0^B, o_0^i), (term_1^B, o_1^i), \ldots, (term_g^B, o_g^i)\}$, $i = 1, \ldots, m$, furthermore, the calculation result in the following function ϕ in (8) can be used to transform $Overall_i$ into the 2-tuple linguistic value on $TERMSET^B$,

$$\phi\left(\{(term_0^B, o_0^i), (term_1^B, o_1^i), \ldots, (term_g^B, o_g^i)\}\right) = \frac{\sum_{l=0}^{g} l o_l^i}{\sum_{l=0}^{g} o_l^i} \tag{8}$$

Denote the calculation result in (8) as d_i. The following discussions focus on transforming d_i into 2-tuple linguistic value on $TERMSET^B$ [3]-[4].

Definition 4 Given a linguistic evaluation set *TERMSET*(*TERMSET*=$\{t_0, t_1, \ldots, t_g\}$), (t_l, λ_l) is called the 2-tuple linguistic value, if λ_l is the symbolic translation about term t_l and satisfies $\lambda_l \in [-0.5, 0.5]$, $l = 0, \ldots, g$.

Usually, the symbol operation result on linguistic evaluation set *TERMSET* (for example, ω) can be transformed into a 2-tuple linguistic value on *TERMSET*, that is,

$$\omega \to t_l \times [-0.5, 0.5) \tag{9}$$

4 Illustration

The illustration data is from [5]: four experts (E_1, E_2, E_3 and E_4) give their preference information on four alternatives (S_1, S_2, S_3 and S_4) in ordinal interval format, as stated in table 1. In addition, suppose the experts are equal important, i.e., w=(0.25,0.25,0.25, 0.25). By employing the proposed approach, the 2-tuple linguistic values of the alternatives are obtained: S_1: (t_3,-0.0912), S_2 : (t_3,0.2673), S_3 :(t_3,0.2050), S_4 :(t_3,0.0873). Thus, the ranking of the alternatives is $S_2 \succ S_3 \succ S_4 \succ S_1$.

5 Summary

With respect to the group decision making problems with preference information on alternatives in the form of ordinal intervals, this paper proposes a fuzzy set approach, which is composed of the following steps: firstly, transform the preference ordering and ordinal intervals information into intervals in [0,1]; secondly, transform the ordinal interval information into the fuzzy set over basic linguistic evaluation set; Thirdly, aggregate the experts' opinions in the form of fuzzy set over the basic linguistic evaluation set; Fourthly, rank the alternatives by comparing the 2-tuple linguistic values of them.

Compared with the current research on the topic focused, the proposed approach in this paper is innovative and simpler that that in [5], and more precise than that in [6]. Furthermore, the proposed approach is readily implemented into a computer-based information system to facilitate the whole process in the group decision making.

Table 1. Matrix of preference in ordinal interval form [5]

S_i	E_1	E_2	E_3	E_4
S_1	[1, 2]	[4, 4]	[2, 4]	[2, 4]
S_2	[1, 4]	[1, 2]	[1, 3]	[2, 4]
S_3	[3, 4]	[1, 2]	[2, 4]	[2, 4]
S_4	[2, 3]	[1, 3]	[1, 3]	[2, 4]

References

1. Hwang, C.L., Lin, M.J.: Group decision making under Multiple criteria: Methods and Applications. Springer, Berlin (1987)
2. Chiclana, F., Herrera, F., Herrera-Viedma, E.: Integrating three representation models in fuzzy multipurpose decision making based on fuzzy preference relations. Fuzzy Sets and Systems 97, 33–48 (1998)
3. Herrera, F., Martinez, L., Sanchez, P.J.: Managing non-homogeneous information in group decision making. European Journal of Operational Research 166, 115–132 (2005)
4. Herrera-Viedma, E., Martínez, L., Mata, F., Chiclana, F.: A Consensus Support System Model for Group Decision-Making Problems With Multigranular Linguistic Preference Relations. IEEE Transactions on Fuzzy Systems 13(5), 644–658 (2005)
5. Gonzle Pachon, J., Romero, C.: Aggregation of partial ordinal rankings: an interval goal programming approach. Computers & Operations Research 28(8), 827–834 (2001)
6. Fan, Z.P., You, T.H.: TOPSIS Method to Solve Group Decision Problems with Preference Information in Ordinal Interval Form. Journal of Northeastern University(Natural Science) 28(12), 1779–1781 (2007)

Semantic Service Discovery Based on Agent and Service Grouping

Lijun Duan[1] and Hao Tian[2]

[1] School of Computer, Hubei University of Education, Wuhan 430205, China
[2] Department of Electronic Engineering, Hubei University of Economics, Wuhan 430205, China
duan_lj@sina.com, th@hbue.edu.cn

Abstract. Many current methods about semantic web service discovery have unsatisfactory precision and low matching efficiency. In order to correct the above-mentioned deficiency, a method of semantic web services discovery based on agent and service grouping is presented in this paper. Firstly, the scheme of web services grouping by agent is introduced. Secondly, a discovery model corresponding to the method is expounded. Finally, the method is tested, and the experiments prove that the method can improve the time efficiency and services precision of service discovery.

Keywords: Semantic, Service Discovery, Agent, Service Grouping.

1 Introduction

Web services discovery is a highly intelligent information processing technology. In its existing theories and technology system, Web service discovery has integrated multiple disciplines such as pattern recognition, natural language processing, machine learning and data mining etc. Its comprehensiveness and interdisiplinarity is very outstanding. Current methods of web services discovery could be divided into two categories: The first kind is service registration and discovery mechanism based on the UDDI (Universal Description, Discovery and Integration), which have poor performance in service description and service precision; Another kind is web service discovery based on semantic knowledge, this type of methods could achieve satisfactory effect by using semantic description technology. Most of current studies use ontology language OWL-S to implement semantic reasoning and service matching. The service rapidity and service precision are the most important performances in web service discovery, and how to improve the two indexes is one of the current research priorities.

In this paper, we propose a semantic service discovery method based on agent and service grouping. We take a web service as an entity with three important performances and adopt agent technology to group the entities dynamically according to semantic characters. The experiment shows that the method can enhance the speed and precision of service discovery.

A. Xie & X. Huang (Eds.): Advances in Electrical Engineering and Automation, AISC 139, pp. 139–144.
springerlink.com

2 Related Works

The research on semantic web services discovery focuses mostly on service matching algorithms. Massimo Paolucci et al. [1] presented a semantic matching algorithm based on web service ability. By using DAML-S [2] description language as a service model, it could implement web service semantic matching through semantic reasoning about the description of the interfaces of concepts in the ontology relation, and the effect is divided into four grades (accurately match, alternative matching, including matching and matching failure). Jeffrey Hau et al. [3] proposed a similarity calculation method based on OWL-S. According to the OWL description language concept and attributes in the inherent semantic relations, the semantic similarity between two OWL object could be calculated based on reasoning information value measurement method and the reasoning information value which are generated by semantic reasoning rules. The disadvantage of these two kinds of methods is the lack of accurate quantitative calculation in semantic matching. In addition, Nayak et al. [4] putted forward a semantic web service clustering method for service grouping. Sriharee [5] brought forward a service services quality rating model based on ontology by adopting flexible matching algorithms find services. There also are many typical semantic service discovery programs. Intelligent Software Agents Group's Matchmaker project [6] of Carnegie Mellon university adopted concentration service discovery framework for multi-level matching; of Information Management Group's Matchmaking Agent project [7] of Manchester university also adopted concentration service discovery framework, but its service discovery strategy is global semantic matching based on Profile; the large distributed information system Lab's METEOR-S project [8] of Georgia university used comprehensive service discovery framework based on P2P, completely semantic support joint registration discovery mechanism for service discovery.

Overall, the existing semantic web services discovery research is focused on service discovery mechanism and semantic matching algorithms. Most of them use the semantic relations between concepts reasoning and realize service matching, and judgment service matching degree. They still belong to a kind of coarse granularity level service matching, could offer only qualitative analysis service. Their semantic description is not meticulous enough, and the recall ratio and the precision of service discovery are not high.

3 Semantic Service Discovery Based on Agent and Service Grouping

If the registered service could be grouped on the basis of semantic similarity before service matching, it is possible to narrow search scope, improve the speed and accuracy of service discovery. Due to agent technology has good dynamic and autonomy, so it can implement real-time monitoring and processing network services. We propose a new method based on agent and service grouping for semantic service discovery.

3.1 Service Description and Grouping

We assume that there already exists clear ontology division, and every web service is corresponds to a certain ontology. Each service is defined by text description, service function parameters, service performance parameters, i.e. the entity of *E = (D, F, P)*. *D* is the parameter of text description, and it contains information including correlation between service and ontology, service type, service function description, etc; *F* is the parameter of service function and describes the service features list, service required input parameters, service output parameters and other information; *P* is the parameter of service performance and includes service time, service cost, service additional conditions and other information.

Based on the above assumptions and definition, we can group web services by agent, as shown in Fig. 1 below.

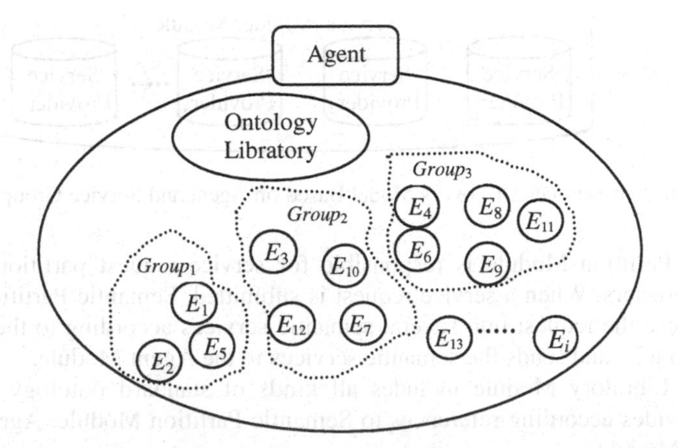

Fig. 1. Scheme of Web Services Grouping by Agent

Fig. 1 describes the principle of services grouping, each agent has good flexibility and autonomy, are responsible for managing an ontology libratory and the corresponding services (entities). When agent receives service request, could group quickly entities according to the service instruction and each entity's parameters. The grouping could change when the web service request changes. Agent also can dynamically manage the behaviors of the services (entities) belonging to the scope of the certain ontology libratory in real time.

3.2 Discovery Model

In order to enhance the speed and precision of service discovery, we propose a new service discovery model based on agent and service grouping, shown as Fig. 2. It is composed of five main modules: Semantic Partition Module, Agent Module, Entities Module, Service Provider Module and Ontology Libratory Module.

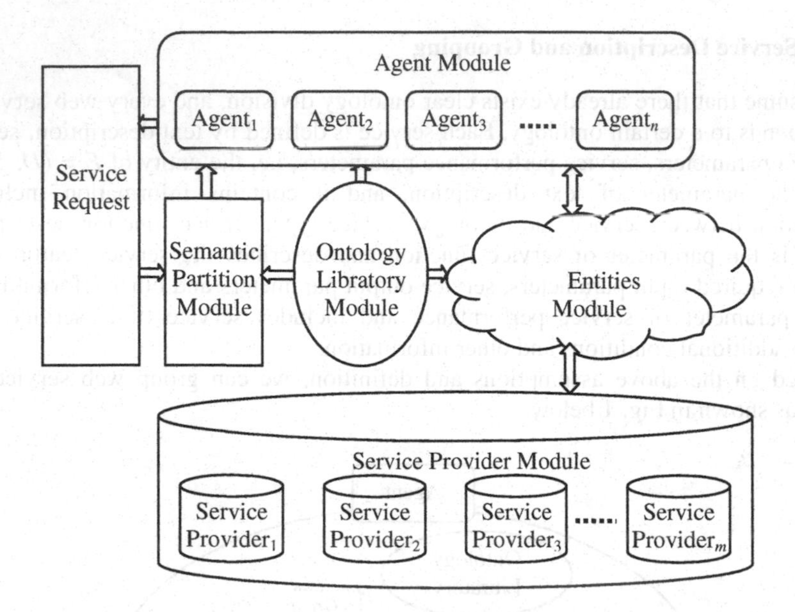

Fig. 2. Web Services Discovery Model Based on Agent and Service Grouping

Semantic Partition Module is responsible for service request partition based on semantic characters. When a service request is submitted, Semantic Partition Module will decompose the request into several semantic services according to the Ontology Library Module, and sends the semantic services to the Agent Module.

Ontology Libratory Module includes all kinds of standard ontology the model needs. It provides according references to Semantic Partition Module, Agent Module and Entities Module.

Service Provider Module is the collection of all web service providers. It contains many heterogeneous and dynamic web resources. All members of Entities Module are from this module.

Entities Module is the collection of all available service. Each service could be taken as an entity defined above in this paper.

Agent Module is composed of multiple agents. When the semantic services are sent to the module, the corresponding agent begins grouping service according to service parameters received. Then every group starts to compare the parameters of every entity according to the user's preference and returns ultimately results to the user.

3.3 Discovery Algorithm

According to the model we can see that, when a request Req_i is submitted, Semantic Partition Module decomposes Req_i into several semantic services $SemSrv$={$SemSrv_1$, $SemSrv_2$, ...$SemSrv_j$}. The module sent $SemSrv$ to Agent Module, thus, Agent Module assigns $SemSrv_j$ to $Agent_n$. $Agent_n$ then starts grouping entities based on the semantic characters of $SemSrv_j$. $Agent_n$ begins to match according to the match threshold value $MatchThresholdValue_i$ and user's preference $UserPf_j$. Finally, Agent Module returns the results to users. We bring out a discovery algorithm which is shown as the following:

```
While (Reqi≠0 ) do {
    MatchResult=0;
    ResultList=0;
    MatchThresholdValue =α;
    SemSrv={SemSrv1, SemSrv2, …SemSrvj};
    For all SemSrv do
    {
        MatchThresholdValuei=Match(Groupk, UserPfj);
        if MatchThresholdValuei≥α
        ResultList = ResultList+ MatchResult;
    }
    Return ResultList;
}
```

4 Experiments

In our experiments, we choose the same experimental environment with [9]. We also use the same tools: Net Beans 5.5.1, Protégé, Jena 2.5.1 and Apache Tomcat 6.0.14. We create services description ontology by Java API of Protégé, query and analyze services description ontology by OWL API of Jena. We also create 1500 semantic service description.

The first experiment tests the speed comparison between our method and the method in [9], and the result is shown as Fig. 3. From Fig. 3 we can see that the method proposed in this paper has better efficiency than the method in [9].

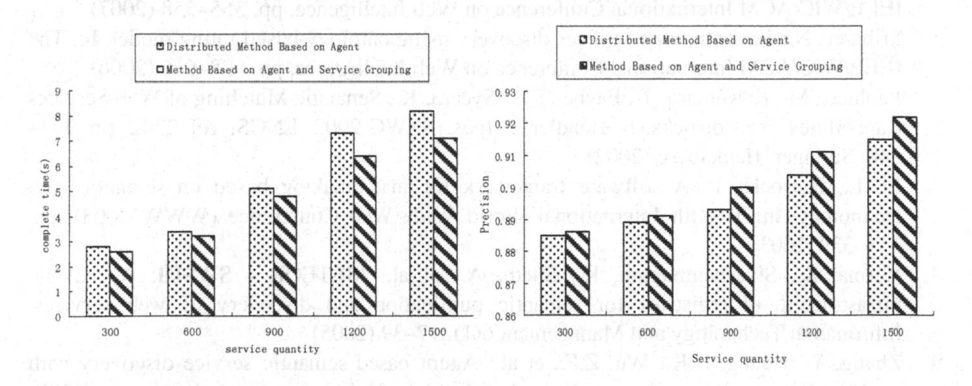

Fig. 3. Comparison of Discovery Efficiency **Fig. 4.** Comparison of Discovery Precision

The second experiment tests the precision comparison between our method and the method in [9], and the result is shown as Fig. 4. Fig. 4 shows that the method proposed in this paper also has better precision than the latter.

5 Conclusion

In this paper, we proposed a new method for semantic service discovery. We adopted agent technology and service grouping in the method. In the first stage we expounded the scheme of web services grouping by agent, and in the second stage we simultaneously built a service discovery model based on the result of the first stage and also gave the discovery algorithm of the model. Finally, we tested the method by the comparison with a distributed agent discovery method cited in [9], and the results prove that the method we brought out is a feasible way for service discovery.

Acknowledgments. This work is supported by the Science and Technology Research Funds of Educational Ministry of Hubei Province, China (Grant No. Q20112206), and the Youth Research Funds of Hubei University of Economics, China (Grant No. XJ201115) .

References

1. Paolucci, M., Kawamura, T., Payne, T.R., Sycara, K.: Semantic Matching of Web Services Capabilities. In: Horrocks, I., Hendler, J. (eds.) ISWC 2002. LNCS, vol. 2342, pp. 333–347. Springer, Heidelberg (2002)
2. Ankolekar, A., Burstein, M., Hobbs, J.R., Lassila, O., Martin, D., McDermott, D., McIlraith, S.A., Narayanan, S., Paolucci, M., Payne, T.R., Sycara, K.: DAML-S: Web Service Description for the Semantic Web. In: Horrocks, I., Hendler, J. (eds.) ISWC 2002. LNCS, vol. 2342, pp. 348–363. Springer, Heidelberg (2002)
3. Hau, J., Lee, W., Darlington, J.: A semantic similarity measure for semantic web services. In: 14th Intl. World Wide Web Conference (WWW 2005), pp. 533–541 (2005)
4. Nayak, R., Lee, B.: Web service discovery with additional semantic and clustering. In: The IEEE/WIC/ACM International Conference on Web Intelligence, pp. 555–558 (2007)
5. Sriharee, N.: Semantic web services discovery using ontology-based rating model. In: The IEEE/WIC/ACM International Conference on Web Intelligence, pp. 608–616 (2006)
6. Paolucci, M., Kawamura, T., Payne, T.R., Sycara, K.: Semantic Matching of Web Services Capabilities. In: Horrocks, I., Hendler, J. (eds.) ISWC 2002. LNCS, vol. 2342, pp. 333–348. Springer, Heidelberg (2002)
7. Li, L., Horrocks, I.: A software framework for matchmaking based on semantic web technology. In: Twelfth International World Wide Web Conference (WWW 2003), pp. 331–339 (2003)
8. Verma, K., Sivashanmugam, K., Sheth, A., et al.: METEOR - SWSDI: A scalable infrastructure of registries for semantic publication and discovery of web services. Information Technology and Management 6(1), 17–39 (2005)
9. Zhang, Y., Huang, H.K., Wu, Z.F., et al.: Agent based semantic service discovery with quality of service measurement. Journal of Nanjing University (Natural Sciences) 44(9), 495–502 (2008)

Design of the Data Mining System Module Based on GT4

Yajuan Lv[1], Shuai Wang[2], Zhigang Liu[3], and Xinyu Chen[3]

[1] Jilin Medical College Computer Office, Jilin 132013
[2] 65127 Army, Jilin Meihekou, 135000
[3] Jilin Medical College Network Management Center, Jilin 132013

Abstract. First, this paper briefly introduces the data mining system structure and the divide basis of system module based on the GT4 platform, detailed introduction the design process: the user interface module, resources registration module, global data mining module, local data mining module.

Keywords: GT4, Data Mining, Web Service, Grid Node.

1 Introduction

Popularization of Computer network technology brought earth-shaking changes to the life of people, and produced a lot of desultory data on the internet. The application of Globus Toolkit 4 (Referred to as GT4) sure to build a practical data mining system based on grid, and help people to find valuable information from the distribution of the cyber source to provide new technology support. Grid technology can make the idle computing resources that geographically dispersed to realize the sharing of resources. Application of GT4 core development kit (Java Web Service Core) and the data mining technology to deal with dispersed data information resources of grid platform just like operating a computer, operating the grid resources just like to use the browser browsing webpage.

2 Design of the System Structure

Design of the system structure is more important in the software engineering, according to user needs and design, it is foundation of the whole data mining system based on grid. This architecture has the following characteristics:

First, it realization the data resource integration and management based on grid platform;

Second, it can deal with local and remote computer data, realization serial / parallel data mining;

Third, system nodes mutual cooperation, adding and deleting nodes are relatively easy to, and system maintainability.

A. Xie & X. Huang (Eds.): Advances in Electrical Engineering and Automation, AISC 139, pp. 145–149.
springerlink.com

2.1 Achieve the Goal of the System

Using of increasingly sophisticated GT4 grid management technique and data mining technique to build a practical distributed data mining system. The main feature of contemporary society: digital, information-based, network-based. Network infrastructure is perfect with each passing day, configuration of software and hardware resources reasonable stability. It is feasibility that using of off-the-shelf cyber source to build a development platform base on grid.

2.2 Divide of the System Module

The data mining system's data source is distributed data source based on GT4, distributed data source refers to the physical distribution and logically centralized data source system. Distributed data source storage characteristics:

(1)Distribution of physical, data stored in more dispersed node of the network,

(2)Logic of overall importance, the data storage in each node is a global data in logical;

(3) Node autonomy, the local user manage the data each node, it has the autonomous.

In the system, each computer is a node of the grid, known as the grid node. In a large number of nodes, there must be a grid node to control and management other node, this node is called grid center control node, the central control node complete of decision support base on grid. If you want to complete a data mining task, the free grid node can complete this node mining task according to the mining needs, the grid center control node to collect each node data mining result. Local grid node management information has limitations, coverage is small, mainly managing single node data, pooled analysis of partial results of data mining. However, these local node data and global node data also has certain connection.

According to the above analysis, the task of data mining base on grid complete by global data mining and local data mining together. The data mining system based on grid mainly include the following several modules: user interface, resources registration, global data mining, and local data mining. The system structure is shown in figure 1:

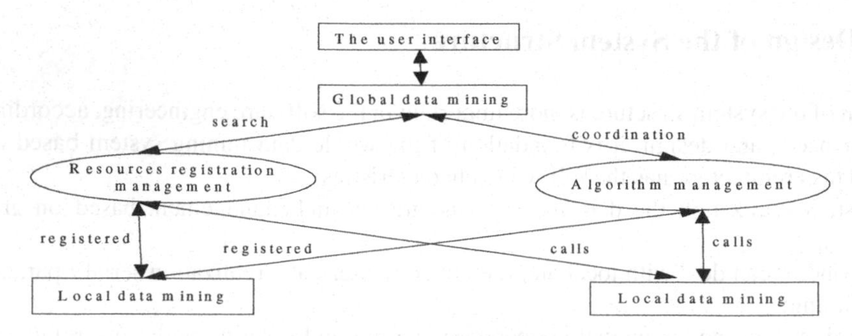

Fig. 1. System structure

2.3 Design Functions and Interface of the System Modules

The system is distributed computing system based on Java_WS _Core (i.e. Web services), each computer in the grid can be as an independent grid node, and it is registered and release as an independent Web Service. Each node has a good independence, operability and portability, in order to better complete the function of the system, it is divided into four modules, the specific functional modules and interface are design as follows:

(1) The user interface module: this module is the interactions between the user and the grid system resources, is the system window, it manage registration resources, global data mining resources and local data mining resources, is the graphical interface, it link the rest of the three modules. In order to operate conveniently, the user interface using a graphic interface of Web way, so it can use the browser to release and find some service for users, and provide a unified interface for search services.

(2) Resources registration module: Free local Web Service resources of the grid system registered in resources registration module. In resources registration center, the global grid node timely increase and delete local Web Service resources.

Local Web Service resource use WSDL standard definition, mainly finish of data mining service providers, describe the specific data mining information. Data mining service providers' descriptive information which includes the grid node computer mainframe models, hard disk size, memory capacity, installed in the operating system, the CPU and memory utilization, and publish library resources registration. Data mining that particular information description includes data mining algorithm examples the address, data distribution examples the address, data mining status, mining directory information such as local files.

(3) Global data mining module: It is the grid data mining system comprehensive service control node. Global data mining module used to unify management data mining based on Web service, in order to achieve the dynamic and heterogeneous Web resources integration. When a local data mining service request, global control node proposed inquires to resources registration center, provide local node basic information, data mining type of data mining service, data collection, data mining algorithm, from many registered Web services to retrieval and meet the requirements of the service, using the grid FTP service transmit Web services to local grid node. Global control node and local node is dynamically-bound, when local node complete mining tasks, returned mining results by grid FTP service to global control node, then global control node summarize and analyze the different local mining results, and submitted the final result to the user interface.

Global data mining module's Web Service resources components include global Web Service interface, global mining resources and their attributes document. Mainly use XML to describe data mining algorithm Web Service example address, local Web Service example address, global data mining status, data mining results etc, and document information of resource attribute. Global Web Service resources mainly operation includes: creating and destroying the examples, executive global mining algorithms, coordinating local Web Service resources.

Global Web Service resources of operation: when local node request data mining, first to create global Web Service resources examples, to build the global calculation containers, submit task (find free resources, allocate data). Performance data for allocating algorithm, monitoring local mining status, task over, collect mining results.

Global data mining a Web Service's main interface as follows:

- Create Web Service resources examples: Create Resource( )
- Destruction Web Service resources examples: Destroy Resource( )
- Start global mining algorithm: Start_QJDM( )
- Receiving algorithm parameters: Receive_ Algorithm parameter( )
- Receiving user parameters: Receive_Userparameter ( )
- Receiving data distribution Web Service resources EPR: Receive_DatamarkEPR( )
- Receiving local mining results: Receive_JBDMResult( )
- Monitoring local mining service status: Ask_JBDMState( )

(4) Local data mining module: It is an indispensable part of the whole system, the local data mining module accept global data mining application of dynamic invocation. According to mining request of user, and calling corresponding of the data mining task, creating a Web service information, and dynamically binding with global control node, receiving global control node of mining Web service information, data resource information, calling a local data mining algorithm, realizing the data mining analysis, using the grid FTP transfer mining results back to global data mining module.

Local data mining module's Web Service resources include local Web Service resources interface, local data mining resources and their attributes document components. Local Web Service resources similar as global Web Service resources, with XML language to describe local mining resources attribute document. This attribute document mainly describe a data mining algorithm Web Service examples , address, data mining type, local data mining status, data mining results and other information. Mainly operation of the local Web Service resources interface includes: performing local data mining algorithm procedures, applying for local data mining data, calls the relevant data, complete local data mining tasks and generate mining results.

Local Web Service resources operational process: first, creating local Web Service resources examples, generating calculation containers, submitting mining task to calculated containers. To start the task, performing local mining algorithms, applying for local mining data, obtained data, completing mining task, returning the results back to the calculation containers, destroyed containers.

Local data mining Web Service main interface as follows:

- ✓ create local Web Service resources examples: CreatResource( )
- ✓ destroy local Web Service resources examples: Destroy Resource( )
- ✓ receiving data distribution Web Service resources EPR: Receive_DatamarkEPR( )

- ✓ receiving mining data: Receive_DMdata( )
- ✓ start local algorithm: Start_JBAlgorithm( )
- ✓ obtain mining data:Get_Data( )
- ✓ began to local mining: Start_JBDM( )

The WSDL file of global and local data mining module not only contains the service interface, also contains the namespace mapping, message element, specific services operation. Message element, defines the basic elements of service use, Service operation provided the main operation of the external services. Other WSDL documents about binding port communication protocol specification, abstract operation specification and other content, also in accordance with the WSDL standard.

3 Conclusion

Building a data mining platform based on grid, provides a new method of data mining, to fill the blank of the data mining system in the domestic field based on GT4, making the distributed data mining application more practical.

References

1. Lei, C.: Medical Data Zmining, pp. 4–5. Higher education press, Beijing (2006)
2. Liang, X.: Distributed Dhain Commercial Data Mining Modeling, pp. 14–16. Aviation industry press, Beijing (2008)
3. Mario, C., Antonio, C., Andrea, P.: Distributed Data Mining on Grids: Services, Tools, and Applications. IEEE Transactions on Systems, Man and Cybernetics (2004)
4. Zheng, Z.: Researeh and Development of Servie-oriented Distributed Data Mining in Grid: (Master thesis). Wuhan university of technology, Wuhan (2009)
5. Rajkumar, B., Manzur, M.: GridSim a toolkit for the Modeling and simulation of distribution resource management and scheduling for Grid computing Concurrency and computation. Practice and Experience (2002)
6. Yan, D.: Data warehouse and data mining technology apply in the hospital information management. South China Defense Medical Journal 22(4), 60–61 (2008)

Comparative Research on Audience's Attitudes toward and Behaviors Caused by Product Placement and Traditional Advertising

Yue Yang[1], Xianliang Wang[2], and Hu Liu[1]

[1] School of Economic& Management, Communication University of China, Beijing, China
[2] Northeastern University at Qinhuangdao, Qinhuangdao, China
{yyqhd,vickywxl}@163.com, liuhu@cuc.edu.cn

Abstract. Product placement is a fast growing multi-billion dollar industry. The investigation is made to compare product placement and traditional advertising from attitudes and behaviors. Findings also reveal differences in both attitudes and behaviors across a range of factors, including watching time, gender, age and education level.

Keywords: Product placement, Traditional advertising, Comparative research, Attitude & behavior.

1 Introduction

Product placement is the inclusion of consumer products or services for promotional purposes in TV programs and movie; it is favored by advertisers and media recently, which has developed into a key component of China's media industry, especially appearing most often in the movies and TV programs. It is very helpful for improving the effectiveness of this kind of adverting to consider the relationship between product placement and audience experience in the advertising operation.

The recent studies on product placement and audience experience mainly focus on the following four dimensions to analyze the trends on the audiences' attitude to product placement and their consumption-related behaviors.

Cultural background of the Audience. The audiences of different ethnicity, different nations, and different educations are different in the degrees for accepting the product placement in movies and TV programs (Federico de Gregorio, 2010). Martin Eisend(2009)indicated that product placement acceptability of ethically charged/controversial products is generalizable over different cultures, but not product placement acceptability of neutral products. The higher the education of audience is, the more negative their attitude tends to be (Federico de Gregorio, 2010;Yin Xiaolin, 2009).

Age and Gender of the Audience. The audience of different age and gender has different attitude toward product placement in movies and TV programs. But results of those researches are somewhat not the same. Some researchers argue that the non-students groups have more positive attitude to product placement than the student groups (Yongjun Sung, 2009). And some researches show that the young people are more inclined to accept product placement (Federico de Gregorio, 2010 ; Cristina, 2010, and Yin Xiaolin, 2009). Some researchers conclude that the female has more positive attitude to product placement than the male (Federico de Gregori, 2010). While, some researchers think that the gender could only in some time affect the audience's acceptance degree of this kind of advertising (Cristina, 2010 and Yin Xiaolin, 2009). It is worth suggesting that Zhao Qian (2010) and Yin Xiaolin (2009) have surveyed the same area in China, but got different conclusions. Zhao Qian concludes that there is no significant difference between the male and the female audiences in degrees of accepting product placement.

Product Category. Not all kinds of products can be accepted to be placed into movies and TV programs by audiences. Ignacio Redond (2008) shows that the most suitable product categories for product placement in movies and TV programs include alcoholic drinks, alcohol-free drinks, groceries, cultural offerings, and personal care product. But Yongjun Sung(2009) finds different results, which reveal that tobacco, firearm, gambling, politically orientated and alcohol brands or products are the most difficult to accept for product placement, and it is easiest to accept car, camera, electronics, sunglasses, charity, and heath care product for product placement.

Form of Product Placement. Although the classification of product placement is slightly different, scholars have similar findings. It is that blunt and highlighting type of product placement is more likely to remember for audience, but is not in favor of audience to accept the ad's content, so it is more difficult to induce the audience to change attitudes and behavior by this type of placement. But the cleverly integrated into the plot and subtle style product placement is more positive, especially in this form of product placement repeated gentle.(Pamela Miles Homer,2009;Eva Van Reijmersdal,2009;Peter Neijens,2009;Moonhee Yang,2007;Nie yanmei,2008 and Dai tingting,2010)

While there has been a growing stream of research focusing on product placement, there has also been a dearth in the use of systematic conceptual groundings to help guide such investigations. Advertising and psychological theories provide valuable explanations for the majority of the effects. However, whether there are some differences in advertising effectiveness between traditional advertising and product placement in China, which will provide some indications for the marketers to choose the advertising media. There has also been very limited analysis of differences between product placement and traditional advertising from consumers' demographic characteristics and the degree of exposure to these advertisements.

2 Research Questions

Back in the 1950s, a lot of scholars studied on the impact of mass media, and they considered the correlation between consumer socialization and viewing time and individual age of movie and TV program. The media exposure has been consistently documented as a significant influencer of their consumer attitudes and behaviors among consumers according to consumer socialization. In our study, we focus on movie and TV program as the socialization agents during the consumer socialization process, and systematically analyze the attitude towards and behaviors caused by product placement and traditional advertising in China. Because our study is focused on product placements in movie and television program, we operationalize media exposure as time of movie and television program watching. And the comparison of influence degree is made between traditional advertising and product placement.

RQ1: There is a significant difference in attitude toward product placement and traditional advertising.

RQ2: There is a significant difference in behaviors caused by product placement and traditional advertising.

If there is a significant difference after the RQ1 and RQ2 analysis, the discussion of what factors and how they impact the attitude and behavior is necessary.

RQ3 : There is a significant difference in relationship of time for movie and TV program watching and consumer's attitude toward product placement and traditional advertising.

RQ4: There is a significant difference in relationship of time for movie and TV program watching and consumer's behaviors caused by product placement and traditional advertising.

RQ5: There is a significant difference in terms of attitude toward traditional advertising and product placement, how do consumers differ in regard to (a) gender, (b) education level and (c) age.

RQ6: There is a significant difference in terms of behaviors caused by traditional advertising, how do consumers differ with regard to (a) gender, (b) education level and (c) age.

3 Method

3.1 Sample

The current study object was people under 45 years old. Trough online questionnaires, the answers were obtained by random respondents. The final sample size was 533. Demographic variables (gender, age, educational level) and time of watching movies and TV programs per week was assessed by a single question. Two questions measured the product placement attitudes toward by 3 scales. Two dichotomous scale questions measured the behaviors in response to traditional advertising and product placement. Demographic characteristics of the respondents are shown in Table.1.

3.2 Variables and Measures

Time（Time of Movie and TV program Watching）. This variable was operationalized as the degree of media exposure. Respondents were asked to indicate how much time they spend watching TV and movie in a week.

A1 (Attitude toward Traditional Advertising). This variable was operationalized to assess attitudes toward traditional advertising by 3 scale in general.

A2 (Attitude toward Product Placement). This variable was operationalized to assess attitudes toward product placement by 3 scale in general.

B1 (Behavior caused by Traditional Advertising). This variable was operationalized to obtain the consumer behaviors caused by traditional.

B2 (Behavior caused by Product Placement). This variable was operationalized to obtain the consumer behaviors caused by product placemen.

GEN, AGE and EDU was operationalized to represent the gender and age of the respondent and whether he (or she) receives a higher education respectively.

Table 1. Demographic Profile of the Sample

		Frequency	Percentage
GEN			
	male	309	58.0
	female	224	42.0
	total	533	
AGE			
	≤18	6	1.1
	19~24	274	51.4
	25~34	195	36.6
	35~44	58	10.9
	total	533	
EDU			
	yes	124	23.3
	no	409	76.7
	total	533	

4 Results of Research Questions

4.1 Results of RQ1 and RQ2

In order to answer ***RQ1*** and ***RQ2***, the ANOVA (analysis of variance) was conducted for the consumer's attitude toward and behavior caused by product placement and traditional advertising (showed in Table 2). As they do not meet the assumption of test of homogeneity of variance (Levene Statistic, $p<0.05$), the Welch and Brown-Forsythe test are used, then we get the probability value less than 0.05 significance level. The analysis suggests that there are significant differences in attitude toward product placement and traditional advertising, the behavior caused by product placement and traditional advertising as well.

Table 2. ANOVA of attitude and behavior

A1 and A2	F	P	B1 and B2	F	P
Levene Statistic	10.204	0.000	Levene Statistic	1278.224	0.000
Welch	4.228	0.015	Welch	273.090	0.000
Brown-Forsythe	4.171	0.016	Brown-Forsythe	273.090	0.000

4.2 Results of RQs 3~6

In order to analyze the impact factors of attitude and behavior, the ANOVA (analysis of variance) was conducted for TIME, GEN, AGE and EDU.

TIME. As showed in Table 3, respondents whose watching time is above 15 hours and who never watch movies and TV program have significantly different attitudes toward traditional advertising and product placement (P<0.05), but respondents whose watching time is in the middle area have no significant difference in attitudes (P>0.05). Further analysis found that respondents who never watch movie and TV hold a negative attitude toward traditional advertising, but are supposed to be basically neutral to product placement. And it is very significant that tolerant attitude toward product placement is held when respondent's watching time for movie and TV program is above 15 hours. Respondents with different watching time for movie and TV program show significantly different consumption-related behaviors (P<0.05).That means respondents' consumption-related behaviors caused by traditional advertising are much more than product placement.

Table 3. ANOVA of attitude and behavior by factors

F/Welch sig		A1& A2	B1 &B2	F/Welch sig		A1& A2	B1 &B2
TIME(hours)	<1	0.003	0.000	GEN	M	0.000	0.000
	1~3	0.787	0.000		F	0.349	0.000
	4~7	0.473	0.000	EDU	Y	0.269	0.000
	7~10	0.104	0.000		N	0.014	0.000
	11~14	0.134	0.000	AGE	<18	-	-
	15~20	0.010	0.025		19~24	0.097	0.000
	>21	0.014	0.000		25~34	0.130	0.000
					35~45	0.119	0.000

GEN. As shown in Table3, the results suggested that male respondents have significantly different attitude toward traditional advertising and product placement (P<0.05), while the female respondents do not (P>0.05). Furthermore, male respondents who hold a negative or neutral attitude toward traditional advertising show more positive attitude toward produce placement on the contrary. However, all the female respondents show more positive attitude toward produce placement no matter how they feel about the traditional advertising. And both male and female respondents' consumption-related behaviors affected by traditional advertising are much more than product placement.

AGE. As shown in Table3, the results suggested that responders of all age groups have significantly different behaviors caused by traditional advertising and product placement. The responders' consumption-related behaviors are mostly caused by traditional advertising. On the contrary, there is no significant difference in attitude in all age groups mentioned above. This result is inconsistent with the conclusion of QsR1~2, because the age group under 18 years old can not be statistically calculated. In addition, because of all age groups' significance tests are also relatively close to 0.1, when the sample size increases, the result will becomes significant.

EDU. As shown in Table3, grouped according to whether respondent receiving higher education, we found that there is no significant difference in attitude toward traditional advertising and product placement (p=0.2690.05) in the higher education group. But, as to the group of not received higher education, there is a significant difference in attitude (p= 0.0140.05), they prefer product placement against traditional advertising. The traditional advertising and product placement affect the respondent's consumption-related behavior significantly differently (p=0.000) no matter they received higher education or not. The behaviors caused by traditional advertising are much more than product placement.

5 Discussion

Our findings indicate the practical implications for placement agents and marketers in China.

First of all, the results suggest that the traditional advertising is more likely cause the consumption-related behaviors, regardless of the responders' age, education level and gender and how long they spent on watching movie and TV program. That means, no matter how novel the product placement is, but its consumption-related behaviors is not better than traditional advertising. That might because the product placement is recessive advertisement and sometimes it is designed to be identity difficultly for the movie's plot. In addition, the main purpose of movie is not product promotion, so it can't deliver product information to consumers directly and completely. Therefore, although product placement is so popular for advertising agents, but it does not show more advantages in consumption-related behaviors. The cost of product placement is expensive, but receives little effect. When advertising agents and marketers place product in movie, they must choose the proper movie and placement pattern.

Second, the analysis result of responder's gender indicated that male responders do not like traditional advertising, but they hold positive attitude toward product placement. If the male consumers are potential target groups for the product, then traditional advertising might lead to ignore or resentment but product placement may be better accepted.

Third, our results provide some indications that respondents who didn't receive higher education prefer product placement. The advertisers would do well to focus their placement efforts on movies popular among this type of consumers in China. These audiences exhibit higher levels of receptiveness to product placement as a practice, and marketers can follow these indications.

All in all, the conclusions on attitude toward product placement and traditional advertising should not be interpreted by advertisers as a blank check to use product placement at their pleasure or to mean that audiences will be willing to tolerate any kind of placement. Traditional advertising is effective—at least for some of the time and is never out of season. Further more, both placement agents and marketers would do well to match up their placement efforts based on a particular target audience of movie and TV programs as to the product placement.

References

1. de Gregorio, F., Sung, Y.: Understanding Attitudes Toward and Behaviors in Response to Product Placement. Journal of Advertising (39), 83–96 (2010)
2. Sung, Y., de Gregorio, F., Jung, J.-H.: Non-student consumer attitudes towards product placement. International Journal of Advertising (28), 257–285 (2009)
3. Redondo, I., Holbrook, M.B.: Illustrating a systematic approach to selecting motion pictures for product placement and tie-ins. International Journal of Advertising 27, 691–714 (2008)
4. Eisend, M.: A Cross-Cultural Generalizability Study of Consumers Acceptance of Product Placements in Movies. Journal of Current Issues and Research in Advertising (31), 15–25 (2009)
5. Cristina, Raluca, Delia: Product placement in Romanian movies produced after 1989. Journal of Media Research (8), 46–73 (2010)
6. Van Reijmersdal, E.: Brand Placement Prominence: Good for Memory Bad for Attitudes. Journal of Advertising Research 6, 151–153 (2009)
7. Yang, M., Roskos-Ewoldsen, D.R.: The Effcetivencess of Brand Placements in the Movies: Levels of Placements, Explicit and Implicit Memory, and Brand-Choice Behavior. Journal of Communication (57), 469–489 (2007)
8. Zhao, Q.: Effect of product placement in film's. China's Collective Economy (2) (2010)
9. Yin, X.: Empirical research on the effect of product placement in films. China's Collective Economy (1) (2009)

Exponential Stability of Numerical Solutions to a Class of Nonlinear Population Diffusion System

Li Wang and QiMin Zhang

School of Mathematics and Computer Science, NingXia University, YinChuan, 750021, P.R. China

Abstract. In recent years, numerical solutions of stochastic differential equations have received much attention. Numerical approximation schemes are very important for exploring their properties. In this paper, a class of nonlinear population diffusion system was introduced and then the exponential stability is obtained by use of Euler method and Barkholder-Davis-Gundy's inequality. It is proved that the numerical approximation solutions converge to the analytic solutions of the equations under the given conditons, where information on the order of approximation is offered.

Keywords: Barkholder-Davis-Gundy's inequality, exponential stability, nonlinear population diffusion system.

1 Introduction

There are many applications of SDE [1,2]. Recently, their numerical solutions have been found much importance in many fields [3,4,5].In this paper, we consider the convergence of population systems with diffusion

$$\begin{cases} \frac{\partial P}{\partial t}+\frac{\partial P}{\partial r}-k\Delta P+\mu_0(r,t,x)P+\mu_e(r,t,x;P)P & \\ = f(r,t,x;P)+g(r;t;x;P)\frac{\partial \omega}{\partial t} & in\ Q \\ P(0,t,x)=\int_0^A \beta(r,t,x;P)P(r,t,x)dr & in\ \Omega_r=(0,T)\times\Omega \\ P(r,0,x)=P_0(r,x) & in\ \Omega_A=(0,A)\times\Omega \\ P(r,t,x)=0 & in\ \sum=(0,A)\times(0,T)\times\partial\Omega \\ P(t,x)=\int_0^A P(t,r,x)dr & in\ \Omega_r \end{cases} \quad (1)$$

where $Q=(0,A)\times(0,T)$, $r\in[0,A]$ denotes the age, $x\in\Omega$ denotes the position variables in space, $\Omega\subset R^N$ denotes a bounded region with smooth bounder $\partial\Omega$; $P(r,t,x)\geq 0$ denotes the age-density function of population at time

A. Xie & X. Huang (Eds.): Advances in Electrical Engineering and Automation, AISC 139, pp. 159–164.
springerlink.com

t age r and station x, $k \geq 0$ is the coefficient of population diffusion; $\mu_e(r,t,x;p) \geq 0$ is the extra death rate function of population. Δ denotes the Laplace operator with respect to the space variable. In this paper, we will develop a numerical approximation method for SDE (1).

2 Preliminaries

Let $O = [0, A] \times \Omega$, and $V \equiv \{\varphi \mid \varphi \in L^2(O), \frac{\partial \varphi}{\partial x_i} \in L^2(O)$, $H = L^2(O)$. We denote by $\|\cdot\|, |\cdot|$ the norms in V, H respectively; and $m > 0$ a constant that $|x| \leq m\|x\|, \forall x \in V$. Let ω_t be a Wiener process defined on complete probability space (Ω, F, P).Let $(F_t)_{t\geq 0}$ be the σ algebras for $\{\omega_s, 0 \leq s \leq t\}$, then ω_t is a martingale relative to $(F_t)_{t\geq 0}$ and $\omega_t = \sum_{i=1}^{\infty} \beta_i(t) e_i$, $\beta_i(t)$ are mutually independent real Wiener processes with incremental covariance $\lambda_i > 0$, $W_{e_i} = \lambda_i e_i$ and $trW = \sum_{i=1}^{\infty} \lambda_i$ (tr denotes the trace of an operator).

Consider the following nonlinear stochastic equation(*):

$$\begin{cases} P_t = P_0 - \int_0^t \frac{\partial P_s}{\partial r} ds + \int_0^t k \Delta P_s ds - \int_0^t \mu_e(r,s,x;P_s) P_s ds \\ -\int_0^t \mu_0 P_s ds + \int_0^t f(r,s,x;P_s) ds + \int_0^t g(r,s,x;P_s) d\omega_s, in \quad Q \\ P(0,t,x) = \int_0^A \beta(r,t,x;P) P(r,t,x) dr \qquad in \quad \Omega_r = (0,T) \times \Omega \end{cases}$$

where $P_t = P(r,t,x), P_0 = P(r,0,x)$

Definition 1. Let $(\Omega, F, \{F_t\}, P)$ be the stochastic basis and ω_t a Wiener process. Suppose that P_0 is a random variable such that $E|P_0|^2 < \infty$. A stochastic process P_t is said to be a strong solution on Ω to the SDE (*) for $t \in [0,T]$ if the following conditions are satisfied:

(1) P_t is a F_t-measurable random variable;

(2) $P_t \in I^p(0,T;V) \cap L^2(\Omega; C(0,T;H))$, $p > 1, T > 0$ where $I^p(0,T;V)$ denotes the space of all V-valued processes $(P_t)_{t\in[0,T]}$ measurable (from $[0,T]\times\Omega$ into V), and satisfying $E\int_0^T \| P_t\|^p dt < \infty$

(3) Eq.(*) is satisfied for every $t \in [0,T]$ with probability one.

If T is replaced by ∞, P_t is called a global strong solution of (*).Let $\Delta t = \frac{T}{N}$,

$$Q_t^{n+1} - Q_t^n - \frac{\partial Q_t^n}{\partial r}\Delta t - k(r,t)\Delta Q_t^n \Delta t + \mu_0 Q_t^n \Delta t + \mu_e(r,t,x) Q_t^n \Delta t$$

$$= f(r,t,x;Q_t^n)\Delta t + g(r,t,x;Q_t^n)\Delta\omega_n$$

Here, Q_t^n is the approximation to $P(r,t_n,x)$.

We first define the step function

$$Z_t \equiv Z(r,t,x) = \sum_{k=0}^{N-1} Q_t^k I_{[k\Delta t, k+1\Delta t]}$$

where I_G is the indicator function for the set G .Then we define

$$Q_t - P_0 + \int_0^t \frac{\partial Q_s}{\partial r} ds - \int_0^t k(r,s)\Delta Q_s ds + \int_0^t \mu_0 Z_s ds + \int_0^t \mu_e(r,s,x) Z_s ds$$

$$= \int_0^t f(r,s,x;Z_s) ds + \int_0^t g(r,s,x;Z_s) d\omega_s$$

with $Q_0 = P(r,0,x), Q_t = Q(r,t,x)$.The objective in this paper is that, we hopefully find a unique process $P_t \in I^p(0,T;V) \cap L^2(\Omega; C(0,T;H))$ such that (*) hold. For this objective, we assume the following conditions are satisfied:

(a.1) $f(r,t,x,0) = 0, g(r,t,x,0) = 0$;

(a.2) (Lipschitz condition) $\exists K > 0$ such that for all $p_1, p_2 \in C$

$$| f(r,t,x;p_1) - f(r,t,x;p_2) | \vee | g(r,t,x;p_1) - g(r,t,x;p_2) | \le K\| p_1 - p_2 \|_C$$

(a.3) $\mu(r,t,x) and\, \beta(r,t,x)$ are continuous in Q such that

$$0 \le \mu_0 \le \overline{\mu} < \infty, 0 \le \beta(r,t,x) \le \overline{\beta} < \infty, k_0 \le k(r,t) \le \overline{k}$$

Definition 2. Suppose that P_0 is a random variable such that $E|P_0|^2 < \infty$. For a given step size $\Delta > 0$, a numerical method is said to be exponentially stable in mean square on Equation (1) if there is a pair of positive constants γ and $\overline{N}$, such that

$$E|Q_t^n|^2 \le \overline{N} E|P_0|^2 e^{-\gamma n \Delta}, \forall n = 0,1,2...$$

3 Stability of Strong Solutions

In this Section, we shall establish some criteria for the exponential stability of SDE (*).we first study the properties of Q_t:

Lemma 1. Under the assumptions of (a.1)-(a.3), $\exists C_{1t}$, for $\forall T > 0$

$$\sup_{0 \le t \le T} E|Q_t|^2 \le C_{1T}$$

Proof. Applying *Itô* formula to Q_t yields

$$|Q_t|^2 = |Q_0|^2 + 2\int_0^t \int_O \left(-\frac{\partial Q_s}{\partial r} + k(r,s)\Delta Q_s - \mu_0 Z_s\right) Q_s \, dxdrds$$

$$-2\int_0^t \int_O \mu_e(r,s,x) Z_s Q_s \, dxdrds + 2\int_0^t \int_O f(r,s,x;Z_s) Q_s \, dxdrds$$

$$\int_0^t \int_O k(r,s)\Delta Q_s Q_s \, dxdrds = -\int_0^t \int_O k(r,s)\Delta \nabla Q_s \nabla Q_s \, dxdrds \le -k_0 \int_0^t \| Q_s\|^2$$

$$E\sup_{0 \le s \le t} |Q_s|^2 \le E|Q_0|^2 + (A\overline{\beta}^2 - k_0 + 1)\int_0^t E\sup_{0 \le s \le t} |Q_s|^2 \, ds$$

$$+(4\overline{\mu} + 2K^2)\int_0^t E|Z_s|^2 \, ds + 2E\sup_{0 \le s \le t} \int_0^s \int_O g(r,\tau,x;Z_s) Q_\tau \, dxdrd\tau$$

By Burkholder-Davis-Gundy's inequality, we have

$$E\sup_{0 \le s \le t} \int_0^s \int_O g(r,\tau,x;Z_s) Q_\tau \, d\omega_\tau \le \frac{1}{4} E\sup_{0 \le s \le t} |Q_s|^2 + K_1 K^2 \int_0^t E|Z_s|^2 \, ds$$

$$E\sup_{0 \le s \le t} |Q_s|^2 \le 2(A\overline{\beta}^2 - k_0 + 1 + 4\overline{\mu} + 2k^2 + k_1 k^2)\int_0^t E\sup_{0 \le s \le t} |Q_s|^2 \, ds + 2E|Q_0|^2$$

Now, Gronwall 's lemma implies the required result, the proof is complete.

Lemma 2. Assume assumptions (a.1)-(a.3),for any T > 0 such that

$$E\sup_{0\le t\le T}|Q_t-Z_t|^2\le C\Delta t\sup_{0\le t\le T}E|Q_t|^2$$

Proof. $$Q_t-Z_t=-\int_{k\Delta t}^{t}\frac{\partial Q_s}{\partial r}ds+\int_{k\Delta t}^{t}k(r,s)\Delta Q_s ds-\int_{k\Delta t}^{t}\mu_0(r,s,x)Z_s ds$$

$$-\int_{k\Delta t}^{t}\mu_e(r,s,x)Z_s ds+\int_{k\Delta t}^{t}f(r,s,x;Z_s)ds+\int_{k\Delta t}^{t}g(r,s,x;Z_s)d\omega_s$$

By the Cauchy-Schwarz inequality and the condition (a.1)-(a.3) we get

$$|Q_t-Z_t|^2\le 5\Delta t\int_{k\Delta t}^{t}|\frac{\partial Q_s}{\partial r}|^2 ds+5\bar{k}^2\Delta t\int_{k\Delta t}^{t}|\Delta Q_s|^2 ds$$

$$+10\bar{\mu}^2\Delta t\int_{k\Delta t}^{t}|Z_s|^2 ds+5k^2\Delta t\int_{k\Delta t}^{t}|Z_s|^2 ds+5|\int_{k\Delta t}^{t}g(r,s,x;Z_s)d\omega_s|^2$$

and $$E\sup_{0\le t\le T}|\int_{k\Delta t}^{t}g(r,s,x;Z_s)d\omega_s|^2\le C_3\int_{k\Delta t}^{t}E\sup_{0\le s\le T}|Z_s|^2 ds$$

$$E\sup_{0\le t\le T}|Q_t-Z_t|^2\le 5(C_4+2\bar{\mu}^2+k^2)\Delta t\sup_{0\le t\le T}E|Q_t|^2$$

Lemma 3. Under assumptions (a.1)-(a.3), for any $T>0$

$$\sup_{0\le t\le T}E|Q_t-P_t|^2\le C_T\Delta t\sup_{0\le t\le T}E|Q_t|^2$$

Proof. Use *Itô* formula and Cauchy-Schwarz inequality and the same method of lemma 3.3 in [6], we can get the conclusion.

Lemma 4. Under assumptions (a.1)-(a.3), the trivial solution of Equation (1) is exponentially stable in mean square . That is , $\exists\lambda, M>0$ such that for any P_0

$$E|P_t|^2\le ME|P_0|^2 e^{-\lambda t},\forall t\ge 0$$

Theorem 1. Under assumption of (a.1)- (a.3), the Euler method applied to Equation (1) is exponentially stable in mean square.

Use lemma 1-lemma 4, we can get the proof of this theorem, and it is also analogous to that of Theorem 2.2 in [7].

Acknowledgement. The author would like to thank the referees for their very comments which greatly improved this paper.

The research was supported by The National Natural Science Foundation (No.11061024)

(China); and was also was supported by Ningxia Natural Science Foundation (No.NZ0936).

References

1. Arnold, L.: Stochastic Differential equations: theory and applications. Wiley (1972)
2. Kloeden, P.E., Platen, E.: Numerical solution of stochastic Differential equations. Springer (1992)
3. Mao, X.: Stochastic Differential equations and applications. Horwood (1997)
4. Platen, E.: An introduction to numerical methods for stochastic Differential equations. Acta Numerica 8, 197–246 (1999)
5. Marion, G., Mao, X., Renshaw, E.: Convergence of the Euler Scheme for a class of Stochastic Differential Equation. International Mathematical Journal 1, 9–22 (2002)
6. Zhang, Q.: Exponential Stability of Numerical Solutions to a Stochastic Age-Structured Population System with Diffusion. Journal of Computational and Applied Mathematics 22, 220–231 (2008)
7. Li, R., Meng, H., Chang, Q.: Exponential Stability of Numerical Solutions to SDDEs with Markovian Switching. Journal of Applied Mathematics and Computation 174, 1302–1313 (2006)

A Study of Multi-human Behavior in Substations' Operation Tickets Processing

Guangwei Yan* and Chao Chen

School of Control and Computer Engineering,
North China Electric Power University,
Beijing, China
yan_guang_wei@126.com, chenchaoany2000@163.com

Abstract. In order to improve the realistic and effects of substations' operation tickets training, an operation tickets simulation system based on multi-human behavior is designed. Main human and equipments of substation are modeled by 3D modeling technology, the multi-human behavior in substations' operation tickets processing is simulated by virtual human animation technology, and a multi-human animation generating method based on location and multi-configure files is proposed, roaming in substation virtual scene is realized by 3D interaction technology, the data processing ,including the creation, examination and the management function is realized by database technology. Final experiment shows a good practical value of the system.

Keywords: Operation ticket, Simulation, Multi-human animation, Roam, Database techniques.

1 Introduction

It is common to make switching operations in substations, and these operations can be described by operation tickets. So it is necessary to make operation processing training for the substation staff, which can keep substations in good status. The current implementation methods of operation processing training have: using database technology with object-oriented technology to create the operation tickets management system [1], using expert system to create the operation tickets management system [2]. However, there are some defects about these methods. First, these methods ignore the training of multi-human behavior in substations' operation tickets processing. Second, these methods can not make realistic training effect for they implement the interactive function only by form of words, tables, planar graph.

In order to improve the realistic and effect of substations' operation tickets training, using virtual reality technology, multi-virtual human animation technology, database technology, an operation tickets simulation system based on multi-human behavior is designed. This system is developed by C++ and C#, implements the multi-human behavior simulation and the operation tickets data processing by OSG platform and .NET platform.

* Corresponding author.

A. Xie & X. Huang (Eds.): Advances in Electrical Engineering and Automation, AISC 139, pp. 165–170.
springerlink.com

2 System Architecture

This system consists of three parts: the creation and optimization of models, the creation of the multi-virtual human animation library and operation tickets database, the realization of roaming scenarios and interactive control. First, found the equipments and virtual human models by 3D modeling technology[3], second, found the multi-virtual human animation library and the operation tickets database by multi-virtual human animation technology[4]-[5]and database technology; at last, the interactive functions are realized by interactive technology. System architecture is showed in Fig. 1.

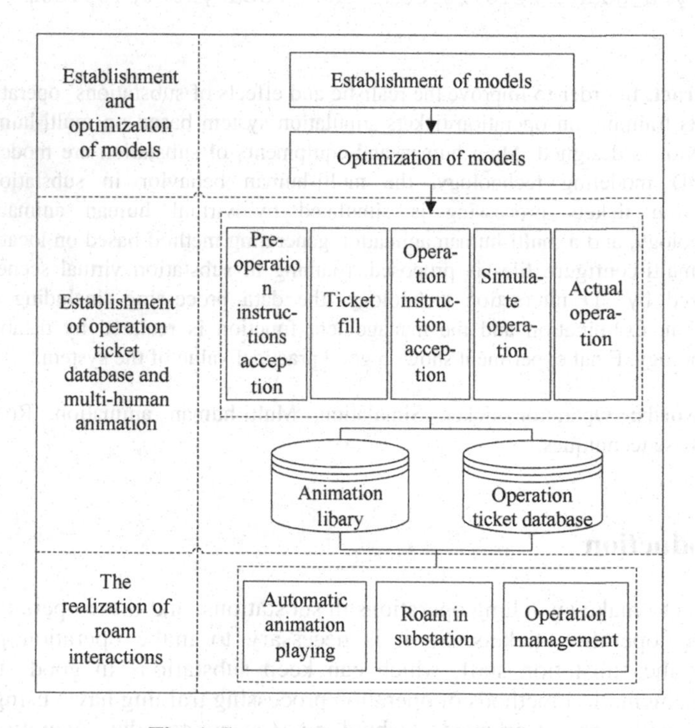

Fig. 1. System function structure diagram

3 Establishment and Optimization of Models

This paper uses 3D modeling technology and tools to simulate objects in reality, making it models that computer can handle, besides, certain methods to optimize the structure and size of model file are used.

A substation is made up of workshop, control room and related work staff. In this paper, the substation models are divided into exterior view model, interior view model and character model.

This paper uses 3dsMax to create substation models, respectively established each model in substation. For each model a texture map is given so that it can better show substation equipment and scene.

Because the number and volume of models in substation is large, so this paper optimizes the models in some degree. The main steps of optimization are as follows:

First, use PolygonCruncher to optimize models. PolygonCruncher is a third party plugin of 3dsMax. It can reduce the number of patches without influencing model fineness.

Second, choose better format files of model. OSG at least can recognize osg format and ive format files .Ive format can reduce file size substantially with nearly no loss in degree of distortion, so this paper only export models except character models to ive format files to improve system speed.

4 Establishment of Operation Ticket Database and Multi-human Animation

4.1 Establishment of Multi-human Animation

Multi-virtual human animation library is system's core simulation data. It constitutes the set of virtual human behavior associated with operation ticket. The establishment of multi-human animation is to cope with behaviors associated with operation ticket, abstract key behavior and simulate it using multi-human virtual technology. Main steps are followings:

First, collection, analysis and abstract of multi-human behavior associated with substation operation tickets. In substation, behavior associated with substation operation tickets is divided into five types according to its process. Table 1 gives out the analysis results.

Table 1. Analysis of people's behavior in substation operation

Process	Behavior description
Pre-operation instructions acception	Guardian calls, receive pre-operation instructions, the turning master prepares the organization of operations
Ticket fill	Write operation of the ticket as requested
Accept the Operating instructions	Guardian calls, receive operation instructions, the master organizes the operation meeting
Simulate operation	Simulate the operation depending on the operating tickets
Practical operation	Practice depending on the operating tickets

Second, according to the results of analysis, we can establish virtual multi-human animation and construct animation library. In the process, bone modeling technology is required. First, import virtual human into 3dMax, at the same time, establish new bone object and adjust it to reality. Then, bind grid and bone, freeze the grid of virtual human model. Last, establish key frame, finish virtual human animation through move and rotate of bone object. That is how we establish the virtual human animation library. Its process is shown in Fig. 2.

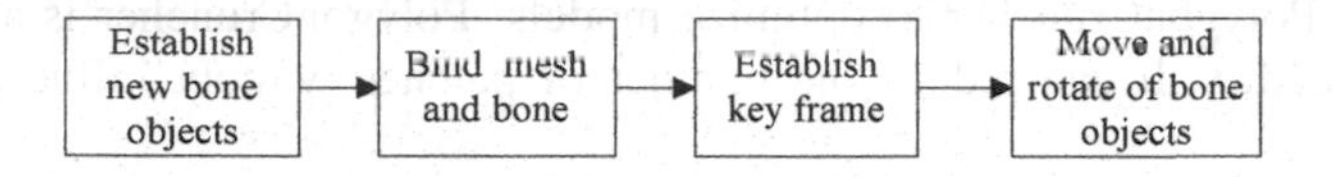

Fig. 2. The establishment of human animation library

4.2 Establishment of Operation Ticket Database

Operation ticket database mainly store data relevant to operation ticket, as source of data to be dealt with. In this paper, we use SQL Server as database building tools. In this software, OPP Test is created. In OPP Test, table Opp_info store operation ticket information, table User_info store user information, table Opp_temp store cache data in memory.

5 The Realization of Roam Interactions

5.1 The Demo of Operation Tickets Processing Animation

Operation tickets' processing animation has simulated multi-human behavior that is associated to the operation tickets, these animation can provide straighter and more realistic experience for the substations' staff. This paper proposes a multi-human animation generating method based on location and multi-config files. This method mainly loads certain animation config files according to the value of view positions. The specific steps are as follows:

1. Create virtual human animation config files. Animation config files describe the virtual human animation. These files organize the skeleton, mesh, texture and animation which the program can load. Main params are showed as table 2.

Table 2. The parameters of animation configuration file

Parameter	Meaning	Unique
Skeleton	Point to the virtual human skeleton files	Yes
Animation	Point to the virtual human animation files	No
Mesh	Point to the virtual human mesh files	No
Material	Point the virtual human material files	No

2. Load global animation parameters file. This file describes the global animation parameters according to which the system loads and processes animation. Main params are showed as table 3.

Table 3. Global animation parameters

Parameter	Meaning
RoutePath	Mark the path file of the path
Position(i)	Position of the viewpoint
Position_Scope(j)	A range of the viewpoint
CFGName	Point to the configuration files of virtual human animation
Result	Mark whether the current animation is complete 0: complete, 1: not completed(Defaults value: 0)

Parameter *RoutePath* describes the path of a text file which records the roaming path. Parameter *Position*(i) is an array whose length is 3, describes the view position.

$$Position(i) = \{ x , y , z \} . \quad (1)$$

Parameter *Positon_Scope(j)* describes a range of view position.

$$Position\_Scope(j) = \{ Position(m) , Position(n) \} . \quad (2)$$

Position_Scope(j) refers the range of Position(m)~Position(n).

1. Get the value of *RoutePath*, if *RoutePath* is null, exit execution, else step 4.
2. Set the roaming path according to the value of *RoutePath*, calculating the value of *Position*(i), when *Position*(i)∈*Position_Scope(j)*, set *CFGName* according to the mapping table, then step 5. The mapping table is showed as table 4.

Table 4. The mapping of view positon and virtual human animation

Position_Scope(j)	Configuration file	Virtual human animation
Position_Scope(1)	Pre_comd_rec.cfg	Accept pre-operation instructions
Position_Scope(2)	Comd_rec.cfg	Accept the Operating instructions
Position_Scope(3)	Sim_Opre.cfg	Simulate operation
Position_Scope(4)	Real_Opre.cfg	Practical operation
Other positions	null	No animation

3. Judge the value of parameter *Result*, if *Result*=0, set *Result* =1, then read the animation config file, run the animation; if *Result*=1, wait until last animation has finished (*Result*=0 now), then read the animation config file, run the animation. After animation, set *Result*=0.
4. Return step 4. The simulation result is showed as Fig. 3.

5.2 The Realization of Roam

In order to correspond with the automatic demonstration, manual roaming is provided by system to meet users' individual needs. Users can roam the scene in the substation, can generate path files by system, and can display different operation tickets processing animation as different roaming positions. The way, which includes camera lens, rotate, push, shake, zoom, and hybrid transformation, is used to achieve the different directions display of virtual scene in substation system roaming.

Fig. 3. The simulation result

5.3 Achievement of Operation Ticket Management Function

In this paper, we achieve some functions such as operation ticket entry, audit and management. Users can demonstrate operation ticket procession in simulation scene.

Microsoft ado.net technology is used in database access. Besides, three layers are needed to construct the program structure. Business logic layer achieve expression function, data access layer achieve functions in business logic layer, data process layer achieve function of access database.

6 Conclusion

The substation operation simulation system is based on virtual human behavior. It not only has common data processing function but also can simulate demo with multi human behavior, making up for the blank of most operation ticket simulation training system. In the following work, in order to achieve better effect of simulation training, we can optimize more in three dimensional driver's performance and virtual character model precision.

References

1. Xu, S.-Z., Zhu, Z.-S., Zhang, J.: The Design of Simulation System of Operation Ticket in Substations. High Voltage Engineering 26(1), 39–41 (2003)
2. Jin, H.-Z., Zhou, L., Zheng, J.: Switching sequence expert system based on online training of simulative substation. Electric Power Automation Equipment 23(1), 45–48 (2003)
3. Tang, Y., Ge, Y.-J., Chen, W.: Research on Human Body Model of Digital Sportsman and Its Simulation. Journal of System Simulation 15(1), 56–58 (2003)
4. Xu, M., Sun, S.-Q., Pan, Y.-H.: The Research of Motion Control in Virtual Human. Journal of System Simulation 15(3), 339–342 (2003)
5. Musse, S.R., Thalmann, D.: Hierarchical model for real time simulation of virtual human crowds. Visualization and Computer Graphics 7(2), 152–164 (2010)

SCARA Robot Control System Design and Trajectory Planning: A Case Study

Guangfeng Chen, Yi Yang, Linlin Zhai, Kun Zou, and Yanzhu Yang

College of Mechanical Engineering, Donghua University
Shanghai 201620, China
chengf@dhu.edu.cn, yycty@sohu.com, zhail@msn.com,
{kouz,yangyz}@dhu.edu.cn

Abstract. This paper conducts the study to construct the control system and trajectory planning for SCARA robot, which is used to pick the biscuits from delivery belt and place on packing line. According to the practical action requirements, defined the desired path in Cartesian space, through calculating the right angle coordinate system of key points in inverse in joint space, choose the trapezoid acceleration curve for joint space trajectory planning, generating a feasible motion control track.

Keywords: SCARA, trajectory planning, mathematical modeling, forward and inverse solutions.

1 Introduction

Different manipulator configurations are available as rectangular, cylindrical, spherical, revolute and horizontal jointed. A horizontal revolute configuration robot, selective compliance articulated robot arm (SCARA) has four degrees of freedom in which two or three horizontal servo controlled joints are shoulder, elbow and wrist. Last. SCARA designed at Japan, is generally suited for small parts insertion tasks for assembly lines like electronic component insertion [1].

In general, traditional SCARA are 4-axis robot arms, i.e., they can move to any X-Y-Z coordinate within their work envelope. There is a fourth axis of motion which is the wrist rotate (Theta-Z). The ‘X’, ‘Y’ and the ‘Theta-Z’ movements are obtained with three parallel-axis rotary joints. The vertical motion is usually an independent linear axis at the wrist or in the base. This paper conducts the study to construct the control system for SCARA robot, which is used to pick the biscuits from delivery belt and place on packing line.

2 SCARA Mechanism and Control Structure

A 4-axis SCARA robot has parallel shoulder, elbow, and wrist rotary joints, and a linear vertical axis through the center of rotation of the wrist. This type of manipulator is very common in light-duty applications such as electronic assembly.

A. Xie & X. Huang (Eds.): Advances in Electrical Engineering and Automation, AISC 139, pp. 171–176.
springerlink.com

The SCARA robot is design to pick and place biscuit on band carrier. In this example, the upper-arm length is L1 350 mm, and the lower-arm length L2 is 250 mm. The shoulder joint, the elbow joint, and the wrist joint have resolutions of 3200000 counts per 360 degrees. Rotation in the positive direction for all 3 joints is counterclockwise when viewed from the top. The vertical axis has a resolution of 100 counts per millimeter, and movement in the positive direction goes up.

The control system hardware structure is illustrated in Fig.1. The control system is consists of SCARA robot, computer, motion controller, servo motor, camera and image grabbing card. The computer captures the image of workpiece and delivers to PC through image grabbing card, the PC analysis the image and computes the position of biscuits on band carrier, and planning the control trajectory for each motors of SCARA and drives the end effector pick and place on predefined position.

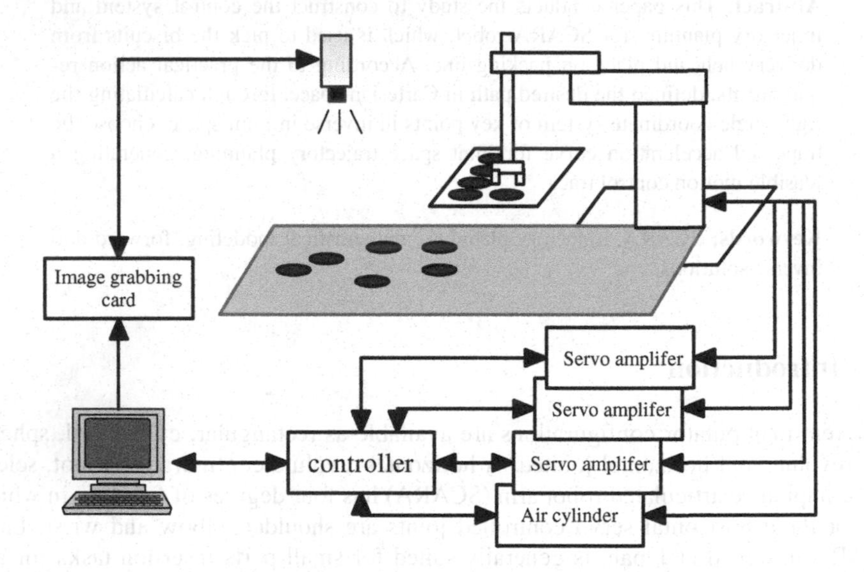

Fig. 1. Hardware structure of SCARA control system

3 Trajectory Planning for Control System

Trajectory planning is the process of defining a desired path in joint or Cartesian space for joints or gripper, respectively [2]. Trajectory planning quality decided the robot system operation performance and efficiency. There are two commonly used trajectory planning methods, one is planning in joint space, and another is planning in Cartesian space.

3.1 SCARA Robot Trajectory Define

In this case, the SCARA is design to pick and place the workpiece on the conveyor belt. The PC gets capture image and get the (x, y, z, δ) with help of image processing technology, the position of forearm end actuator is at P_0 (which is equal to P_6 of last trajectory planning). For the band carrier is keep on moving at a constant speed, the PC estimated time and position P_2 of end actuator could arrival to pick the workpiece. Then the end actuator move to P_1 which is exactly above P_2, and down to P_2, pick the workpiece, move to P_3, move to P_4 which is right above destination point P_5, then down to P_5 and release the workpiece, and move the end actuator to P_6. In the trajectory P_4=P_6 and P_1=P_3.

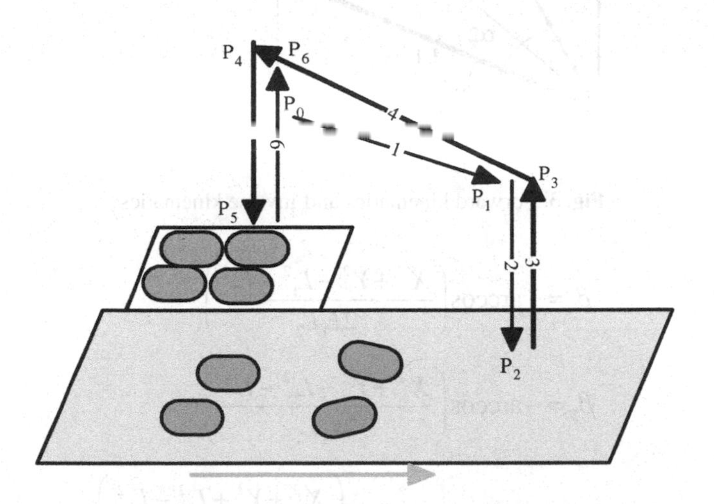

Fig. 2. Trajectory planning optimization schemes

3.2 Forward and Inverse Kinematics

Kinematics is the process of calculating the position in space of the end of a linked structure, given the angles of all the joints. It is easy, and there is only one solution. Inverse Kinematics does the reverse that given the point of the end actuator get the angles do the joints need to be in. It can be difficult, and there are usually many or infinitely many solutions. To generate the trajectory planning, the forward and inverse and kinematics solution must be got first. For SCARA robot, there are four parameters could be control as (x, y, z, δ), but z and δ could be easier got. This diagram shows a top view of the mechanism. Z-axis motion is into and out of the page. As Fig. 3 shown is the forward-kinematic equations are:

$$X = L_1\cos(\alpha) + L_2\cos(\alpha + \beta) \tag{1}$$

$$Y = L_1\sin(\alpha) + L_2\sin(\alpha + \beta) \tag{2}$$

Through the geometry calculation we could get: Joint angle of upper-arm are β_1 and β_2, and lower-arm are α_1 and α_2 as (3) to (6).

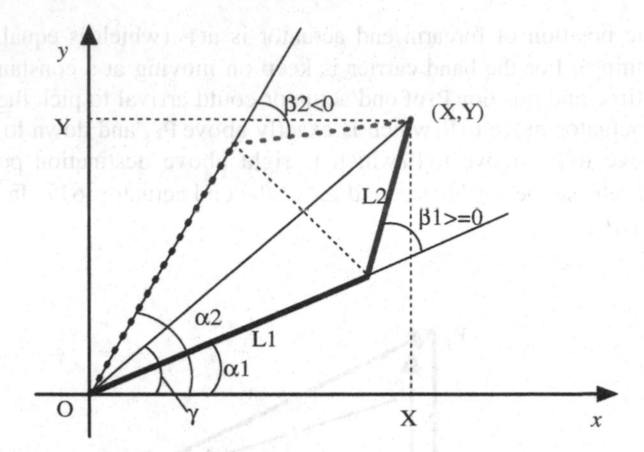

Fig. 3. Forward kinematics and inverse kinematics

$$\beta_1 = +\arccos\left(\frac{X^2+Y^2-L_1^2-L_2^2}{2L_1L_2}\right) \tag{3}$$

$$\beta_2 = -\arccos\left(\frac{X^2+Y^2-L_1^2-L_2^2}{2L_1L_2}\right) \tag{4}$$

$$\alpha_1 = \arctan(Y/X) - \arccos\left(\frac{X^2+Y^2+L_1^2-L_2^2}{2L_1\sqrt{X^2+Y^2}}\right) \tag{5}$$

$$\alpha_2 = \arctan(Y/X) + \arccos\left(\frac{X^2+Y^2+L_1^2-L_2^2}{2L_1\sqrt{X^2+Y^2}}\right) \tag{6}$$

Since in joint space trajectory planning, there might be multiple solutions inverse kinematics, this involves joints optimization principles. In general, the premise of collisions should follow the "multiple mobile small joints, less move large joints" principle. As Fig.4 shown, from point A move to point B, the starting point A and end point B exist two and unequal inverse solutions. If current posture is in $OC_{1R}A$, the destination point B has two feasible postures are $OC_{2L}B$ and $OC_{2R}B$. In accordance with the above principles, select the optimal solution as starting point A posture $OC_{1R}A$ to destination posture $OC_{2L}B$.

3.3 Motion Curve Selection

The choice of a particular trajectory has direct and relevant implications on several aspects of the design and use of a robot, like the dimensioning of the actuators and of the reduction gears, the vibrations and efforts generated on the machine and on the load, the tracking errors during the motion execution. The detail information co

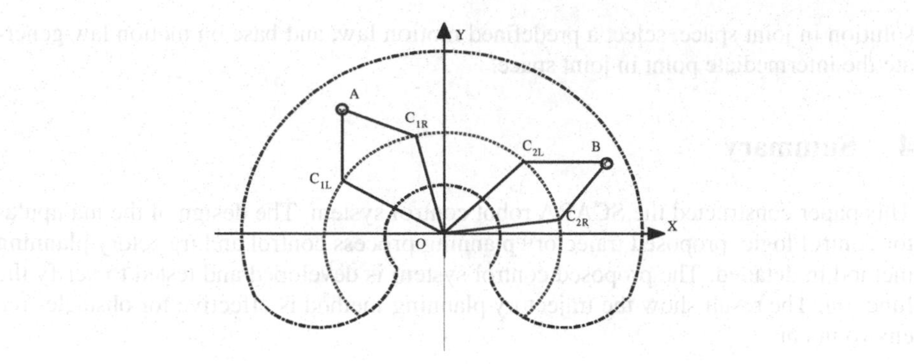

Fig. 4. Trajectory planning optimization schemes

3.4 Flowchart for Trajectory Generation

As Fig.5 illustrated is the flowchart for trajectory planning. The planning process starts from the input of start and end points in Cartesian space and get its inverse

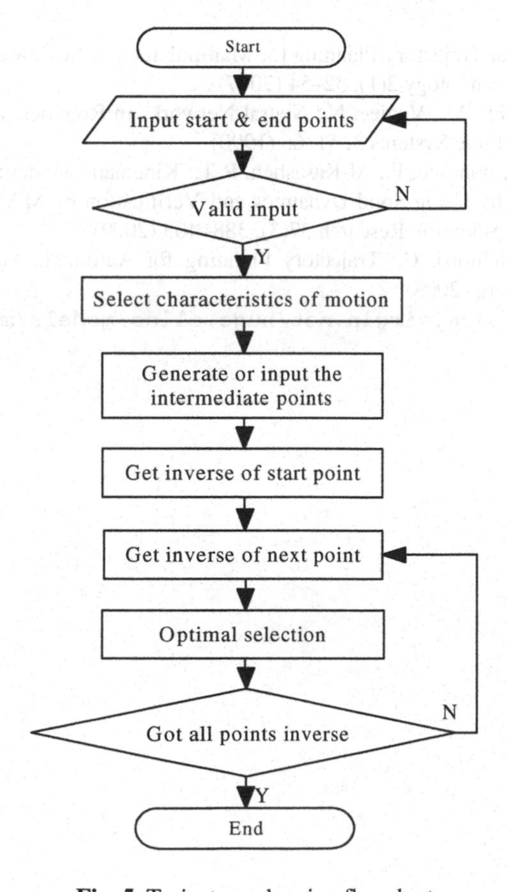

Fig. 5. Trajectory planning flowchart

solution in joint space, select a predefined motion law, and base on motion law generate the intermediate point in joint space.

4 Summary

This paper constructed the SCARA robot control system. The design of the manipulator control logic, proposed trajectory planning process control and trajectory planning method in detailed. The proposed control system is developed and tested to verify the function. The result show the trajectory planning method is effective for obstacle-free environment.

Acknowledgements. This work was supported by Shanghai Natural Science Foundation (No.10ZR1401500) and the Fundamental Research Funds for the Central Universities (9D10307).

References

1. Ata, A.A.: Optimal Trajectory Planning for Manipulators: A Review. Journal of Engineering Science and Technology 2(1), 32–54 (2007)
2. Horne, B., Jamshidi, M., Vadiee, N.: Neural Networks in Robotics: A Survey. Journal of Intelligent and Robotic Systems 3, 51–66 (1990)
3. Alshamasin, M.S., Ionescu, F., Al-Kasasbeh, R.T.: Kinematic Modeling and Simulation of a SCARA Robot by Using Solid Dynamics and Verification by MATLAB/Simulink. European Journal of Scientific Research 37(3), 388–405 (2009)
4. Biagiotti, L., Melchiorri, C.: Trajectory Planning for Automatic Machines and Robots. Springer, Heidelberg (2008)
5. `http://freespace.virgin.net/hugo.elias/models/m_ik.htm`

A Robust Relay Selection Scheme for Cooperative Communication Network

Quan Zhang, Liusheng Huang, and Hongli Xu

School of Computer Science and Technology, University of Science and Technology of China, Hefei, China
zqlol@mail.ustc.edu.cn, {lshuang,xuhongli}@ustc.edu.cn

Abstract. Cooperative relay selection strategy design, as one of the fundamental issues in cooperative communication, has attracted dramatic attention in recent years. However, many of the previous works have been done under the assumption that the transceiver nodes were fixed. In this paper, we address the robust relay selection problem that the active transceiver nodes are dynamic. First we use "*capacity gain*" as the metric for relay selection strategies and formally define the robust relay selection problem. Then we propose a *robust relay selection scheme* called RRS which aims to maximize the transmission capacity gain in the worst case. Finally, our simulations have shown the RRS scheme can improve about 50% minimum capacity gain than the random select algorithm.

Keywords: cooperative communication, relay selection, submodular.

1 Introduction

For cooperative communication, relay selection strategy plays a very important role in improving the network performance. Thus a lot of research has been done on relay selection strategies. In [1], Bletsas et al. introduced a distributed relay selection scheme based on the instantaneous channel state information (ICSI) for just a single source-destination pair. Centralized algorithm for a single source-destination pair selecting multiple relays was proposed in [2]. Cai et al. designed a semi-distributed algorithm which considered both relay selection and power allocation in [3]. Y. Shi et al. [4] investigated a relay selection algorithm for multiple source-destination pairs by using "linear marking" mechanism to achieve Max-Min fairness. But in [5], Xu et al. showed that this relay selection strategy may drastically reduce the aggregate throughput of the network. Recently, mobility in cooperative communication networks has been studied in [6]. However, their algorithms only considered fixed transmission pairs. Consider a communication network which consists of four nodes as shown in Fig. 1, we can see the active source-destination pairs change over time. In this scenario, the relays we selected which are the best for certain transmission pairs could be the worst for others.

A. Xie & X. Huang (Eds.): Advances in Electrical Engineering and Automation, AISC 139, pp. 177–184.
springerlink.com

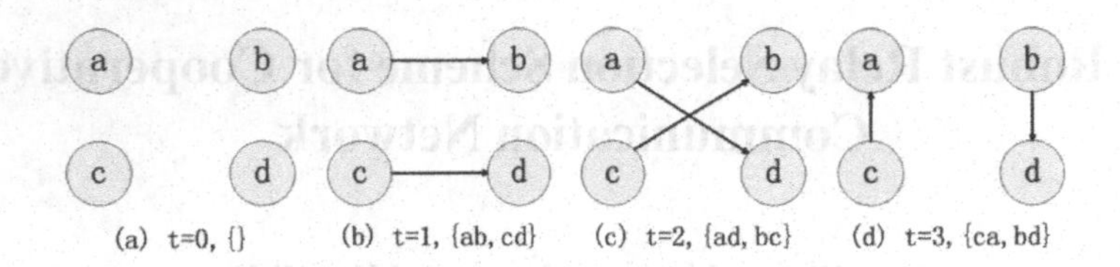

Fig. 1. Active source-destination pairs at different time intervals

Our main contributions are as follows:

- We present a new metric, called Capacity Gain, which shows the robust level of a relay selection strategy.
- We define and formalize the robust relay selection problem, and by reducing the problem to submodular set covering problem, we design a robust relay selection scheme called RRS which maximizes the capacity gain in the worst case when there is a constraint on budget. Moreover, it is a predetermined relay selection strategy, and the required information is easy to get, such as statistical channel state information (SCSI).
- Simulations also show the *RRS* scheme can improve about 50% minimum capacity gain than random select algorithm.

2 Preliminaries

2.1 Cooperative Communication Model

The simplest cooperative communication model includes a source node, a destination node and a relay node. The data transmission from the source node to the destination node is done by using a frame-by-frame mechanism, and the time of transmitting one frame is divided into two time slots as shown in Fig. 2.

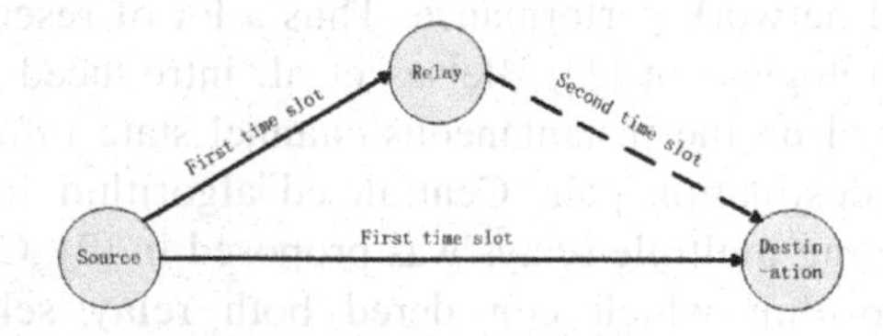

Fig. 2. An example of the simplest cooperative communication model

There are two main forwarding methods for cooperative communication, detect-forward mode (DF) and amplify-forward mode (AF) [7]. For simplicity, we only discuss the AF mode in this paper, since our work can be easily generalized to the DF mode. Let *s* denotes the source node, *d* denotes the destination node and *r* denotes the relay node. Under the AF mode, the maximum average mutual information or capacity from the source node to the destination node is given by [8]:

$$I_{AF}(SNR_{sd}, SNR_{sr}, SNR_{rd}) = \frac{1}{2}\log_2\left(1 + SNR_{sd} + \frac{SNR_{sr}\cdot SNR_{rd}}{SNR_{sr}+SNR_{rd}+1}\right). \quad (1)$$

Where SNR_{sd} is the signal-noise-ratio from *s* to *d*. When the source node *s* transmits directly, the capacity is computed by [8]:

$$I_D(SNR_{sd}) = \log_2(1 + SNR_{sd}) \ . \tag{2}$$

In this paper we assume one source-destination pair can only choose one relay, while one relay node could serve more than one source-destination pairs. This is reasonable, the source nodes and the destination nodes could be personal users' cell phones or other communication devices, the relay nodes may represent the APs which have numbers of sub-channels so that one AP can serve more than one communication pairs. We also assume each AP has unlimited sub-channels, since it is not likely all the source-destination pairs would select the same AP.

2.2 Problem Definition and Formalization

Definition 1 (Communication Topology). A Communication Topology is the collection of source-destination pairs which are active at a certain time point. We assume that a node can either transmit to or receive data from another node, but not simultaneously. So in a Communication Topology, a node can only appear once. For example, there are four different communication topologies in Fig. 1.

Now assume there is a node set $N=\{n_1,n_2,\ldots,n_t\}$ and an AP set $A=\{a_1,a_2,\ldots,a_r\}$, each of the APs has a cost $w(a_i)$. And any node in the node set is able to communicate with any other node of the set as long as they want, we use $\mathbb{T}$ to denote the collection of these communication topologies. To evaluate the performance of a relay selection scheme, we define the **capacity gain** of a communication topology *T* which is denoted as η_T as the fraction of capacity increment on *T* which is served by an AP set *H*, where $H \subseteq A$. η_T can be computed as follows:

$$\eta_T(H) = \frac{C(T,H)-C(T,\emptyset)}{C(T,\emptyset)} \ . \tag{3}$$

Where $C(T,H)$ is the maximum capacity of communication topology *T* when *T* is served by AP set *H*. So the problem is: Given a budget *B*, find an AP set which maximizes the minimum capacity gain among all communication topologies and the total cost does not exceed *B*. Formally,

$$\max_{H \subseteq A} \min_{T \in \mathbb{T}} \eta_T(H) subject\ to\ w(H) \leq B \ . \tag{4}$$

Where *H* is the collection of the selected APs and *w(H)* is the total cost of *H*. Note that if there is no constraint on the budget, one can just select all the APs.

3 Submodularity Analysis

To show Problem (4) is an instance of the submodular set covering problem, we need to show the objective function $\eta_T(\cdot)$ is *monotonic, normalized* and *submodular*. As it is a function of $C(\cdot,\cdot)$, we will show how to compute $C(\cdot,\cdot)$ at first.

Let there be a communication topology *T* that includes $|T|$ source-destination pairs, the *i*th source-destination pair is denoted as t_i, $1 \leq i \leq |T|$. And suppose there is an AP set $H \subseteq A$, the *j*th AP in *H* is denoted as h_j, $1 \leq j \leq |H|$. We also set a virtual AP called h_0 which means any source-destination pair that selects h_0 will use direct transmission, and whenever an AP set $H \subseteq A$ (including $\emptyset$) is selected, we add h_0 to *H*. Let x_{t_i,h_j} represents the relationship between source-destination pair t_i and AP h_j, obviously, $x_{t_i,h_j} = 0$ *or* 1. Since each source-destination pair can only select at most one AP, we have $\sum_{h_j \in H \cup h_0} x_{t_i,h_j} = 1$. Then the capacity of each source-destination pair denoted as c_{t_i,h_j} can be computed by $c_{t_i,h_j} = \sum_{h_j \in H \cup h_0} I_X \cdot x_{t_i,h_j}$, $1 \leq i \leq |T|$ and $1 \leq j \leq |H|$, where I_X can either be I_D or I_{AF}, depending on which AP is selected. So the total capacity of communication topology *T* can be obtained as follows:

$$C(T,H) = \max\{\textstyle\sum_{t_i \in T} c_{t_i,h_j}\} = \sum_{t_i \in T} \max c_{t_i,h_j} \,. \tag{5}$$

That is selecting the best AP for each source-destination pair in the communication topology. Then for function $\eta_T(\cdot)$, we have the following lemmas.

Lemma 1. $\eta_T(\cdot)$ is monotonic.

Proof. For any $H_1 \subseteq H_2 \subseteq A$, $\eta_T(H_1) - \eta_T(H_2) = \frac{C(T,H_1) - C(T,H_2)}{C(T,\emptyset)}$, it is not difficult to see that $C(\cdot,\cdot)$ is nondecreasing, so $\eta_T(H_1) - \eta_T(H_2) \leq 0$.

Lemma 2. $\eta_T(\cdot)$ is normalized.

Proof. $\eta_T(\emptyset) = \frac{C(T,\emptyset) - C(T,\emptyset)}{C(T,\emptyset)} = 0$.

Lemma 3. $\eta_T(\cdot)$ is submodular.

Proof. First we give the definition of submodular [15]: a set function $F : 2^S \longrightarrow \mathbb{R}$ is called submodular, if it holds that $F(A \cup B) + F(A \cap B) \leq F(A) + F(B)$ for all subsets $A, B \subseteq S$. And there is another definition which is equal to the above one, that is: F is called submodular, if it holds $F(A \cup \{x\}) - F(A) \geq F(B \cup \{x\}) - F(B)$ for all $A \subseteq B \subseteq S$ and $x \in S - B$. We use the second definition when proving lemma 3.

Let subsets $H_1 \subseteq H_2 \subseteq A$ and $h \in A - H_2$, we have:

$$\eta_T(H_1 \cup \{h\}) - \eta_T(H_1) = \frac{C(T,H_1 \cup \{h\}) - C(T,H_1)}{C(T,\emptyset)} \,. \tag{6}$$

$$\eta_T(H_2 \cup \{h\}) - \eta_T(H_2) = \frac{C(T,H_2 \cup \{h\}) - C(T,H_2)}{C(T,\emptyset)} \tag{7}$$

As $C(\cdot,\cdot)$ is monotonic, then $C(T,H_1) \leq C(T,H_2)$, $C(T,H_1) \leq C(T,H_1 \cup \{h\})$ and $C(T,H_2) \leq C(T,H_2 \cup \{h\})$. We discuss the following cases:

1) $C(T,H_1) = C(T,H_1 \cup \{h\})$, $C(T,H_2) = C(T,H_2 \cup \{h\})$

It means the new AP *h* cannot improve the capacity of *T* in both situations. Hence we get $\eta_T(H_1 \cup \{h\}) - \eta_T(H_1) = 0$, $\eta_T(H_2 \cup \{h\}) - \eta_T(H_2) = 0$. So $\eta_T(H_1 \cup \{h\}) - \eta_T(H_1) = \eta_T(H_2 \cup \{h\}) - \eta_T(H_2)$.

2) $C(T,H_1) = C(T,H_1 \cup \{h\}), C(T,H_2) < C(T,H_2 \cup \{h\})$

This case is impossible. If adding h to H_2 can improve the capacity of T, then adding h to H_l can improve the capacity of T too.

3) $C(T,H_1) < C(T,H_1 \cup \{h\}), C(T,H_2) = C(T,H_2 \cup \{h\})$

This shows h is better than some of selected APs in H_l, but not than those in H_2. Clearly, we have $\eta_T(H_1 \cup \{h\}) - \eta_T(H_1) > \eta_T(H_2 \cup \{h\}) - \eta_T(H_2)$.

4) $C(T,H_1) < C(T,H_1 \cup \{h\}), C(T,H_2) < C(T,H_2 \cup \{h\})$

Let T_l denotes the collection of source-destination pairs that will replace their selected APs in H_l by h, let H_1^* denotes the collection of selected APs of T_l. Similarly, we use T_2 and H_2^* for H_2, and we have $H_1^* \subseteq H_1$, $H_2^* \subseteq H_2$ and $T_2 \subseteq T_1 \subseteq T$. Hence the capacity of T under H_l is the sum of T_l and $T - T_1$, and similarly for H_2. So we get $C(T,H_1 \cup \{h\}) - C(T,H_1) = C(T_1,\{h\}) + C(T-T_1,H_1) - C(T_1,H_1^*) - C(T-T_1,H_1) = C(T_1,\{h\}) - C(T_1,H_1^*)$, $C(T,H_2 \cup \{h\}) - C(T, H_2) = C(T_2,\{h\}) - C(T_2,H_2^*)$. Since $T_2 \subseteq T_1 \subseteq T$, the capacity of T_l is the sum of T_2 and $T_1 - T_2$. Then $C(T_1,\{h\}) - C(T_1,H_1^*) = C(T_2,\{h\}) + C(T_1 - T_2,\{h\}) - C(T_2,H_1^*) - C(T_1 - T_2,H_1^*)$. Obviously $C(T_1 - T_2,\{h\}) - C(T_1 - T_2,H_1^*) > 0$ and $C(T_2,H_1^*) \leq C(T_2,H_2^*)$. Thus $C(T,H_1 \cup \{h\}) - C(T,H_1) > C(T,H_2 \cup \{h\}) - C(T,H_2)$. So $\eta_T(H_1 \cup \{h\}) - \eta_T(H_1) > \eta_T(H_2 \cup \{h\}) - \eta_T(H_2)$.

4 Robust Relay Selection Scheme

Our approximation algorithm is mainly based on the solution framework in [10]. Consider the following variant of problem (4):

$$\max_{H \subseteq A} \ \lambda \ subject\ to\ \eta_{T \in \mathbb{T}}(H) \geq \lambda\ and\ w(H) \leq B\ . \tag{8}$$

Where λ denotes the capacity gain. It is easy to see that problem (8) is equal to (4). Now we relax the budget constraint, define problem (9) as follows:

$$\max_{H \subseteq A} \ \lambda \ subject\ to\ \eta_{T \in \mathbb{T}}(H) \geq \lambda\ and\ w(H) \leq \alpha B\ . \tag{9}$$

Note when $\alpha = 1$, problem (9) is the same as (8). Further, we eliminate the budget constraint completely and fix the capacity gain as μ, define problem (10) as:

$$H_\mu = \arg\min_H \ w(H)\ subject\ to\ \eta_{T \in \mathbb{T}}(H) \geq \mu\ \cdot \tag{10}$$

Suppose we have obtained a solution of problem (10), if the total cost of the solution is less than αB, then μ is a feasible solution for (9) and is a lower bound of the optimal solution of (9), if the total cost is higher than αB, then μ is an upper bound. Thus, a binary research can be used to improve μ step by step. Since $\eta_T(\cdot)$ is nondecreasing, we initially set $\mu_{min} = 0 \leq \min_{T \in \mathbb{T}} \eta_T(\emptyset)$ and $\mu_{max} = \min_{T \in \mathbb{T}} \eta_T(A)$, and parameter δ is set to control the number of iterations.

Now we show how to solve problem (10) for a given capacity gain μ. First we truncate the function $\eta_T(H)$ to $\eta_T^\mu(H)$, where $\eta_T^\mu(H) = \min\{\eta_T(H), \mu\}$. Let

$\eta^{\mu}(H)$ denotes the sum of the capacity gain of all communication topologies in $\mathbb{T}$, that is $\eta^{\mu}(H) = \sum_{T \in \mathbb{T}} \eta^{\mu}_T(H)$. As $\eta_T(H)$ is submodular and monotonic, $\eta^{\mu}_T(H)$ is also submodular and monotonic [11]. Moreover, the sum of submodular functions is also submodular and monotonic, thus $\eta^{\mu}(H)$ is submodular and monotonic. Note that $\eta_T(H) \geq \mu$ for $T \in \mathbb{T}$ if and only if $\eta^{\mu}(H) = |\mathbb{T}|\mu$. Because $\eta^{\mu}(H)$ is monotonic, if $\eta^{\mu}(H) - |\mathbb{T}|\mu$, we have $\eta^{\mu}(H) = \eta^{\mu}(A)$. So problem (9) is equal to:

$$H_{\mu} = \arg\min_{H} \; w(H) \; subject \; to \; \eta^{\mu}(H) = \eta^{\mu}(A) \cdot \tag{11}$$

However (11) is also a NP-hard problem [12]. Fortunately, a greedy algorithm can be used to approximate the optimal solution [13]. Algorithm 1 shows how to solve problem (11), and the whole algorithm for problem (9) is shown in Algorithm 2.

Remember that we have relaxed the budget constraint by using α , if $\alpha < 1 + \log(\max_{h \in A} \sum_{T \in \mathbb{T}} \eta_T(\{h\}))$, though we can still get a solution , it does not guarantee its performance as good as the optimal solution [10].

Algorithm 1: GreedyMinimumCostAPSet(GMCAS)

```
Input: η_T(·), T, A, μ.
Output: An AP set H ⊆ A
begin
    H ⟵ ∅ ;
    h* ⟵ NULL ;
    while η^μ(H) < η^μ(A) do
        ForAll h ∈ A − H;
        h* ⟵ arg max{η^μ(H ∪ h) − η^μ(H)} ;
        H ⟵ H ∪ h* ;
    end
    return H ;
end
```

Algorithm 2: RobustRelaySelection(RRS)

```
Input: η_T(·), T, A, α, B, δ
Output: An AP set H_best ⊆ A
begin
    μ_min ⟵ 0;
    μ_max ⟵ min_{T∈T} η_T(A);
    H_temp ⟵ ∅;
    H_best ⟵ ∅;
    while μ_max − μ_min > δ do
        μ ⟵ (μ_max + μ_min)/2;
        H_temp ⟵ GMCAS(η_T(·), T, A, μ);
        if w(H_temp) > αB then
            μ_max ⟵ μ;
        end
        else
            μ_min ⟵ μ;
            H_best ⟵ H_temp;
        end
    end
    return H_best;
end
```

Fig. 3. Algorithm 1and Algorithm 2

5 Simulations

We consider a wireless communication network with 50 transceiver nodes, and assume every two of them can communicate with each other. For each possible source-destination pair, we initialize it with a random capacity and set the maximum capacity achieved by selecting AP does not exceed twice of the initial capacity. Thus we have obtained a matrix, the rows represent different source-destination pairs and the columns represent different APs. To make our simulation more generalized, the size for the collection of communication topologies is random, and in each communication topology, the number of source-destination pairs is also random with an upper bound 25 since one node only appears in a communication topology once.

For comparison, we also set a baseline algorithm called Random Select algorithm, initially an empty set, the algorithm will select an AP randomly from A to its final AP set as long as the total cost of the selected APs does not exceed the budget.

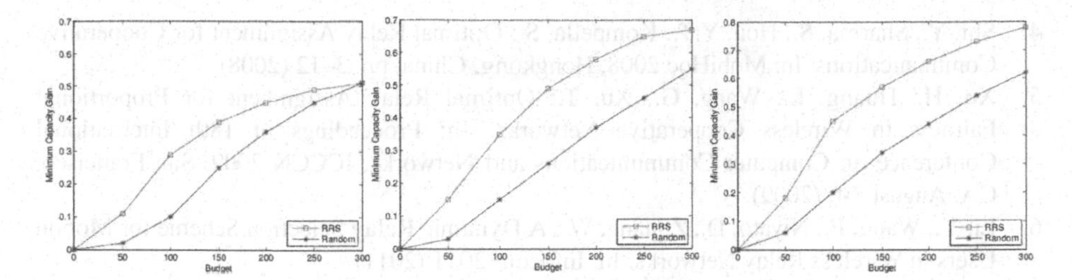

Fig. 4. Minimum capacity gain over various sizes of AP set and budgets (50 nodes, AP=5,15,25)

Fig. 4 shows the minimum capacity gain achieved by RRS scheme and Random Select algorithm under various sizes of AP set and budgets. From Fig. 4, we can see when the budget is low, the minimum capacity gain is low too for both of the two algorithms, then if we increase the budget, the capacity gain will grow fast, but when the budget is large enough, the growing speed becomes low. That is because the benefits by selecting APs are almost saturated. The simulation results show the RRS can improve about 50% minimum capacity gain than the random select algorithm.

6 Conclusion

This paper has studied the relay selection problem in cooperative communication networks, taking dynamic transmission pairs into consideration. We proved the relay selection problem in this scenario could be reduced to the submodular set covering problem and proposed an approximation algorithm RRS. Our simulations also show that the RRS algorithm can improve about 50% minimum capacity gain than the random select algorithm.

Acknowledgments. This paper is supported by the National Grand Fundamental Research 973 Program of China under Grant No. 2011CB302-905, the National Science Foundation of China under Grant No. 61170058, National Science and Technology Major Project under Grant No. 2011ZX03005-004-04.

References

1. Bletsas, A., Khisti, A., Reed, D., Lippman, A.: A Simple Cooperative Diversity Method Based on Network Path Selection. IEEE Journal on Selected Areas in Communications 24(3), 659–672 (2006)
2. Madan, R., Mehta, N., Molisch, A., Zhang, J.: Energy-Efficient Cooperative Relaying over Fading Channels with Simple Relay Selection. IEEE Transactions on Wireless Communications 7(8), 3013–3025 (2008)
3. Cai, J., Shen, S., Mark, J.W., Alfa, A.S.: Semi-distributed User Relaying Algorithm for Amplify-and-forward Wireless Relay Networks. IEEE Transactions on Wireless Communications 7(4), 1348–1357 (2008)

4. Shi, Y., Sharma, S., Hou, Y.T., Kompella, S.: Optimal Relay Assignment for Cooperative Communications. In: MobiHoc 2008, Hongkong, China, pp. 3–12 (2008)
5. Xu, H., Huang, L., Wang, G., Xu, T.: Optimal Relay Assignment for Proportional Fairness in Wireless Cooperative Networks. In: Proceedings of 18th International Conference on Computer Communications and Networks, ICCCN 2009, San Francisco, CA, August 3-6 (2009)
6. Li, Y., Wang, P., Niyato, D., Zhuang, W.: A Dynamic Relay Selection Scheme for Mobile Users in Wireless Relay Networks. In: Infocom 2011 (2011)
7. Nosratinia, A., Hedayat, A.: Cooperative Communication in Wireless Networks. IEEE Communications Magazine, 74–80 (October 2004)
8. Laneman, J.N., Tse, D.N.C., Wornell, G.W.: Cooperative diversity in wireless networks: Efficient protocols and outage behavior. IEEE Transactions on Information Theory 50(12), 3062–3080 (2004)
9. Zhao, Y., Adve, R.S., Lim, T.J.: Improving Amplify-and-forward Relay Networks: Optimal Power Allocation Versus Selection. In: Proc. IEEE International Symposium on Information Theory, Seattle, July 9-14, pp. 1234–1238 (2006)
10. Krause, A., McMahan, B., Guestrin, C., Gupta, A.: Robust Submodular Observation Selection. Journal of Machine Learning Research (JMLR) 9, 2761–2801 (2008)
11. Nemhauser, G.L., Wolsey, L.A.: Integer and Combinatorial Optimization. Wiley-Interscience (1988)
12. Feige, U.: A threshold of ln n for approximating set cover. J. ACM 45(4) (1998)
13. Wolsey, L.A.: An analysis of the greedy algorithm for the submodular set covering problem. Combinatorica 2, 385–393 (1982)
14. Zheng, Z., Lu, Z., Sinha, P., Kumar, S.: Maximizing the contact opportunity for vehicular Internet access. In: Infocom 2010 (2010)
15. Nemhauser, G., Wolsey, L., Fisher, M.: An analysis of the approximations for maximizing submodular set functions. Mathematical Programming 14, 265–294 (1978)

Wireless Network Transmission Fault Tolerance Strategy Based on Flexible Macroblock Ordering and Intra Refreshment

Shuanghong Liu and Yongling Yu

Zhengzhou Institute of Aeronautical Industry Management, ZhengZhou, 450015, China
{cloudyni, yuyongling}@zzia.edu.cn

Abstract. A hybrid fault-tolerant method was proposed by employ both the flexible macroblock order (FMO) and intra refreshment. First of all, the problems and factors during network transmission in wireless video surveillance system were presented. Based on analysis, an intra refreshment method was proposed within the framework of rate distortion optimization. Experimental results show that the proposed hybrid method outperforms any individual error resilience tool.

Keywords: flexible macroblock ordering (FMO), wireless transmission, intra refreshment, fault tolerance.

1 Introduction

As the Internet and wireless mobile communication technology of rapid development, and video communication gradually become the main business of network communication。 Frame refresh macro blocks is a kind of improving streaming video of robustness effective technology, but too much frame macro block of code rate will make the increased dramatically and influence coding efficiency. Therefore, in the bit rate limited channel, how to determine the number of macro blocks frame and the position to become frame macro block of brush the new study key.

2 Analysis of Commonly Used Fault-Tolerant Techniques

H. 264 provide some fault-tolerant tools, but they have different purposes and purpose, that is, in different occasions need to choose different combination to use. Error concealment will be able to use the data that they receive to restore the lost data, so usually used in the decoder. In the wireless network environment, the ability of the decoder is particularly important [1]. Because wireless network environment, many RTP packet transmission in the gateway or router discarded, and these lost data and must be in the decoder according to the correlation of space and time to recover.

A. Xie & X. Huang (Eds.): Advances in Electrical Engineering and Automation, AISC 139, pp. 185–190.
springerlink.com

2.1 Intra Block Refreshment

Because frame coding do not depend on the time frame of data, so the adjacent frame coding piece can effectively prevent because packet loss even frame and cause error propagation loss. For dialogue for the business of video in real-time, and I frame refresh high frequency is low, so can use frame coding to part of the role of the frame instead of I. H.264 / AVC offer two frame coding piece of refresh mode[2]; Among them, a kind is random model, that is, the user can choose frame coding, and the number of the piece by random decide which position encoder the macro block of code a frame; Another kind is line refresh mode, namely the encoder in the image in a choice of rows of frame coding.

2.2 Parameter Sets

H. 264 standard cancelled Sequence layer and image layer, will originally belong to Sequence and the image of the head of separate element syntax forming Sequence Parameter Set SPS (Sequence Parameter Set) and image Parameter Set PPS (Picture Parameter Set). Sequence parameter set includes and an image sequence for all relevant information, such as the grade of the code used image size and level, such as used in video sequences,. Image parameter set contains belong to one of the image of all the information, such as primogeniture coding method, FMO, macro blocks to the mapping of the way of group, a sequence of a video application or some independent image. Several different sequence parameter set and image parameter set was decoder received correctly, stored in different position, decoder has coding based on the first of every has coding location of choose the right image parameter set to use.

2.3 Flexible Macroblock Ordering

FMO technology achieves through the piece of group technology. Piece by a group of composed in a series of each usually include the macro blocks. The video coding for the benefits of the FMO can make for channel transmission and erroneous scattered. Specific implementation method is, the frame graph of the macro blocks may form a or a few piece of groups and each a piece of group alone, when a piece of transmission of missing, can use happened with the near has been received correctly to another piece of the group of the macro block effective error cover up[3]. Group composed of way can be rectangular way or regular decentralized way, also can be completely randomly scattered way.

3 Wireless Video Transmission Fault-Tolerant Control Simulation System

A complete wireless video network transmission system included video acquisition, video coding, transmission control protocol processing, communication network and video decoding , which provided video application service in communication network based on TCP/IP with the characteristics of random time delay and packet loss. The principle was shown in Fig. 1.

The framework of wireless video transmission system was provided for the analysis of proceeding video stream in the system, and the fault-tolerant techniques for all links in wireless network transmission were proposed.

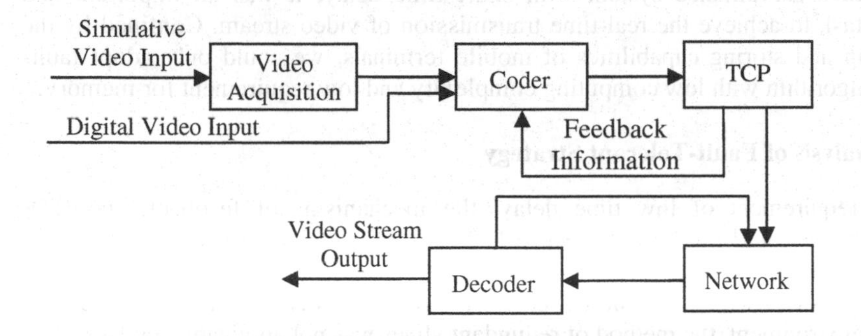

Fig. 1. Framework of Video Network Transmission System

In the system shown in Fig. 1, processing and transmission of the whole video stream was described as follows: at the sending end, simulative video was sampled to get digital video and then it was coded or the digital video input was coded directly to generate video bit stream for network communication that was adapt to the transmission; according to the feedback information, the available transmission bandwidth for network was estimated and the output rate of the coder was regulated adaptively (including the regulation of source and channel rates), so that the video bit stream could satisfy the limit of available bandwidth for current network transmission; at the receiving end, the received video stream was decoded, the video signals were reconstructed, the parameters of current network transmission were calculated (e.g. the packet loss rate in transmission, etc.) and the feedback control information was sent out.

The network for wireless video surveillance system was error-prone, in which data were easy to be transferred incorrectly, such as packet loss, bit inversion, etc. Since the error resilience technique in some video surveillance systems was very sensitive to bit error in channels, even a single primary error might result in rapid deterioration of video quality; and it was more intolerable to lose packet in IP network, which might result in the collapse of decoding terminal if severely. In particular, when the packet that was transmitting error was regarded as the reference frame, the image based on it would also go wrong, and it might attribute to the wrong transmission. In the transmission of wireless network, TCP protocol could ensure a reliable transmission, but its retransmission mechanism brought about long time delay, which was unacceptable in real-time video application [4]. The UDP protocol could reduce time delay, but it was of high packet-loss rate without the guarantee of QOS; especially when it was applied in wireless network environment, the high rate of packet loss had exerted great influence on the quality of video. Based on the overall consideration of above factors, UDP protocol might be used in transport layer, while in the top layer, a proper fault-tolerant strategy was necessary to reduce the rate of packet loss and effective measures were adopted to recover lost data, so as to minimize the visual influence by transmission error.

4 Optimizing Fault-Tolerance Strategy during Wireless Network Transmission

In the video surveillance system with short time delay, it was an important and difficult task to achieve the real-time transmission of video stream. Confined by the computing and storing capabilities of mobile terminals, we could only adopt fault-tolerant algorithm with low computing complexity and low requirement for memory.

4.1 Analysis of Fault-Tolerant Strategy

For the requirement of low time delay, the mechanisms of feedback and RTP retransmission could not be used. Since H.264 used in the surveillance system was baseline profile, which did not support data partition, so data partition and the unequal error protection based on it could not be used, either. In order to reduce data traffic in wireless environment, the method of redundant slices was not applicable, either.

In consideration of the above factors, block coding could be used during the network transmission in wireless video surveillance system, together with the fault-tolerant strategy such as FMO structure and intra block refreshing.

Slice coding might reduce the length of each packet, so that the probability of packet loss was reduced accordingly; and even in case of packet loss, it would be easier for error resilience [5]. There were 3 types of data block in H.264: head information, intra data block and inter data block. Head information included the type of macro block, quantitative parameters, motion vector, etc. which was the most important and should be provided with maximum protection, so these blocks were regarded as Class A in H.264. Intra data block included intra CBPs and intra coefficient, which could prevent spread of error bits effectively, and they belonged to Class B. Inter data block included inter CBPs and inter coefficient, which was the least important, so they were regarded as Class C. FMO mode (interleaving mode) could reduce bit rate, as well as make full use of spatial information for error correction. FMO classified macro blocks in the same frame into different slice groups, so the error was concealed effectively by means of acquiring the information of macro blocks adjacent to the lost one from other slice groups that were receiving correctly, when the error occurred in a macro block or several macro blocks [6]. In this way, some adjacent macro blocks might be distributed in different slices, so it helped to restore the wrong image better, and avoided spreading the error as well.

Intra block refreshing was used to eliminate the incorrect transmission by the loss of inter prediction macro blocks. By means of scattering adjacent macro blocks into different slices groups, FMO helped the receiving end improve its property of error concealment, utilizing the spatial correlation. In spite of that, it was still difficult to avoid the network error, such as packet loss and its result of the mismatch of reference frames at the encoding and decoding ends, and the influence by accumulated spatial errors was severe, so it could only be restored rapidly by encoding the image region as intra blocks [7]. However, full Frame I would not be inserted in the application of video phone, for the instant bit rate and its introduction would increase time delay dramatically. Therefore, the coding tool FMO for error resilience, which could improve the property of error concealment at the decoding end effectively, should be combined with intra block refreshing to get a more effective scheme of mixed error resilience.

4.2 Experiment of Fault-Tolerant Strategy

The fault-tolerant strategy would be tested. Table 1 showed several testing conditions in wireless network that was proposed in Reference, and the first condition was adopted in this paper, i.e. 64 kbps bit rate, 0.051% error rate; as the reference software of JVT's H.264/AVC, JM 9.2 was adopted for encoding and decoding, which had patched up some bugs. IPPP⋯IPPP (one Frame I from every other 150 frames) was adopted for testing video sequence; RTP packet was output from the encoder, the packet damage was simulated with general testing software in off-line environment and the file after damage should be decoded by the decoder directly. In the decoder, if some frame could not be restored, then it would copy the previous frame directly to ensure the integrity of video sequence. The testing results could be evaluated by peak-value single-to-noise ratio of intensity.

Table 1. Testing Environment in Wireless Network

No.	Bit Rate (kbps)	Video Length (s)	Error Rate (%)	Application Environment
1	64	60	0.049	PCS
2	64	60	0.031	PCS
3	64	60	0.05	PCS
4	128	180	0.67	PCS
5	128	180	0.81	PSS
6	128	180	0.44	PSS

Other parameters for the encoder and decoder included: ProflleDC=66, LevelDC=l0, COnstralnedSetlflag=1, NUmberReferenceFrames=3, the bit rate was within 64kbps, Loop Filter, Symbolmode=UVLC, outFileMode=RTP were not adopted; decoder: considering that there were many slices in Frame I, the MAX-NUM-SLICES in the program should be given a high value (200 or so).

The testing results were shown as Table 2.

Table 2. Comparison of Results before and after the Fault-tolerant Strategy

Sequence	Format	Frame Rate (fps)	Frames	Error-Resilient Strategy	RTP Packets	Packets Lost	Min Y	Avg Y
Carphone	QCIF	15	380	Yes	2526	32	32.08	25.01
Carphone	QCIF	15	380	No	2325	29	30.67	22.53
Carphone	QCIF	15	380	Yes	2448	33	32.14	26.33
Forenlan	QCIF	15	380	Yes	2511	29	30.57	23.41
Forenlan	QCIF	15	380	No	2384	31	26.42	21.22
Forenlan	QCIF	15	380	No	2364	32	30.81	24.36

It could be seen from the experimental results that the application of slice coding, together with the fault-tolerant strategy of FMO and intra block refreshing during the network transmission in wireless video surveillance system would improve the quality of video and the average intensity PSNR could be improved 2dB or so.

5 Conclusion

In this paper, it proposed the fault-tolerant strategy of combining FMO and intra block refreshing based on slice coding. The adoption of complex macro blocks reduced distortion that was related to error concealment, so the quantity of intra blocks in inter encoding frames decreased, thus effectively improved the efficiency of encoding as well as the quality of video.

References

1. Xie, L., Zhang, Y.: Joint FMO and Rate-distortion Optimized Mode Selection for H. 264 Fault-tolerant Video Coding. Journal on Communications 28, 128–132 (2007)
2. Xu, Y., Ji, W., Chen, Y., Zhou, J.: Real-Time Wireless Video Transmission Fault-Tolerant Control Algorithm Based on NS2. The Computer Simulation 28, 150–153 (2011)
3. Yu, H.-B., Yu, S., Wang, C.: Highly Efficient, Low Delay Architecture for Transporting H.264 Video over Wireless Channel. Signal Processing: Image Communication 19, 369–385 (2010)
4. Wenger, S.: H. 264/AVC over IP. IEEE Transactions on Circuits and Systems for Video Technology 13-7, 645–656 (2010)
5. Xie, Z., Zheng, G., He, G.: H.264 Error Resilient Coding Scheme Based on Flexible Macroblock Ordering. Computer Engineering and Applications 31, 200–202 (2006)
6. Jang, S.H., Jayant, N.: An efficient error resilient technique for applications of one way video using transcoding and analysis by synthesis. In: Proceedings of the IEEE Global Telecommunications Conference Workshops, Texas, USA, pp. 428–432 (2004)
7. Chiou, H.-J., Lee, Y.-R., Lin, C.-W.: Content aware fault-tolerant transcoding using prioritized intra-refresh for video streaming. Journal of Visual Communication and Image Representation 16-4, 563–588 (2005)

Research of Algorithm Based on Improved Self-Organization Neural Network of Fuzzy Clustering

Xu Zhang, LongZhen Duan, Yunqiao Zhan, and Guifeng Wang

Department of Computer Science and Technology, Nanchang University,
Nanchang, China

Abstract. This paper introduced an improved fuzzy SOFM clustering algorithm. In the initialization phase, by the way of subtractive clustering, optimized initial weights of network and determined the number of clusters. To verify the effectiveness of the algorithm, this algorithm will be applied to web log mining. Experimental results showed that the improved fuzzy SOFM neural network training speed and convergence results have improved to some extent, and for a variety of users interested in mining provides a feasible approach.

Keywords: Web log mining, Fuzzy clustering, SOFM neural network.

1 Introduction

Web log mining is the current frontier of computer science research focus, application of neural network technology can enhance the intelligence of Web log mining.

Proposed in recent years the use of self-organizing feature map neural network, fuzzy clustering method, clustering can be achieved quickly. Using self-organizing feature map neural network fuzzy clustering method because it can maintain the data in the space of topological properties, able to quickly bring high-dimensional data is mapped to two-dimensional plane, can solve the problem of high dimensional data clustering. However, there are some problems in this method, such as the initial cluster centers sensitive, can only achieve a rough clustering, the clustering of non-spherical family is not satisfied and so on.

Data sets for the topology of real data sets in order to preserve the structural information, we propose a data set can reflect the basic structure of a hierarchical fuzzy approach to adaptive fuzzy-level growth of self-organizing network, but it's still sensitive to the initial clustering [1].

In the fuzzy Kohonen neural network, based on the fuzzy output neuron bias degrees, lateral inhibition and the weighted coefficient of fuzzy membership to improve in three areas, effectively avoid the appearance of dead neurons and improve the network clustering performance, but the number of neurons in the output settings cannot be set dynamically, based on the known number of clusters in the case set [2].

Rough sets and fuzzy Kohonen effective converged network integration, multi-sensor data fusion has obvious advantages, but its existence is difficult to summarize the network structure of the shortcomings of integration of the sample [3].

A. Xie & X. Huang (Eds.): Advances in Electrical Engineering and Automation, AISC 139, pp. 191–197.
springerlink.com

Based on SOFM neural network model of the advantages of clustering, after the subtraction result as a combination of clustering fuzzy C-means approach the initial value of SOFM neural network to optimize the initial parameters of the network, and applied to Web log mining.

2 Clustering Analysis

2.1 Fuzzy C-Means Clustering Algorithm (FCM)

FCM is a fuzzy clustering method. The algorithm uses the membership function of each sample point belongs to a calculation of the degree of clustering, according to the fuzzy set theory, vectors with different degree of membership attributable to different sets of clustering, according to the size of the membership categories to determine the ownership of the sample [4][5]. FCM objective function through the establishment, and so that it minimized to achieve clustering, the establishment of the following objective function.

$$J_m = \sum_{i=1}^{N} \sum_{j=1}^{c} u_{ij}^{m} \|x_i - c_j\|^2 \quad 1 \le m < \infty \tag{1}$$

Where m is the weighted index, x_i belongs to category j for the degree of membership, x_i is the i measured d-dimensional data, c_j is the class j of the cluster center, the objective function J shows that various types of data to the corresponding cluster center and the square of the weighted distance.

Fuzzy C-Means clustering algorithm specific steps:

Step 1: Initialization, given the number of clusters c, $2 \le c \le n$, 11, is the number of data, set the iteration stopping threshold ε, $0 < \varepsilon < 1$, initialize the cluster prototype model c_j, set the iteration counter $k = 0$.

Step 2: Using equation (2) Calculate or update the partition matrix

$$u_{ij} = \left[\sum_{k=1}^{c} \left(\frac{\|x_i - c_j\|}{\|x_i - c_k\|} \right)^{\frac{2}{m}-1} \right]^{-1} \tag{2}$$

Step 3: Using equation (3) Calculate or update the cluster prototype model matrix

$$c_j = \frac{\sum_{i=1}^{N} u_{ij}^{m} x_i}{\sum_{i=1}^{N} u_{ij}^{m}} \tag{3}$$

Step 4: To meet the conditions $\max_{ij}\{\|u_{ij}^{k+1} - u_{ij}^{k}\| < \varepsilon\}$, The iteration stops.

The algorithm is essentially a gradient descent optimization algorithm, the time when a large sample data, can easily fall into local minimum, but not the global optimal solution, but also extremely sensitive to initial values.

2.2 Subtractive Clustering

Subtraction clustering method is an input-output training data by the detection of clustering to automatically generate the model inference system technology [6]. It is R. Yager mountain clustering method into - further expansion. A subtraction clustering each - data points are as - a possible cluster center, and is based on the data points around this point to calculate the density of points as cluster centers this possibility[7].

For the sample data points $X = \{X_1, X_2, \ldots, X_n | X_i \subseteq R^m\}$, N is the number of samples, m is the dimension of sample data, subtract the clustering process steps are as follows:

Step 1: Each point on the X in X_i. Calculate the density.

$$d_i = \sum_{j=1}^{N} \exp\left[\frac{-\|X_i - X_j\|^2}{(0.5\gamma_a)^2}\right] (\gamma_a \text{ is parameter}) \tag{4}$$

The highest density of data points X_{c1} is the first cluster center.

Step 2: Set X_{ck} Selected for the k-th cluster center, the density of cluster center is d_{ck}. The density of each data point be amended by the following formula.

$$d_i = d_i - d_{c_k} \sum_{j=1}^{N} \exp\left[\frac{-\|X_i - X_{c_k}\|^2}{(0.5\gamma_\beta)^2}\right] (\gamma_\beta \text{ is parameter}) \tag{5}$$

Find the remaining N-1 data point density, the highest density of points selected as the new cluster center. To avoid close proximity of the cluster center, take $\gamma_\beta = 1.5\gamma_a$.

Step 3: Determine the new cluster centers corresponding density targets are met.

$$\frac{d_{c_{k+1}}}{d_{c_1}} < \delta (\delta < 1) \tag{6}$$

If conditions are met, then go to step 2, if the conditions are met out of circulation, no new cluster center. δ is the scaling parameter to control the number of clusters.

2.3 SOFM Neural Network Overview

SOFM neural network is a two-tier feed-forward neural network, namely the input layer and output layer [8]. Number of neurons in input layer is equal to the dimension of the sample; the output layer is also a competitive level. Neurons in input layer weights vector to the outside world through the information into the output layer of neurons, through the input mode of self-organization learning, classification results in the output layer will be represented. SOFM neural network input patterns through repeated study, the spatial distribution of the weight vector connecting the input mode of the probability density and distribution of consistent, that is, the connection weights vector can reflect the spatial distribution of the statistical characteristics of input patterns [9].

Self-organizing feature map neural network has a strong parallel computing, self-organizing, self-regulating ability, but it’s still the following weaknesses:

(1) Weight vector of the network updating strategy makes the end result depends on the vector of the input sample sequence.

(2) Defined by the network learning rate is decreasing as a function of the number of iterations to ensure convergence of workers forced to terminate the network strategy, this can only amend the rules to ensure a sufficient number of iterations the

weight vector after the correction is very small, in order to ensure optimal weight vector.

(3) Different initial conditions often lead to different results. In addition, the correction field size and learning rate selection strategy to deepen the learning algorithm of SOFM is essential for assembly of different parameters produce different results.

(4) A pre-fixed number of competitive layer neurons, we can have a maximum number of classes, which for some applications, especially in advance how many do not know the number of categories, and it cannot be applied.

3 An Improved Fuzzy Clustering Algorithm SOFM

Subtraction clustering algorithm using the initial sample concentration and the number of clusters of cluster centers, and take them as the parameters of SOFM neural network initialization conditions. In the SOFM neural network learning mechanism in the integration of fuzzy C-means algorithm (FCM), proposed a modified fuzzy SOFM clustering neural network (FSOFM). Network structure of two layers, the first layer is input layer, neurons containing P, P is the input data dimension; second layer of competition in the output fuzzy neurons, their number is equal to the number of class c, the state value of [0 , 1] continuum, when the i-th output neuron. The input sample X k, k ∈ {1,2, ..., n} belongs to i-th class of the membership degree u_{ik}. The algorithm steps are as follows:

Step 1: Given a sample space x = {x 1, x 2... x 1}, the number of clusters is c, $2 \leq c \leq n$, the error threshold $\varepsilon > 0$.

Step 2: Using Subtractive Clustering algorithm (detailed steps as shown in section 2.2), to be cluster centers set {M 1, M 2, ..., M k}, then the number of clusters c = k.

Define the neighborhood distance $r(t) = \frac{k-1}{2}\left(1 - \frac{t}{t_{max}}\right)$.

Step 3: Initialize the network weights W = {W 1, W 2... W c}, initialize the network weights set of cluster centers, $W_i = M_i$, i= 1, 2... c. Given fuzzy parameters $l_0 > 1$, the maximum number of iterations t max, the initial iteration is t = 0.

Step 4: According to equation (7) iteratively update the fuzzy membership functions {u ik}, and according to (8) calculate the learning rate { α ik}.

$$u_{ik} = \frac{1}{\sum_{j=1}^{c}\left(\frac{\|X_k - W_i\|}{\|X_k - W_j\|}\right)^{\frac{2}{m-1}}} \tag{7}$$

$$\alpha_{ik}(t) = \left(u_{ik}(t)\right)^{l_m} \qquad l_m = l_0 - \frac{t}{t_{max}}(l_0 - 1) \tag{8}$$

Step 5: Adjust the network weights $W_i(t)$.

$$W_i(t) = W_i(t-1) + \frac{\sum_k^n \alpha_{ik}\left(X_k - W_i(t-1)\right)}{\sum_k^n \alpha_{ik}(t)} \tag{9}$$

Step 6: Calculate the energy function $E(t)=\sum_{i=1}^{c}\|W_i(t)-W_i(t-1)\|^2$, If t> t_{max} or E (t) <ε, stop the iteration, otherwise t = t +1 go to step 4.

The improved fuzzy SOFM clustering algorithm (FSOFM) analysis, its advantages are the following:

(1) Subtraction clustering algorithm using the initial sample concentration and the number of clusters of cluster centers, and as a SOFM neural network parameters to initialize the condition that the number of competitive layer neurons is initialized to the number of clusters, to avoid for any given cluster configuration error caused the number of categories; initialize the network weights set for the cluster center, to ensure the weight of discrete, making the weight distribution and the probability density distribution of the input sample is similar to enhance the operation of the network capacity and improve network convergence.

(2) Learning rate adaptively according to the adjustment of membership, to [illegible] neighborhood selection and adjustment.

(3) Into the fuzzy C-means algorithm, the degree of membership into SOFM, in the output neurons can be achieved by setting the threshold number of outputs, a sample will be assigned to multiple classes to achieve a fuzzy clustering.

4 Experimental Results

The above model is applied to Web log mining process, through the fuzzy SOFM neural network clustering of users. This choice of Jiangxi Xinhua Bookstore Online Shopping Mall in December 2010 as an experimental server log data sets to experiment. After pre-processing, it selects 920 records, a total of 105 user sessions, 203 different pages. After subtraction to get the initial cluster center of cluster number c = 5, and will be set as initial cluster centers initialization of fuzzy SOFM network weights. After the iterative algorithm, and finally obtained five clusters, as shown in Table 1.

Table 1. Clustering Results

Cluster center number	Users belonging to the cluster center number	Content Summary	User Number
6	6,11,12,19,21,28,33,41,69,81, 96,103,100,101,102,105	Computer Class	17
12	4,9,11,30,32,35,36,41,42,49,53,56,60,72,82,87,96,104	Economics and Management	18
29	1,3,5,7,8,11,22,26,27,29,31,34,35,38,40,44,45,47,50,51,52, 57,59,61,62,64,80,82,85,93,94,98,102,104	Foreign language	35
51	2,10,13,14,16,20,23,37,48,58, 60,68,83,84,86,87,90,91,92,97,99,101,102,103	Literature	24
88	4,8,15,17,18,24,25,39,43,46, 54,55,63,65,88,89,95,100	Art	16

Seen from Table 1, numbers of computer users for the 11 class, economics and management, more interested in foreign language category pages; number 8 of the users of foreign languages, more interested in art pages. To the proposed improved fuzzy SOFM algorithm (FSOFM) with the literature [10] proposed the fuzzy Kohonen algorithm (FKNN) compared in Table 2.

Table 2. Comparison of algorithms

Heading level	Example	Font size and style
Iterations	309	105
Running time	176 seconds	211 seconds
The number of clusters	Determined in advance	Adaptive

As FSOFM algorithm FKNN algorithm based on the initial optimization of the network to do so in the network convergence is better than FKNN algorithm, clustering capacity. On the number of clusters, FSOFM adaptive algorithm can effectively change the data set, and therefore have certain robustness. However, due to the optimization process is also a process of clustering the original data set, the running time is still lacking.

5 Conclusion

In this paper, subtractive clustering and distribution of the initial cluster centers to get the number of clusters to optimize the fuzzy SOFM neural network the number of output neurons and the network initial weights, the algorithm not only reduces the number of iterations, accelerate the network convergence, and the effect of clustering algorithm has been some improvement, and improved fuzzy SOFM neural network algorithm is applied to the Web log mining test analysis, the proposed algorithm is an effective method of cluster analysis , convergence results have greater degree of improvement, and to solve the multi-class clustering problem, but because some of the parameters in the algorithm has a certain empirical running time is not ideal. In future work, will work to continue to improve the algorithm, the set of parameters, further improving the quality of clustering.

References

1. Barreto-Sanz, M., Pérez-Uribe, A., Peña-Reyes, C.-A., Tomassini, M.: Fuzzy Growing Hierarchical Self-Organizing Networks. In: Kůrková, V., Neruda, R., Koutník, J. (eds.) ICANN 2008, Part II. LNCS, vol. 5164, pp. 713–722. Springer, Heidelberg (2008)
2. Xu, M., Tan, X.: Kohonen clustering network based on improved fuzzy algorithm. Computer Simulation 26(4), 228–232 (2009)
3. Liu, H., Chen, X., Liu, Y.: Kohonen based on rough sets and fuzzy clustering network of multi-sensor data fusion. Electronic Measurement and Instrument 23(3), 218–223 (2009)
4. Kang, family silver, Ji, Z.-C., Gong, C.: A nuclear fuzzy C-means algorithm and its application. Journal of Scientific Instrument 31(7), 1657–1663 (2010)

5. Meng, L., Song, Y., Zhu, F.: Space-based neighborhood weighted fuzzy C-means clustering and its application. Computer Science 27(10), 2968–3973 (2010)
6. Gu, L., Wu, H.: A method of subtraction based on genetic algorithm clustering. Pattern Recognition and Artificial Intelligence 21(6), 758–762 (2008)
7. Yu, D., Li, Y.: Based on Subtractive Clustering improved fuzzy c-means fuzzy clustering algorithm. Microcomputer and Applications 29(16), 14–20 (2010)
8. Kohonen, T.: Self-organized Formation of Topologically Correct Feature Maps. Biological Cybernetics 43, 59–69 (1982)
9. Ding, C., Yeung: SOFM network based on improved K-means clustering algorithm. Technology Review 27(10), 61–63 (2009)
10. Tsao, E.C., Beulek, J.C., Pal, N.R.: Fuzzy Kohonen Clustering Network. Pattern Recognition 27(5), 757–764 (1994)

Fast Mining Algorithm of Global Maximum Frequent Itemsets

Bo He

School of Computer Science and Engineering, ChongQing University of Technology, 400054 ChongQing, China
heboswnu@sina.com

Abstract. As far as we know, a little research of global maximum frequent itemsets had been done. Therefore, a fast mining algorithm of global maximum frequent itemsets was proposed, namely, FMAGMFI algorithm. Firstly, each computer nodes compute local maximum frequent itemsets with DMFIA algorithm and local FP-tree. Secondly, the center node combined local maximum frequent itemsets. Finally, global maximum frequent itemsets were gained by the searching strategy of top-bottom. Adopting FP-tree structure, FMAGMFI algorithm greatly reduces runtime compared with Apriori-like algorithms. Theoretical analysis and experimental results suggest that FMAGMFI algorithm is efficient.

Keywords: DMFIA algorithm, Global maximum frequent itemsets, FP-tree, FP-growth.

1 Introduction

There are some mining algorithms[1] of maximum frequent itemsets, such as DMFIA[2] and CD[3], etc. However, these mining algorithms do not suit mining of global maximum frequent itemsets. Aiming at these problems, a fast mining algorithm of global maximum frequent itemsets was proposed, namely, FMAGMFI algorithm.

2 Related Description

2.1 Mining of Global Maximum Frequent Itemsets

The global transaction database is *DB*, the total number of tuples is *M*. Suppose P_1, P_2,..., P_n are *n* computer nodes, there are M_i tuples in DB_i. if DB_i (i=1,2,...,n) is a part of *DB* and stores in P_i, then $DB = \bigcup_{i=1}^{n} DB_i$, $M = \sum_{i=1}^{n} M_i$.

Mining of global maximum frequent itemsets can be described as follows.

A. Xie & X. Huang (Eds.): Advances in Electrical Engineering and Automation, AISC 139, pp. 199–204.
springerlink.com

Each node P_i deals with local database DB_i, and communicates with other nodes. finally, global maximum frequent itemsets of global transaction database are gained.

2.2 Related Definition

Definition 1. For itemsets *X*, the number of tuples which contain *X* in local database $DB_i(i=1,2,\ldots,n)$ is defined as local frequency of *X*, symbolized as $X.si$.

Definition 2. For itemsets *X*, the number of tuples which contain *X* in global database is global frequency of *X*, symbolized as $X.s$.

Definition 3. For itemsets *X*, if $X.si \geq min\_sup*M_i(i=1,2,\ldots,n)$, then *X* are defined as local frequent itemsets of DB_i, symbolized as F_i. If $|X|=k$, then *X* symbolized as F_i^k , and *min_sup* is the minimum support threshold.

Definition 4. For itemsets *X*, if $X.s \geq min\_sup*M$, then *X* are defined as global frequent itemsets, symbolized as *F*. If $|X|=k$, then *X* symbolized as F_k.

Definition 5. For global frequent itemsets *X*, if all superset of *X* are not global frequent itemsets, then *X* are defined as global maximum frequent itemsets, symbolized as *FM*.

2.3 Related Theorem[4]

Theorem 1. If itemsets *X* are local frequent itemsets of DB_i, then any nonempty subset of *X* are also local frequent itemsets of DB_i.

Corollary 1. If itemsets *X* are not local frequent itemsets of DB_i, then the superset of *X* must not be local frequent itemsets of DB_i.

Theorem 2. If itemsets *X* are global frequent itemsets, then *X* and all nonempty subset of *X* are at least local frequent itemsets of a certain local database.

Theorem 3. If itemsets *X* are global frequent itemsets, then any nonempty subset of *X* are also global frequent itemsets.

Corollary 2. If itemsets *X* are not global frequent itemsets, then superset of *X* must not be global frequent itemsets.

Theorem 4. If itemsets *X* are global maximum frequent itemsets, then *X* must be global frequent itemsets.

Theorem 5. If itemsets *X* are global maximum frequent itemsets, then *X* and all nonempty subset of *X* are at least local frequent itemsets of a certain local database.

Corollary 3. If itemsets *X* are not local frequent itemsets of any local database, then *X* must not be global maximum frequent itemsets.

Theorem 6. If itemsets *X* are global maximum frequent itemsets, then *X* are at least the subset of local maximum frequent itemsets in a certain local database.

Corollary 4. If itemsets X are not the subset of local maximum frequent itemsets of any local database, then X must not be global maximum frequent itemsets.

Theorem 7. If item x_i is not global frequent item, and $\{x_i\} \subseteq X$, then itemsets X must not be global frequent itemsets, and must not be global maximum frequent itemsets.

3 FMAGMFI Algorithm

3.1 Design Thoughts of FMAGMFI Algorithm

Each node adopts DMFIA algorithm and FP-tree[5] to compute local maximum frequent itemsets in FMAGMFI.

FM' are pruned by the searching strategy of top-bottom.

The searching strategy of top-bottom is described as follow.

(1) P_0 confirms the largest size k of itemsets in *FM'*;

(2) P_0 collects global frequency of all k-itemsets in *FM'* from other nodes;

(3) If k-itemsets Q are global frequent itemsets, then $FM=FM \cup \{Q\}$, and P_0 deletes Q and any nonempty subset of Q from *FM'*;

(4) If k-itemsets Q are not global frequent itemsets, then P_0 deletes Q from *FM'*. For all item $x \in Q$, $FM' = FM' \cup \{Q - \{x\}\}$; Turn to (1).

3.2 Description of FMAGMFI Algorithm

Alogrithm. FMAGMFI

Input: The local transaction database DB_i which has M_i tuples and $M = \sum_{i=1}^{n} M_i$, n nodes $P_i(i=1,2,\ldots n)$, the center node P_0, the minimum support threshold *min_sup*.

Output: The global maximum frequent itemsets *FM*.

Methods: According to the following steps.

Step1. /*each node adopts DMFIA[2] algorithm and FP-tree to produce local maximum frequent itemsets*/

for(i=1;i<=n;i++) /*gaining global frequent items*/

{Scanning DB_i once;

computing local frequency of local items E_i;

P_i sends E_i and local frequency of E_i to P_0;}

P_0 collects global frequent items E from E_i;

E is sorted in the order of descending support count;

```
P0 sends E to other nodes Pi; /*transmitting global frequent items to other nodes Pi */
for(i=1;i<=n;i++)
{creating the FP-tree^i;   /*FP-tree^i represents FP-tree of DBi */
 FMi =DMFIA(FP-tree^i, min_sup); /* node adopts DMFIA algorithm to produce local
        maximum frequent itemsets FMi aiming at FP-tree^i*/}
step2./* P0 gets the union of all local maximum frequent itemsets */
for(i=1;i<=n;i++)
 Pi sends FMi to P0; /* FMi represent local frequent itemsets of Pi */
P0 combines FMi and produces FM'';   /* FM' = ∪(i=1..n) FMi , represent the union of all local
maximum frequent itemsets*/
step3./*P0 gets global maximum frequent itemsets according to the searching strategy
of top-bottom */
FM=∅ ;
while FM'≠∅
{P0 confirms the largest size k of itemsets in FM';
 for all itemsets Q ∈ local frequent k-itemsets in FM'   /* P0 collects global frequency
 of Q from other nodes Pi;*/
 if Q are not the subset of any itemsets in FM
{ P0 broadcasts Q;
    Pi sends Q.si to P0;   /* Pi computes local frequency Q.si of Q according to FP-tree^i
    */
    Q.s= Σ(i=1..n) Q.si ;   /* Q.s represents global frequency of Q*/
    if Q.s>=min_sup*M   /* Q are global maximum frequent itemsets*/
    { FM=FM∪{Q};
     P0 deletes Q and any nonempty subset of Q from FM';}
    else   /* Q are not global maximum frequent itemsets*/
    { P0 deletes Q from FM';
     for all item x∈ Q
      if Q−{x} are not the subset of any itemsets in FM
       FM'= FM' ∪ {Q−{x}};
    }}}
```

4 Experiments of FMAGMFI

Comparison experiment: It is a way of changing the minimum support threshold while adopting fixed number of nodes. FMAGMFI compares with CD in terms of runtime. The results are reported in Fig.1.

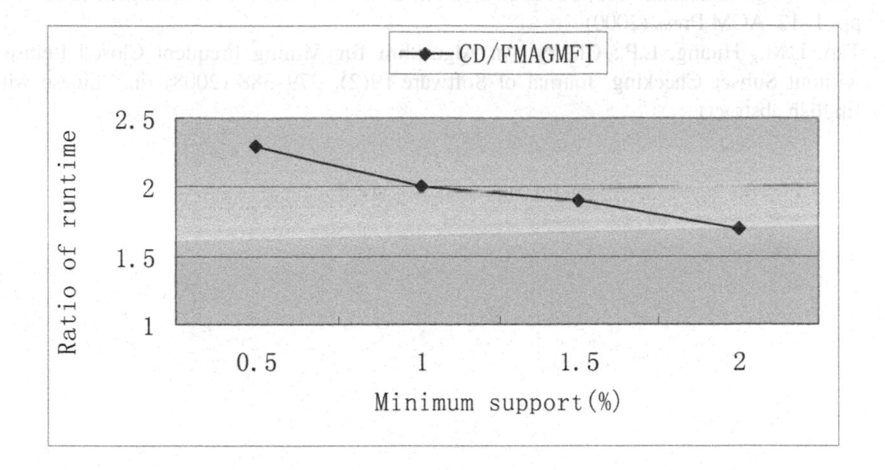

Fig. 1. Comparison of runtime

The comparison experiment results indicate that under the same minimum support threshold, runtime of FMAGMFI decreases while comparing with CD.

5 Conclusion

FMAGMFI gains global maximum frequent itemsets by the searching strategy of top-bottom. It lessens runtime. Experimental results suggest that FMAGMFI is fast and effective.

Acknowledgment. The paper is supported by the ministry of education humanity and social scientific research items grant No. 09yjc870032 and the national natural science foundation of China under Grant No. 61173184.

References

1. Chen, Z.B., Han, H., Wang, J.X.: Data warehouse and data mining. Tsinghua University Press, Beijing (2009) (in Chinese)
2. Song, Y.Q., Zhu, Z.H., Chen, G.: An algorithm and its updating algorithm based on FP-tree for mining maximum frequent itemsets. Journal of Software 14(9), 1586–1592 (2003) (in Chinese with English abstract)

3. Agrawal, R., Shafer, J.C.: Parallel mining of association rules. IEEE Transaction on Knowledge and Data Engineering 8(6), 962–969 (1996)
4. He, B., Wang, H.Q., Liu, Z., Wang, Y.: A fast and parallel algorithm for mining frequent itemsets. Journal of Computer Applications 26(2), 391–392 (2006) (in Chinese with English abstract)
5. Han, J.W., Pei, J., Yin, Y.: Mining frequent patterns without Candidate Generation. In: Proceedings of the 2000 ACM SIGMOD International Conference on Management of Data, pp. 1–12. ACM Press (2000)
6. Tao, L.M., Huang, L.P.: Cherry: An Algorithm for Mining Frequent Closed Itemsets without Subset Checking. Journal of Software 19(2), 379–388 (2008) (in Chinese with English abstract)

A Study on Web2.0 Blending Learning

Rui Cheng

Education School, JiangHan University, Wuhan, China
carycheng78@gmail.com

Abstract. An instructional model of online blending learning based on web2.0 tools has been presented with this study. Compared with other blending learning, the advantages and deficiencies of the model have been analyzed and detailed. General guidelines for applying the model to instructional practice are also discussed based on an experiment.

Keywords: blending learning, web2.0, learning model.

1 Introduction

Blending learning is a learning manner integrating diverse technologies and pedagogical approaches for instructional purposes. In the web era, it is endowed with new concepts as follows:

For one thing, it is to combine or mix different web-based technologies and various pedagogical or instructional approaches; for another, it is to combine F2F (face-to-face) instructor-led training with online self-driven learning manner.

Consequently, web2.0 blending learning (blending learning under the web2.0 environment) is supposed to comprise two imperative parts -- online learning based on web2.0 technologies and F2F training led by the instructor.

2 Web2.0 Blending Learning

2.1 Web2.0 Online Learning

According to formats, learning materials can be categorized into text, image, audio, video, animation and so on. There are 3 kinds of web2.0 tools -- blog, video website and web bookmark can be utilized for the storage, transference and retrieval of the materials. The instructor may apply several tools simultaneously to online instruction.

As is shown in fig 1, various blog sites can be used for storing and transferring text or image contents because they support materials of these kind well; video websites can provide us with space and convenience to store and manage multimedia learning contents such as video, audio or animation. Recently, some blog websites begin to support multimedia contents too, though they don't provide space as big as video websites do in general. For learning contents with little multimedia, the instructor may just use a blog website to distribute learning materials.

A. Xie & X. Huang (Eds.): Advances in Electrical Engineering and Automation, AISC 139, pp. 205–211.
springerlink.com

Web bookmark sites such as Delicious make it easy for us to register and retrieve useful web resources by providing us a page to log hyperlinks. With its help, instructors may register web resources for future use anytime and anywhere; the other way round, students can find useful information from these refined sites conveniently for either homework or further learning.

In addition, some micro blog sites such as Twitter and so on could be used for brief talk or notice publication.

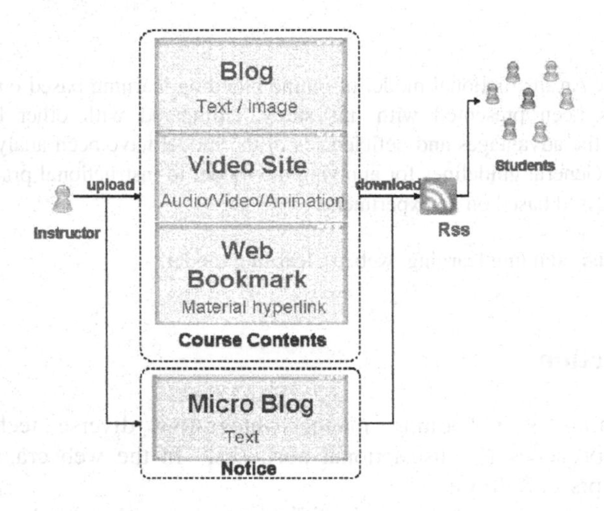

Fig. 1. Web2.0 online learning

Hence, the instructor just need to upload learning contents to different web2.0 sites according to their forms, and subsequently the students can carry out self-driven or collaborative learning online. In view of assembling learning materials dispersed in different web2.0 sites is troublesome and time-consuming, rss tools such as GoogleReader and the like are supposed to be used by students.

Furthermore, feedbacks and interactivities between the instructor and students or among students are necessary too. Many tools may be employed for these purposes.

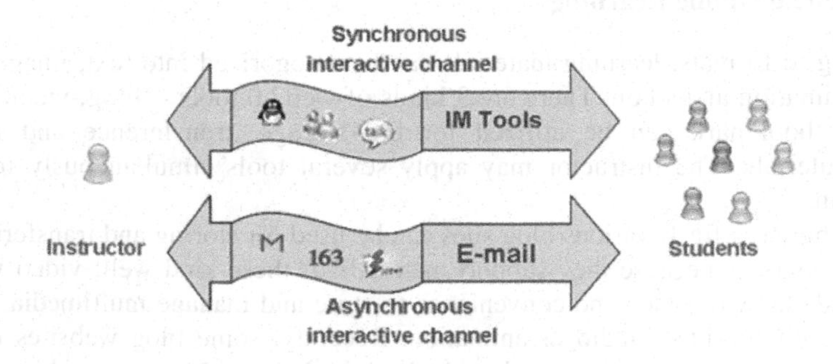

Fig. 2. Interactive channels

As shown in fig 2, the IM (instant message) tools can be utilized as synchronous interactive channel, and the E-mail tools are liable to make up the asynchronous ones.

2.2 Web2.0 Blending Learning Model

As demonstrated in fig 3, the learning model comprises 3 stages, 2 branches and several steps.

The first is instructional design stage, in which the instructor prepares for the work to be carried out in both class room settings and online environments. In this period, the fundamental tasks include: analyzing learning needs, analyzing learners, analyzing learning contents, setting down learning objectives, developing learning strategies and approaches and selecting evaluation instruments.

The second is performance stage. According to the difference of instruction delivery contexts, it is divided into two learning branches -- F2F instruction and web2.0 learning. Generally, the former is used for lecture or class discussion and the latter is applied to online self-driven learning or review. Sometimes, the web2.0 learning can be used for task-driven learning or collaborative learning, and then the students are supposed to be assigned some tasks or asked to communicate with each other.

From the perspective of instructor, he or she may carry out instruction via either branch in the first place according to his or her preference, arrangement or schedule, and then turn to the other one to blend. The 2 instruction manners are performed alternately to achieve the mutual goals.

The third is evaluation stage, which consists of formative evaluation and summative evaluation. With results of formative evaluation, inappropriate instructional strategies, approaches or tools can be identified and revised so as to improve performance in the rest learning periods. Summative evaluation is used for judging whether the learning objectives have been achieved and weighing the learning effectiveness.

2.3 Advantages of the Model

Low cost: compared with the other kind of blending learning model, this one may be the cheapest. Because most web2.0 tools are free of charge, instructors can apply them to online instruction without caring about the expenses. It is the primary feature of the model, which may be crucial for clients, especially for those who are short of funds.

Ease of use: most web2.0 tools are designed user-friendly, neither the instructor nor the students need preliminary training. They can develop learning materials, lead online instruction or participate in online collaboration and the like with various web2.0 tools immediately. Most functions of these tools will be mastered in a few minutes in authentic jobs. Conversely, it will save much time for both the instructor and the students.

Functioning diversity: web2.0 tools are cheap but not primitive or crude. Most of them provide us with diverse powerful functions, which can be applied to online learning. Many of them may be equal or even better than the modules comprised in a specialized platform in function. The only defect is the deficiency of interconnection between different tools, which is prone to be conquered by some means.

Relative independence: because of the abundance and diversity of web2.0 tools, we may find different tools providing us with similar functions. It means that we may have several choices to fulfill the same work in our blending learning. In other words, the web2.0 blending learning model is independent from some given tool or platform. Instructors or students are prone to use different tools for instruction or learning.

Additionally, this model has many mutual advantages of all blending learning models as well -- for example, the instructor and students tend to acquire more interactions in blending learning than in traditional classroom training; the F2F instruction and online learning can reinforce learning effect and supplement deficiency for each other etc.

3 An Experiment

In accordance with the model, an experiment last for 1 semester has been carried out in our university, and its effectiveness was proved positive.

3.1 Performance of the Experiment

Considering that the instruction is supposed to make full use of online interactivities, so we chose the course of photography as our instructional contents, which may need lots of discussions or team work. Two similar class of the same grade were selected as our experimental subject -- one was fulfilled traditional F2F training, and the other was fulfilled web2.0 blending learning.

In the process of web2.0 learning, several web2.0 tools were applied to our work.

The first one is a picture blog website known as HaoKanBu, in which we accommodated the instructional contents and photo cases for learning and discussion.

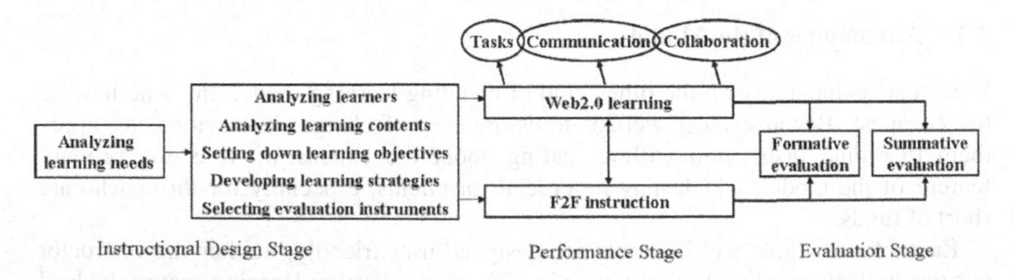

Fig. 3. Blending learning procedure

The second is YouTube, the most popular video website, in which we held some instructional videos.

The third is a web bookmark sites called Delicious, in which the instructor registered useful hyperlinks to other web resources for students' consultation.

An Rss tool called FriendFeed has been used for subscribing materials dispersed in HaoKanBu and YouTube and an instant message tool called QQ and Emails were used for exchanging ideas between the instructor and students.

The performance of the experiment complied with the model aforementioned completely. Meanwhile, the instructor has assigned independent shooting tasks for students to inspire their study, asked them to carry out online learning and discussion of cases, provided online direction for both learning and discussion, and joined the other online interactivities.

Finally, a classroom lecture or summarization was held to finish a learning unit.

Sometimes the instructor may ask students to turn in their photo works or commentary papers on some cases. There are several ways are capable of achieving that objective: (1) Students send their works to instructor via Emails or instant messages. (2) Students publish their photos or papers on their own blog pages, and then the instructor visit these pages to score them. (3) Students upload their works to a given mutual net space, for example: a network hard disk or an FTP server. (4) Students turn in their works in classroom face to face.

3.2 Result of the Experiment

After the experiment, we have compared the learning effect of both classes from different viewpoints.

Firstly, a final exam was carried out, which showed that students of the two different classes grasped the course knowledge equally well. It means that the web2.0 blending learning is not better than traditional F2F instruction absolutely in effectiveness; at least it appeared that way.

But a questionnaire to students of both classes revealed more information that proved the value of web2.0 blending learning model.

First and foremost, students in the class fulfilled web2.0 blending learning were more active in participating online interactions, and their learning interest on the course was maintained higher than students of the other class.

In the second place, most students of the blending learning class agreed that the new learning model had improved learning efficiency, enhanced understanding, broaden their vision on the subject, and strengthened communication between students and the instructor.

In the third place, most students thought web2.0 tools were easier to use than a specialized online learning platform or a learning management system.

In the fourth place, many students appreciated the weekly classroom instruction which was blended with continual web2.0 learning. They said that the F2F instruction could solve some problems which could not be solved via online communication and spur them to keep pace with instructional scheme. Additionally, they thought the F2F instruction could make the learning activities more real and concrete; the online learning environments are virtual and isolated after all, which may make them feel lonely and depressed.

According to these facts, the model is proved effective.

4 General Guidelines

4.1 Blend Rather Than Duplicate

The most common mistake made by instructors is to duplicate their lecture contents and classroom activities to web2.0 learning environments.

Students may feel boring to study the same contents or participate in the same activities twice, and they will lose interest and passion to study soon. Hence, careful consideration should be made to ensure the activities and learning contents of each setting are different, and the instructor should blend rather than duplicate them in each.

4.2 Select Proper Tools Just Enough for Work

Through our study and experiment, we found that some web2.0 tools are likely to surpass the other ones in some functions or details, which may facilitate instructors' work or students' study. Accordingly, tools applied to web2.0 learning are supposed to be selected carefully by the instructor and recommended to students, or the following work will be harder.

Furthermore, the tools to be used had better be just enough for the work. Superabundant tools may cost more time of both instructors' and students', and mislead their attention to the tools themselves instead of the course contents. Therefore, instructors should control the number of tools that to be applied to their instruction strictly.

4.3 Find the Right Balance

Some students who have participated in the experiment addressed repeatedly that the web2.0 learning activities cost their more time in study than before. Though the learning via web2.0 tools is more interesting and attractive, without clear requirements or detailed recommendations they had to visit those websites for learning up to themselves frequently. As a result, they spent more time on study per week generally, especially when there are some tasks or discussions to be fulfilled online. They often get exhausted in participating online activities and preparing works or assignments for class and the like.

Hence, the instructor is supposed to direct students to find the right balance in their study activities, for both web2.0 environments and classroom settings, or the learning manner will be worthless.

4.4 Avoid Excessive Burden

Supporting web2.0 learning besides F2F instruction may double the workload for instructors, especially when there are not appropriate instructional schemes.

Some instructors may be keen on task driven learning or collaborative learning. They often assign several tasks to students simultaneously regardless of the work to be performed online by themselves.It will turn to heavy burdens for them soon, and then they will lose interest and motivity of applying this model to their courses.

5 Conclusions

As a brand-new instructional model, web2.0 blending learning emerges numerous advantages and immense value for schools, instructors and students.

Though the instructor and students may face challenges in varied areas such as technology, logistics, organization, management and the like participating in web2.0 learning, they would not deny the attraction of the model.

More and more professionals and instructors are supposed to make use of web2.0 tools as substitutes for specialized online learning platforms, and perform their online instruction without caring about the expense. The web2.0 blending learning model is sure to be popular in schools in the future.

References

1. Driscoll, M.: Blended Learning: let's go beyond the hype. E-learning, 54 (March 2002)
2. Singh, H.: Building Effective Blended Learning Programs. Educational Technology 43, 51–54 (2003)
3. Osguthorpe, R.T., Graham, C.R.: Blended Learning Systems: definitions and directions. Quarterly Review of Distance Education 4, 227–234 (2003)
4. Cheng, R.: Research of blending learning making use of web2.0 tools. In: 2009 WASE International Conference on Information Engineering, vol. 2, pp. 466–469 (July 2009)
5. Henrich, A., Sieber, S.: Blended learning and pure e-learning concepts for information retrieval: experiences and future directions. Information Retrieval 12, 117–147 (2008)
6. Stacey, E., Gerbic, P.: Teaching for blended learning–Research perspectives from on-campus and distance students. Education and Information Technologies 12, 165–174 (2007)

A Normalization Method of Converting Online Handwritten Chinese Character to Stroke- Segment-Mesh Glyph

HanQuan Huang, Min Lin, and MaiKu Zhang

Computer & Information Engineering College, Inner Mongolia Normal University, Huhhot, China
hhq04012412@163.com, cieclm@imnu.edu.cn,
20094019001@mail.imnu.edu.cn

Abstract. Based on Stroke-Segment-Mesh model, a method of normalization converting online handwriting Chinese characters to Stroke- Segment-Mesh Glyph is presented. Firstly, normalize the key points of every stroke. And then normalize relative position relationships of strokes. Using the method of normalizing sequential key points improves the normalization method of handwriting Chinese character, and thus figure out the problem of stroke segment translocation when normalizes complex strokes. The method of normalizing relative position relationships of strokes is to ensure relative position relationships consistency among the strokes of mesh glyph and the handwritten Chinese character by moving, extending or shortening the corresponding strokes. Experimental results indicate that the presented method has much better converting effect and sets foundation for stroke- segment-mesh glyph recognition and the Chinese characters quality evaluation.

Keywords: Handwritten character, mesh glyph, relative position relationship, normalization.

1 Introduction

Stroke-segment-mesh consists of N*N grid small squares. A grid small square contains 18 stoke segments which are the four sides of square and two diagonals, the each of four vertexes of square lines to the each of corresponding midpoints of the opposite sides. Stroke segments are divided into horizontal strokes, vertical strokes and oblique strokes (as in Fig.1). We can see from the Fig.1, line segment (basic stroke) with a specific stroke type ("heng" "shu" "pie" "na") and specific direction(the specific angle of line segment) can be combined into arbitrarily complex strokes and the Chinese characters. Stroke-Segment-Mesh Depiction of Chinese Character Glyph [1] (mesh glyph) can depict input of any glyph and computation for comparing either the whole or part of glyphs. We have two methods for inputting Chinese character glyph. One is depiction on the stroke-segment-mesh; The other is written in the writing tablet. However, due to restrictions of the stroke-segment-mesh model, the former method is not natural nor convenient. The second method is more conforming to writing habits,

A. Xie & X. Huang (Eds.): Advances in Electrical Engineering and Automation, AISC 139, pp. 213–219.
springerlink.com

so writing is natural and convenient and efficient, just can make up for the deficiency of the former method.

The purpose of the paper is to find an effective normalization method converting the handwritten Chinese character to Stroke-segment-mesh glyph , and to retain as much of the original features of the handwriting character, including error information and non-standard features, the most important feature is the relative position between strokes, and lay the foundation for the Stroke-segment-mesh glyph recognition and the Chinese characters quality evaluation.

2 The Description of the Normalization Method

According to the feature of the Stroke-segment-mesh glyph, the normalization method is presented. The first step is that normalize each of strokes in sequence of writing order. The next step is that normalize the relative position relationships of strokes by moving, extending or shortening relative strokes, in order to ensure relative position relationships consistency among the strokes of mesh glyph and the handwritten Chinese character.

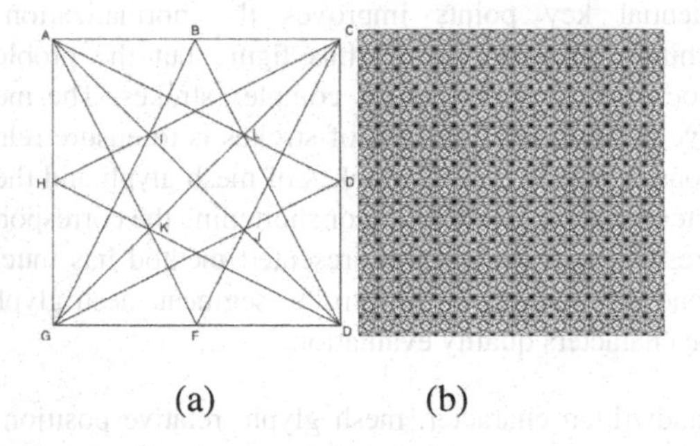

(a) (b)

Fig. 1. Stroke-segment-mesh model. (a) grid small squares. (b) Stroke-segment-mesh.

2.1 Corresponding Definitions

For the sake of the description more efficient, the corresponding definitions are as follows:

Definition 2.1. A stroke is defined as s<start, K, end>. The s is the name of the stroke; start is starting point of the stroke and end is ending point of the stroke, they are called endpoint. the key points set represented as K which contains endpoint set represented as E and character point(s) set represented as C, if C is nonempty set, each element of the set is different.

A stroke is either simple stroke or complex stroke.

Definition 2.2. If C is empty set, s is a simple stroke which represented as simple_stroke< s<start, end>>, or SS<start, end>.

Definition 2.3. If C is non-void set, s is a complex stroke which represented as:complex_stroke<SS_1<start,c_1>,SS_2<c_1,c_2>,…,SS_n<c_{n-1},c_n>,SS_{n+1}<c_n,end>>, c_1,…, c_n are the elements of the C.

Definition 2.4. The vertex point and midpoint of grid small square are called valid point.

Hypothesize that there are two strokes, s_i<start1, C_i, end1> and s_j<start2, C_j, end2>. Their key points sets are the K_i and K_j.

Definition 2.5. Meet point: If s_i and s_j intersect a point, and this point is one of elements of the K_i or K_j, this point is meet point.

Definition 2.6. s_j meet with s_i: If meet point is one of elements of K_j, s_j meet with s_i.

Definition 2.7. s_j and s_i meet with each other: If s_j meet with s_i and s_i meet with s_j, s_j and s_i meet with each other.

Definition 2.8. s_j disjoint s_i: disjoint point is one of elements of K_j and this point to s_i distance is shortest distance. If C_i is empty set, the line segment that disjoint point belongs to is non-parallel s_i and all elements of K_j are on the same side of s_i; If C_i is nonempty set, and there is a simple stroke s_k which belongs to s_i and the line segment that disjoint point belongs to is non-parallel s_k and all elements of K_j are on the same side of s_k. We call s_j disjoint s_i.

Definition 2.9. s_j and s_i disjoint each other: If s_j disjoint s_i and s_i disjoint s_j, s_j and s_i disjoint each other.

Definition 2.10. Moved stroke and reference stroke: If meet point or disjointed point is one of elements of K_j, s_j is moved stroke and s_i is reference stroke, or confirm the reference stroke and moved stroke in sequence of writing order.

Definition 2.11. Cover joint point: If three strokes at least intersect at the same point, the point is defined as cover joint point, these strokes are the relevance strokes of the point.

2.2 Stroke(s) Normalization

Normalize a stroke is to point to put the stroke's key points on The vertex point of grid small square.

The complex stroke normalization method of [2] is that normalized the endpoints firstly and then other key points. It is possible that the method bring about the problem of stroke segment translocation(Fig.2) because key point(s) is(are) not vertex point. In order to figure out the problem, it is necessary each of key points is normalized to vertex point because only the vertex point can line to other valid points. Using method of normalizing sequential key points improves the normalization method of handwriting Chinese character, and thus figure out the problem. The basic idea is stroke's direction is invariable and its length is variable. First, normalized start point to the nearest vertex point of stroke-segment-mesh. And then according to the direction and length of the line which belongs to the point to normalize the next key point, and so on.

2.3 Description of the Relative Position Relationships of Strokes

The relations stable principle of Chinese characters[4] point that in Chinese character scripts, the attributes(such as direction, length, position) of stroke primitives are not stable, but the relations between strokes primitives are stable, the relations between primitives, which are the main part of Chinese character scripts information, reflect the cssential of Chinese character scripts. According to the principle, it is necessary to ensure relative position relationships consistency among the strokes of mesh glyph and the handwritten Chinese character.

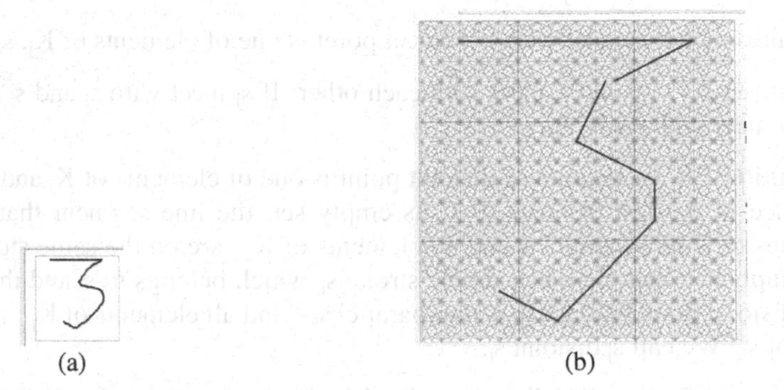

Fig. 2. Stroke segments translocation. (a) handwritten character. (b) corresponding mesh glyph

According to the references[3-7] involved the research of the relative position relationships of strokes, the relative position relationships of strokes are divided into two types. One, according to intersection relationship, is disjoint(l), meet(a), intersection(j). The other is according to the projection position of moved stroke's end point on the reference stroke.

(1). Disjoint is divided three types, disjoint point are start(s), end(e) and character point(c). Similarly meet is divided three types, meet point are start, end and character point. Projection position of Disjoint point, meet point or intersection point on the reference may be start, end, middle point(m), left point(u) which is between start and middle point, right point(d) which is between middle point and end.

(2). According to the relations stable principle of Chinese characters, the length of stroke is not stable, it is rarely effect to many Chinese characters recognition, for example, changed the length of the Chinese character's ("feng") third "heng stroke". But taking into account point of view of writing standardization Chinese character, the character("feng") is not a standardization character if the two stroke's length is longer than three strokes. The non-standardization feature should be saved to the Chinese characters quality evaluation. For the better relative position relationships of strokes, the projection position of the end points of moved stroke should be taken into account. They are the start, end, middle point, left point and right point of the reference stroke.

The relationships(relationship string) represented as a string consisted 6 letters. It is as the follows:

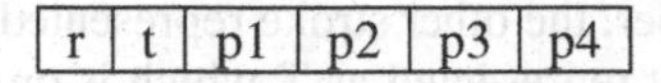

r	t	p1	p2	p3	p4

r represented as relationships type and r∈{l, a, j, #} ; t represented as the type of meet point or disjoint point and t∈{s, e, c, #} ; p1(p2) represented as the position of meet point or intersection point on the reference stroke(moved stroke) or the projection position of disjoint point on the reference stroke(moved stroke); p3(p4) represented as the start(end) of moved stroke on the reference stroke. p1, p2,p3,p4∈{s, m, u, d, e,#}. "#" represented as nothing. r, t, p1 and p2 confirm the position of disjoint point, meet point or intersection point. p3 and p4 confirm the position of moved stroke's end points.

For example, "asu##u" represented as two strokes meet, meet point is the start of moved stoke and left point of the reference stroke, the end of moved stroke is left point of the reference stroke. "j#umud" represented as two strokes intersect, intersection point is the left point of reference stroke and the middle point of moved stroke, the projection position of the start(end) of moved stroke is the left point(right point) of the reference stroke.

Firstly, according to the corresponding information of disjoint point, meet point or intersection point confirm r, t, p1, p2. if disjoint and meet, p2 is '#'. And then according the following rules to confirm p3 and p4. (1)count the projection position of the other end point on the reference stroke if disjoint point or meet point is end points. (2)count the projection position of the two end points on the reference stroke if disjoint point or meet point is character point or two strokes are parallel. r, t, p1 and p2 are '#' and according to (2) to confirm p3 and p4 if two strokes are parallel.

The relative position relations of the handwritten strokes are stored in a two-dimensional array represented as RB[n][n], n is the number of simple line segments, the row represent sequence number of moved strokes; the column represent sequence number of reference strokes. As an example, RB[i][j]= "asu##u", this indicates that s_i meet s_j, meet point is the start of moved stoke and left point of the reference stroke, the end of moved stroke is left point of the reference stroke.

2.4 The Relative Position Relationships of Strokes Normalization

Because the relationships of original handwritten character strokes may be changed after stroke(s) normalization, it is necessary to normalize the relationships.

For Chinese character which are whole-surround structure, only just move the associated strokes can guarantee a consistent relationship of intersection and meet; otherwise extend or shorten the related strokes is essential. In order to normalize relationships more convenient, move associated strokes depending on the first four letter of relationship string at the first round of circulation; at the second round of circulation, extend or shorten the stroke-related depending on the whole relationship string. To determine whether the strokes in the final position, with a fixed stroke type confirmType that is initialized to 0, that has not been fixed, the other value indicates the location of stroke has been fixed.

The step of moving a stroke as follows:

Step1. get the stroke represented as s_i which will be moved depending on their confirmtype and order number, the other stroke represented as s_j.

Step2. get the target point represented as T which is on the s_j depending on the relationship string represented as r.

Step3. get the moving distance of s_i represented as D: if r[0]= 'a' or r[0]= 'l', get the meet point or disjoint point depending on r[1]; otherwise, get a point of s_i depending on r[3]. The point gotten represented as P. D=T-P.

Step4. move the s_i a distance of D, and assign 1 to its confirmType if it is not 1.

Step5. move the stroke(s) which is(are) the relevance strokes of the cover point joint except s_i and s_j to T

The step of normalize relationships as follows:

Step1. get the relationships stored in a two-dimensional array represented as RB.

Step2. assign 1 to all strokes' confirmType.

Step3. confirm the first stroke.

Step4. move stroke(s):

Step4.1. When the i-th stroke is normalized, find relationships among the stroke and front of i-1 strokes.

Step4.2. move relative stroke(s) depending on the front four letters of the relationship string, and assign 1 to their confirmType.

Step4.3. jump to Step5. if all strokes are normalized, otherwise, jump to step4.1.

Step5. extend or shorten stroke.

Step5.1. assign 0 to all strokes' confirmType.

Step5.2. When the i-th stroke is normalized, find relationships among the stroke and front of i-1 strokes, and assign 1 to their confirmType.

Step5.3. extend or shorten relative stroke(s) depending on the front four letters of the relationship string.

Step5.4. jump to Step6. if all strokes are normalized, otherwise, jump to step5.2.

Step6. normalize end point position of strokes depending on last two letters of relationship string.

3 Experiment and Result

700 handwritten Chinese character of 70 Chinese character are selected depending on the structure and number of strokes of Chinese characters, a part of experimental result is illustrated in Fig.3. The structure of Chinese characters is divided into the following categories: Left-Right Structure, Left-Middle-right structure Top-Bottom Structure, Top-Middle-Bottom structure, Single-body structure, Four-part or multipart structure, Encircling or half-encircling structure. The number of strokes can be divided into three categories :1-10 ; 11-20 ; more than 21 . 10 characters are selected whose number of strokes are different in the each of structure. Each of the different characters are written five times by two peoples. So there are a total of 700 handwritten Chinese characters. The experimental results indicate that this method can better maintain the relationship and get a higher goodness of fitting.

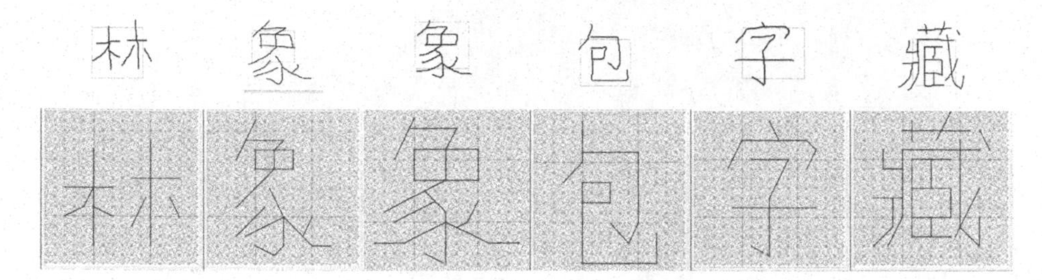

Fig. 3. A part of experimental result

4 Conclusion

The method of stroke(s) normalization figure out the problem of the stroke segment translocation when normalizes complex strokes. Relationship normalization resolves basically the distortion of relationship and get a higher goodness of fitting of relationship. The experimental results indicate that this method can better maintain the relationship and get a higher goodness of fitting. The work make the input of mesh glyph more convenient and efficient; and maximize the retention of the original features of handwritten character; and lay the foundation of mesh glyph recognition and the Chinese characters quality evaluation.

Acknowledgments. The preferred Project Supported by National Natural Science Foundation(60863007).

References

1. Lin, M., Song, R.: Stroke-Segment-Mesh Depiction of Chinese Character Glyph and Algorithm for Glyph Comparing. Journal of Computer-Aided Design & Computer Graphics 21(9) (September 2009)
2. Zhao, H., Lin, M.: Character Glyph Analysis Oriented On-line Handwritten Chinese Character Stroke Recognition. Journal of Inner Mongolia Normal University (Natural Science Edition) 37(6) (November 2008)
3. Allen, J.F.: Maintaining Knowledge about Temporal Interval. Communication of the ACM 26(11), 832–843 (1983)
4. Wang, K., Wang, Y.: The relations stable principle of Chinese characters script. Journal of Chinese Information Processing 10(4), 24–31 (1996)
5. Tan, C.K.: An Algorithm for On-line Strokes Verification of Chinese Characters using Discrete Features. In: Eight International Workshop on Frontiers in Handwriting Recognition, Niagara-on-the-lake, Ontario, Canada, pp. 339–344 (2002)
6. Zhang, W.-Y., Sun, X.-M., Zeng, Z.-B., Wu, J.-Z.: Automatic Generation of Mathmatical Expression of Chinese Characters. Journal of Computer Research and Development 41(5), 848–852 (2004)
7. Lin, M.: The Study of Formal Description of Chinese Character glyph and Application, pp. 89–93. Beijing University of Technology, Beijing (2009)

Fig. [illegible]. A part of experimental result

4 Conclusion

The method of stroke(s) normalization figure out the problem of the stroke segment translocation when normalizes complex strokes; Relationship normalization resolves basically the distortion of relationship and get a higher goodness of fitting of relationship. The experimental results indicate that this method can better maintain the relationships and get a higher goodness of fitting. The work make the input of mesh glyph more convenient and efficient, and maximize the retention of the original features of handwritten character, and lay the foundation of mesh glyph recognition and the Chinese characters quality evaluation.

Acknowledgements. This project: Project Supported by National Natural Science Foundation(60863007).

References

1. Lin, M., Song, K.: Stroke-Segment-Mesh Depiction of Chinese Character Glyph and Algorithm for Glyph Computing. Journal of Computer-Aided Design & Computer Graphics 21(9) (September 2009)
2. Chao, H., Lin, M.: Character Glyph Analysis Oriented On-line Handwritten Chinese Character Stroke Recognition. Journal of Inner Mongolia Normal University (Natural Science Edition) 37(6) (November 2008)
3. Allen, J.F.: Maintaining Knowledge about Temporal Interval. Communication of the ACM 26(11), 832–843 (1983)
4. Wang, K., Wang, Y.: The relations stable principle of Chinese characters script. Journal of Chinese Information Processing 10(4), 24–31 (1996)
5. Tan, C.K.: An Algorithm for On-line Strokes Verification of Chinese Characters using Discrete Features. In: Eight International Workshop on Frontiers in Handwriting Recognition, Niagara-on-the-lake, Ontario, Canada, pp. 339–344 (2002)
6. Zhang, W.-Y., Sun, X.-M., Zeng, Z.-B., Wu, J.-Z.: Automatic Generation of Mathematical Expression of Chinese Characters. Journal of Computer Research and Development 41(5), 848–852 (2004)
7. Lin, M.: The Study of Formal Description of Chinese Character glyph and application, pp. 89–91. Beijing University of Technology, Beijing (2009)

Housing Problem Research for the College Graduates Based on Interim Model–The Case of Shanghai*

Xiaolong Hu[1], Xuezhen Wang[2], and Zhihao Lin[3]

[1] Institute of Real Estate, Shanghai University,
Shanghai, China
huxiaolongsh@yahoo.com.cn

[2] Ningbo Institute of Material Technology &Engineering, CAS,
Ningbo, China
duozhiyin@shu.edu.cn

[3] Shanghai 3rd Reeducation through Labor Management Station,
Shanghai, China
linzhihao01@sina.com

Abstract. The college graduates is the main power of improving Chinese modernization. But most of them are unable to obtain a satisfactory living environment when they make important economic contributions. On the other hand, Chinese Housing-for-all program is designed for urban residents who have local census register and pay for the enough social security, most of the graduates can not match them. How to resolve this problem is very important. This paper uses shanghai's data as an example. The survey found the housing problem of college graduates in Shanghai, propose college graduates interim housing model

Keywords: college graduates in Shanghai, housing problems, housing security, policies, housing interim mode.

1 Introduction

The college graduates, as the main power of improving modernization in shanghai, couldn't afford the rising price of apartments. Moreover, a large number of college graduates are outside the Housing-for-all program which is designed for urban residents in Shanghai. Because these graduates have no census register of shanghai and do not pay for the enough social security. Therefore, most of them are unable to obtain a satisfactory living environment when they make important economic contributions to shanghai. This kind of situation seriously damages the identity of this city for the graduates, also it will be inevitably bring about a series of social problem, which led to the city of an unstable social order. So, research for the apartment problem of college graduates in shanghai is not only the need of the city's development but also the need of improve people's livelihoods and promote social harmony.

* Humanities & Social Science Planning Fund of MOE: "The research for the new provide mode of Chinese public rent apartment", No: 10YJA630065.

Real Estate Institute of Shanghai University project : The research of the housing security policies for the low-income college graduates at shanghai.

A. Xie & X. Huang (Eds.): Advances in Electrical Engineering and Automation, AISC 139, pp. 221–227.
springerlink.com

This paper investigated the college graduates' housing situation and Questionnaire is the main means, taking direct questionnaires, web surveys and field visits in three ways at the same time, focus on analysis of the housing situation of college graduates, living patterns, and internal and external factors [1]. Specifically, the object of this research - college graduates in Shanghai is not participated in work groups, including college students, bachelor (full time), master (full-time). This is a very important part of the population in common is that graduation is basically no deposit.

2 The Basic Status of Housing of College Graduates in Shanghai

321 questionnaires are designed and distributed, 267 of them are valid. The site visits, Web searches, phone interviews are also used for investigation. The results are as following:

1. The per Capita Area is Small, The Rent-Based

According to the survey data, per capita housing area of 45% respondents is less than 30 m2. 53 % respondents rent the apartments, and 64% of them are shared rent groups. Through investigation, Most of them rent instead of buying apartments, because the income level can not match the rising prices of apartments. Relative to the purchase, the financial burden of renting is light, and it can adapt to strong liquidity of the work of college graduates who recently graduated.

2. The Economic Burden of Housing is Large

Survey of college graduates in Shanghai focused on the persons that their annual income is less than RMB 50,000 (79%), 48% of the respondents' annual income is less than RMB 20,000. In the rental group, there are 45% person that will pay 30% income for their rent. In the purchase group, loan ratio of home buyers(77%) is 70% - 80% and more than 37% of them repay the pressure index above 0.5.

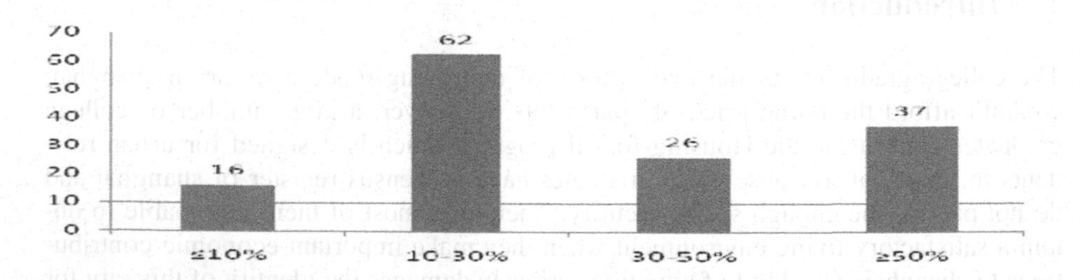

Fig. 1. Rent Income Proportion the rent of respondents

3. Low Level of Residential Satisfaction

A survey shows that most college graduates aren't satisfied with their residential conditions, especially about habitable area and traffic convenience. 34% of surveyed people think their habitable area is generally low and 32% think the traffic is inconvenient (58% of surveyed people spend 1-3 hours to commute every day).

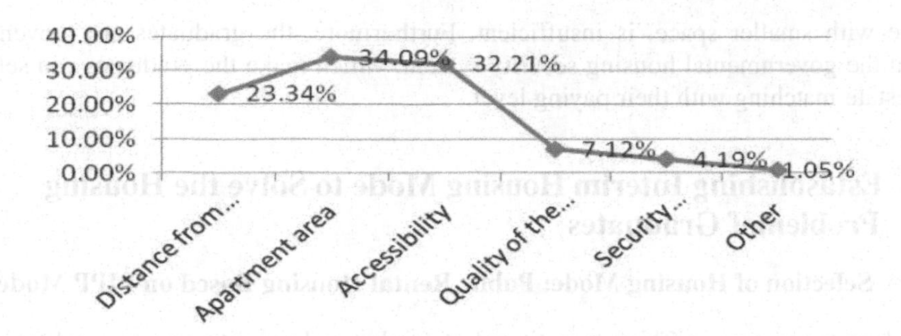

Fig. 2. Rent Income Proportion the rent of respondents

3 Analysis of the Issue of House Buying and Renting of College Graduates in Shanghai

A classification analysis of research results by clustering method shows that almost every group of people, no matter how high or how low their income is, have their own ways to solve the housing problem. [2]To solve their housing problem, people with high-income or come from a wealthy family buy houses, native people with middle income live with their parents to avoid rent spending, while migrant people with low- and middle-income usually face both the high pressure of large amount of rent and low level of housing condition. The main reasons for this include:

1. A Coverage Lacking of Housing Security System in Shanghai

According to the average wage in shanghai, quite a large proportion of college graduates in shanghai belong to low- and middle-income people. However, the house security system fails to include college graduates, especially those without a local census register, into the system, while public rental housing, which includes the college graduates, just started and has not yet provided any real accommodation houses for them.

2. Bad Employment Situation for College Graduates in Shanghai

The continuous increasing population of college graduates makes a surplus of supply of college graduates than the demand in the labor market. Global economy downturn affects the growth of China economy, and thus slows the demand increasing for college graduates. Meanwhile, Shanghai, as a fast-developing metropolitan, draws more and more college graduates from all around China. All these dim the employment situation for college graduates in Shanghai and cause the wage paid to college graduates remain stagnant, a sharp contrast to the sharply-increasing house price and rent in Shanghai.

In summary, on one hand, graduates in Shanghai who have just start working have a weak purchasing power and there are differences in their consumption concepts. On the other hand, housing price in Shanghai has already been at a high level and is rising at a rapid speed which is much faster than the rising speed of the graduates' income. In addition, the supply of estates catering the graduates, e.g. condominium at low price with smaller space, is insufficient. Furthermore, the graduates are prevented from the governmental housing security system, which make the graduates can select no estate matching with their paying level.

4 Estasblishing Interim Housing Mode to Solve the Housing Problcmof Graduates

4.1 Selection of Housing Mode: Public Rental Housing Based on PIPP Mode

The housing system in China, constituted of market and security, can be used for reference in the research of the housing problem of graduates. As is analyse above, most of the graduates are in the middle-and-low-income groups as soon as they leave school and then, with the promotion of proficiency and post, their income level will climb and they themselves will step in middle or even high-income groups. Thus, the greatest difference between graduates and those others who need social security lies in that the graduates have good capacity of self-improvement and that they just need help for temporary troubles. At this moment, it is required that the government and the society offer the graduates some help of transition. Under such train of thought, this paper holds that public rental housing is the solution. Because public rental housing can provide the youngsters, who have just started their careers, with vastly covered dwelling which can be rented for long term at a comparatively low price. This policy can help the young get over the period of saving for purchasing estate.

At present, the construction of public rental housing in Shanghai is operated under the marketing system: with part of the capital, resources and preferential policy provided by the government, specialized institutions invest, operate and manage the public rental housing, according to the marketization standards and the principle of breaking even with little profit and aiming at the sustainable operation[3].

Considering the problems of the public rental housing in cities in China, such as insufficiency of supply, shortage of capital and difficulties in successive management, this paper holds that it is ought to build youth apartments for graduates to meet the housing demand of them and to form a long-term feasible virtuous cycle which meets the interests of different parties. The so-called youth apartment is run under the PIPP mode of public rental housing supply, which is based on the practical capacity of public rental housing supply of the local authority in Shanghai, macro-controlled by government, constructed by private section and coordinated by mid-organization. There are different specific modes, including purchase-transfer-operate, construct-rent-transfer, transfer-operate, etc. The purchase-transfer-operate mode means that government purchase the estates, which are being constructed by private organizations or are finished for sale, fitting for public dwelling, at a negotiated price, which is certainly lower than the price for individual buyers. Then the estate will be transferred to mid-organizations, who will arrange the renting and transfer of the estate reasonably. This mode is a good choice for Shanghai, a city confronted with the scarcity of land. The construct-rent-transfer mode refers to that private organizations rent the public housing to government after finishing construction. The estate is operated and managed by the mid-organization and the ownership does not belong to the government until the end of the authorizing period. This mode can release the financial pressure of the local authority on the building of public dwelling.[4]

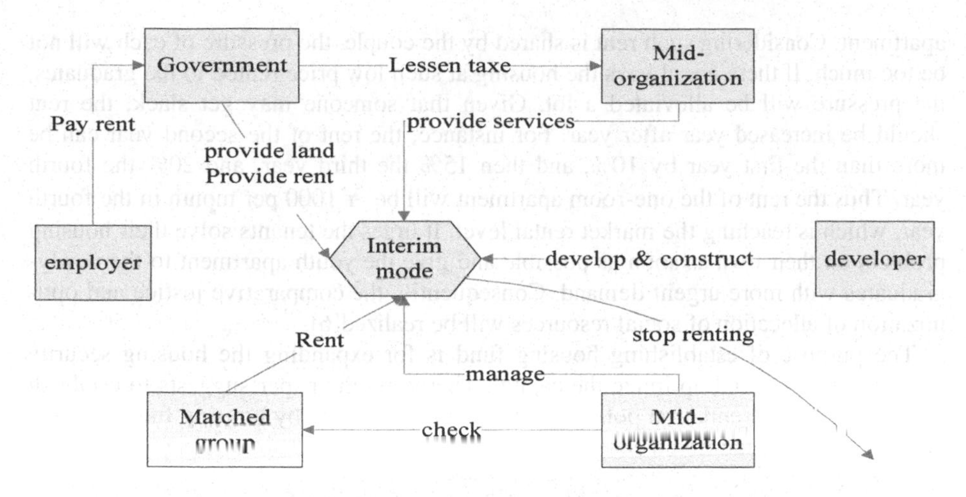

Fig. 3. Relationship of Interim Mode Based on Public Rental Housing

4.2 The Different Youth Apartment rent

The youth apartment is a kind of public rent apartment, which is provided for the low-income college graduates. So the rent should be considered whether the tenants can pay for and social funds want to attend or not. The government should use all kinds of policy instruments to support it and handle the relationship among market rent, cost rent and afford rent. In other words, the rent of youth apartment is decided by the marketing rent, government finance, tenants' income and other objective factors. The different rents are fit for all kinds of tenants. The model is as following [5]:

$$\sum_{t=0}^{T}((\sum_{j=1}^{J} KP_j S_j - C)_t)[1+(f_0+I)]^{-t} = 0$$

Therein, *T* stands for the operation time of youth apartment, *J* stands differentiation demand modulus of elasticity for college graduates, *Kj* stands differentiation rent level modulus within the same area, *Pj* stands demands modulus of elasticity for the J's rent, *Sj* stands the demands modulus of elasticity for the J's apartment area, *C* stands the cost of apartment, *f0* stands the cost of funds for investor, I stands the rate of return for the youth apartment.

According to the result of survey and the statistics of file, it is appropriate that the rent take up 10%~15% of the income. As the lowest income per graduate is no more than ￥2000 (esp. those who start their career within one year), the proper rent should be ￥200~300 per graduate. It is certain that different type of estate requires different rent. The condition of dwelling co-rented by two people is comparatively the worst and the space of such type is the smallest, so the rent is comparatively low. The rent of the dwelling providing individual room and washroom is comparatively high, i.e. ￥400 per month. An apartment of 40~50m2 is enough for the temporary demand of newly-weds and the rent can be higher, i.e. ￥600~700 per month for a one-room apartment. Considering such rent is shared by the couple, the pressure of each will not be too much. If there are always the housing at such low price rented to the graduates, the pressure will be alleviated a lot. Given that someone may get slack, the rent should be increased year after year. For instance, the rent of the second year can be more than the first year by 10%, and then 15% the third year, and 20% the fourth year. Thus the rent of the one-room apartment will be ￥1000 per month in the fourth year, which is reaching the market rental level. It urges the tenants solve their housing problem on their own as soon as possible and give the youth apartment to the newly-graduates with more urgent demand. Consequently, the comparative justice and optimization of allocation of social resources will be realized[6].

The purpose of establishing housing fund is for expanding the housing security policies' function and optimize the capital. Therefore, the paper suggests to establish some kind of differentiation policy that the rent can be paid by housing fund. So that, all kind of saving can find a correct consumption mode.

4.3 Reasonable Selection of Housing Environment, Site and Rent

The present housing demand of the graduate is the basic dwelling demand instead of the improving one. Given that graduates mostly have low income and remain single or keep a family of two, it is required to provide them with a diversity of dwelling. For instance, the site of public dwelling should be separated in different locations to cut the commuting cost of graduates as much as possible.

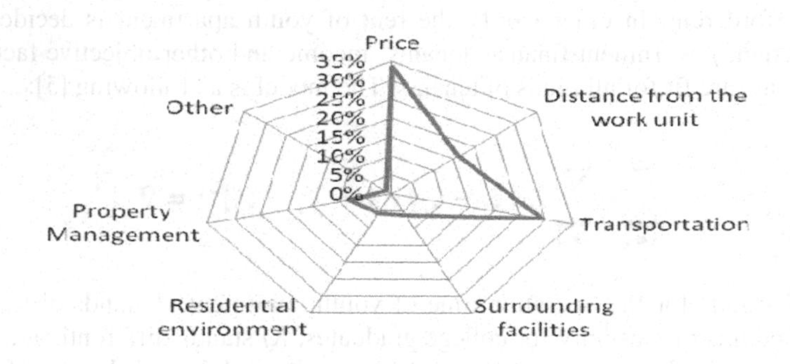

Fig. 4. Location factor

4.4 Reasonable Selection of Housing Environment, Site and Rent

Having graduated not long, without marriage, most of the graduates are still single, so they can apply for public dwelling as individuals rather than families. The department in charge may check the applicants of the aspects, such as: Diploma, Native or not, Number of years set for career, Monthly income.

The real estate administration is supposed to check the tenants' situation regularly. One can be disqualified of the entry of youth apartment, if he/she behaves like following:

1. Lend or rent the rental apartment to others
2. Change the use of the apartment without permission
3. Have not lived in the rental apartment for more than 6 months successively

As youth apartment is supported by governmental subsidy and has taken up a lot of social resources, the tenants should agree with the local authority on working there for a while to contribute to the local economic development.

References

1. Yang, D., Ma, F.: The research of housing problem for HangZhou's college graduates. Co-Operative Economy & Science, 126–128 (2008)
2. Zhang, J.-K., Yao, Y.: A research on housing problem of graduates and countermeasures —on the basis of situation in Nanjing. Journal of Southeast University (Philosophy and Social Science) 127, 35–38 (2009)
3. Shao, Z.: The research for the application of PPP new financing model, http://www.fdrc.com.cn/PaperView.aspx?paperID=6cd83bbd-c894-4a61-873d-86cc51b228a6
4. Tian, Y.: The Research of PIPP Pattern in Public Housing Security System (2008)
5. Xu, L., Liang, C., Qin, Y.: The Monthly Runoff Prediction of Zi Ping-pu by Improved Elman Model. Journal of Sichuan University(Engineering Science Edition), 38–42 (2006)
6. Zhuang, Z.: Rent pricing model for pubulic rent house. CO-Operative Economy & Science, 15–19 (2008)

The Impact of Entrepreneurial Learning on Entrepreneurial Performance: The Mediating Role of Entrepreneurial Competencies

Xiaoxia Zhang

School of Economics and Management,
ZhengZhou University of Light Industry
zhangxiaxiamail@163.com

Abstract. Entrepreneurial learning has emerged as an important area of enquiry in relation to both the academic study of entrepreneurship and the practical development of new entrepreneurs, yet it is insufficiently understood area. This paper develops new understanding in this area from entrepreneurial competencies and constructs a tentative framework of entrepreneurial competencies as a mediator between entrepreneurial learning and entrepreneurial performance. This establishes the foundation for the future empirical research.

Keywords: entrepreneurial learning, entrepreneurial competency, performance.

1 Introduction

Entrepreneurship is an inter-related process of creating, recognizing and acting on opportunities, combining innovating, decision making and enaction. To keep the business progressing, an entrepreneur has to assume the role of innovator, manager, mall business owner, division vice president at various stages of business and each role requires unique set of skills whose possession and application translate into a unique learning exercise. In this way, from opportunity recognition to actual creation of organization, the entrepreneur is involved in the variety of learning cycles which further adds to his/her experience. Entrepreneurial learning has emerged as an area of scholarly interest and academic inquiry. Entrepreneurial learning has been explored by a number of researchers including Young & Sexton (1997), Deakins & Freel (1998), Cope & Watts (2000), and Rae & Carswell (2001), yet it is an area which is not well understood (Deakins, 2000).

The existing research, paying attention to the concept, the feature description and its constitution, is superficial, a fraction of the research focuses on the model and the effect of entrepreneurial learning on entrepreneurial psychology and entrepreneurial behavior. However, that is just a preliminary research.

At the same time, in 1973, MacMillan first proposed the concept of the competency, based on the previous studies [1], Chandler and Hanks (1993) put forward the concept of entrepreneurial competencies [2]. It has become a hotspot of many scholars and enterprises to study the characteristics of entrepreneurs by using the methodology applied in conducting research on entrepreneurial competency. Based

A. Xie & X. Huang (Eds.): Advances in Electrical Engineering and Automation, AISC 139, pp. 229–234.
springerlink.com

on entrepreneurial competencies, the majority thinks that entrepreneurs need to combine environment factors with organizational factors to choose the strategy which fit with organizational development and environment change and construct unique enterprise dynamic capability to gain sustainable competitive advantage and get the good performance.

According to the above analysis, the existent researching paradigm is the mechanisms either entrepreneurial learning on entrepreneurial performance or entrepreneurial competencies on entrepreneurial performance. In order to better understand the specifics of this relationship, we have to take a closer look at the processes of entrepreneurial learning and entrepreneurial competencies, based on the analysis, this paper constructs a tentative framework of entrepreneurial competencies as a mediator between entrepreneurial learning and entrepreneurial performance.

2 Entrepreneurial Learning and Entrepreneurial Competencies

2.1 Entrepreneurial Learning

Research on learning processes in entrepreneurial ventures is still in an early stage (Minniti and Bygrave, 2001) [3]. Although the importance of learning process in discriminating successful ventures from unsuccessful ones is widely acknowledged. Entrepreneurial learning is often described as a continuous process that facilitates the development of necessary knowledge for being effective in starting up and managing new ventures.

That is not to disregard the contribution of economics, for although Schumpeter (1934) and Kirzner (1973) observed the importance of learning in the entrepreneurial process [4], Binks and Vale (1990) commented on the limitations of economic theory in understanding the human, sociological and psychological aspects of entrepreneurial behavior [5]. To understand these activities and offer understanding to it, many theorist have offered understanding from variety of angles. Cognitive theorists, which concentrates on the individual acquisition and comprehension of knowledge, has dominated the study of learning, but has limitations in using the metaphor of man as computer as a means of understanding the human mind, the ability to learn, and social interaction. The current research, however, aims to offer understating beyond cognitivism. The cognitivisim, having more of an entitative stance, looking at learning as happening in vacuum and isolation from external factors and predominantly overlooks the role of personal experience. However, the more recent turn within this stream of research is that of social learning which offer an understanding from a social and collaborative process of learning (Rae, 2001, 2004, 2006) [6]. Wenger (1998) developed a comprehensive social and behavioral theory of learning as a transformational process of identity creation, including dimensions of meaning, practice, identity and community [7]. This provides a conceptual foundation for understanding learning that accommodates social participation and human action as well as cognition, enabling advanced learning theory to be applied to the subject of entrepreneurship. Rae (2006) constructed a framework based on Wenger's social theory of learning, emphasizing the creation, recognition and development of opportunities and proposed a framework for entrepreneurial learning that was based on social constructionist, narrative and antecedent theories such as pragmatism. The framework includes three propositions. First, that the development of entrepreneurial identity is the outcome of personal and social emergence. Second, that the recognition and enaction of opportunities in specialized situations is an outcome of a process of contextual learning. Third, that the enaction and growth of a business venture is an outcome of negotiated enterprise.

Tell (2008) built on the notion of the network even more explicitly in his study of university-facilitated monthly learning groups of SMEs [8]. He found that the entrepreneurs used their learning network as a reflective tool, which enabled both single-loop and double-loop learning to take place. This is another moment when the entrepreneurial learning literature comes very close to the insights of the ANT and STS-inspired works of economic sociology, such as e.g. Beunza and Stark's (2004, 2009) [9][10] study of financial models as social devices of reflexivity.

2.2 Entrepreneurial Competencies

A competency is defined as an underlying characteristic of a person, which results in effective and/or superior performance in a job. Entrepreneurs' jobs should not be understood in the traditional sense, instead, as those tasks involved in pursuing and running a new business (Bird, 2002) [11]. The roles and tasks performed by entrepreneurs are those that are relevant for their personal and venture success.

The competency approach has become an increasingly popular means of studying entrepreneurial characteristics. While competency can be studied from its inputs (antecedents to competencies), process (task or behavior leading to competencies), or outcomes (achieving standards of competence in functional areas) (Mole et al., 1993) [12], moreover, in terms of casual relationship, behaviors are closer to performance than other entrepreneurial characteristics like personality traits, intentions or motivations Gartner and Starr, 1993) [13]. According to Bird (1995) [14], competencies are seen as behavioral and observable characteristics of an entrepreneur. Man et al. (2000) held that entrepreneurial competencies included six competency areas: opportunity competencies, organizing competencies, strategic competencies, social competencies, commitment competencies, and conceptual competencies [15].

Entrepreneurial competencies, therefore, include those clusters of related knowledge, attitudes, and skills which an entrepreneur must acquire through managerial training and development to enable him produce outstanding performance, and maximize profit, while managing a business venture or an enterprise. Although the conception of a competency has been used as the guiding principle of analysis (Chandler and Jansen, 1992; Man and Lau, 2000), studies have been mainly oriented to link managerial or entrepreneurial competencies with firm-level performance.

3 The Impact of Entrepreneurial Learning on Entrepreneurial Performance: The Mediating Role of Entrepreneurial Competencies

Man (2002) held that entrepreneurial learning could be seen as the entrepreneurial core competency and could be interpreted using competency approach [16], according to the above discussion, as shown in Fig.1, this paper constructs a tentative framework of entrepreneurial competencies as a mediator between entrepreneurial learning and entrepreneurial performance (EP).

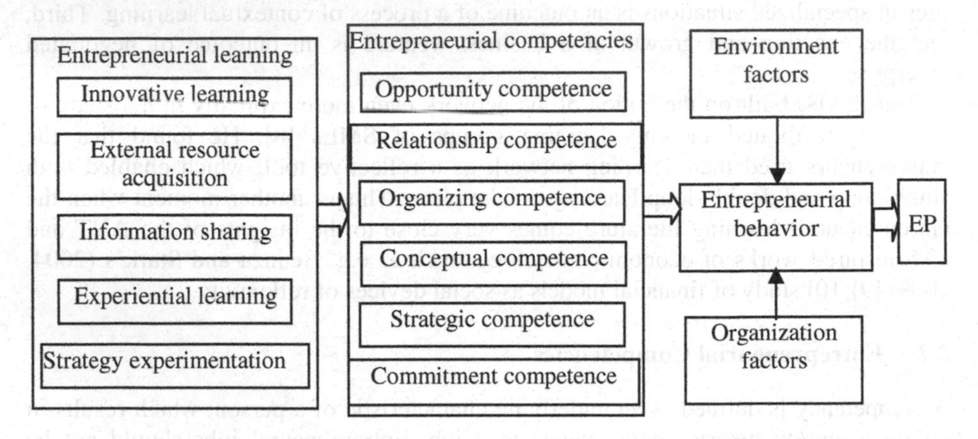

Fig. 1. Impact of entrepreneurial learning on entrepreneurial performance

3.1 Dimensions of Entrepreneurial Learning

Deakins (1998) identified these critical factors of entrepreneurial learning in SMEs: to network in their sector at an early stage, to assimilate experience and opportunity, to reflect on past strategy, to recognize mistakes, to access resources, and to bring in external members as part of the "entrepreneurial team" [17].

Man (2002), based on a competency framework of entrepreneurial learning, using semi-structured interviews to 12 entrepreneurs found that there were six behavioral patterns of entrepreneurial learning, including: actively seeking learning opportunities, learning selectively and purposely, learning in depth, learning continuously, improving and reflecting upon experience, and successfully transferring prior experience into current practices.

Chen Wenting and Li Xinchun (2010), based on the research of Man and Deakins, taking Chinese entrepreneurial firms as research objectives, using Nvivo8.0 software and context analysis method, pointed out the core dimensions of entrepreneurial learning that could be used in the future empirical study, the results found that entrepreneurial learning could be divided into five dimensions: innovative learning, external resource acquisition, information sharing, experiential learning and strategy experimentation [18].

3.2 Dimensions of Entrepreneurial Competencies

Man's (2002) in his paper "the competitiveness of small and medium enterprises: a conceptualization with focus on entrepreneurial competencies", by drawing upon the concept of competitiveness and the competency approach, a conceptual model was developed to link the characteristics of small and medium-sized enterprises' (SMEs') owner–managers and their firms' performance together. The model consisted of four constructs of competitives cope, organizational capabilities, entrepreneurial competencies and performance. Major areas of entrepreneurial competencies are distinguished, including opportunity, relationship, conceptual, organizing, strategic, and commitment competencies. Wang Chongming (2003, 2004, 2006) developed and modified the structure of the entrepreneurial competency in China based on the academic and practical research, and found that the entrepreneurial competency was multi-dimensional including opportunity competence, relationship competence, conceptual competence, organizing competence, strategic competence and commitment competence [19].

3.3 Environment Factors and Organization Factors

In order to obtain sustainable competitive advantage, entrepreneurs need according to the change of external environment adjust the strategy timely. Porter's Five Forces Model and W. Chan Kim's Blue Ocean Strategy in Strategic Management are distinguished methods of environmental analysis. For the new firms, the core environment factors include the business opportunities and the market environment.

Resources are the basis of long-range strategic plans in the enterprises. According to the resource-based theory, the enterprise is the unique combination of visible and invisible resources. Entrepreneurs need to combine environment factors with organizational factors to choose the strategy which fitts with organizational development and environment change and construct unique enterprise dynamic capability to gain sustainable competitive advantage in the competitive environment.

4 Conclusion

All in all, entrepreneurial learning is still a relatively new area of study, and its emergence may owe something to the revival of interest in all things entrepreneurial since the 1980s. At the same time, entrepreneurial competencies were investigated quite intensively in recent years. However, it remains unclear the relationship between entrepreneurial learning and entrepreneurial competencies. The primary focus of this paper is on entrepreneurial competencies as a mediator between entrepreneurial learning and entrepreneurial performance.

References

1. McClelland, D.C.: Testing for Competency Rather Than Intelligence. American Psychologist 28, 1–14 (1973)
2. Chandler, G.N., Hanks, S.H.: Market attractiveness, resource-based capabilities, venture strategies, and venture performance. Entrepreneurship Theory and Practice 9, 331–349 (1993)
3. Minniti, M., Bygrave, W.: A dynamic model of entrepreneurial learning. Entrepreneurship Theory and Practice, 5–16 (2001)
4. Kirzner, I.M.: Competition and entrepreneurship. University of Chicago Press (1973)
5. Binks, M., Vale, P.: Entrepreneurship and Economic Change. McGraw-Hill, Maidenhead (1990)

6. Rae, D.: Entrepreneurial learning: A Conceptual Framework for Technology-based Enterprise. Technology Analysis & Strategic Management 18(1), 39–56 (2006)
7. Wenger, E.: Communities of Practice: Learning, Meaning and Identity. Cambridge University Press, Cambridge (1998)
8. Tell, J.: The Emergent Nature of Learning Networks. In: Harrison, R.T., Leitch, C. (eds.) Entrepreneurial Learning: Conceptual Frameworks and Applications, pp. 272–287. Routledge, London (2008)
9. Beunza, D., Stark, D.: 'Tools of the Trade: The Socio-Technology of Arbitrage in a Wall Street Trading Room. Industrial and Corporate Change 13(2), 369–400 (2004)
10. Beunza, D., Stark, D.: Looking out, Locking. Financial Models and the Social Dynamics of Arbitrage Disasters (2009) (Unpublished manuscript)
11. Bird, B.: Learning Entrepreneurship Competencies: The Self-Directed Learning Approach. International Journal of Entrepreneurship Education 1, 203–227 (2002)
12. Mole, V., Dawson, S., Winstanley, D., Sherval, J.: Researching managerial competencies. Paper Presented to the British Academy of Management Annual Conference, Milton Keynes (September 1993)
13. Gartner, W.B., Starr, J.A.: The nature of entrepreneurial work. In: Birley, S., MacMillan, I.C. (eds.) Entrepreneurship Research: Global Perspectives, pp. 35–67 (1993)
14. Bird, B.: Towards a theory of entrepreneurial competency. Advances in Entrepreneurship. In: Firm Emergence and Growth, vol. 2, pp. 51–72. JAI Press, Greenwich (1995)
15. Man, T.W.Y., Lau, T.: Entrepreneurial Competencies of SME Owner/Manager in the Hong Kong Services Sector: A Qualitative Analysis. Journal of Enterprising Culture 8, 235–254 (2000)
16. Man, T.W.Y., Lau, T., Chan, K.F.: The competitiveness of small and medium enterprises: A conceptualization with focus on entrepreneurial competencies. Journal of Business Venturing (17), 121–142 (2002)
17. Deakins, D., Freel, M.: Entrepreneurial Learning and the Growth Process in SMEs. The Learning Organization 5(3), 144–155 (1998)
18. Chen, W., Li, X.: Entrepreneurial Learning in Chinese Firms: Dimensions and Empirical Test. Economic Management (8), 63–72 (2010) (in Chinese)
19. Miao, Q., Wang, C.: A Conceptual Model: Entrepreneurial Competence Based on Enterprise Competitiveness. Journal of China University of Geosciences (3), 18–24 (2003) (in Chinese)

Study on the Influence of Corporate Knowledge Integration on Operational Performance-Using Trust as a Moderator Variable

Chiung-En Huang and Chih-Chung Chen

Aletheia University,
70-11, Beishiliao, Madou Dist., Tainan City, 72147, Taiwan (R.O.C.)
a3126747@gmail.com, jason556@mail.au.edu.tw

Abstract. This study aims to determine if knowledge integration of the financial industry in Taiwan influences operational performance through the mediating effect of trust. Hierarchical regression analysis was conducted on questionnaire survey. A total of 500 questionnaires were distributed, and 315 valid questionnaires were retrieved, with a valid return rate of 63%. The results showed that trust reveals mediating effects between knowledge integration and operational performance.

Keywords: Knowledge Integration, Operational Performance, Trust.

1 Introduction

With the coming of a knowledge economy era, knowledge gradually replaces traditional resources (e.g., capital, land, and labor) to become the niche of firms survival and maintain competitiveness. In the trend of financial freedom and internationalization, new financial products are continuously introduced. As there are various types of businesses and competition is severe, the financial industry must integrate internal professional knowledge and external market information in order to immediately and effectively respond to external environments.

How does corporate knowledge effectively integrate and lead to more competitiveness in dynamic environments? This study aims to determine if corporate knowledge integration can effectively respond to environmental change through the mediating effects of trust, thus, enhancing operational performance.

2 Literature References

[1] Treated knowledge integration as the continuous process of collective construction, expression, and re-definition of common beliefs, which was accomplished by social interaction among organizational members. Knowledge integration leads to employee negotiations and cooperation by the design of appropriate organizational frameworks, which integrate empowerment and communication. Thus, employees' knowledge and abilities will be fully utilized, and individual Specialized Knowledge within an organization can be integrated in order to enhance organizational competitiveness [2].

A. Xie & X. Huang (Eds.): Advances in Electrical Engineering and Automation, AISC 139, pp. 235–239.
springerlink.com

[3] Suggested that if organizational management considers members, the members will have the intention to share knowledge or opinions through trust, which would positively influence opinion sharing and knowledge integration. When employees are committed to and trust the firms and supervisors, they will enjoy duties assigned by supervisors, rendering them more willing to share their knowledge and experience, and learn. According to [4], system trust is the justice of organizational procedures. For instance, fair and reasonable human resources procedures are positively related to employees' attitude, and will enhance employees' communication with an organization.

[5] Suggested that when employees have stronger trust in an organization, they are more willing to share knowledge. In other words, when employee trust in system is high, knowledge sharing will be more frequent.

[6] Suggested that the development and integration of new knowledge and vision and would improve performance. Therefore, new knowledge created by organizational knowledge integration will enhance performance. According to [7], if corporate members can construct common goals, cognition, and behavioral regulations, they can thus avoid communication difficulties and enhance their intention to share experience and knowledge.

[8] Suggested that by integrating related resources and developing rapid product innovation, firms can be successful in global competitive environments. Competence of resource integration and innovation is based on knowledge. When organizational members have higher degrees of interaction, exchange, and trust, strong ties will be established. Upon the effect of strong ties, individuals' knowledge exchange and flow will be effective. [9] indicated that in an increasingly competitive industrial environment, integration between knowledge and technique is the expected trend, and firms with higher degrees of knowledge integration will demonstrate more effective R&D, and better products and services.

[10] Pointed out that knowledge integration means knowledge sharing and upgrading, which combines original knowledge and creates new knowledge. Their market knowledge, service quality, general knowledge, and performance would be enhanced. Therefore, this study infers that knowledge integration will influence operational performance through the mediating effect of trust.

3 Research Method

3.1 Sample

This study treated the finance industry in Taiwan as the subject, and distributed 500 questionnaires. A total of 315 valid samples were retrieved, with a valid return rate of 63%.

3.2 Measurement

The data were analyzed by SPSS18.0. Scales of knowledge integration, trust, and operational performance were measured upon a Likert-type 7-point scale, where 7 indicates "strongly agree" and 1 means "strongly disagree". *Cronbach's* α is used to measure reliability of scales. *Cronbachs'* α of scales of knowledge integration, trust, and operational performance are 0.9257, 0.969, and 0.915, respectively, which are higher than 0.9, and demonstrates a high degree of internal consistency of the scales. Accumulated explained variances are 75.283%, 77.372%, and 90.457%, respectively.

4 Research Result

By hierarchical regression analysis, two-dimensional interaction of influence of knowledge integration on operational performance upon trust was validated. Multiplying items of variables included in regression analysis usually lead to collinearity, which might affect the precision of statistical analysis. Thus, according to the suggestion of [11], this study conducted the analysis after normalizing all variables.

Analysis of interactive effects of knowledge integration and trust on operational performance is shown in Table 1. The control variable is included in Model 1. Direct effect variables, knowledge integration, and trust, are included in Model 2 (β=0.743, *p*<0.001; β=0.082, *p*>.05); two-dimensional interactive effect of knowledge integration and trust is included in Model 3 (β=0.113, *p*<.05). The outcomes demonstrate significant influence.

In three regression models, the VIF of independent variables are 1.000~2.329. According to the suggestion of [12], VIF must be lower than 10, which means there is no high degree of collinearity in this regression model. Thus, interaction between knowledge integration and trust can significantly predict operational performance. The F value of R^2 is 38.247 and *P*<.001. In other words, interaction between knowledge integration and trust will significantly predict operational performance. According to the results of Table 1, the simplified regression equation (insignificant predictors are not included in the equation) is: "operational performance" =-0.159+ 0.776 "knowledge integration" +0.113 "knowledge integration × trust". With fixed gender, marital status, and post, the operational performance of a high degree of trust (higher than the mean of trust) is higher than that of the low degree of trust (lower than the mean of trust).

Table 1. Interaction effect analysis of knowledge integration and trust on operational performance

Variables	Operational performance		
	Model 1	Model 2	Model 3
Control variables			
Gender	0.047	0.073	0.078
Marital status	-0.013	0.017	0.061
Post	-0.022	-0.021	-0.020
Direct effect variables			
Knowledge integration		0.743***	0.776***
Trust		0.082	0.064
Interaction effect			
Knowledge integration×Trust			0.113*
F-value	0.673	44.455***	38.247***
R^2	0.006	0.418	0.427
$\triangle R^2$		0.412***	0.011*

***P<.001, **P<.01, *P<.05

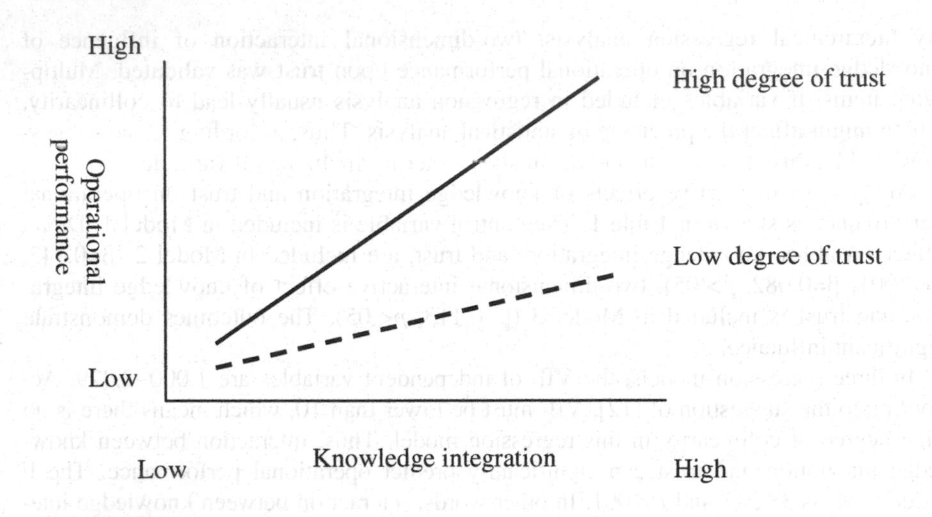

Fig. 1. Interaction effect of knowledge integration and trust on operational performance

4 Conclusions

According to research findings, a high degree of influence on trust knowledge integration and operational performance is more significant than the low degree of trust. When employees have more trust in colleagues, they believe that their colleagues are willing to share resources; thus, they will have the intention to exchange knowledge with them [13], particularly when colleagues are competent, honest, and reliable. Employees will realize that the sharing of valuable knowledge will not be their loss of valuable organizational resources, and their contribution and exchange of valuable resources will result in the return of others' valuable knowledge. Hence, employees will obtain more knowledge resources, leading to greater knowledge competence.

If employees can share information and knowledge with different departments, selfish departmentalism will be avoided and overall trust in the firms will be enhanced. Once the employees trust a corporate system and policies, they will be willing to be integrated with different departments. Employees' trust in overall corporate system is significantly reflected on employees' performance.

When the employees are confident in the competence of the colleagues and supervisors, they will have the intention to learn from others' shared experiences and knowledge. Care and respect among the employees will enhance their willingness to share their personal knowledge and experience. Thus, once the firms construct complete policies and processes, knowledge integration will be enhanced, the members' will engage in learning, experience sharing, and knowledge upgrading, thus strengthening corporate business capabilities and service quality, as well as corporate performance.

References

1. Huang, J.C., Newell, S.: Knowledge Integration Processes and Dynamics within the Context of Cross-Functional Projects. Int. J. Proj. Manag. 21(3), 167–176 (2003)
2. Grant, R.M.: Prospering in Dynamically-competitive Environment: Organizational Capability as Knowledge Integration. Organ. Sci. 7(4), 375–387 (1996)
3. Clegg, C., Unsworth, K., Epitropaki, O., Parker, G.: Implicating Trust in the Innovation Process. J. Occup. Organ. Psychol. 75(4), 409–422 (2002)
4. Pearce, J., Branyiczki, I., Bigley, G.: Insufficient Bureaucracy: Trust and Commitment in Particularistic Organizations. Organ. Sci. 11(2), 148–162 (2000)
5. Mayer, R.C., Davis, J.H.: An Integration Model of Organizational Trust. Acad. Manage. Rev. 20(3), 709–735 (1995)
6. Slater, S.F., Narver, J.C.: Market Orientation and the Learning Organization. J. Marketing 59(3), 63–74 (1995)
7. Dyer, J.H., Nobeka, K.: Creating and Managing a High-Performance Knowledge-Sharing Network: The Toyota Case. Strategic Manage. J. 21(3), 345–367 (2000)
8. Teece, D.J.: Research Directions for Knowledge Management. Calif. Manag. Rev. 40(3), 289–292 (1998)
9. Iansiti, M., West, J.: Technology Integration: Turning Great Research into Great Product. Harvard Bus. Rev. 75(3), 69–79 (1997)
10. Sabherwal, R., Becerra-Fernandez, I.: Integrating Specific Knowledge: Insights from the Kennedy Space Center. IEEE T. Eng. Manage. 52(3), 301–315 (2005)
11. Aiken, L.S., West, S.G.: Multiple Regression: Testing and interpreting interactions. Sage, Newbury Park (1991)
12. Hair, J.F., Anderson, R.E., Tatham, R.L., Black, W.C.: Multivariate Data Analysis, 5th edn. Prentice Hall, New Jersey (1998)
13. Mayer, R.C., Davis, J.H.: The Effect of the Performance Appraisal System on Trust for Management: A Field Quasi-experiment. J. Appl. Psychol. 84(1), 123–136 (1999)

The Game of Efficiency-Wage and the System Design

Jianlin Qiu

Guangxi University of Finance and Economics, Nanning, P.R. China, 530001
qiujianlin@163.com

Abstract. The role of high-quality personnel is increasingly being recognized in the process of business development by Chinese enterprises, and Chinese enterprises play an increasing emphasis on incentive pay for personnel. High level of pay has become the effective means to attract, retain and motivate highly qualified personnel. Efficiency-wages as a new incentive system is widely used in many companies. This article discusses the game theory and process of efficiency-wage, and propose some principles of designing efficiency-wage system.

Keywords: Personnel-incentive, Efficiency-wages, Process of game, System-design.

1 The Concept of Efficiency-Wages

Efficiency-wages is a pay system that enterprises to pay high wages to stimulate high-quality personnel highly qualified personnel to work, mainly for enterprises to attract, motivate, discipline the high-quality personnel and reduce the role of monitoring costs, so as to create greater value, maintain business continuity, stability and development.

2 The Game of Efficiency-Wage

The goal of enterprises efficiency-wage l is to reduce the total efficiency units of labor costs, making it a true sense of "the efficiency-wage" that subject to a series of constraints and constraints.

2.1 The Theory of Efficiency-Wage Game

The assumption of the economic game of efficiency-wage: Through higher levels of Efficiency-wages to attract highly qualified personnel. For high-quality personnel to accept efficiency wages, it must be efficient to pay the utility is greater than the other to get the reservation utility (opportunity cost), be to meet the conditions of participationconstraint; At the same time, companies make high-quality personnel motivated to choose the actions that the companies want them to choose that promote high-quality talent to work hard to obtain high-yield. companies must make high-quality personnel in the choice of the desired corporate action (work) be to get the expected utility of not less than choose other actions (such as lazy, part-time, etc.) to get the expected

A. Xie & X. Huang (Eds.): Advances in Electrical Engineering and Automation, AISC 139, pp. 241–245.
springerlink.com

utility, that meet the conditions of incentive-compatibilityconstraint; In addition, while the high-quality personnel in an enterprise to another enterprise for the poor performance, its reputation will be damaged. If the business and high-quality personnel are between the infinitely repeated games, and take the "cold-blooded strategy", then the personnel will pay a price for its dismissed reputation. its difficulty to find a new job and the cost will increase dramatically. Work hard to get a high level of efficiency-wage is the optimal choice for high-quality personnel; Once the company found that employees do not work hard (such as lazy, part-time, etc.), then punish those lazy (dismissal), is the company's optimal choice. Therefore, the establishment of reciprocal relationship between high-quality personnel and the enterprises is the basic assumptions of efficiency-wage working.

2.2 The Information Transmission of Efficiency-Wage Game

The request of efficiency wage under the asymmetric information. Only the high-quality personnel in the information superiority know their ability and quality of their work effort, the enterprise is not entirely clear of the information disadvantage. Therefore, there is asymmetric information both. Only in this way, companies can through the control of the wage level of the signal to identify the level of quality talent and the work effort : If companies choose to low-wage, low quality will be the ability of staff to come to every one, may form a "pooling equilibrium" and difficult to distinguish between good and bad people. If companies choose to high wage (efficiency-wage), high-quality personnel will be attracted and come to work-hard, and those low quality people because they do not have the necessary qualities and can not afford to work pressure, but not the job candidate. Therefore, a high level of efficiency wages must be high enough to cause separation of the imbalance, so that high-quality personnel and low quality of people's choice behavior is not the same; At the same time, high-quality talent also delivered high-quality signal to the business (eg, highly educated, high-grade and previous work experience and achievements, etc.). Of course, the efficiency wage determination has a certain subjective, it depends on the identity of both companies and talent, the good reputation and the reputation of the actual effectiveness and efficiency-wages.

2.3 The Game Process of Efficiency-Wage

The theory of efficiency-wage incentive systems. Because the company designed a high level of efficiency wage, high-quality companies don’t choose the actions enterprises don’t want them to (such as lazy or part-time) in order to maintain their maximum effectiveness. Due to the higher wages enterprises reduces the demand for employees, meanwhile employees increase the cost of looking for work again, so that it reserves to pay down, this double role of incentives will encourage more employees to work hard. Therefore, "hard work" is the best choice of action rational high-quality personnel; Meanwhile, the enterprises of the efficiency-wage, lower demand for employees, which really reduces the efficiency of the total unit labor costs. Therefore, the implementation of a high level of efficiency wages is the reason companies’ best option for action.

The following is the net pay matrix during the game :

	Action	High-quality personnel	
		Not work hard	**Work hard**
Enterprises	**Implementation of the efficiency-wage**	e−A, a+C	E−A, A
	No implementation of the efficiency-wage	e−a, a+C	E−a, a

In the above pay matrix: e—the enterprises proceeds under the action of "Not work hard"; E—the enterprises proceeds under the action of "Work hard"; a—Low efficiency-wage; A—High efficiency-wage; C—The opportunity proceeds of high-quality personnel under the action of "do not work hard".

The explanation for the above that the enterprises proceeds is "e−a " instead of " E −a "under the condition of "work hard and do not implement the efficiency wage": Even high-quality personnel select the action of "hard work", if the enterprises will choose "not to implement efficiency wage" action, rational high-quality people will choose "does not work hard" to response, thus resulting in the enterprises net income for the "e−a ".

When the enterprises choose to "the implementation of efficiency wages" action, if A> a + C, the best action for high-quality personnel to choose is "hard work"; When high-quality personnel choose to "work hard" action, if E-A>e-a, the enterprises will choose the action of " Implementation of the efficiency wage "..

3 The Design Ideas of Efficiency-Wage-System

3.1 Establish the Integrity of the Mechanism Is the Premise of Effective Implementation of Efficiency-Wage

Establish the integrity of the talent market is the fundamental way to implement efficiency wage. Efficiency-wage system requirements enterprises and personnel must comply with the credibility , and establish relationship of labor contracts under each other's past experience and reputation. From the fundamental point of view, efficiency-wage is effective because of both sides provide a reliable signal in the condition of incomplete information and asymmetric to help both enterprises or personnel to make screening and selection. Signaling effectiveness and success rate depends on the level of trust between the company and the personnel. Both sides believe the other is the emphasis on reputation, and thus comply with the principle of reciprocity.

3.2 Establish a Comprehensive Employee Performance Appraisal System Is the Basis for the Implementation of Efficiency-Wages

China's labor market development is far from perfect, especially the corresponding market order is not perfect, there is no scientific employee performance evaluation system, resulting in the talent market in the situation of "pass off the fish eyes as pearls", and thus undermine the integrity of the entire principle of the talent market, the occurrence of a crisis of confidence. Therefore, the establishment of a comprehensive employee performance appraisal system, can effectively identify the performance of efficiency wage, the same employees also need to establish a comprehensive assessment system of their own value, to scientifically assess the effectiveness of efforts have been chosen. This is the basis for the implementation of efficiency-wage system.

3.3 Establish a Sound Mechanism Is the Indemnification to Implement the Efficiency-Wage

If We want to give full play to the role of high-quality personnel, we must start from the market mechanism, to introduce the competition \ incentives systems, to implement the reasonable price of labor distribution system of "competitive prices", to give full play to the price regulation mechanism, to promote the rational flow of trained personnel to enhance the intrinsic value, to make a reasonable allocation of human resources, thereby improving the job market. To this end, to break the non-market behavior's shackle which is both denied the employer's bargaining autonomy and the main supply of human resources offer autonomy, and negate the market's effect in the formation process of human capital price, we have to do the following two aspects: First, establish a market regulation system of Talent Price, through the supply and demand mechanisms and competition to the formation and regulation of the talent market, creating a good price-oriented mechanism; Second, establish a negotiated bargaining system of Talent Price of supply and demand sides. On the one hand, enterprises as the demand side, according to the human resources supply and demand conditions, trends, ability to pay wages, national wage policy and other companies with the industry wage levels and other factors to determine their level of efficiency wages; the other hand, high-quality personnel as the supply side, according to their strength, supply and demand conditions .etc to determine the expected level of efficiency-wages. supply and demand efficiency wages, both in their expected level of efficiency wages based on the two sides through equal consultations to reach acceptable efficiency wages. The two sides through equal consultations to reach a mutually acceptable efficiency wage based on their expected efficiency-wages.

4 Conclusion

Efficiency-wages to motivate the role of high-quality personnel of Chinese enterprises is obvious. Implementation of the effective-wage system is the inevitable choice of Chinese enterprises during their development process. To explore the problems and difficulties in the implementation process of effective-wage system, to solve these problems and difficulties, and to develop he effective-wage system in Chinese enterprises is the Chinese enterprise managers and researchers'obligation.

References

1. Du, Z.Y., Wang, X.: Corporate compensation design and management. Guangdong Economic Press, Guangdong (2001)
2. Zhang, J.H.: Thoughts Of the Salary Management. J. Economist (8) (2002)
3. Yu, P., Pu, J.Y.: Analysis of companies and candidates Game. J. Kong Economy (1) (2007)
4. Zhang, W.Y.: Game Theory and Information Economics. Shanghai People's Publishing House (2004)

Combination of Fuzzy C-Means and Particle Swarm Optimization for Text Document Clustering

Jiayin Kang[1] and Wenjuan Zhang[2]

[1] School of Electronics Engineering, Huaihai Institute of Technology, 222005, Lianyungang, China
[2] School of Computer Engineering, Huaihai Institute of Technology, 222005, Lianyungang, China
{jiayinkang,zhangwenjuan2009}@gmail.com

Abstract. Document clustering, an important tool for document organization and browsing, has become an active field of research in the machine learning community. Fuzzy c-means (FCM), a powerfully unsupervised clustering algorithm, has been widely used for categorization problems. However, as an optimization algorithm, it easily leads to local optimal clusters. Particle swarm optimization (PSO) algorithm is a stochastic global optimization technique. This paper presents a hybrid approach for text document clustering based on fuzzy c-means and particle swarm optimization (PSO-FCM), which makes full use of the merits of both algorithms. The PSO-FCM not only helps the FCM clustering escape from local optima but also overcomes the shortcoming of the slow convergence speed of the PSO algorithm. Experimental results on two commonly used data sets show that the proposed method outperforms than that of FCM and PSO algorithms.

Keywords: document clustering, fuzzy c-means, particle swarm optimization, global optimization.

1 Introduction

With the rapid growth of the Internet, there has been ever expanding amount of electronic information available. How to explore and utilize the huge amount of text documents is major question in the areas of information retrieval and text mining. Document clustering is the task of assigning a given text document to one or more predefined categories. It is an important research field, and attracts increasing attention and popularity due to the increased availability of documents in digital forms and the following need to access them in flexible ways [1-3].

Clustering is an unsupervised learning technique that has been widely used in categorization problems. It groups the input space into C regions based on some similarity or dissimilarity metric. The partition is done such that patterns within a group are more similar to each other than patterns belonging to different groups [4]. Some of the more familiar clustering methods are: partitioning algorithms based on dividing entire data into dissimilar groups, hierarchical methods, density and grid based clustering, some graph based methods, etc [5]. Perhaps the most influential fuzzy clustering

A. Xie & X. Huang (Eds.): Advances in Electrical Engineering and Automation, AISC 139, pp. 247–252.
springerlink.com

algorithm is fuzzy C-means (FCM) originally proposed by Bezdek [6]. In a simple form, it selects C points as cluster centers and assigns each data point to the nearest center. The updating and reassigning process can be kept until a convergence criterion is met.

Although FCM algorithm is easy to be implemented and works fast in most situations, it suffers from two major drawbacks that make it inappropriate for many applications. One is sensitivity to initialization and the other is convergence to local optima. To deal with the limitations that exist in traditional FCM clustering algorithm, recently, new concepts and techniques have been incorporated into the classic FCM to satisfying need for the text document clustering. One of these techniques is optimization method that tries to optimize a pre-defined function [5]. Particle swarm optimization algorithm [7], one of the optimization methods, has gained much attention and has been successfully used in optimization problems, presenting several advantages with respect to traditional optimization techniques [8, 9, 10]. In this paper, we investigate the application of PSO to help the FCM algorithm escape from local minima. A hybrid clustering algorithm for text document clustering based on FCM and PSO , called PSO-FCM, is proposed. The experimental results on two commonly used data sets show that the FCM-PSO algorithm is superior to the FCM algorithm and the PSO algorithm.

2 Document Representation

For most existing document clustering algorithms, documents are represented by using the vector space model [1, 5]. In this model, each document d is considered as a vector in the term-space and represented by the term frequency (TF) vector:

$$d_{tf} = [tf_1, tf_2, \ldots, tf_t] \tag{1}$$

Where $tf_p (p = 1, 2, \ldots, t)$ is the frequency of term p in the document, and t is the total number of unique terms in the text database. A widely used refinement to this model is to weight each term based on its inverse document frequency (IDF) in the document collection. The idea is that the terms appearing frequently in many documents have limited discrimination power, so they need to be deemphasized [1]. This is commonly done by multiplying the frequency of each term p by $\log(n / df_p)$, where n is the total number of documents in the collection, and df_p is the number of documents that contain term p (i.e., document frequency). Thus, the TF-IDF representation of the document d is:

$$d_{tf-idf} = [tf_1 \log(n / df_1), tf_2 \log(n / df_2), \ldots, tf_t \log(n / df_t)] \tag{2}$$

Where n is the total number of documents in the collection, and $df_p (p = 1, 2, \ldots t)$ is the number of documents that contain term p , i.e., document frequency.

3 Brief Reviews of FCM and PSO

3.1 FCM Clustering Algorithm

The FCM algorithm assigns documents to each category by using fuzzy memberships:

Let $D=\{d_i, i=1,2,\cdots,n \mid d_i \in \mathbb{R}^t\}$ denotes a document collection with n documents to be partitioned into c clusters, where d_i represents features data. The algorithm is an iterative optimization that minimizes the objective function defined as follows:

$$J_m=\sum_{k=1}^{c}\sum_{i=1}^{n}u_{ki}^m \left\|d_i - v_k\right\|^2 \tag{3}$$

With the following constraints:

$$\{u_{ki} \in [0,1] \mid \sum_{k=1}^{c} u_{ki}=1, \forall i, 0<\sum_{i=1}^{n} u_{ki}<n, \forall k\} \tag{4}$$

Where u_{ki} represents the membership of document d_i in the kth cluster, v_k is the kth class center; $\|\cdot\|$ denotes the Euclidean distance, $m>1$ is a weighting exponent on each fuzzy membership. The parameter m controls the fuzziness of the resulting partition. The membership functions and cluster centers are updated by the following expressions:

$$u_{ki}=\sum_{l=1}^{c}\left(\frac{\left\|d_i - v_k\right\|}{\left\|d_i - v_l\right\|}\right)^{-2/(m-1)} \tag{5}$$

And

$$v_k=\sum_{i=1}^{n}(u_{ki}^m d_i) / \sum_{i=1}^{n}(u_{ki}^m) \tag{6}$$

3.2 Particle Swarm Optimization Searching Algorithm

The PSO algorithm was originally introduced by Kennedy and Eberhart [7], and has been successfully applied to many fields. In PSO, the solution space of the problem is formulated as a search space. Each position in the search space is a correlated solution of the problem. Particles cooperate to find the best position (best solution) in the search space (solution space). Each particle moves according to its velocity.

Let $x_i=(x_{i1},x_{i2},\ldots,x_{in})$ represents the position of the ith particle in the $n-$dimensional space, and $v_i=(v_{i1},v_{i2},\ldots,v_{in})$ denotes the velocity of ith particle. During the search process, the particle successively adjusts its position according to their personal best position (pbest) and the global best position (gbest). The personal best position $p_i=(p_{i1},p_{i2},\ldots,p_{in})$ is the best previous position of the ith particle that gives the optimal fitness value. The global best position $p_g=(p_{g1},p_{g2},\ldots,p_{gn})$

is the best particle among all the particles in the population. Each particle updates its position and velocity as follows:

$$x_{ij}(t+1) = x_{ij}(t) + v_{ij}(t+1) \quad j = 1,2,\ldots,n \tag{7}$$

$$v_{ij}(t+1) = wv_{ij}(t) + c_1r_1(p_{ij}(t) - x_{ij}(t)) + c_2r_2(p_{gj}(t) - x_{ij}(t)) \quad j = 1,2,\ldots,n \tag{8}$$

Where w is a intertia weight scaling the previous time step velocity; c_1 and c_2 are two acceleration coefficients that scale the influence of the best personal position of the particle (pbest) and the best global position (gbest); r_1 and r_2 are elements from two uniform random sequences in the range $[0,1]$.

4 PSO-FCM Algorithm

Fuzzy c-means clustering is an effective algorithm, but it easily convergence to local optima. We combine FCM with PSO to form a hybrid clustering algorithm called PSO-FCM, which maintains the merits of FCM and PSO. We summarize the PSO-FCM algorithm as follows:

Step 1: Set the initial parameters including the maximum iterative number $n_{\max}$; weighting exponent m; clusters' number c; termination threshold ε (a very small value, and $\varepsilon > 0$); the population size p_{size}; intertia weight w; acceleration coefficients c_1 and c_2.

Step 2: Initialize a population of size p_{size}.

Step 3: Set iterative count $Gen_1 = 0$.

Step 4: Set iterative count $Gen_2 = Gen_3 = 0$.

Step 5 (PSO Method)

Step 5.1: Apply the PSO operator to update the p_{size} particles.

Step 5.2: $Gen_2 = Gen_2 + 1$. If $Gen_2 < 8$, go to Step 5.1.

Step 6 (FCM Method) For each particle i do

Step 6.1: Take the position of particle i as the initial cluster centers of the FCM algorithm.

Step 6.2: Recalculate each cluster center using the FCM algorithm.

Step 6.3: $Gen_3 = Gen_3 + 1$, If $Gen_3 < 4$, go to Step 6.2.

Step 7: $Gen_1 = Gen_1 + 1$. If $Gen_1 < n_{\max}$, go to Step 4.

Step 8: Assign data point d_i to cluster j with the biggest u_{ki}.

5 Experimental Results and Analysis

In order to verify the performance of proposed algorithm, we carried out experiments on two commonly used data sets: one is Reuters-21578 data set [11], and another is 20Newsgroup data set [12]. Moreover, tests on two data collections using traditional FCM and PSO algorithm are conducted for the purpose of comparison. Fig. 1 shows the F -measure values [13] of the clustering results of all the algorithms on two data collections, and Table 1 displays the purity values [1] of the clustering results using three algorithms on two data collections. In general, the larger the F -measure is, the better the clustering result is; the larger the purity value is, the better the clustering result is.

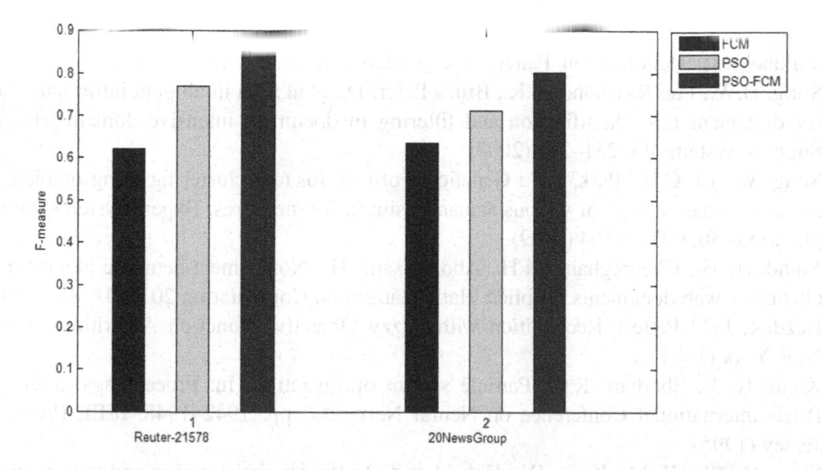

Fig. 1. Results of F-measure values using three algorithms on two data collections

Table 1. Purity values of three algorithms on two data collections

Collection	FCM	PSO	PSO-FCM
Reuter-21578	0.6409	0.7885	0.8601
20NewsGroup	0.6633	0.7416	0.8319

As shown in Fig. 1, the PSO-FCM algorithm obtains significant improvements over the FCM and the PSO algorithm in terms of the F -measure value both on the Reuters-21578 data collection and on the 20Newsgroups data collection. Furthermore, from Table 1, we can see that the PSO-FCM algorithm really improves the purity values over the FCM algorithm and the PSO algorithm on two date collections.

In summary, it is clear that the overall performance of our approach is better than the performance of both the traditional FCM and the PSO algorithm.

6 Conclusion

In this paper, we proposed a hybrid method for clustering of text document using FCM combined with PSO algorithm. We compare its performance with the other document clustering methods via experiments on two widely used datasets. Experimental results show that the presented approach outperforms than other two algorithms, traditional FCM algorithm and PSO algorithm.

References

1. Luo, C.N., Li, Y.J., Chung Soon, M.: Text document clustering based on neighbors. Data & Knowledge Engineering 68, 1271–1288 (2009)
2. Kumar, M.A., Gopal, M.: A comparison study on multiple binary-class SVM methods for unilabel text categorization. Pattern Recognition Letters 31, 1437–1444 (2010)
3. Song, D.W., Lau Raymond, Y.K., Bruza Peter, D., et al.: An intelligent information agent for document title classification and filtering in document-intensive domains. Decision Support Systems 44, 251–265 (2007)
4. Song, W., Li, C.H., Park, S.C.: Genetic algorithm for text clustering using ontology and evaluating the validity of various semantic similarity measures. Expert Systems with Applications 36, 9095–9104 (2009)
5. Mahdavi, M., Chehreghani, M.H., Abolhassani, H.: Novel meta-heuristic algorithms for clustering web documents. Applied Mathematics and Computation 201, 441–451 (2008)
6. Bezdek, J.C.: Pattern Recognition with Fuzzy Objective Function Algorithms. Plenum, New York (1981)
7. Kennedy, J., Eberhart, R.C.: Particle swarm optimization. In: Proceedings of the 1995 IEEE International Conference on Neural Networks, pp. 1942–1948. IEEE Press, New Jersey (1995)
8. Shen, Q., Shi, W.M., Kong, W.: Hybrid particle swarm optimization and tabu search approach for selecting genes for tumor classification using gene expression data. Computational Biology and Chemistry 32, 53–60 (2008)
9. Yang, F.Q., Sun, T.L., Zhang, C.H.: An efficient hybrid data clustering method based on K-harmonic means and particle swarm optimization. Expert Systems with Applications 36, 9847–9852 (2009)
10. Huang, K.Y.: A hybrid particle swarm optimization approach for clustering and classification of datasets. Knowledge-Based Systems 24, 420–426 (2011)
11. http://www.daviddlewis.com/resources/testcollections/reuters21578/
12. http://people.csail.mit.edu/jrennie/20newsgroups/
13. Zhang, W., Yoshida, T., Tang, X.J., et al.: Text clustering using frequent itemsets. Knowledge-Based Systems 23, 379–388 (2010)

A Survey on Unicast Routing Protocol for Vehicular Ad Hoc Networks

Yuanzhen Li[1], Yiqiao Yu[1], and Shuzhen Li[2]

[1] School of Computer Science, Liaocheng University, Shandong Liaocheng, 252059
[2] Linqing City Office State Administration Taxation, Linqing, Shandong, 25600
li_yuanzhen@hotmail.com

Abstract. VANET(Vehicular ad hoc networks), in which vehicles directly exchange information through wireless communication with the dual purpose of increasing road safety and comfort in driving, have recently received great attention. Because of high dynamic topology and partitioned networks in VANET, an important problem in VANET is to find efficient ways to disseminate information on the target areas before the vehicle entering into another road. To aims to make a summary, comparison and analysis of the mainstream methods and the latest progress in the field, expecting to be helpful to the future research, the unicast routing strategy for VANET(Vehicular Ad Hoc Networks) as well as some closely related topics was stated. The problems of current research are discussed, and finally the future directions are pointed out.

Keywords: Vehicular Ad Hoc Networks, Mobile Ad Hoc Networks, Inter-vehicle communication; unicast routing.

1 Introduction

Mobile Ad hoc NETworks (MANETs) which lack a network infrastructure, such as base stations, are comprised of self-organizing mobile nodes. This technology has been proposed as a complementary to the fourth-generation wireless networks. And in recent years, there is a growing interest on the research and deployment of Mobile Ad hoc Networks (MANET) technology for vehicular communication. VANET(Vehicular Ad Hoc Networks)) is a specific type of self-organizing mobile ad hoc network, which is formed by vehicles that are equipped with wireless communication devices. The prospective applications of VANETs are categorized into two groups as comfort and safety applications. The first group is expected to improve the passengers comfort and optimize traffic efficiency, whereas the second one improves driving safety.

The remainder of this paper is organized as follows: Section 2 describes the characteristics of vehicular ad hoc networks. Section 3 presents the importance of routing in vehicular ad hoc networks and routing strategy for vehicular ad hoc networks is surveyed in Section 4. Finally, Section 5 provides our conclusion.

A. Xie & X. Huang (Eds.): Advances in Electrical Engineering and Automation, AISC 139, pp. 253–258.
springerlink.com

2 Characteristics of Vehicular Ad Hoc Networks

In addition to its similarities to ad hoc networks, VANET is distinguished from other kinds of ad hoc networks in the following aspects[1]: (1) highly dynamic topology. Due to high speed of movement between vehicles, the topology of VANET is always changing; (2) Frequently disconnected network (Intermittent connectivity). Due to the same reason, the connectivity of the VANET could also be changed frequently especially in spare VANET; (3) sufficient energy and storage. Nodes in VANET have ample energy and computing power (e.g. storage and processing); (4) mobility predication. The vehicle's movement is constrained by prebuilt highways, roads and streets, and its future position can be predicated given the speed and the street map; (5) delay constraints. Some applications (e.g. emergency alarm) in VANET require strict delay constraints but not high data rates. (6)On-board Sensors. Nodes are generally assumed to be equipped with sensors to provide information useful for routing purposes. (7) Various communications environments. VANETs are usually operated in two typical communications environments: in highway traffic scenarios or in city conditions. (8) Geographical type of communication. The VANETs often have a new type of communication which addresses geographical areas where packets need to be forwarded (e.g., in safety driving applications).Thus, when designing a routing algorithm for VANET, we must take into account these factors.

3 Routing in Vehicular Ad Hoc Networks

Routing is crucial to the success of VANET applications, which is complicated by the fact that VANETs are highly mobile, and sometimes partitioned. Unicast routing is a fundamental operation for vehicle to construct a source-to-destination routing in a VANET. Multicast is defined by delivering multicast packets from a single source vehicle to all multicast members by multi-hop communication. Geocast routing is to deliver a geocast packet to a specific geographic region. Broadcast protocol is utilized for a source vehicle sends broadcast message to all other vehicles in the network.

Unicast routing in VANET has been studied recently and many different protocols were proposed. Several papers surveyed the state-of-the-art research on unicast routing for VANET(Vehicular Ad Hoc Networks), but few of them discuss this question in mainstream methods [1-3]. This article attempts to classify the routing algorithms according to the key technologies of each kinds of algorithms. The progress of the routing algorithm was described based the development of key technologies used in the algorithm.

4 Unicast Routing Strategy for Vehicular Ad Hoc Networks

The main goal of unicast routing in VANETs is to transmit data from a single source to a single destination via wireless multi-hop transmission or carry-and-forward techniques. In the wireless multi-hop transmission, the intermediate vehicles in a routing path should relay data as soon as possible. In the carry-and-forward technique, source vehicle carries data as long as possible to reduce the number of data packets.

The delivery delay-time cost by carry-and-forward technique is normally longer than wireless multi-hop transmission technique. Unicast routing algorithm for VANET can be classified into several types, including Topology-based routing, Position-based routing, Anchor-based routing, Traffic-aware routing, Link-state-based routing, Knowledge-Based Routing, Node-aided-based routing, Contention-Based routing, Carry-and-forward strategy based on the main technology used.

4.1 Topology-Based Routing

These topology-based routing protocols which can further be divided into proactive (table-driven)[4] and reactive (on-demand) routing[5], use links' information that exists in the network to forward packet. Proactive topology-based routing [4] preserves the routing information such as the next forwarding hop regardless of communication requests, while reactive topology based routing[5] opens a route only when it is necessary for a node to communicate with another node.

In proactive routing, control packets are constantly broadcast and flooded among nodes to maintain the paths or the link states between any pair of nodes even though some of paths are never used. While in reactive routing maintains only the routes that are currently in use, thereby reducing the burden on the network.

4.2 Position-Based Routing

In position-based routing[6,7], three types of information need to know: 1 the position of the node itself which can be obtained by the Global Position System (GPS), 2 the position of the node's one-hop neighbor which can be obtained by the beans sent periodically, 3 the position of the destination which can be obtained by locations services[8]. A node makes the forwarding decision based on the position of a packet's destination and the position of the node's one-hop neighbors. Packet will be forwarded to the node which is closest to the destination. The forwarding strategy will fail if no neighbor is closer to the destination than the node itself. In this case, we say that the packet has reached the local maximum at the node since it has made the maximum local progress at the current node. The routing protocols in this category have their own recovery strategy to deal with such a failure.

4.3 Anchor-Based Routing

Anchor-based routing is similar to the source routing of DSR. In anchor-based routing the source node includes into each packet a list of anchors or fixed geographic points that the packet must be forwarded along. Between anchors, greedy position-based routing is employed. The fundamental principle in the greedy approach is that a node forwards its packet to its neighbor that is closest to the anchor. Routing algorithms in [9] are examples of a scheme that employs anchor-based routing to forward packets to remote destinations.

In anchor-based routing, on the one hand, packets are routed from one intersection to another intersection. On the other hand, packets are routed from one vehicle to another vehicle.

4.4 Traffic-Aware Routing

In traffic aware routing [10,11], road traffic information are utilized to identify the anchor path. The roads where there are more vehicles were used to forward the packets.

The anchor-based routing is based on the map information, according to the physical length of the street to select the street when the data packet forwarding, without considering the actual density of vehicles on the streets or called traffic flow information. Traffic-aware routing is based on the physical length of the street, the street traffic flow information on a comprehensive selection of the appropriate path of macro-level packet forwarding, as the anchor sequence. Traffic flow routing is based on the macro level, in fact, the process in the selection anchor the flow of information using a vehicle.

4.5 Link-State-Based Routing

In the link-state-based routing[12,13], packet will be forwarded through the most stable and reliable link which can be evaluated by the speed, direction, position of nodes.

4.6 Knowledge-Based Routing

In knowledge-based routing algorithm[14], it is assumes that the movement knowledge of vehicles can obtained. The movement knowledge of vehicles includes the destination of a vehicle, the trajectory of a vehicle. Knowledge-based routing algorithm makes the routing decision based on the movement knowledge of vehicles.

Suppose the network movement information of all vehicles, including the current node position, speed, movement direction, driving route and the destination of vehicles is known, knowledge-based routing algorithm can be theoretically optimal routing. Routing algorithm based on knowledge, knowledge of vehicle movement synchronization and optimal routing method remains to be further studied. At the same time knowledge-based routing algorithm, requires the user to provide information on vehicle movement, undermining the privacy of user information.

4.7 Node-Aided-Based Routing

Node-aided-based routing assumed that there are some auxiliary nodes in the networks which help forward the packets. This type of routing algorithm can further be divided into static node-aided-based routing[15] and dynamic node-aided-based routing[16] based on the auxiliary nodes' movement. The message ferry proposed by [16] belongs to this type routing algorithm, which utilizes a set of special mobile nodes called "message ferries" to provide communication service for nodes in the deployment area. The main idea behind this approach is to introduce non-randomness in the movement of nodes and exploit such non-randomness to help deliver data.

4.8 Contention-Based Routing

Contention-Based Forwarding (CBF) [17] is a geographic routing protocol that does not require proactive transmission of beacon messages. Data packets are broadcast to all direct neighbors and the neighbors decide if they should forward the packet. The actual forwarder is selected by a distributed timer-based contention process which allows the most-suitable node to forward the packet and to suppress other potential forwarders. Receivers of the broadcast data would compare their distance to the destination to the last hop's distance to the destination. The bigger the difference, the larger is the progress and shorter is the timer.

4.9 Carry-and-Forward Strategy

To deal with partitioned networks, researchers adopt the idea of "carry and forward"[18-20], where nodes carry the packet when routes do not exist, and forward the packet to the new receiver that moves into its vicinity. These protocols can be classified into two categories: one is to follow the traditional ad hoc network literature and add no control on mobility; the other is to control the mobility of node to help packet forwarding. The strategy "carry and forward" can deal the partition networks effectively.

5 Conclusion and Future Work

Vehicle Ad Hoc network can effectively improve the urban traffic safety. Routing is the core of the vehicular ad hoc networks and many researchers have proposed a number of routing algorithms for vehicular ad hoc networks. This paper classified the routing algorithms according to the key technologies of each kinds of algorithms. The progress of the routing algorithm was described based the development of key technologies used in the algorithm.

References

1. Fan, L., Yu, W.: Routing in vehicular ad hoc networks: A survey. IEEE Veh. Technol. Mag., 12–22 (2007)
2. Bernsen, J., Manivannan, D.: Unicast routing protocols for vehicular ad hoc networks: A critical comparison and classification. Pervasive and Mobile Computing. 1–18 (2009)
3. Chang, C., Xiang, Y., Shi, M.: Development and status of vehicular ad hoc networks. Journal on Communications, 116–126 (2007)
4. Perkins, C.E., Bhagwat, P.: Highly dynamic Destination-Sequenced Distance-Vector routing (DSDV) for mobile computers. ACM SIGCOMM Computer Communication Review, 234–244 (1994)
5. Johnson, D.B., Maltz, D.A.: Dynamic source routing in ad hoc wireless networks. Kluwer International Series in Engineering and Computer Science, pp. 153–179 (1996)
6. Karp, B., Kung, H.T.: GPSR: greedy perimeter stateless routing for wireless networks. In: Mobile Computing and Networking, pp. 243–254. ACM Press, Boston (2000)
7. Lochert, C., Mauve, M., Ler, H.F., Hartenstein, H.: Geographic routing in city scenarios. ACM SIGMOBILE Mobile Computing and Communications Review, 69–72 (2005)

8. Xiang-yu, B., Xin-ming, Y., Jun, L., Hai, J.: VLS: A Map-Based Vehicle Location Service for City Environments. In: IEEE International Conference on Communications(ICC), pp. 1–5. IEEE Press, Dresden (2009)
9. Lochert, C., Hartenstein, H., Tian, J., Fussler, H., Hermann, D., Mauve, M.: A routing strategy for vehicular ad hoc networks in city environments. In: IEEE Intelligent Vehicles Symposium, pp. 156–161. IEEE Press, Heidelberg (2003)
10. Mouzna, J., Uppoor, S., Boussedjra, M., Pai, M.M.M.: Density aware routing using road hierarchy for vehicular networks. In: IEEE International Conference on Service Operations, Logistics and Informatics (SOLI), pp. 443–448. IEEE Press, Etienne (2009)
11. Jerbi, M., Senouci, S.M., Meraihi, R., Ghamri-Doudane, Y.: An Improved Vehicular Ad Hoc Routing Protocol for City Environments. In: IEEE International Conference on Communications (ICC), pp. 3972–3979. IEEE Press, Lannion (2007)
12. Taleb, T., Sakhaee, E., Jamalipour, A., Hashimoto, K., Kato, N., Nemoto, Y.: A Stable Routing Protocol to Support ITS Services in VANET Networks. IEEE Trans. Veh. Technol., 3337–3347 (2007)
13. Namboodiri, V., Gao, L.: Prediction-based routing for vehicular ad hoc networks. IEEE Trans. Veh. Technol.,s 2332–2345 (2007)
14. Leontiadis, I., Mascolo, C.: GeOpps: Geographical Opportunistic Routing for Vehicular Networks. In: IEEE International Symposium on a World of Wireless, Mobile and Multimedia Networks (WoWMoM), pp. 1–6. IEEE Press, London (2007)
15. Yong, D., Li, X.: SADV: Static-Node-Assisted Adaptive Data Dissemination in Vehicular Networks. IEEE Trans. Veh. Technol., 2445–2455 (2010)
16. Zhao, W., Ammar, M., Zegura, E.: A message ferrying approach for data delivery in sparse mobile ad hoc networks. In: Mobile ad Hoc Networking and Computing, pp. 187–198. ACM Press, Tokyo (2004)
17. Yuanzhen, L., Jianxin, L., Tonghong, L., Xiaomin, Z.: A Contention-Based Forwarding Routing Protocol for Vehicular Ad hoc Networks in City Scenarios. Acta Electronica Sinica, 2639–2645 (2009)
18. Jing, Z., Guohong, C.: VADD: Vehicle-Assisted Data Delivery in Vehicular Ad Hoc Networks. IEEE Trans. Veh. Technol., 1910–1922 (2008)
19. Cheng, P., Lee, K., Gerla, M., Härri, J.: GeoDTN+Nav: Geographic DTN Routing with Navigator Prediction for Urban Vehicular Environments. Mobile Netw. Appl., 61–82 (2010)
20. Burgess, J., Gallagher, B., Jensen, D., Levine, B.N.: MaxProp: Routing for Vehicle-Based Disruption-Tolerant Networks. In: 25th IEEE International Conference on Computer Communications (INFOCOM), pp. 1–11. IEEE Press, Barcelona (2006)

Generating Inductive Invariants for Petri Nets

Bin Wu[1,3] and YongJun Fu[2,*]

[1] Department of Basic Course,
Shanghai University of Finance and Economics Zhejiang College,
Jinhua 321019, China
binwu.cs@gmail.com
[2] Jinhua College of Vocation and Technology, Jinhua, 321000, China
jhsfyj@126.com
[3] Shanghai Key Laboratory of Trustworthy Computing,
East China Normal University, Shanghai 200062, China

Abstract. Petri nets play a central role in the analysis of discrete, parallel and distributed systems. Reachability problem is one of the analysis of Petri nets. This paper address the analysis of reachable marking of petri nets using invariants. We translate Petri nets as transition systems. Then, through computing vanishing ideal of sample points, we present a new approach for constructing polynomial equation invariants of Petri nets. Our approach avoids first-order quantifier elimination and cylindrical algebraic decomposition. From the preliminary experiment results, we demonstrate the feasibility of our approach.

Keywords: Petri nets, Reachability, Inductive Invariants, Vanishing Ideals.

1 Introduction

For representing and analyzing discrete, parallel and distributed systems, Petri nets are formal models and first introduced by Petri in 1962 [8,9]. Since then, the study of Petri net is becoming increasingly popular. Reachability problem is one of the most central problem in the analysis of Petri nets. It is well known that the reachability problem is decidable [5], but its bottleneck is at least exponential complexity in the general case. A lot of effort has been devoted by Petri net researchers in order to obtain more efficient algorithm for solve it.

In this paper, for analyzing the reachability marking of Petri nets we present a new method to generate polynomial invariants of petri nets based on computing vanishing ideal of the associated transition systems sample points. In general, abstract interpretation as a traditional method is applied on invariant generation of Petri nets [3]. In addition, some other methods have been proposed to construct invariants of Petri nets. In [10], authors analyzed the reachability problem of Petri nets using Farkas' Lemma to generate invariants. In [1], Presburger

* Corresponding author.

A. Xie & X. Huang (Eds.): Advances in Electrical Engineering and Automation, AISC 139, pp. 259–266.
springerlink.com

arithmetics and real arithmetic have been used to represent the state space of Petri nets, and linear inequality invariants were generated.

This paper presents a technique to generate the polynomial invariants of Petri nets. Given a Petri net $\mathcal{P}$, we model the net $\mathcal{P}$ as the associated transition system Ψ. Then we get finite sample points by recording the values of system Ψ. Next we apply Buchberger-Möller algorithm to compute the vanishing ideal of these sample points as candidate invariants (candidate may not be real invariant), and BM-algorithm remains the amazing polynomial time complexity. Subsequently, we verify the candidate invariants based on **polynomial-scale consecution**. Finally, we can either generate the polynomial invariants or conclude that the polynomial invariants with degree $\leq e'$ do not exist, where e' is the minimal degree of the polynomials in the vanishing ideals.

The rest of the paper is organized as follows. In Section 2, we recall the notions of vanishing ideals for finitely many points, hybrid systems and (inductive) invariants. In Section 3, we present an efficient method to generate polynomial equation invariants for hybrid systems and an example is given. We conclude our results in Section 4.

2 Preliminaries

This section contains a collection of definitions and facts about vanishing ideals of finitely many points, Petri nets and transition systems, which will be needed later. Then we represent Petri nets as transition systems.

Throughout this paper, let $\mathcal{K}$ be a (commutative) field of characteristic zero, $\mathcal{K}[x_1, \ldots, x_n]$ be the ring of polynomials in n indeterminates $x_1, \ldots, x_n$ over $\mathcal{K}$, the monomial order σ in $\mathcal{K}[x_1, \ldots, x_n]$ be the graded lexicographic order, and $\deg(f)$ denote the total degree of a polynomial $f \in \mathcal{K}[x_1, \ldots, x_n]$.

2.1 Vanishing Ideals of Finitely Many Points

Definition 1 (Ideal of Polynomials). *A set $\mathcal{I} \subseteq \mathcal{K}[x_1, \ldots, x_n]$ is an* ideal *in $\mathcal{K}[x_1, \ldots, x_n]$ if for any $f, g \in \mathcal{I}$ and $h \in \mathcal{K}[x_1, \ldots, x_n]$, we have $f + g \in \mathcal{I}$ and $f \cdot h \in \mathcal{I}$.*

For $h_1, \ldots, h_r \in \mathcal{K}[x_1, \ldots, x_n]$ we denote by $\langle h_1, \ldots, h_r \rangle$ the smallest ideal containing $h_1, \ldots, h_r$, i.e.

$$\langle h_1, \ldots, h_r \rangle = \left\{ \sum_{i=1}^{r} f_i h_i \mid f_i \in \mathcal{K}[x_1, \ldots, x_n] \right\}.$$

If $\mathcal{I} = \langle h_1, \ldots, h_r \rangle$, we say that $\mathcal{I}$ is an ideal generated *by $h_1, \ldots, h_r$ and that $h_1, \ldots, h_r$ is a basis of $\mathcal{I}$.*

By Hilbert' Basis Theorem, any ideal in $\mathcal{K}[x_1, \ldots, x_n]$ has a *finite basis.* We can compute Gröbner bases of ideals using Buchberger's algorithm [2].

Definition 2 (Vanishing Ideals of Finitely Many Points). *Let A be a finite set of $\mathcal{K}^n$. The* vanishing ideal *of the point set A is the ideal*

$$\mathcal{I}(A) = \{f \in \mathcal{K}[x_1, \ldots, x_n] \mid f(a) = 0, \text{ for all } a \in A\}$$

of all the polynomials vanishing on each point in A.

Buchberger and Möller presented an algorithm [7] to compute the reduced Gröbner basis of the vanishing idea for finite many points, based on Gaussian elimination on a generalized Vandermonde matrix.

Remark 1. Stated in [6], Buchberger-Möller algorithm is of polynomial time complexity $O(n^2 s^4)$, where n is the dimension of the affine space and s is the number of points.

2.2 Petri Nets and Invariants

Definition 3 (Petri net). *A* Petri net $\mathcal{P}$ *is a tuple $\langle P, T, F, \omega, \mu^{(0)} \rangle$, where*

- *P is the set $\{p_1, p_2, \ldots, p_n\}$ of* places*;*
- *T is the set $\{t_1, t_2, \ldots, t_m\}$ of* transitions*;*
- *$F \subseteq (P \times T) \cup (T \times P)$, the input-output relation;*
- *$\omega : F \mapsto \mathbb{N}^+$ is an assignment of weights $\omega(p,t)$ or $\omega(t,p)$ to arcs $(p,t) \in F$ or $(t,p) \in F$. The value $\omega(p,t)$ represent the token of an input-arc from input-place p to transition t, and $\omega(t,p)$ represent the token of an output-arc from t to output-place p;*
- *$\mu^{(0)} : P \mapsto \mathbb{N}$ is an initial marking.*

A transition $t \in T$ is *enable* in some marking μ if and only if $\forall p \in I(t)$, $\mu(p) \geq \omega(p,t)$ in t, where $I(t)$ denote the set of input-places of $t \in T$. When a transition $t \in T$ fires in μ, it create a new marking μ', such that $\forall p \in I(t) \cup O(t)$, $\mu'(p) = \mu(p) - \omega(p,t) + \omega(t,p)$, where $O(t)$ denote the set of output-places of $t \in T$. The firing of t in the marking $\mu^{(i)}$ is denoted by $\mu^{(i)} \xrightarrow{t} \mu^{(i+1)}$.

Definition 4 (Transition System). *A* transition system Ψ *is a tuple $\langle V, \mathcal{T}, \Theta \rangle$, where*

- *$V = \{x_1, \ldots, x_n\}$ is a set of state variables;*
- *$\mathcal{T}$ is a set of transitions, where each transition $\tau \in \mathcal{T}$ is a tuple*

$$\langle V, V', \rho_\tau, g_\tau \rangle,$$

  such that
  - *ρ_τ is the* transition relation*, a first-order formula, over $V \cup V'$, where V represents the current-state variables and its prime version $V' := \{x'_1, \ldots, x'_n\}$ represents the next-state variables. Here x'_i is the new variable introduced to stand for the values of x_i after the assignment;*
  - *g_τ is the guard of the transition τ. Only if g_τ holds, the transition can take place.*
- *Θ is a first-order formula over V that represents the initial condition.*

A sequence of state $s_0, s_1, \ldots$ is a *run* of a transition system $\Psi : \langle V, \mathcal{T}, \Theta \rangle$, if (1) the initial state satisfies the initial condition, i.e., $s_0 \models \Theta$, and (2) for each pair of consecutive states s_i, s_{i+1} follows a transition $\tau \in \mathcal{T}$ that leads from s_i to s_{i+1}, that is, $s_i \wedge s_{i+1} \wedge g_\tau \models \rho_\tau$.

Definition 5 (Petri nets as Transition Systems). *Given a Petri net $\mathcal{P} : \langle P, T, F, \omega, \mu^{(0)} \rangle$, the transition system $\Psi : \langle V, \mathcal{T}, \Theta \rangle$ is called the associated transition system of $\mathcal{P}$ if*

- *for each $p \in P$ there exists a variable x_p in V;*
- *for each $t \in T$ there exists a transition $\tau \in \mathcal{T}$ with transition relation*

$$x'_p = x_p - \omega(p,t) + \omega(t,p)$$

with the transition guards $\bigwedge_{p \in P} \geq \omega(p,t)$.

- $\Theta = \bigwedge_{p \in P} (x_p = \mu^{(0)}(p))$.

Sankaranarayanan et al. concluded that the conversion preservers reachable marking in [10].

Theorem 1 ([10]). *Let Ψ be the associate transition system of Petri nets $\mathcal{P}$. Then a marking μ is reachable in $\mathcal{P}$ if and only if the corresponding variable assignment is reachable in Ψ.*

Example 1. Let a Petri net show in Figure 1. Its associated transition system is as follow.

$$V = \{x_1, x_2, x_3\};$$
$$\mathcal{T} = \{\tau_1, \tau_2\};$$
$$\Theta = (x_1 = 1, x_2 = 2, x_3 = 3);$$

$$g_{\tau_1} = (x_1 \geq 1 \wedge x_2 \geq 2 \wedge x_3 \geq 2), \qquad \rho_{\tau_1} : \begin{bmatrix} x'_1 = x_1 - 1 \\ x'_2 = x_2 + 1 \\ x'_3 = x_3 - 2 \end{bmatrix};$$

$$g_{\tau_2} = (x_1 \geq 0 \wedge x_2 \geq 2 \wedge x_3 \geq 2), \qquad \rho_{\tau_1} : \begin{bmatrix} x'_1 = x_1 + 1 \\ x'_2 = x_2 - 2 \\ x'_3 = x_3 - 2 \end{bmatrix}.$$

By an *assertion*, we mean a first-order formula over the program variables. We use the notation $s \models \varphi$ to denote that a state s satisfies an assertion φ. We will also write $\varphi_1 \models \varphi_2$ for two assertions φ_1, φ_2 to represent that φ_2 is true at least in all the states in which φ_2 is true.

Next, we introduce the notions of (inductive) invariants for transition systems.

Definition 6 (Invariant). *An* invariant *of the Petri net Ψ is an assertion η over V which holds at all reachable states of Ψ.*

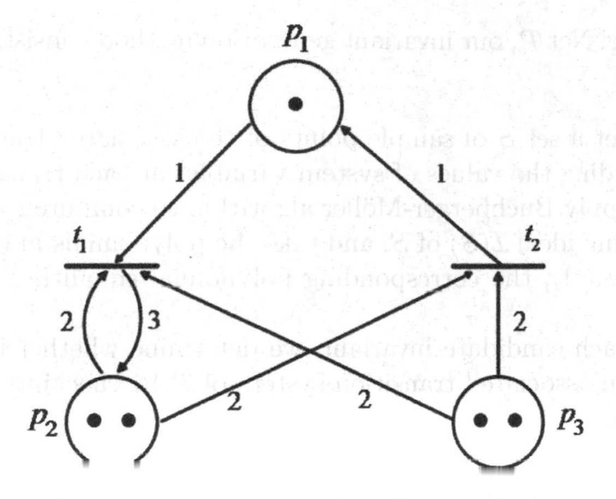

Fig. 1. Petri net of Example 1

Definition 7 (Inductive invariant). *We say that an assertion* η *is* inductive *for a transition system* $\Psi : \langle V, \mathcal{T}, \Theta \rangle$ *if and only if the* Initiation and *Consecution conditions hold:*

Initiation: *η holds initially,* $\Theta \models \eta(l_0)$;

Consecution: *if η holds prior to a transition being taken then it holds in any state obtained after the transition is taken, that is, for all* $\tau \in \mathcal{T}$, $\eta \wedge g_\tau \wedge \rho_\tau \models \eta'$.

It is a well-known result [4] that if η is an inductive assertion map then $\eta(l)$ is an invariant at location l for each $l \in L$. However, an invariant assertion is not necessarily inductive.

In this paper, we are interested in finding inductive invariants of the form $p(V) = 0$, where $p(V)$ is a polynomial over the state variables. For brevity, we shall use η to denote both the assertion $p = 0$ and the polynomial p.

In the sequel, we will use the following stronger but more practical discrete consecution condition defined in [11] or [12].

Definition 8 (Polynomial-Scale Consecution). *Let* τ *be a transition and* η *be an assertion map. We say* η *satisfies* polynomial-scale *(PS) consecution for* τ *if there exists a polynomial* q *over* V *such that*

$$\rho_\tau \models (\eta' - q \cdot \eta = 0).$$

In particular, if $\deg(q) = 0$*, polynomial-scale consecution reduces to constant-scale consecution.*

From Definition 8, to verify whether η satisfies polynomial-scale consecution, it suffices to check whether $\eta \mid \eta'$.

3 Invariant Generation

In this section, we present an approach to generate polynomial equation invariants for Petri nets by computing vanishing ideals of sample points.

Given a Petri Net $\mathcal{P}$, our invariant generation method consists of the following steps:

Step 1: We get a set S of sample points of the associated transition system of $\mathcal{P}$ by recording the values of system variables at each transition τ.

Step 2: We apply Buchberger-Möller algorithm to compute a Gröbner basis of the vanishing ideal $\mathcal{I}(S)$ of S, and take the polynomials in the basis of $\mathcal{I}(S)$, or more exactly, the corresponding polynomial equalities, as candidate invariants.

Step 3: For each candidate invariant, we determine whether it is an invariant of the given associated transition system of $\mathcal{P}$ by checking polynomial-scale consecution.

3.1 Getting a Set of Sample Points

The first step in our method is to get a set S of sample points of the associated transition system of $\mathcal{P}$. Let Ψ be the associated transition system of $\mathcal{P}$ with m transitions and n state variables. Given an upper bound e for the total degree of its potential polynomial invariants. By a classical result from combinatorics, we need at most $\binom{e+n}{n}$ sample points to determine a candidate invariant in n variables with total degree bound e. For each transition τ_i, $i = 1, \ldots, m$, we can get a set S_{τ_i} containing no more than N_{τ_i} sample points by executing consecution. Finally, we will obtain a finite set $S = \bigcup_{i=1}^{m} S_{\tau_i}$ containing no more than $\binom{e+n}{n}$ sample points.

Example 2. For the system introduced in Example 1, set the degree bound of the polynomial invariants to be $e = 1$, we need at most $\binom{e+n}{n} = 4$ sample points. By running the assignment at each transition τ_1, τ_2 with the initial values $(x_1, x_2, x_3) = (1, 2, 2)$, we get sample points:

$$S_{\tau_1} = \{(1,2,2),(0,3,0) \quad \text{and} \quad S_{\tau_2} = \{(1,2,2),(2,0,0)\},$$

$$S = \{(1,2,2),(0,3,0),(2,0,0)\}.$$

3.2 Computing Vanishing Ideal

The second step in our technique is computing the vanishing ideal

$$\mathcal{I}(S) = \langle \eta_1, \ldots, \eta_r \rangle$$

of the point set S as candidate invariants, where $\eta_1, \ldots, \eta_r$ is a Gröbner basis of the ideal $\mathcal{I}(S)$. In addition, we can get the minimal degree $e'(\leq e)$ of all the polynomials in $\mathcal{I}(S)$.

Example 3. Apply Buchberger-Möller algorithm to get a vanishing ideal $\langle \eta_1, \eta_2, \eta_3, \eta_4 \rangle$ of 3 sample points in Example 2 with respect to the graded lexicographical ordering $x_3 \prec x_2 \prec x_1$, where

$$\begin{aligned}\eta_1 &= -2x_3 + x_2x_3\\ \eta_2 &= x_3^2 - 2x_3\\ \eta_3 &= x_3 - 3x_2 + x_2^2\\ \eta_4 &= 6x_1 - x_3 + 4x_2 - 12\end{aligned}$$

3.3 Verifying the Candidate Invariants

The third step in our approach is verifying whether each candidate in $\{\eta_1 = 0, \ldots, \eta_r = 0\}$ is an actual invariant. Clearly, all the η_i satisfy **Initiation** condition, since they belong to the vanishing ideal of sample points of the associated transition system of $\mathcal{P}$. Therefore, the remaining task is to determine whether η_i satisfies **polynomial-scale consecution**. According to Definition 8 we only need check whether $\eta_i \mid \eta_i'$.

As a result, we can either generate the polynomial invariants of total degree $\leq e$ or conclude that polynomial invariants with degree $\leq e'$ do not exist.

Example 4. Consider the **polynomial-scale consecution**, we determine whether $\eta_i(x_1, x_2, x_3) \mid \eta_i(x_1', x_2', x_3')$ for $i = 1, 2, 3, 4$. We only get that $\eta_4(x_1, x_2, x_3) \mid \eta_4(x_1', x_2', x_3')$.

We obtain the invariant of the Petri Net $\mathcal{P}$:

$$6x_1 - x_3 + 4x_2 - 12 = 0.$$

Moreover, we conclude if $\eta(x_1, x_2, x_3) = 0$ is an invariant of the Petri Net $\mathcal{P}$, the minimal degree of the polynomials $\eta(x_1, x_2, x_3)$ is $e' = 1$.

4 Conclusions

In this paper, we present a new method to generate polynomial equation invariants for Petri nets. We first translate Petri nets as transition systems and compute vanishing ideals of transition system sample points to get candidate invariants, then check whether the candidates satisfy **polynomial-scale consecution** based on Definition (8). Finally, we obtain the polynomial equation invariants of Petri nets. Our approach avoids first-order quantifier elimination and cylindrical algebraic decomposition as well as they do not depend on any abstraction interpretation methods. However, the inequality invariants can not be constructed by our approach. How to further improve the efficiency of our method is still a big challenge and our main future work.

Acknowledgments. We are so grateful to Dr. Min Wu and Dr. Zhengfeng Yang for their contribution to the previous work as well as lots of fruitful discussions on this work and valuable comments on the draft of this paper. This research was partly supported by the NSFC projects No.90718041.

References

1. Bérard, B., Fribourg, L.: Reachability Analysis of (Timed) Petri Nets Using Real Arithmetic. In: Baeten, J.C.M., Mauw, S. (eds.) CONCUR 1999. LNCS, vol. 1664, pp. 178–193. Springer, Heidelberg (1999)
2. Buchberger, B.: Gröbner-bases: An algorithmic method in polynomial ideal theory. In: Multidimensional Systems Theory - Progress, Directions and Open Problems in Multidimensional Systems, pp. 184–232. Reidel Publishing Company, Dodrecht (1985)
3. Clarisó, R., Rodríguez-Carbonell, E., Cortadella, J.: Derivation of Non-structural Invariants of Petri Nets Using Abstract Interpretation. In: Ciardo, G., Darondeau, P. (eds.) ICATPN 2005. LNCS, vol. 3536, pp. 188–207. Springer, Heidelberg (2005)
4. Floyd, R.W.: Assigning meanings to programs. In: Schwartz, J.T. (ed.) Proceedings of Symposia. Applied Mathematics on Mathematical Aspects of Computer Science, vol. 19, pp. 19–32. American Mathematical Socity, Rhode Island (1967)
5. Kosaraju, S.R.: Decidability of reachability in vector addition systems. In: STOC 1982: Proceedings of the Fourteenth Annual ACM Symposium on Theory of Computing, pp. 267–281. ACM Press, New York (1982)
6. Marinari, M.G., Möller, H.M., Mora, T.: Gröbner bases of ideals defined by functionals with an application to ideals of projective points. Applicable Algebra in Engineering, Communication and Computing 4(2), 103–145 (1993)
7. Möller, H.M., Buchberger, B.: The construction of multivariate polynomials with preassigned zeros. In: Calmet, J. (ed.) ISSAC 1982 and EUROCAM 1982. LNCS, vol. 144, pp. 24–31. Springer, Heidelberg (1982)
8. Petri, C.A.: Kommunikation mit Automaten. Bonn: Institut fü Instrumentelle Mathematik, Schriften des IIM Nr. 2 (1962)
9. Petri, C.A.: Fundamentals of a theory of asynchronous information flow. In: Proc. of IFIP Congress, vol. 62, pp. 386–390. North Holland Publ. Comp, Amsterdam (1963)
10. Sankaranarayanan, S., Sipma, H., Manna, Z.: Petri Net Analysis Using Invariant Generation. In: Dershowitz, N. (ed.) Verification: Theory and Practice. LNCS, vol. 2772, pp. 682–701. Springer, Heidelberg (2004)
11. Sankaranarayanan, S., Sipma, H., Manna, Z.: Constructing Invariants for Hybrid Systems. In: Alur, R., Pappas, G.J. (eds.) HSCC 2004. LNCS, vol. 2993, pp. 539–554. Springer, Heidelberg (2004)
12. Sankaranarayanan, S., Sipma, H.B., Manna, Z.: Non-linear loop invariant generation using gröbner bases. In: POPL 2004: Proceedings of the 31st ACM SIGPLAN-SIGACT Symposium on Principles of Programming Languages, pp. 318–329. ACM Press, New York (2004)

Contrasting Accuracies of Regions' Statistical Data with B-F Problem

Minqing Gong[1,2], Mingwei Sun[1], and Mingzhong Jin[2]

[1] Science College, Guizhou University, Guiyang, 550025 China
[2] Guizhou University for Nationalities, Guiyang, 550025 China
http://www.springer.com/lncs

Abstract. In this paper we evaluated and analyzed the differences of the accuracy qualities of statistical data from different regions with the Behrens[1]–Fisher[2] problem and applied Scheffe transformation[3] to solve the several-sample B-F problem. We selected the indexes' data samples dated from 1999 to 2008 from three regions, contrasted the accuracy qualities of the data and ranked the qualities. The method here could also be utilized in the other aspects of data qualities, which makes it possible to construct the data evaluation and comparison system from the eight aspects of the data quality.

Keywords: Quality of data, B-F problem, Empirical analysis, Test.

1 Introduction

Behrens[1]-Fisher[2](B-F) problem tests the equality of the mean vectors of two normal populations with unknown and unequal covariance structures and requires to give the interval estimation. For the univariate B-F problem, Scheffe[3] transformed the original two-sample to a one-sample problem. Welch[4] proposed an approximate degrees of freedom method based on Student's. For the multivariate, Bennett[5] extended Scheffe's method[3]. Anderson[6] and Welch's[4] also extended. Zhang, J. etc.[7] researched the k-sample high-dimensional B-F, and proposed L^2−test in the situations T^2−test could not be utilized.

About region governmental data assessments Yu, G.[8] presented a review for a data quality seminar in China. Xian, Z.[8] said internationally data quality was evaluated mainly from eight aspects of applicability, timeliness, accuracy, comparability, connectivity, accessibility, transparency and effectiveness. Feng, L.[8] highlighted heteroscedasticity difficulty. Yu, F.[9] also studied the evaluation system abroad. Zhang, F.[10] made an evaluation indicator system. Lu, E.[8] evaluated GDP data quality via Robust MM estimates[11]. Liu, H. etc.[12] made use of combined models to time series, analyzed the difference between predicted and actual value. Jin, Y. etc.[13] summarized domestic researches in governmental data quality. Sun, M. etc.[14] constructed B-F problem of regional data quality of accuracy through modeling the change of time series data and tested significant difference between the qualities. As the successive of paper[14] the essential here lies in the method to contrast and rank the accuracies qualities of different regions' data.

A. Xie & X. Huang (Eds.): Advances in Electrical Engineering and Automation, AISC 139, pp. 267–273.
springerlink.com

2 Data Evaluation with Extension of Scheffe's Method[7]

With the method of extension of Scheffe's transformation[3,7] we may reduce multi-samples B-F to one-sample, the distribution of one-sample problem is

$$z_1, z_2, ..., z_n \ iid \sim \ N_q(\mu, \Sigma), \tag{1}$$

where $N_q(\mu, \Sigma)$ denotes a q -dimensional normal distribution with mean vector μ and covariance Σ, q is very large, μ and Σ are unknown. The hypotheses are:

$$H_0 : \mu = \mu_0, \ \ H_1 : \mu \neq \mu_0, \tag{2}$$

where μ_0 is a certain $q-$dimensional vector. Considering the multi-samples B-F as: assuming that we have the following k independent normal samples $x_{lj}, j = 1, 2, ..., n_l \sim N_p(\mu_l, \Sigma_l), l = 1, 2, ..., k$. The multi-samples B-F refers to the problem of testing whether the k mean vectors are equal each other:

$$H_0 : \mu_1 = \mu_2 = ... = \mu_k, \tag{3}$$

without assuming that the covariance $\Sigma_l, l = 1, 2, ..., k$ are equal.

Then we will present Scheffe's method[3] which transforms two normal samples into one. Without loss of generality, we assume that $n_1 \leq n_2 \leq ... \leq n_k$. Denote $\bar{x}_l(m) = m^{-1} \sum_{j=1}^{m} x_{lj}$ as the partial sample mean of the first m observations of the $l-th$ sample. Obviously, $\bar{x}_l(m)$ is an unbiased estimator of μ_l. Define

$$y_{lj} = [x_{1j} - \bar{x}_l(n_l)] + \sqrt{\tfrac{n_1}{n_l}}[x_{lj} - \bar{x}_l(n_1)], \ j = 1, 2, ..., n_1, \ l = 2, ..., k.$$

Combining the resulting $k-1$ samples into one via above Scheffes[3], we have:

$$z_j = [y_{2j}^T, y_{3j}^T, ..., y_{kj}^T]^T, \ \ j = 1, 2, ..., n_1; \ \ z_1, z_2, ..., z_{n_1} \ iid \ \sim N_q(\mu, \Sigma). \tag{4}$$

where

$$q = (k-1)p, \ \ \mu = [\mu_1^T - \mu_2^T, \mu_1^T - \mu_3^T, ..., \mu_1^T - \mu_k^T]^T,$$

$$\Sigma = J_{k-1} \otimes \Sigma_1 + n_1 diag(\frac{\Sigma_2}{n_2}, \frac{\Sigma_3}{n_3}, ..., \frac{\Sigma_k}{n_k}). \tag{5}$$

$A \otimes B$ denotes the Kroneck product of two matrices A and B, J_{k-1} the $(k-1) \times (k-1)$ matrix with all entries being 1, $diag(A_1, A_2, ..., A_l)$ the block diagonal matrix with entries $A_1, A_2, ..., A_l$. In this way, the variables in original $k-$sample B-F problem (3) are reduced to one-sample variable based on the iid (independent identically distributed) normal sample (4), hence the testing method of one-sample problem can be used. From Sun, M. etc.'s [14], it can be seen that when q is much smaller than n_1, the problem can be tested by the Hotelling's T^2-test. The variables in (4) are independently normally distributed with mean vector $\mu = [\mu_1^T - \mu_2^T, \mu_1^T - \mu_3^T, ..., \mu_1^T - \mu_k^T]^T$, which indicates that $\mu = \mathbf{0}$ if and only if $\mu_1 = \mu_2 = ... = \mu_k$. Therefore, testing the null hypothesis of the original $k-$sample B-F problem (3) is equivalent to testing the null of the one-sample problem (2). μ can be estimated by the associated sample mean vector

$\bar{z} = [\bar{x_1}^T - \bar{x_2}^T, ..., \bar{x_1}^T - \bar{x_k}^T]^T$. $\bar{z}$ is invariant to the orders of the k−sample and is most relevant to the tests of (3) and (2).

For time series data it is applicable to model according to the change of the data, fit the trend of the sequence, and then assess the quality of statistical data by comparing the actual value with the predicted value of the model. As for a non-stationary time sequence X_t, it can be described using the following model[12]: $X_t = f(t) + \nu_t$, where $f(t)$ denotes the means of X_t over time which is the deterministic part of the sequence, ν_t denotes the rest of X_t.

Statistical data of good quality can reflect the real situation of the corresponding areas accurately, and can be used to predict the corresponding future values effectively controlling the error in the effective range. The quantitative analysis of the accuracy of the statistics is to model by using the changes of statistical indicators' data over time, making a series of statistical tests for the model, comparing the values fitted by the model and the real values on the basis of model getting through a variety of tests and having property of giving good statistical predictions, calculating the deviations. We will present relative error method in paper[12] to assess the data quality.

Comparing statistical indicator's real value with the corresponding predicted value calculated according to the model to calculate the relative error ε, the $n-th$ relative error is:

$$\varepsilon_n = \frac{X_n - \widehat{X}_n}{X_n}. \tag{6}$$

For a given permissible error range, if $|\varepsilon_n|$ is in the error range, then the statistical indicators' corresponding data can be initially judged reliable. We will fit the selected statistics data of regions by trend evaluation method, compare the fitted values with the actual values, calculate the absolute value of the relative error to assess the quality, and build multi-sample B-F problem of the quality of regional statistics data. Then we will test if there are significant differences between the qualities, and rank the qualities in order.

3 Positive Analysis

We selected three indicators, Gross Domestic Product (GDP), Fixed Asset Investment (FAI) and Urban Residents Final Consumption Expenditure (FCE). The data were not converted by the constant prices conversion coefficients since the current price had no effect on our problems. And the data were taken from 3 administrative regions in China, referred to as B, G and Y region respectively, from 1999 to 2008[15]. Then we fitted the trends of the indicators' data using trend fitting assessment method, calculated the fitted values of corresponding years respectively by fitting equations and got their absolute values of relative errors that was used to measure data quality, built multi-sample B-F problem of data quality, tested whether there were significant differences among the accuracies of regions' data, ordered and comparably tested their accuracy qualities. When 3 regional statistical data plotted on time series, the three indicator sequences were non-smooth as most other economic series and rose exponentially approximately. So we chose exponential model to fit the growth trend.

First, with Matlab we calculated out the fitting equations between time t and the actual observations of the three indicators respectively from B and Y region. Then the corresponding years t were generated into the equations, calculating the fitted values of every indicator of the corresponding year. After that we put the corresponding fitted values and the actual observed values into the formula (6), calculated the absolute values of relative errors of the corresponding index for that year, whose values reflected the accuracy quality of corresponding data for that year in the region. For instance GDP of B region increased exponentially with the rising time t from 1999 to 2008. With Matlab, it was calculated that the GDP fitting formula of B region was as: $\hat{G} = 1.1688e^{7.7462(t-1998)}$. Then the value of year t was put into the above equation and the GDP fitted value of B region in year t was obtained. According to B region's corresponding GDP actual observations G which was put into the formula (6) with the fitted value $\hat{G}$, the corresponding relative error ε could be calculated, and whose absolute value was obtained. With this method we calculated the absolute values of relative errors (AVRE) of the data of three indicators from B, Y and G region, dated from 1999 to 2008[15] respectively, and we contained the AVREs in Table 1.

Table 1. Absolute Value of Relative Errors(AVRE) of GDP, FAI and FCE in 3 regions

	AVRE of B Region:			AVRE of Y Region:			AVRE of G Region:		
Year	GDP AVRE	FAI AVRE	FCE AVRE	GDP AVRE	FAI AVRE	FCE AVRE	GDP AVRE	FAI AVRE	FCE AVRE
1999	.0096	.0042	.0520	.1003	.1998	.1728	.0373	.0887	.0542
2000	.0004	.0524	.0093	.0381	.0533	.0419	.0327	.0068	.0061
2001	.0047	.0289	.0804	.0239	.0685	.0586	.0069	.0462	.1082
2002	.0032	.0017	.0085	.0714	.1805	.0370	.0475	.1047	.0814
2003	.0043	.0417	.0044	.0972	.1718	.0063	.0415	.0314	.0228
2004	.0269	.0499	.0088	.0299	.1058	.0977	.0206	.0130	.0934
2005	.0010	.0183	.0065	.0344	.0209	.0582	.0043	.0311	.0459
2006	.0238	.0271	.0166	.0147	.0396	.0234	.0001	.0104	.0022
2007	.0069	.0516	.0117	.0295	.0631	.0725	.0163	.0093	.0063
2008	.0096	.1224	.0107	.0851	.0832	.0458	.0305	.0113	.0414
(sum)	(.0904)	(.3982)	(.2089)	(.5245)	(.9865)	(.6142)	(.2377)	(.3412)	(.4619)

Data refer to Yearbook-(2009)[15]: B means Beijing, Y Yunnan and G Guangdong

Then the problem could be researched by the methods of two-sample B-F problem which can be transformed into single sample problem by the method of extension of Scheffe's transformation, which was utilized to solve B-F problems described in section 2. The one-sample problem was as: $z_1, z_2, ..., z_{10},\ iid \sim N_q(\hat{\mu}, \hat{\Sigma})$ approximately(similarly hereinafter), where the sample size $n = 10$ and dimension $q = 3$. The estimating values of μ and Σ were as follows:

$\hat{\mu} = (-.0434 - .0588 - .0405)$, $\hat{\Sigma} = \begin{pmatrix} .0011 & .0017 & .0003 \\ .0017 & .0052 & .0007 \\ .0003 & .0007 & .0028 \end{pmatrix}$. We tested μ by nonnormal Hotelling (similarly hereinafter)T^2−test: $H_0 : \mu = \mu_0, H_1 : \mu \neq \mu_0$ (*here* $\mu_0 = \mathbf{0}$), took the test statistics: $T^2 = n(\bar{z}-\mu_0)^T \Sigma^{-1}(\bar{z}-\mu_0)$, calculated that $T^2 = 20.1067$, obtained by $T^2_{\alpha,n} = \frac{(n-1)q}{n-q} F_{\alpha,q,n-q}$ and $F = \frac{n-q}{(n-1)q} T^2 \doteq 5.2128$. For the given significance level $\alpha = 0.05$, by the Matlab p value could be calculated out: $p = P\{F \geq 5.2128\} = .0148$, where the test statistic $F \sim F(3,7)$. As $p = .0148 < .05 = \alpha$, H_0 should be refused, means there was significant difference between the accuracies' qualities of B and Y regions' data. To rank the accuracies of the three indicators' data for the two regions, we proposed hypothesis comparing the mean values of the samples of AVREs as: $H_0 : \mu_1 < \mu_2$, $H_1 : \mu_1 \geq \mu_2$, where μ_1 and μ_2 denoted the means of relative errors' absolute values of indicators' data of B and Y region respectively. For a given level of significance $\alpha = .05$, using Matlab to calculate, we introduced parameters $[H,p]$ in the results. Where $H = 0$ denoted that the original hypothesis could not be rejected under level α , $H = 1$ denoted be rejected. p was the probability to obtain the observed values when H_0 was true.

Suppose the original hypothesis was the quality of data of B region was better than that of Y region: $H_0 : \mu_{Bi} < \mu_{Yi}$, $H_1 : \mu_{Bi} \geq \mu_{Yi}$ $(i = 1,2,3)$. (By Table 1 each sum of AVREs of B region < that of Y region, we can use $H_0 : \mu_{Bi} = \mu_{Yi}$, $H_1 : \mu_{Bi} > \mu_{Yi}$ $i = 1,2,3$ instead of above hypothesis. We will treat the similar test situation below as the way here.) From T^2−statistic the result of hypothesis test were as follows: $[H_1, p_1; H_2, p_2; H_3, p_3] = [0, 0.9978; 0, 0.9764; 0, 0.9929]$. As all H_i equaled to 0 and the values of p_i were close to 1, the qualities of statistical data of B region was better than Y region's.

Similarly, we compared the accuracies of regional data of B and G region. Also first we calculated the AVREs in Table 1. Then we transformed the two-sample B-F problem constructed by the AVREs of B and G into single sample problem by the method of extension of Scheffe's transformation: $z_1, z_2, ..., z_{10}$ *iid* $\sim N_q(\hat{\mu}, \hat{\Sigma})$ approximately where $\hat{\mu} = (-.0147\ .0045\ -.0253)$ and $\hat{\Sigma} = \begin{pmatrix} .0004 & .0003 & .0000 \\ .0003 & .0024 & .0004 \\ .0000 & .0004 & .0021 \end{pmatrix}$. Further, we tested its mean vector by the nonnormal Hotelling's T^2−test: $H_0 : \mu = \mu_0$ $H_1 : \mu \neq \mu_0$ (here $\mu_0 = \mathbf{0}$). Similar to the analysis of B and Y region, we calculated the test statistic: $T^2 = 11.5566$, $F = 2.9961$. For the level of $\alpha = .05$ as $p = P\{F > 2.9961\} = .0637 > .05 = \alpha$, H_0 was accepted. Namely, there was no significant difference between the accuracies of the quality of the data from B and G region. But if $\alpha = 0.1$, H_0 was refused. Further we decided and tested the ranking of the data quality accuracies. We compared and tested the mean of the samples of AVREs of the indicators respectively, payed attention to FAI AVRE of B > that of G a little bit. H_0 means that the accuracy quality of data from B region was better than that from G region: $H_0 : \mu_{Bi} < \mu_{Gi}$, $H_1 : \mu_{Bi} \geq \mu_{Gi}$ $(i = 1,2,3)$ (Treat here as in Note above, just pay attention to FAI AVRE of B > that of G a little bit). For the given level $\alpha = .05$, as the above analysis of data qualities of B and Y region, with nonnormal Hotelling T^2−distribution the result of the test was calculated as: $[H_1, p_1; H_2, p_2; H_3, p_3] = [0, 0.9713; 0, 0.4132; 0, 0.9803]$. It could be seen that the quality of data from B region was better than that from G region. That FAI AVRE of B > that of G a little bit has relation with 0.4132 here.

And then we analyzed the accuracy qualities of data from G and Y region, obtained the estimated values of μ and Σ of the single sample problem transformed from the two-sample B-F problem constructed by AVREs of G and Y as: $\hat{\mu} = (-.0287\ -.0633\ -.0152)$, $\hat{\Sigma} = \begin{pmatrix} .0013 & .0020 & .0003 \\ .0020 & .0053 & .0016 \\ .0003 & .0016 & .0036 \end{pmatrix}$. Also tested its mean vector by the nonnormal Hotelling's T^2−test: $H_0 : \mu = \mu_0$ $H_1 : \mu \neq \mu_0$ (here $\mu_0 = \mathbf{0}$). The test statistics were calculated: $T^2 = 8.0856$, $F = 2.0963$. We computed that $p = P\{F \geq 2.0963\} = .1342$. If the level $\alpha = .05$, $p = .1342 > .05 = \alpha$ means H_0 was accepted. Namely there was no significant difference. Even if $\alpha = .10$, H_0 was also accepted. But 0.1342 was very closed to 0.1 it was hard to accept H_0 and was also difficult to refuse the test result that there was no significant difference.

As in the note above we decided and tested the ranking of the data quality accuracies, compared and tested the mean of the samples of AVREs of the three indicators respectively. H_0 was that the quality of data from G region was better than that from Y region: $H_0 : \mu_{Gi} < \mu_{Yi}$, $H_1 : \mu_{Gi} \geq \mu_{Yi}$ $(i = 1, 2, 3)$ (Treat here as in the Note above). For level $\alpha = .05$ We obtained the test result by Matlab as follows: $[H_1, p_1; H_2, p_2; H_3, p_3] = [0, 0.9989; 0, 0.9992; 0, 0.8186]$. From the result we can see that the quality of data from G region was better than that from Y region.

4 Discussion

We researched the accuracy qualities of regional data on indicators GDP, FAI and FCE from B, G and Y region, constructed the B-F problem of the data, tested the differences of the accuracies, obtained the ranking of the accuracies qualities of the data from the three regions: B region's was better than both G region's and Y region's, and G region's was better than Y region's. Further this method can be utilized to compare the accuracies of data qualities of different historical periods in a region. With the method the data qualities of different regions can be assessed into several ranks. Further more by quantitative way it is possible to apply this method in the qualities of statistical data quality's eight aspects, thus, this method is also of a certain meaning and effect in the other aspects of improving statistical data qualities besides accuracy.

Acknowledgment. NNSF of China (111061008/A0112), Guizhou Dept. of Sci. and Tech.(QianKeHe GYZi[2011]3055; QianKeHeWaiGZi[2010]7011; QianKeHe-JZi[2010]2136), Guiyang Bureau of Sci. and Tech.(ZhuKeHeTong [2011101]), Bijie District and Guizhou Univ.:(BiXun ZhuanHeZi(2010) SK003), Guizhou Univ.

for Nationalities.(State Ethnic Affairs Commission Funding 2010, Guizhou Governor Funding 2010).

References

1. Behrens, B.V.: Ein Beitrag zur Fehlerberechnung bei wenige Beobachtungen. Landwirtch. Jb. 6, 807–837 (1929)
2. Fisher, R.A.: Fiducial Argument in Statistical Inference. Annals of Eugenics 6, 141–172 (1935)
3. Scheffe, H.: On solutions of the Behrens-Fisher problem, based on the t-distribution. Ann. Math. Statist. 14, 35–44 (1943)
4. Welch, B.L.: The generalization of Student's problem when several different population variances are involved. Biometrika 34, 28–35 (1947)
5. Bennett, B.M.: Note on a solution of the generalized Behrens-Fisher problem. Ann. Inst. Stati. Math. 2, 87–90 (1951)
6. Anderson, T.W.: Introduction to Multi. Stati. Analysis, p. 178. Wiley, N.Y (1984)
7. Zhang, J., Xu, J.: On the k-sample Behrens-Fisher problem for high-dimensional data. Science in China Series (A) Mathematics 6, 1285–1304 (2009)
8. Yu, G.: Review and Summary of 2010 National Government Statistics Data Quality Seminar in China. Statistical Research 27(9), 109–112 (2010)
9. Yu, F.: Quality Implication of Foreign Statistics Data and Relevant Management and Suggestions for Our Country. Statistical Research 2, 26–29 (2002) (in Chinese)
10. Zhang, F.: Establishment of Evaluation Index System of Governmental Statistical Data Quality. Statistics and Decision 3, 53–54 (2007) (in Chinese)
11. Lu, E.: Review of Statistical Data Quality Assessment methods. Statistics and Decision 12, 70–71 (2006) (in Chinese)
12. Liu, H., Huang, Y.: Study on Evaluation Methods of Quality of Statistics Data in China-Trend. Simul. and Appl. Statistical Research 8, 17–21 (2007) (in Chinese)
13. Jin, Y., Tao, R.: Statistical Data Quality Theory and practice in China. Statistical Research 1, 62–67 (2010) (in Chinese)
14. Sun, M., Gong, M., Jin, M.: The Research on Behrens-Fisher Problem of Regional Statistical Data quality (2010) ISBN:9781935068426 US-SRP Pub. 1926-29
15. National Bureau of Statistics: China Statistical Yearbook-(2009). China Statistics Press, Beijing (2009)

Research on Heterogeneous Wireless Network Identity Based on IHS

ZuoBin Yang and Xin Yang

Message Control Department, Air Defense Forces Academy,
Zhengzhou, 450002, China
ggbbx@sohu.com

Abstract. Aiming at the low efficiency and the safety short-board effects by different authentication mechanism in vary wireless networks, the paper proposes IHS mechanism, and the research of the identifier mapping in this mechanism has been take, which provides network with the global memory, mapping and inquires of all the identifier information. It is important to communicate safely in heterogeneous wireless network.

Keywords: IMS, HIS, Identity.

1 Introduction

The integration of heterogeneous networks is the development trend of future networks. Currently vary networks have their own architectures and access technologies, if in the network integration process the same users still has multiple identities, and different network maintain sets of authentication facilities simultaneity, it is bound to greatly reduce the efficiency of heterogeneous integrated network access, and reduce the integration network security for short-board effect of several sets of authentication mechanism. The dual role of IP led to some problems difficult to solve, and constraint its performance improved. So it appears the idea of the separation of identity and location, such as HIP [1] (Host Identity Protocol), Enke's LISP [2] (Locator / ID Separation Protocol) and so on.

Communicating parties initiate communication with identity after separation technology has been taken, for the user identity does not contain the information similar to the location of the network prefix, so it needs the complete set of location management solution that enables users to access the router according to the identity information to obtain the partner location. In integration of heterogeneous wireless networks, it needs deploy identity mapping server in each district for Identity Hiding and Substitution (IHS) mechanism to store the mapping between access identifier and location of terminals, provide the inquiry service from the identity to the location information. When the terminal removes, to update the information stored in the server will be need to avoid a communication failure by error location information.

A. Xie & X. Huang (Eds.): Advances in Electrical Engineering and Automation, AISC 139, pp. 275–280.
springerlink.com

2 Identity Hidden Substitution Mechanism (IHS)

In heterogeneous wireless Integration networks, IHS mechanism applicants, the domain server implements the user temporary identity distribution, local mapping memory, mapping analysis and other functions; IMS provides global memory, analysis and other relevant functions of all identity mapping.

To improve the query efficiency of mapping, the cache table to quickly find mapping is set in the certification entity server, it stores the corresponding mapping between the entities. By querying and response messages the entities complete the query process. The design of identity mapping information memory system has an important impact to establish the correctness of communication.

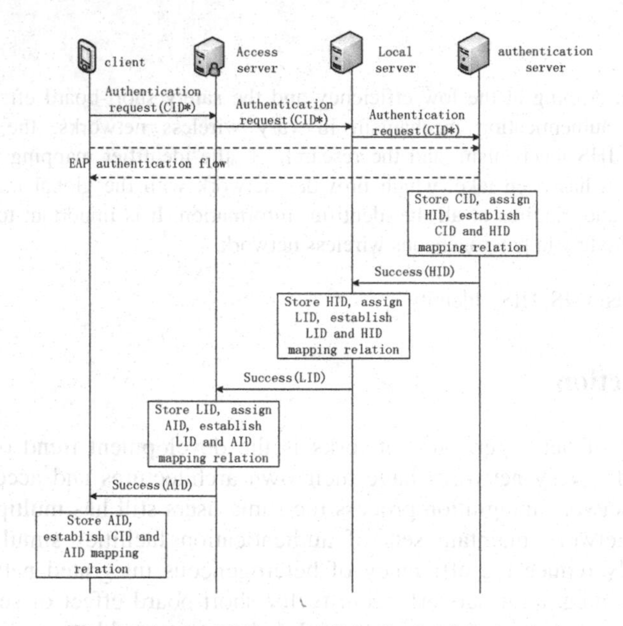

Fig. 1. Identity Hiding and Substitution Mechanisms

3 Identity Mapping Information

In heterogeneous wireless Integration networks, IHS mechanism applicants, the domain server implements the user temporary identity distribution, local mapping memory, mapping analysis and other functions; IMS provides global memory, analysis and other relevant functions of all identity mapping.

To improve the query efficiency of mapping, the cache table to quickly find mapping is set in the certification entity server, it stores the corresponding mapping between the entities. By querying and response messages the entities complete the query process. The design of identity mapping information memory system has an important impact to establish the correctness of communication.

3.1 Identity Mapping Server

3.1.1 Memory Content

IMS (Identifier Mapping Server) stores the mapping relationship of the external permanent identity MSISDN, user home address and current local address of various user terminals in heterogeneous network. It is the most important memory entity in communication network.. When a user initiates communication, he obtains the remote user's home address and current address by querying the MSISDN in the IMS table, then sent the packet with partner's HID to the his local server to communicate , and update the partner memory table in time in the initiator and the partner's home server.

Table 1. Identifier mapping server memory unit

Client Communication Identifier	Home Realm Finger	Local Realm Finger
MSISDN	Home_REG	CUR_REG

Table 1 is the identifier mapping server memory cell, it shows the user identity and location information need to preserve. MSISDN is the client external permanent identity; Home_REG is the client home server address; CUR_REG is the client current geographical server address. We can get an client current address in the network by these info.

According to the different home domain pointer field, the mapping information memory table has three types: owner region memory table, outside region memory table and temporary Memory Table.

3.1.2 Identifier Memory Mode

In resolving the identity mapping server memory issues, drawing on existing methods and, As Chord protocol [3] in the P2P network has some advantages as a good distribution, load balancing, and high query efficiency and so on, we use for reference of existing memory method, and propose the identity mapping memory method based on IHS mechanism, that is the DHT algorithm, which construct the Chord ring with a fast query capabilities.

To map key identifier to the node identifier, and which corresponding the node stores the key identifier information. All the (key, value) constitute a large file index hash table. As the standard DHT, the table stores the mapping information in the mapping server which closest to its key value. When searching, to calculate each key of keyword identifier with the same hash algorithm, and get corresponding memory address according key identifier, which can quickly locate the resources.

3.2 Domain Server

3.2.1 Memory Content

Every domain server is an important function entity in HIS mechanism, they need to complete the function of the part mapping of the local identity distribution and memory. To set the local cache in server is to improve the efficiency of mapping query. The client identity memory in server could be divided into two parts:

1) Local Client Mapping Table

The identity mapping information of the client which has login is stored in local mapping table in each server, and in each domain the corresponding server distribute appropriate temporary identity, so there are (UID, HID), (HID, LID), (LID, AID), then append which in the local client mapping table to substitute the identity on data packet facilitate, and to improve the efficiency of identity substitution.

As well as the local access server, shows in Table 2.

Table 2. Mapping Server temporary table memory unit

Access Identity	Local Identity			Overdue Time	
AID	LID	FLAGS	N-LID	Time Update	Time Delete
				UPDATETIME	DELTIME

2) Partner Mapping Table

When the communication is connected, both of the local server store part of the partner mapping information in its own partner mapping table. In addition, in order to achieve the support of terminal mobility, how to quickly find the partner local identity is also very important, so which is stored in partner mapping table can better improve the efficiency of packet replacement operations.

Table 3. Home server partner mapping table memory unit

Communication Identifier	Local Identity	Overdue Time		Local Finger
		Time Update	Time Delete	
AID	LID	UPDATETIME	DELTIME	CUR_REG

Table 3 shows the local partner memory unit, MSISDN and LID are user ID, UPDATETIME and DELTIME are timing updating and deletion unit, CUR_REG is the local address of the partner.

3.2.2 Server Memory Mapping Table

To consider from the communication performance, the query of identity should be quick and accurate in communication process of identity substitution transmission. The mapping table memory and research efficient have a crucial effect foe identity substitution speed.

In communication connection, a client may be the caller or the responder. For example, in access server the caller query LID by AID, and it is opposite when it s the responder, so one-way hash can not meet the needs of the system, we proposes D-Hash (Double Hash) method, show as in Figure 2. It adopts the two-way hash to the memory and search.

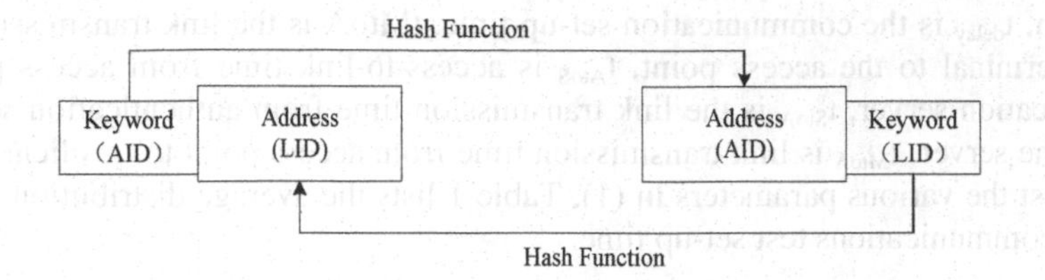

Fig. 2. Access server memory mapping table

4 Identity Mapping Information Experimental Test

Test environment build to test the unified authentication platform established in heterogeneous wireless integration network to verify whether user could authenticate and access normally the heterogeneous network. And Connectivity test and access delay analysis have been take to verify the feasibility of the unified authentication platform.

Figure 3 shows the testing network topology model, including user terminals, an important part of the access points and authentication servers.In test process, the system runs well, the function of all network elements are normal, all messages exchange successfully. In the Models, the terminal of 3G, WLAN and WiMAX are test on the network access authentication. The terminals could transmit and receipt data normally, which verifies the feasibility of unification authentication of the heterogeneous wireless network access authentication.

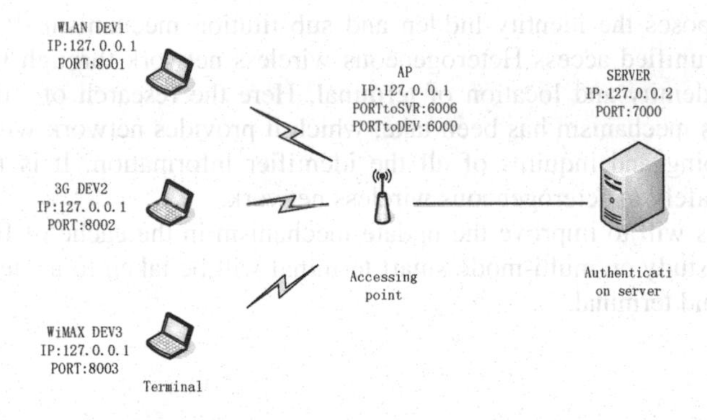

Fig. 3. Network topology model of test environment

Heterogeneous network terminal needs to switch data format to access; we define $t_{MStoAPData}$ as data conversion delay. Communication set-up time can be described as (1):

$$t_{delay} = t_{MtoA} + t_{MStoAPData} + t_{AtoS} + t_{StoA} + t_{MStoAPData} + t_{MtoA} \quad (1)$$
$$= 2(t_{MtoA} + t_{MStoAPData} + t_{AtoS})$$

In which, t_{delay} is the communication set-up time, tMtoA is the link transmission time of the terminal to the access point, t_{AtoS} is access-to-link time from access point to authentication server, t_{StoA} is the link transmission time from authentication server to access the server, t_{MtoA} is link transmission time from access point to client terminal. To test the various parameters in (1), Table 1 lists the average distribution of numbers of communications test set-up time.

Table 4. Communication set-up time distribution (ms)

Parameter	Values	Percentage
$2t_{\text{MtoA}}$	0.189	19.05%
$2t_{MStoAPData}$	0.656	66.13%
$2t_{StoA}$	0.174	14.82%
t_{delay}	1.984	100%

Results show that the link transmission time between terminals and access server, between access server and authentication server is almost negligible;The total delay time of packet conversion is 66.13%, from the results we can see that the experimental communication time is only milliseconds, it does not have any impact on the normal traffic, and it is feasible to increase the part of the network overhead appropriately for improving universal service.

5 Summary

The paper proposes the identity hidden and substitution mechanism. It implements vary terminals unified access Heterogeneous wireless network through the identifier separation of identity and location of terminal. Here the research of the identifier mapping in this mechanism has been take, which it provides network with the global memory, mapping and inquires of all the identifier information. It is important to communicate safely in heterogeneous wireless network.

Future works will to improve the update mechanism in the cache of IHS mechanism, and more study on multi-mode smart terminal will be taken to achieve the separation of user and terminal.

References

1. Moskowitz, R., Nikander, P., Henderson, T.: Host Identity Protocol (EB/OL) (2005), `http://www.ietf.org/internet-drafts/draft-ietf-hip-base-04.txt`
2. Farinacci, D., Fuller, V., Oran, D.: Locator/ID Separation Protocol(LISP)[S], IETF (January 2007)
3. Stocia, I., Morris, R., Kargeret, D.: al: Chord: A scalable Peer-to-Peer lookup service for internet application. In: Proceedings of the 2001 ACM SIGCOMM Conference, pp. 149–160 (2001)

The Optimized Design of E-Yuan Multimedia System

Weichang Feng

School of Computer and Communication Engineering,
Weifang University, Weifang, 261061, China
fweichang@163.com

Abstract. E-Yuan multimedia system is developed for the rich audio and video resource on the Internet and on its server side, it can automatically search and integration of network video and audio resources, and send to the client side for the user in real-time broadcast TV viewing, full use of remote control operation, Simply it's a very easy to use multimedia system. This article introduces its infrastructure, main technical ideas and you can also see some details about server side and client side. At the same time, the improvement on how to collect and integrate video resources is comprehensively elaborated.

Keywords: Linux, Embedded system, Web 2.0, Server, Client, Video search.

1 Introduction

E-Yuan multimedia system is developed for the rich audio and video resource on the Internet and on its server side, it can automatically search and integration of network video and audio resources, and send to the client side for the user in real-time broadcast TV viewing, full use of remote control operation, Simply it's a very easy to use multimedia system. System is divided into server side and client side. On the server side, for different network multimedia resources, we developed different search branch, through the unified data format, every parts of the system could use and generate data for others and finally generate the data for client side. Server's framework is extensible, if there is a new network media resource, we just dynamic increase a appropriate search branch for it, and no change in the overall framework. On the server side, the content for the client side is filtered to ensure that it's health and effectiveness. On the client side, beside the network resource browsing, also supports browsing the most mainstream formats of video, audio and pictures in the removable storage media.

2 Server Side Architecture

Server is used as the data collection and analysis center. It's framework is extensible and according to using a unified XML format for data resources, every parts of server could share data, and the speed & quality of the collection and analysis are more efficiency and accuracy. Server side architecture shown in Fig. 1.

A. Xie & X. Huang (Eds.): Advances in Electrical Engineering and Automation, AISC 139, pp. 281–286.
springerlink.com

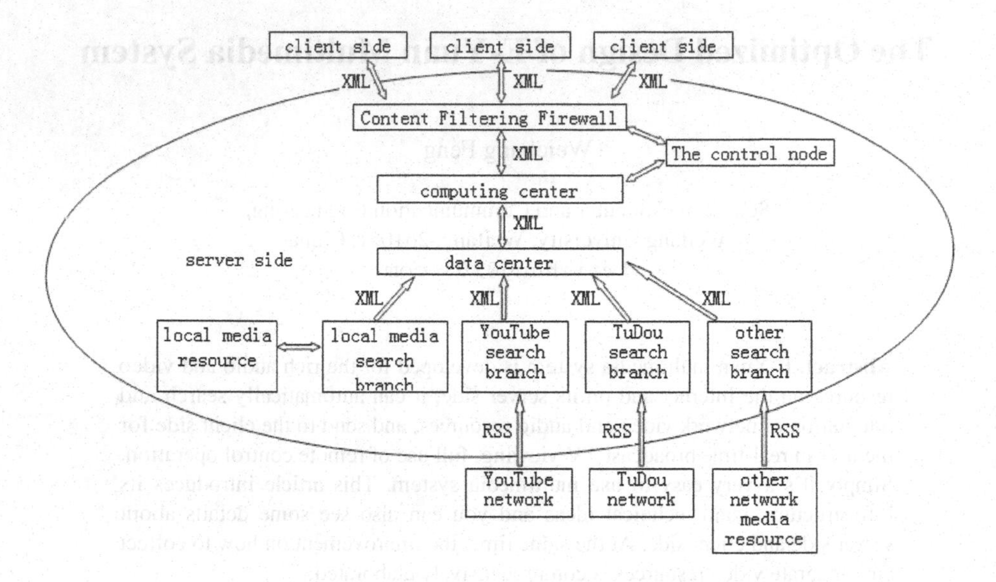

Fig. 1. Sever side architecture

(1) Through several machine-level network protocols, server do the data exchange with the well-known network resources, and get the media information quickly and accurately.
(2) By using a unified XML format for data resources, all the search branches store the network multimedia resources to the data center, the data center through the inspection, alignment and other methods to ensure the effectiveness of multimedia resources, and for storage.
(3) According to the data in the data center, computing center generate the page and information for the client side. and complete the instructions according to user-oriented needs.
(4) Through the control nodes, we can manually adjusted the server's parameters and the pages for client side, and fix the computing center's result if need.
(5) through the content filtering firewall, we can ensure the health and effectiveness of the content, and will make reasonable control of the customer feedback to the computer center to complete the whole system interaction.

3 Main Technical Characteristic of Server

By developing different search branch for different network multimedia resource, server architecture achieve a good scalability.

For example, the search branch for YouTube, it does the following work. First, it get the latest multimedia information through the RSS protocol and then transformed it into the unified XML format, and put it into the data storage center for the entire system.

The unified XML multimedia data format helps every parts of sever to share data, do the collection and analysis automatically, and make the entire system work efficiently and orderly, also it's a good foundation for further server expansion.

4 Client Side Introduction

E-Yuan client side based on embedded systems , because it wants to minimize the costs.

Embedded system is undoubtedly the most popular areas of IT applications, especially with the intelligent household electrical appliance. As we are usually familiar with mobile phones, electronic dictionaries, video phone, MP3 players, digital video (DV), high-definition television (HDTV), game consoles, smart toys are typical of embedded systems.

E-Yuan client side separates into two parts (Hardware part and software part).

4.1 Client Hardware Structure

Client hardware includes: motherboard, power supply, memory, hard drive, infrared remote control and the system enclosure.

The major part of system hardware is an industrial motherboards. Usually an Industrial motherboard designs for an fixed use and not easy to expend its function. With the market requirements of the board become more complex, it requires industrial motherboard has a strong expansion capabilities to meet different customer needs.

This system picks up J7F2WE1G5D embedded high-performance motherboard.

4.2 Client Software Structure

Client software includes the embedded Linux operating system and its applications. Linux operating system is the most successful applications in the embedded field. After the hardware design is complete, we need software to achieve every kind of functions. The value-added of embedded system largely depends on the level of the embedded software application. That is the intelligence level of products is determined by the software.

Software system includes the base subsystem, multimedia subsystem, and human-computer interaction subsystems. Functional description of each subsystem are as follows:

Base subsystem: Contains the basic library and some base functions. Other subsystems depends on this subsystem.

Multimedia Subsystem: The core subsystem, including all media-related input and output functions.

HCI subsystem: allows the users to quickly and easyly interact with the system.

System data flow control principle shown in Fig. 2. (MMS means Multimedia Subsystem)

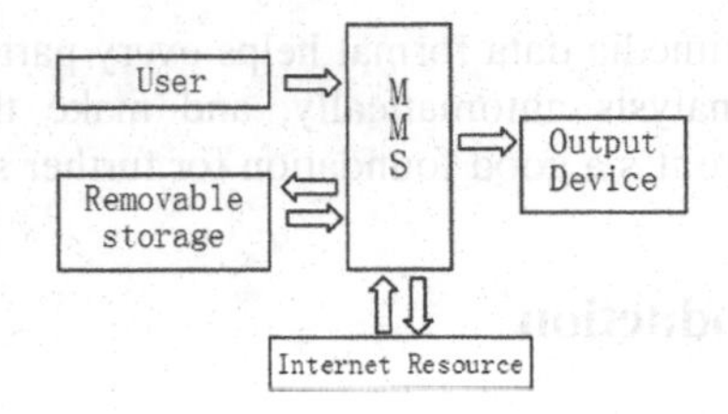

Fig. 2. System data flow control principle

5 Improvement Project on Collecting and Integrating Video Resources

The design project of collecting and integrating audio and video resources automatically is changed. Instead of using RSS subscription functionality originally, we designed a special video vertical search engines to realize the site directional search. Moreover, we designed a web video retrieving central database to achieve the purpose of integrating high quality web video resource. Thus more rich and the broadcast program source can be provided for the embedded online play equipment E-Yuan which is developed independently.

The improved project of collecting and integrating video resources can be divided into four modules: video collecting module, video data updating module, the terminal interface module and client player interface supporting moduleA. Adjusting instruction architecture according to the cpu on the motherboard.

5.1 Video Collecting Module

The main functions of the video collecting module are as fellow : Developing and improving the customization Nutch worms, capturing video website data, preprocessing the website data, judging the effectiveness of resource and saving the valid data into website video searching centre database, as shown in Fig. 3.

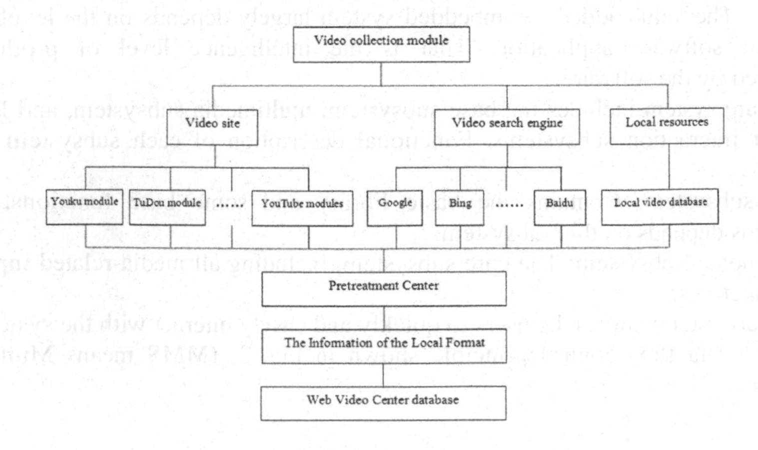

Fig. 3. Web video collection module

5.2 Video Data Updating Module

The video data updating module implements the function of data update timely which is the important indicator of the precise online playing on the client side.

Although the Nutch worms limited the search scope, but in order to guarantee the data accuracy of the database centre, the capture interval is designed short, such as a day or even shorter time to update a database to ensure the correctness of client play resources data.

The video data updating module can carry out independently to specific video website. It can update and maintain regularly according to the column setting style of different website and generate the XML file used by the web pages and terminals, to ensure the synchronism and accuracy of display list of client data. As this module has heavy workload, so it needs more manpower. The follow chart of the video data updating module is shown in Fig. 4.

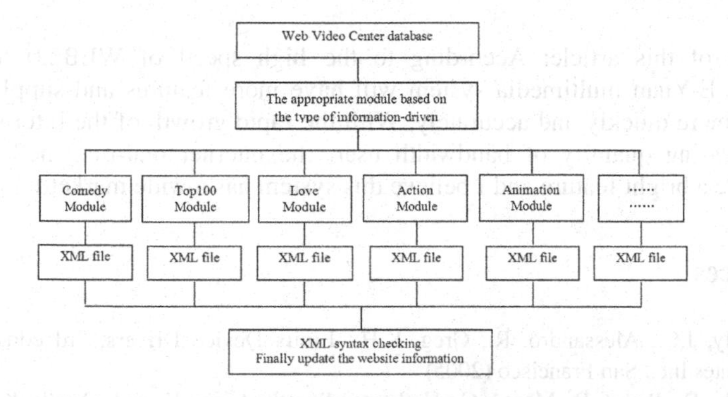

Fig. 4. Web video data update module

5.3 The Terminal Interface Module

According to the peculiarity that the E-Yuan uses the remote operating, the client interface adopts the tree list to display the programme source information so as to the location and search of remote.

5.4 The Client Player Interface Supports Module

Some famous video sites such as Youku and Tudou adopt customized player to broadcast video program. It is very easy to download and install a video player on the computer, and each video website supports download and installation. But it is not so easy to install a video player designed by a web site in the embedded multimedia device. The system has been achieved to supply interface in the E-Yuan client to play Youku, ppstream, Google and many other video resources by using their own custom player.

Parts of the codes of Youku's cross-platform video accessing component are listed as fellows:

```
int main()
{           HINSTANCE hDLL =
LoadLibrary("D:\\libyouku.dll");
            if(hDLL)
            { HFUNC hFun = (HFUNC)GetProcAddress(hDLL,
"request");
              if (hFun)
              { printf("%s\n","ok");
printf( "%s\n" ,hFun("1302167461","XMjUzODUyMDYw","WE1U
VXdNQT09","WTJJek1ERXdaRFkxTVRjMllUbGxOVE0yTkRWaU9HUTVa
V1JqTmpZME1EST0=") );
              } }}
......
```

6 Conclusion

Innovation of this article: According to the high speed of WEB2.0 technology increasing, E-Yuan multimedia system will have more features and supply the data user need more quickly and accurately. With the rapid growth of the Internet and the rapid increasing quantity of bandwidth user, the internet real-tine media industry should have a bright feature and I believe this system has a wide marketing prospect.

References

1. O'Reilly, J.C., Alessandro, R., Greg, K.H.: Linux Device Drivers, 3rd edn. Oreilly & Associates Inc., San Francisco (2005)
2. O'Reilly, D., Bovet, P., Marco, C.: Understanding the Linux Kernel. Oreilly & Associates Inc., San Francisco (2007)
3. Mymediasystem.org.My Media System, `http://mymediasystem.org/`
4. `http://www.w3.org.XML`. Extensible Markup Language, `http://www.w3.org.XML`
5. Teng, C., Huang, B., Ma, X.: Research and application of file system in embedded Linux. Microcomputer Information 11(2), 88–90 (2008)
6. Xiao, G., Li, J., Chang, W., Ding, Z.: The realization of USB driving program in embedded Linux. Microcomputer Information 9(2), 86–88 (2007)
7. `http://www.python.org`. Python Programming Language, `http://www.python.org/`
8. Lombardo, J.: Embedded Linux. SAMS Technical Publishing, Indianapolis (2001)

A New Reputation Model Based on Trust Cluster in Peer-to-Peer Networks

Yu Jin and Hongwu Zhao

School of Computer Science, Wuhan University of Science and Technology,
Wuhan, 430081, China
{jinyu,zhaohongwu}@wust.edu.cn

Abstract. How to build trust relationship between nodes to restrict malicious behaviors and improve the quality of service is becoming a crucial problem in the peer-to-peer systems, and the reputation model is the primary solution. However current P2P reputation model have defects in reputation data management. So pointing to these problems, this paper presents a new reputation data management method that can ensure the security of reputation information. By analysis, TLT can protect malicious peers from tampering, forging and replacing the reputation voters and keep this information secret.

Keywords: peer-to-peer network, trust, reputation, security, trust cluster.

1 Introduction

In P2P systems peers might maliciously behave and harm others in the system [1]. Therefore, one of the fundamental challenges for open and decentralized P2P systems is the ability to manage risks involved in interacting and collaborating with previously unknown and potentially malicious peers. Reputation-based trust management systems can successfully mitigate this risk by computing the trustworthiness of a certain peer from that peer's behavior history.

Currently almost all of Reputation systems for P2P applications [2] were purely decentralized. They had defects such as slow trust convergence speed, low scalability and reputation data management. Point to these problems, we propose TLT, a new trust-cluster-based reputation model for P2P applications. In our previous work, we solve the problem of recent reputation models in the aspects of trust convergence speed and scalability. In this paper we present a new reputation data management method to keep these data secure when being transmitted.

2 Related Work

P2PRep [3] was the representative of purely distributed reputation model in P2P system, proposed for Gnutella, where peer kept track and shared with others reputation of their peers. Reputation sharing was based on a distributed polling protocol. P2PRep protocol was very simple and could be easily integrated into Gnutella 0.4 protocol. However, In P2PRep, the minimum unit of trust evaluation was single node, the

A. Xie & X. Huang (Eds.): Advances in Electrical Engineering and Automation, AISC 139, pp. 287–292.
springerlink.com

convergence speed of the trust was slow and the trust management was complexity. Moreover the cost of network in this system was overwhelming because flooding was the method of forwarding messages.

[4] Discussed the effect of reputation information sharing on the efficiency and load distribution of a peer-to-peer system, in which peers only had limited (peers share their opinions) or no information sharing (peers only used their local ratings).While this approach did not distinguish the ratings for service (reliability) from ratings for voting (credibility). [5-7] propose a global reputation metric regarding performance of each peer was calculated. To this end, each peer's local beliefs on the performance of other peers were weighted by the others' beliefs on his own performance. [8] proposed a Bayesian network-based trust model in peer-to-peer networks that provided a flexible method to represent differentiated trust and combine different aspects of trust. However when computing integrated trust value in a given peer, other peers' ratings were weighted by their own service performance.

Except P2PRep, those models mentioned above did not take security of reputation data into consideration. Although P2PRep provided secure assurance to reputation data, it was not adaptive to hierarchical P2P systems. SupRep [9] was the extension of P2PRep in Gnutella 0.6 (hierarchical P2P). So it also kept reputation data secure in certain degree. However it could not resist the proxy attack exerted by the malicious super-nodes that are analogous to cluster headers in TLT. while TLT is adaptive to both the purely distributed and hierarchical P2P systems. Moreover there is a vote evaluation mechanism in our model that can resist malicious attack launched by super-nodes.

3 Reputation Model

TLT has mainly the following work steps:

1. Joining the trust cluster

1.1 The weak node (such as N_j) sends JCRM(Joining Cluster Request Message) to a cluster header (such as N_i) and becomes its member.

JCRM(SID_j|IP_j|$port_j$|PK_j)

1.2 N_j sends a series of RSRIM(Reporting Sharing Resource Index Message) to N_i to report its sharing resources.

RSRIM (SID_j|Length|Index|RI)

Length: the number of sharing resources;

Index: the index of the current resource in the total sharing resources;

RI: the resource index, in TLT, RI=hash (Resource Name)

2. Searching the resource

2.1 N_j sends a RRqM (Resource Request Message) to N_i;

RRqM (Hop|SID_j|SID_c|RI)

Hop: the number of message being forwarded, now hop=0;

SID_c: the identity number of the cluster header forwarding this message, now SID_c=null;

2.2 N_i examines whether N_j is in the punished period of sending request according to the Algorithm1 in [9], if it is, N_i discards this request; Otherwise modifies the SID_c field and forwarding this request to the system, namely SID_c=SID_i;

2.3 After receiving RRqM, the cluster header (such as N_k), according to the Algorithm 2 in [9], sends RRsM message and chooses member (such as N_v) that has the resource being requested and was not be punished to provide service to respond this request in the honor of a cluster;

$RRsM(SID_k|IP_k|port_k|SID_v|IP_v|port_v|RI)$

3. Polling the inter-cluster reputation

3.1 N_i creates a pair of session keys PK_{poll},SK_{poll};

3.2 N_i sends CRPM (Cluster Reputation Polling Message) and polls those the reputation of responding cluster;

$CRPM(SID_{x1}|SID_{x2}|\ldots|SID_{xi}|\ldots|SID_{xn}|PK_{poll})$

3.3 When receiving CRPM, the cluster header (such as N_u) sends CRRM (Cluster Reputation Responding Message) to respond this inter-cluster reputation polling. The members of N_s had transactions with the members of the cluster header (such as SID_{xi})in the CRPM.

$CRRM((IP_s|port_s|DT_1(N_s,N_{xi})|SID_s|PK_s|SID_{xi}|$
$(IP_s|port_s|DT_1(N_s,N_{xi})|SID_s|SID_{xi})_{SKs})_{PKpoll})$

4. Evaluating the votes

4.1 N_i discards those suspicious votes in which IP address are similar;

4.2 N_i randomly chooses a few votes to evaluate their reality and then sends RVEM (Reputation Vote Evaluation Message) to the voter(such as N_s);

$RVEM(SID_s \mid SID_{xi} \mid DT_1(N_s,N_{xi}))$

4.3 N_s returns a RVERM (Reputation Vote Evaluation Response Message) to report the true vote;

$RVERM(SID_s \mid SID_{xi} \mid DT_1(N_s,N_{xi}))$

4.4 According to those valid votes, N_i evaluates the inter-cluster trust in responding cluster;

4.5 According to the result of inter-cluster trust evaluation, N_i chooses a response cluster (such as N_k) as service providing cluster.

5. Having the transaction

5.1 N_j sends FSM message to N_v to request sharing resource;

$FSM(SID_j|IP_j|port_j|RI)$

5.2 N_v sends sharing resource to N_j

6. Reporting and validating the feedbacks

6.1 After transaction, both N_j and N_v send FRM (Feedback Reporting Message) to its cluster header to report the result of this transaction;

$FRM(SID_j|IP_j|port_j|PK_j|SID_v|R_j|(SID_j|IP_j| port_j|SID_v| R_j)_{SKj}|RI)$

R_j: the intra-cluster rating reported by N_j

6.2 N_i sends FVM(Feedback Evaluation Message) to N_j to validate the feedback;

$FVM(SID_j \mid SID_v \mid R_j)$

6.3 N_j returns a FVRM (Feedback Evaluation Response Message)to report the true intra-cluster rating;

$FVRM(SID_j \mid SID_v \mid R_j)$

6.4 N_i sends ICRM(Inter-Cluster Rating Message) to N_k; $ICRM(SID_i |IP_i|port_i|PK_i|FT_j|SID_j|IP_j|port_j|PK_j|SID_v|SR_i|SR_j|RI)$

FT_j: the current feedback trust value of N_i in N_j

SR_j: the signed intra-cluster feedback of N_j, $SR_j = (SID_j|IP_j|port_j|SID_v|R_j)_{SKj}$

SR_i: the signed inter-cluster rating, $SR_i = (SID_i|IP_i|port_i|SID_j|FT_j|SR_j)_{SKi}$

6.5 N_k sends FVM to N_v to validate the reality of the feedback;

$FVM(SID_v | SID_j |R_v)$

R_v: the intra-cluster rating by N_v

6.6 N_v returns a FVRM to report the true intra-cluster rating;

$FVRM(SID_v | SID_j |R_v)$

6.7 N_k sends two ICREM(Inter-Cluster Rating Evaluation Message) to validate the inter-cluster rating. One is sent to N_i, another is to N_j;

$ICREM (SID_i|SID_v|R_i)$

6.8 N_i and N_j separately return a ICRERM (Inter-Cluster Rating Evaluation Response Message)to report the true inter-cluster rating;

$ICRERM (SID_i|SID_v|R_i)$

7. Processing Feedbacks and Updating Trust Value

7.1 N_k, according the two intra-cluster feedbacks, updates the intra-cluster service trust in N_v and, if necessary, begins to punish N_v to provide service as described in [9];

7.2 N_i, according the two intra-cluster feedbacks, updates the intra-cluster feedback trust in N_j, and if necessary, begins to punish N_j to request service as described in [9];

7.3 According to the result of the transaction, N_j updates the proxy trust in N_i; Likewise N_v updates the proxy trust in N_k as depicted in [10];

7.4 N_i updates the inter-cluster trust in N_k according to these two feedbacks as given in [9].

So the rest parts of our model are arranged in the order mentioned above. Also we take the example mentioned above to introduce the details of our model.

4 Security Analysis

In this section, we will observe TLT from the angles of security. In our model, reputation data mainly include feedback and reputation poll response. So keeping them secure is the goal of the reputation data management.

4.1 Feedback Security Analysis

Suppose Alice and Bob have a transaction. Alice is the consumer and its cluster header is Mallory; Bob is the provider and its cluster header is Carl. Moreover David is the other member of Mallory and Mike is the other cluster header between Mallory and Carl. Now Alice reports the service result of the Bob to Mallory. Then according to this report Mallory sends an inter-cluster rating to Carl. Our model can achieve the following security objectives:

1) The feedbacks' integrity and non-repudiation. According the section of 3, Alice uses FRM message signed by its private key to report service result of Bob to

Mallory. So Mallory can not tamper Alice's rating. Furthermore Mallory also uses ICRM message signed by its own private key to report inter-cluster rating to Carl. So the other cluster header (such as Mike) can not tamper this rating. Moreover Carl can ensure the resource of ICRM is Mallory and the Alice is the consumer. Namely Mallory and Alice is non-repudiation to this ICRM because in ICRM there are the signature of Alice and Mallory.

2) The feedbacks' non-forgery. Malicious peers can forge feedbacks in three ways:

A) Mike creates IP_{fake}, por_{fake}, PK_{fake}, SK_{fake} and SID_{fake} and uses them to report false or inexistent inter-cluster rating under the name of Mallory. Consequently it can have influence on the intra-cluster service reputation of Bob.

B) David creates $IP_{fake'}$, $port_{fake'}$, $PK_{fake'}$, $SK_{fake'}$ and $SID_{fake'}$ and uses them to report false intra-cluster rating on the honor of Alice.

C) Mallory can use other feedback to replace that of Alice. For example, Mallory is the competitor of Bob. It can substitute unsatisfactory feedback of Alice for positive feedback to decrease the reputation of Bob.

In our model, the above attacks can be prevented. As described in the section of 3, after Mallory receives the feedback reported by Alice, it probably uses FVM messages to validate this feedback according to a given probability. Then Alice returns a FVRM message to report the authentic intra-cluster rating. So the attack like B) can be resisted. After Carl receives the inter-cluster rating submitted by Mallory, it probably sends two ICREM messages to validate this rating according to a given probability. One is sent to Alice, another is to Mallory. Both Mallory and Alice use an ICRERM message to submit the authentic feedbacks. If these two feedbacks are consistent, the inter-cluster rating is valid; if not, this inter-cluster rating will be discarded. So the attacks depicted in the A) and C) can be prevented.

4.2 Reputation Poll Security Analysis

Suppose Mallory polls the inter-cluster reputation of Carl. Tom is the cluster header that responses this poll. Mike is other cluster header between Mallory and Carl.

According to the methods in the 3.2.2, when sending CRRM message to report the inter-cluster reputation of Carl to Mallory, Tom uses its own private key to encrypt this message to ensure its integrity. So other peers (such as Mike) can not tamper this vote. Moreover Mallory can ensure this rating is from Tom, namely Tom is non-repudiate to this vote. Besides this message is also encrypted by PKpoll created by Mallory, so this vote is secret and invisible for other peers (such as Mike).

If Mike creates $IP_{fake''}$, $port_{fake''}$, $PK_{fake''}$, $SK_{fake''}$ and $SID_{fake''}$ and wants to use them to report false CRRM in the honor of Tom. Our model can also prevent this attack. After receiving this vote, Mallory will use RVEM message to evaluate the authenticity of this inter-cluster reputation voter.

5 Conclusions and Future Work

This paper proposes a new reputation data management for P2P application. By analysis our new reputation data management method in our model that can protect malicious peers from tampering, forging and replacing the reputation voters and keep this information secret.

We may carry out the future work in two directions: 1) the behavior model of the peers. Although our work analyzes a few kinds of behaviors (such as service, feedback and proxy behaviors), it does not consider the all behaviors of peers; 2) prototype implementation. We also wish to design and implement a prototype of our model and apply it to an actual P2P system.

Acknowledgments. This work was supported by Foundation of Hubei Educational Committee (Grant No: Q20111110).

References

1. Liang, J., Kumar, R., Xi, Y., Ross, K.: Pollution in P2P File Sharing Systems. In: Proceeding of the IEEE Infocom 2005, pp. 1174–1185 (2005)
2. Damiani, E., Vimercati, S.D.C., Paraboschi, S., Amarati, P.S.: Managing and Sharing Servants' reputation in P2P Systems. IEEE Transactions on Knowledge and Data Engineering 15(4), 840–854 (2003)
3. Marti, S., Garcia-molina, H.: Limited Reputation Sharing in P2P Systems. In: Proceedings of ACM EC (2004)
4. Kamvar, S., Schlosser, M., Garcia-Molina, H.: The EigenTrust Algorithm for Reputation Management in P2P Networks. In: Proceedings of the Twelfth International World Wide Web Conference (May 2003)
5. Dou, W., Wang, H., Jia, Y., Zou, P.: A Recommendation-Based Peer-to-Peer Trust Model. Journal of Software 15(4), 571–583 (2004)
6. Tian, C., Zou, S., Wang, W., Cheng, S.: A New Trust Model Based on Recommendation Evidence for P2P Networks. 31(2), 270–281 (2008)
7. Wang, Y., Vassileva, J.: Bayesian Network Trust Model in Peer-to-Peer Networks. LNCS (LNAI), pp. 23–34 (2004)
8. Chhabra, S., Damiani, E., di Vimercati, S.D.C.: A Protocol for Reputation Management in Super-Peer Networks. In: DEXA 2004, pp. 979–983 (2004)
9. Jin, Y., Gu, Z.-M., Gu, J., Zhao, H.: A New Reputation-Based Trust Management Mechanism Against False Feedbacks in Peer-to-Peer Systems. In: Benatallah, B., Casati, F., Georgakopoulos, D., Bartolini, C., Sadiq, W., Godart, C. (eds.) WISE 2007. LNCS, vol. 4831, pp. 62–73. Springer, Heidelberg (2007)
10. Jin, Y., Gu, Z., Gu, J., Zhao, H.: Two-Level Trust Model Based on Mutual Trust in Peer-to-Peer Networks. Journal of Software 20(7), 1909–1920 (2009)

Drafting and Retrieving Standards with XML and Data Mining

Wenhong Yang, Ping Cao, Liang Zhu, and Lei Xing

China Aero-polytechnology Establishment,
100028 Beijing, China
wenhongyang@yahoo.com.cn

Abstract. With increasing in number of standard documents and its applications, the user's needs are becoming various. Sometimes, the user doesn't need a standard document, but a parameter, a term, a kind of specification or a method. On the other side, sometimes the user needs a series of standards as directives of activities. In general, a standard application system is expected to understand and focus on the user need quickly and exactly. The combination of XML and data mining will make this expectation true. This paper discusses how standards drafting and retrieval benefit from XML and data mining, and proposes a uniform standard model with which a variety of XML schemas can be designed. In the end, a schema for general layer of the structured standard model is proposed and verified.

Keywords: drafting standards, standards retrieval, structured standards, XML, data mining.

1 Introduction

The standard is playing an important role in many industry fields. More and more organizations, corporations, manufactories pay attention to the standard drafting and using. However, there are many different structures of standards due to different drafting and publishing rules, which confuse users. In additional, with increasing in number of technical standards, how to find out the useful information is becoming difficult for users.

XML and data mining technology will change traditional manners people draft standards, and improve the efficiency of information retrieval service greatly [1, 2]. In the standard field, now most of the standard retrieval services are based on the standard title, the document number, the publish organization etc. If users don't have enough experience, the result may not the real concern. Obviously, this way can't meet the needs of users. Generally, users don't need a whole of standard, but some of these content, such as a term, a request, a chart, a parameters and so on. On the other hand, sometimes users need a series of standards, such as users search for RFID standards; he needs a series of standards to understand requirements including air interface parameters, labels, readers, coding rulers and so on.

In short, it is necessary for users to focus on the really need technical content rapidly, and this demand gave birth to the research of standards drafting and retrieving based on XML and data mining in this paper.

A. Xie & X. Huang (Eds.): Advances in Electrical Engineering and Automation, AISC 139, pp. 293–298.
springerlink.com

2 Requirements and Benefits

2.1 Drafting Standards with XML

When a standard is drafted, firstly, the drafter has to look through the existed similar standards, so as to ensure there is no technology conflict. Secondly, some of elements of the standard have drafting rules specified in some directives documents, such as the cover, foreword, title, scope and so on. But there is no fixed rule for the structure of the normative technical element due to diversity of many kinds of technologies, so the drafter presents the technical content in an undefined manner, which will bring inconvenience to standard users. Thirdly, the drafters often spend a lot of care on the format of a standard document, in order to make it ready to publish.

If using XML and data mining technology, firstly, because of the high information search efficiency, the drafters can find out whether and how a term, a parameters, or a method has been specified, he/she can find out the related content rapidly, rather than to look through all of the existed similar standards. Secondly, we can specify a schema for a kind of standard, for example, the structures of the test purpose, conditions, steps in the test method standard are specified, make it easy to express the technical content and to understand by users. Thirdly, the drafter does not have to spent effort on the document format; the XML style document will be helpful.

2.2 Retrieving Structured Standards with Data Mining

The precondition of standard using is to get the really need technical content from a great many standards documents. The traditional manner of retrieving standards can't dig out the core knowledge, already can't satisfy users needs. After the standard is structured with XML, data mining would enable users to access to the really need technical content rapidly and accurately.

If users don't have enough experience, the result maybe nothing due to the absence of relationship for most of the current standards searching systems. Sometimes, some of standards compose a resolution for users, such as some standards which are the parts of the same subject. With data mining, the system is supposed to give out a series of standards, in order to provide a resolution for a technical issue.

The existing standards retrieval generally searches keywords from the standard number, title, organizations which defined as informative preliminary elements [3-5]. In fact, the technical content is presented with normative technical elements, in addition, some technical reports are related with the standard, will also help users to understand some content of the standard. With data mining, users will dig out useful information from all of the elements of the standard, technical reports and other resources.

Currently, what users get is a whole of the standard document, but most of the time, it is not necessary since users may only need to know parts of the standard, such as a parameter, a term, and a kind of specification or a test method. A system with XML standards and data mining technology can provide structured knowledge for users, so users can get key content easily without any irrelevant information.

3 A Model for Structured Standards Documents

In order to structured standards documents, we would design one or more schemas. IETM (Interactive electronic technical manual) is a good reference for the schema design [6], in which eight schemas are designed for technical manuals of equipments.

When refer to schemas for standards, the first question is how to divide all of content of a standard document. We analyzed ISO/IEC Directives, Part 2, GB/T 1.1-2009, GJB 6000-2001 and GJB 0.1-2001, which provide guidelines for the standard structure and drafting. Although the cover, preface and structure vary with different standards, but the standard always contains four classes of elements: informative preliminary element, normative general element, normative technical element and informative supplementary element, and these elements play different roles.

The role of informative preliminary elements is basically to identify a standard or describe its status, such as the element contained in cover and preface. The role of informative supplementary elements is to provide additional information to assist the understanding or use of the standard document. Normative general elements and normative technical elements describe the scope of technical content. According to roles of elements, we designed a model for structured standards, as shown in Fig.1.

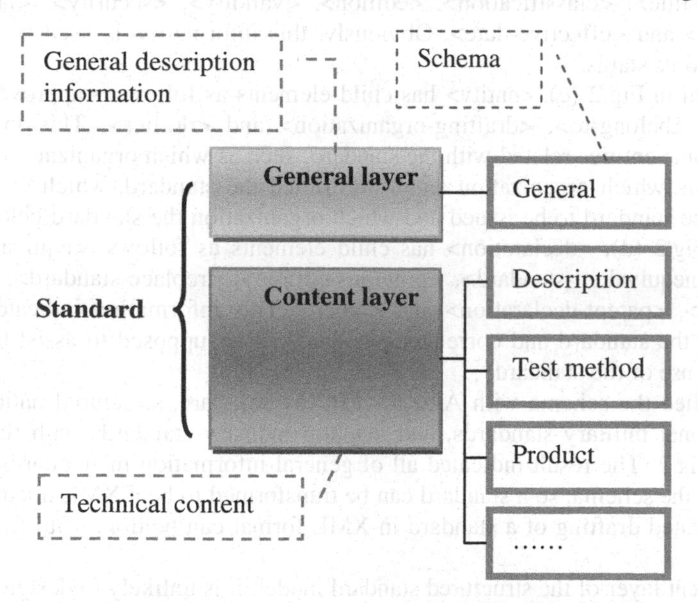

Fig.1. A uniform model for standards

As illustrated in Fig.1, the model contains two layers: general layer and content layer. General layer contains informative preliminary elements, and content layer contains normative elements and supplementary elements.

Information included in general layer is informative preliminary element defined in the traditional standard [1-3]. This information exists in most of standards, and the way of description is fixed when refer to a kind of standard, so we can design a schema for the general layer. But it is not the technical expertise; its purpose is not to solve some problems, but to identify a standard and to support standards management.

Information included in content layer is technical element or normative element defined in the traditional standard [1-3]. This information has no fixed format, the expression and the structure of this information is closely related to technical problems involved in the standard [7], so it's better to design different schemas for different kind of standards. Its purpose is to provide rules, guidelines or characteristics for activities or their results.

4 The XML Schema Analysis

We designed a schema for general layer of the standard model shown in Fig.1, the element is presented with "<>". As shown in Fig.2 (a), general layer is composed of three classes of elements: <identification>, <entity> and <declaration>. Each of elements has several child elements, details are shown in Fig.2 (b)-(e), the element with "n1…n2" is repeatable; the element with the dotted line frame is optional.

As shown in Fig.2 (b), <identification> has child elements as follows: <document number>, <title>, <classification>, <edition>, <validity>, <security>, <language>, <issue-date> and <effective-date>. Obviously, this information is used to identify a standard and its status.

As shown in Fig.2 (c), <entity> has child elements as follows: <approver>, <proposed-by>, <belong-to>, <drafting-organization> and <drafter>. This information indicates some entities related with the standard, such as which organization proposed this program, which organization and who drafted the standard, which organization approved the standard to be issued and which organization the standard belong to. As shown in Fig.2 (d), <declaration> has child elements as follows: <equivalent standard>, <nonequivalent-standard>, <previous-edition>, <replace-standard>, <series>, <other-part>, <patent-declaration> and <other>. This information indicates patents involves in the standard and correlated standards. It is supposed to assist the understanding or use of the standard.

We verified the schema with Altova XMLSpy software, structured national standards, national military standards, and aviation industry standards with the schema shown in Fig.2. The result indicated all of general information in standards could be reflected in the schema, so a standard can be transformed to be a XML document, and it also indicated drafting of a standard in XML format can be done with filling some forms.

For content layer of the structured standard model, it is unlikely to design a schema for all of standards, because key technical content varies with different standards, such as the structure and the express manner of the technical content within the interface standard and the test method standard are totally different. It is necessary to design different schemas for the content level of different kinds of standards, which is the future work of this research program.

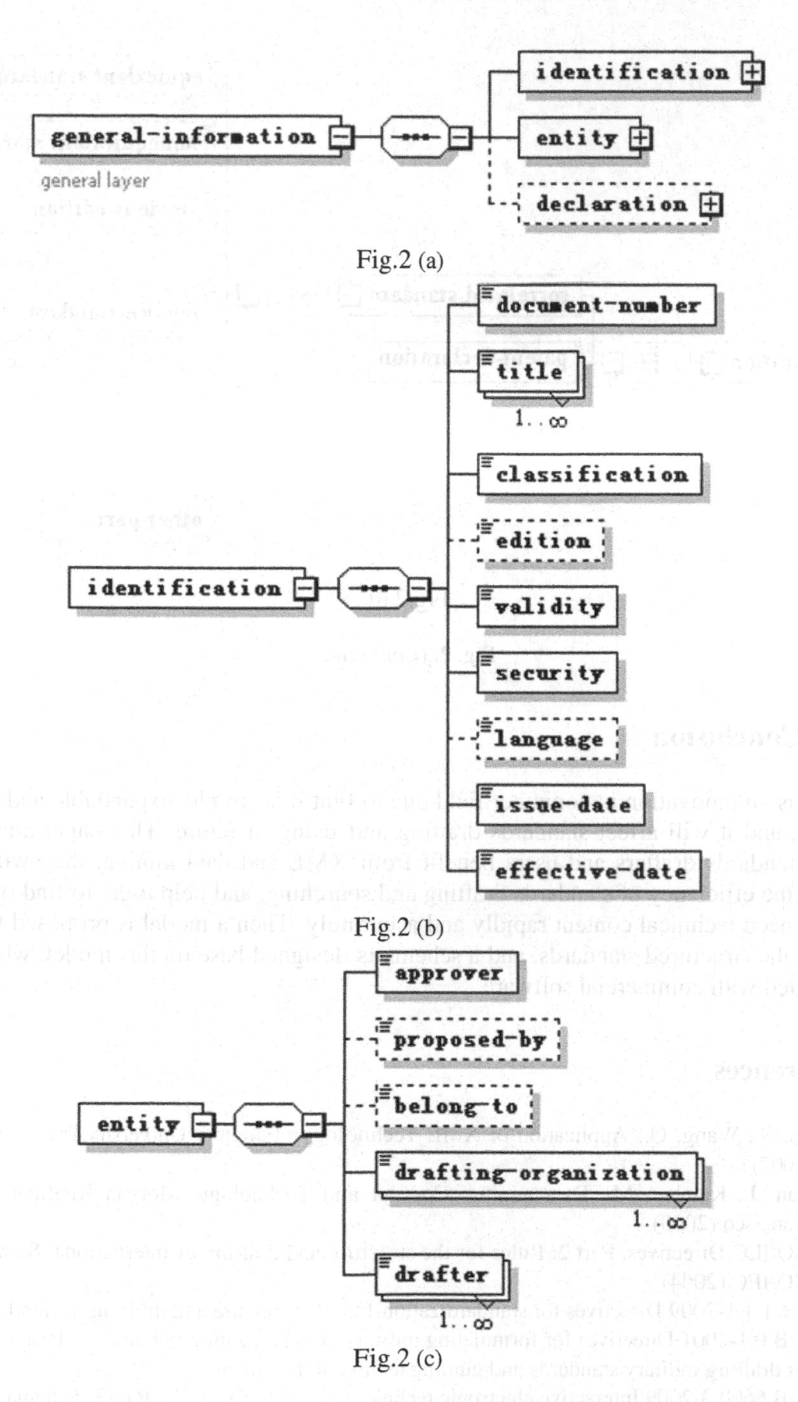

Fig. 2. A schema for general layer of the standard model

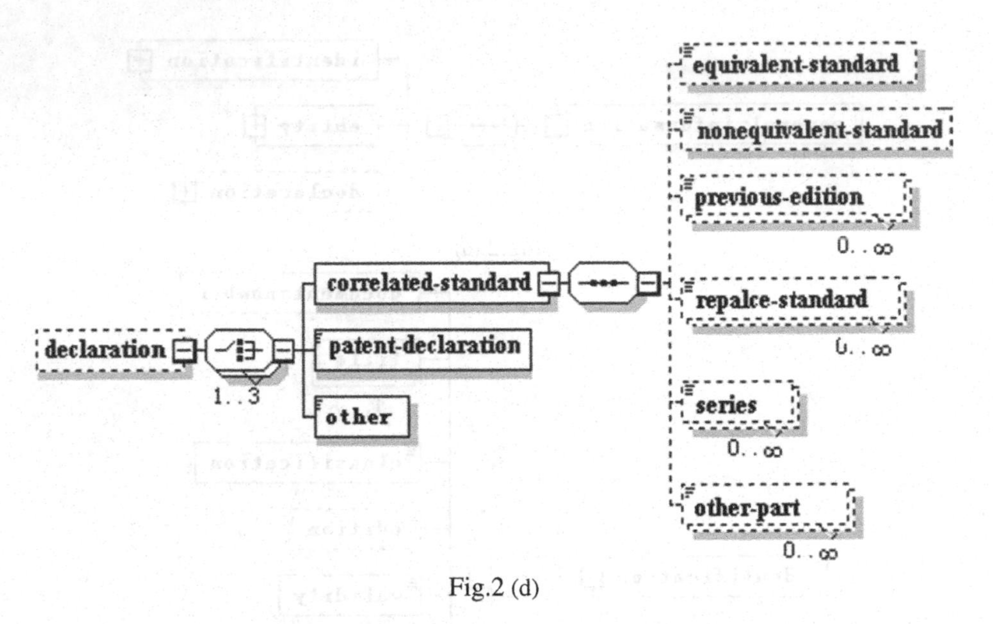

Fig.2 (d)

Fig. 2. (*Continued*)

5 Conclusion

XML is an innovation in computer field due to that it is simple, expandable and interactive, and it will affect standards drafting and using in future. This paper analyzes how standards drafters and users benefit from XML and data mining, they will improve the efficiency of standards drafting and searching, and help users to find out the really need technical content rapidly and accurately. Then a model is proposed to describe the structured standards, and a schema is designed base on this model, which is validated with commercial software.

References

1. Jia, S., Wang, Q.: Application of XML Technology. Tsinghua University Press, Beijing (2007)
2. Han, J., Kamber, M.: Data mining: Concept and Technology. Morgan Kaufmann, San Francisco (2000)
3. ISO/IEC Directives, Part 2: Rules for the structure and drafting of International Standards. ISO/IEC (2004)
4. GB/T 1.1-2009 Directives for standardization-Part 1: Structure and drafting of standards
5. GJB 0.1-2001 Directives for formulating military standardization documents- Part 1: Rules for drafting military standards and guiding technical documents
6. GJB 6600.3-2009 Interactive electronic technical manual of materiel Part 3: Schema
7. Bai, D.: Guides for the drafting of standards. Standards Press of China, Beijing (2009)

The Indices Analysis of a Repairable Computer System Reliability

Xing Qiao[1], Dan Ma[1], Zhixin Zhao[2], and Fengzhi Sun[1]

[1] Department of Mathematics, Daqing Normal University, Daqing, 163712, China
[2] Beijing Institute of Information and Control, Beijing, 100037, China
xiaoqiao1502@163.com

Abstract. In paper, a mathematical model was developed by the supplementary variable method, which is consisted of hardware and software in series. The hardware is repaired to be as good as new; while the software is repaired periodically with decreasing life time and after a period of time, it is repaired as a new one. Under the assumption that the life times of hardware and software both follow exponential distribution and the repair times are subject to general distribution, we take the reliability indices of a repairable computer system.

Keywords: Overhaul, Minimal repair, Availability analysis, Failure frequency.

1 Introduction

Reliability is an important concept at the plan, design and operation stages of various complex systems. The need of obtaining highly reliable system has been more and more important with the development of the modern technology. The high degree of reliability is usually achieved by introducing redundancy or repairman (e.g.[1-4]) or applying preventive maintenance (e.g.[5-6]), optimal inspection plans (e.g.[7-9]) or optimal replacement policy (e.g.[10-11]). The aim is to increase the performance of the system by reducing the downtime or the maintenance and inspection cost of the system.

Repairable system is a kind of important system discussed in reliability theory and is also one of the main objects studied in reliability mathematics. In the general reliability analysis of the computer system, most of authors singly discussed the hardware or software. It is rare to analyze synthetically[12]. Because of different characteristics of the hardware and software, we can not simply take the hardware and software as a unit or two different types of units[13]. With the passage of the using time and the number of failures increasing, the reliability of the hardware would descend and the repair time would be longer[14]. During the software debugging and testing stage, as the failures occur, potential software error is discovered and corrected constantly. These make the software reliability grow[15]. Because the hardware failure or software failure lead to the entire computer system, the computer system can be described as a series system with hardware and software (namely hardware and software in series).

In the reliability analysis of repair system, it usually assumed that the repaired units which compose system is as good as new, and the failed units be repaired immediately. But the reality usually is not the case. In real life, it is possible that the reliability reduce after the software failure each time. That is the condition (3)(see.[16])

A. Xie & X. Huang (Eds.): Advances in Electrical Engineering and Automation, AISC 139, pp. 299–305.
springerlink.com

$F_S^{(n)}(t)=F_S(a^{n-1}t)=1-e^{-a^{n-1}\lambda_s t}, t\geq 0, \lambda_s>0$, the coefficient $a>1$. With the number of repair times increasing, the failure rate is increasing gradually. In view of the aging and accumulative wear, the repair time will become longer and longer and tend towards infinity, i.e. finally, the system is non-repairable. As so, we assume that the software is overhauled to be as good as new after the $(N-1)$ th minimal repair. Then, we discuss the number of minimal repair before overhaul repair is more appropriate. And we will study how its reliability will be affected by the number of minimal repair and overhaul. In Ref.[16], the author assumed that the software can not be repaired as good as new and utilize the geometric process and supplementary variable technique to analyze the system reliability. However, the life and repair time of the hardware and software are assumed that they follow exponential distributed. In this paper, we established a mathematical model of the computer system by the supplementary variable method, which is composed of the hardware and software in series. The hardware is repaired to be as good as new, the software is repaired periodically, and restored software life decreases. After a period of time, to overhaul to make it as a new unit. Under assumption that the life time of the hardware and software follow exponential distribution and repair time subject to the general distribution, we discuss reliability indices and numerical example.

2 Formulation of Mathematical Model

The mathematical model is described specifically as follows:

(1) The computer system is composed of hardware H and software S in series.
(2) The distribution function of the life time X_H of hardware H is

$F_H(t)=1-e^{-\lambda_h t}, t\geq 0, \lambda_h>0$.

(3) The distribution function of the life time $X_S^{(n)}$ of software S during its n th period (e.g. the time between the completion of its $(n-1)$ th repair and that of the n th repair) is $F_S^{(n)}(t)=F_S(a^{n-1}t)=1-e^{-a^{n-1}\lambda_s t}, t\geq 0, \lambda_s>0, a>1, n=1,2,\cdots,N$.
(4) Let Y_1 be the repair time of hardware H, Y_2 be the repair time of software S after its n th failure ($n=1,2,\cdots,N-1$), Y_3 be the repair time of software S after its N th failure, respectively. Their distribution functions are

$G_j(t)=\int_0^t g_j(x)dx=1-e^{-\int_0^t \mu_j(x)dx}$ and $E[Y_j]=1/\mu_j, j=1,2,3$, where $\mu_3(x)>\mu_2(x), \forall x\geq 0$.

(5) The hardware H is repaired as good as new. The software S is performed by a minimal repair (e.g. a maintenance action performed on a failed system by which its survival time is decreasing) during its n th ($n=1,2,\cdots,N-1$) period and a overhaul repair (e.g. a maintenance action performed on a failed system by which it is

repaired as good as new) during its N th period. The above stochastic variables are independent with each other.

Let $N(t)$ be the system state at time t and assume all the possible states as below:

$0i$ $(i=1,2,\cdots,N)$ the system is working.

$1i$ $(i=1,2,\cdots,N)$ the system is failed because the failure hardware H is being repaired in the i th time.

$2i$ $(i=1,2,\cdots,N)$ the system is in failure state because the failure software S is being repaired minimally in the i th time($i=1,2,\cdots,N-1$) and the software S is being overhauled in the N th time.

Then the system state space is $E=\{0i,1i,2i\}$ $(i=1,2,\cdots,N)$; where the work ing state space is $W=\{0i\}$ $(i=1,2,\cdots,N)$ and the failure state space is $F=\{1i,2i\}$ $(i=1,2,\cdots,N)$.

When $N(t)=ji$ ($j=0,1,2$; $i=1,2,\cdots,N$), supplement variable $Y_j(t)$ $(j=1,2,3)$ which denote the elapsed repair time of hardware H, the elapsed minimal repair time of software S and its elapsed overhaul time at time t, respectively. Then $\{N(t),Y_j(t)\}$ constitutes a matrix Markov process whose state probabilities are defined as follows:

$$P_{0i}(t)=P\{N(t)=0i\}, i=1,2,\cdots N$$

$$P_{1i}(x,t)dx=P\{x<Y_1(t)\le x+dx, N(t)=1i\}, i=1,2,\cdots N$$

$$P_{2i}(x,t)dx=P\{x<Y_2(t)\le x+dx, N(t)=2i\}, i=1,2,\cdots N-1$$

$$P_{2N}(x,t)dx=P\{x<Y_3(t)\le x+dx, N(t)=2N\}$$

Then by using the method of probability analysis, the system can be formulated as below:

$$\frac{dP_{01}(t)}{dt}=-(\lambda_h+\lambda_s)P_{01}(t)+\int_0^\infty P_{11}(x,t)\mu_1(x)dx+\int_0^\infty P_{2N}(x,t)\mu_3(x)dx \quad (1)$$

$$\frac{dP_{0i}(t)}{dt}=-(\lambda_h+a^{i-1}\lambda_s)P_{0i}(t)+\int_0^\infty P_{1i}(x,t)\mu_1(x)dx+\int_0^\infty P_{2,i-1}(x,t)\mu_2(x)dx, \quad i=2,\cdots,N \quad (2)$$

$$\frac{\partial P_{1i}(x,t)}{\partial x}+\frac{\partial P_{1i}(x,t)}{\partial t}=-\mu_1(x)P_{1i}(x,t),\ i=1,2,\cdots,N \quad (3)$$

$$\frac{\partial P_{2i}(x,t)}{\partial x}+\frac{\partial P_{2i}(x,t)}{\partial t}=-\mu_2(x)P_{2i}(x,t),\ i=1,2,\cdots,N-1 \quad (4)$$

$$\frac{\partial P_{2N}(x,t)}{\partial x}+\frac{\partial P_{2N}(x,t)}{\partial t}=-\mu_3(x)P_{2N}(x,t) \quad (5)$$

$$P_{1i}(0,t)=\lambda_h P_{0i}(t),\ i=1,2,\cdots,N \quad (6)$$

$$P_{2i}(0,t)=a^{i-1}\lambda_s P_{0i}(t),\ i=1,2,\cdots,N \quad (7)$$

$$P_{01}(0)=1, P_{0i}(0)=0, i=2,\cdots,N; P_{1i}(x,0)=P_{2i}(x,0)=0,\ i=1,2,\cdots,N \quad (8)$$

3 Conclusions

Theorem 1 The steady-state availability of the mathematical model after the N th repair is

$$A_N=\sum_{i=1}^{N}a^{i-1}\Big/\sum_{i=1}^{N}a^{i-1}\left[1+\lambda_h E(R_1)\right]+a^{N-1}\lambda_s\left[(N-1)E(R_2)+E(R_3)\right] \quad (9)$$

Where $E(R_i)=\int_0^\infty e^{-\int_0^x \mu(s)ds}dx\ (i=1,2,3)$.

Proof. Solving equations (1)—(8)($t\to\infty$), we get

$$P_{01}^*=a^{N-1}P_{0N}, i=1,2,\cdots,N,\ P_{1i}^*=\lambda_h P_{0i}e^{-\int_0^x \mu_1(y)dy}, i=1,2,\cdots,N$$

$$P_{2i}^*=a^{i-1}\lambda_s P_{0i}e^{-\int_0^x \mu_2(y)dy}, i=1,2,\cdots,N-1,\ P_{2N}^*=a^{N-1}\lambda_s P_{0N}e^{-\int_0^x \mu_3(y)dy}$$

Setting $P_{ji}^*=\int_0^\infty P_{ji}^*(x)dx, j=0,1,2; i=1,2,\cdots,N$.

Let $M=\sum_{j=0}^{2}\sum_{i=1}^{N}P_{ji}^*\triangleq\sum_{i=1}^{N}P_{0i}^*+\sum_{i=1}^{N}\int_0^\infty P_{1i}^*(x)dx+\sum_{i=1}^{N}\int_0^\infty P_{2i}^*(x)dx$

$$=\sum_{i=1}^{N}a^{i-1}[1+\lambda_h E(R_1)]+a^{N-1}\lambda_s[(N-1)E(R_2)+E(R_3)]$$

Hence,

$$A_N=\sum_{i=1}^{N}a^{i-1}\Big/\sum_{i=1}^{N}a^{i-1}\left[1+\lambda_h E(R_1)\right]+a^{N-1}\lambda_s\left[(N-1)E(R_2)+E(R_3)\right]$$

where $E(R_i)=\int_0^\infty e^{-\int_0^x \mu_i(s)ds}dx(i=1,2,3)$.

From the above availability expression, the mathematical model steady-state availability decreases with the number of the minimal repair times increasing. So, the mathematical model steady-state availability is gradually decreasing (for $a>1$).

Theorem 2 The steady-state failure frequency of the mathematical model after the N th repair is

$$M_N = \frac{\lambda_h \sum_{i=1}^{N} a^{i-1} + \lambda_s N a^{N-1}}{\sum_{i=1}^{N} a^{i-1}[1+\lambda_h E(R_1)] + a^{N-1}\lambda_s[(N-1)E(R_2)+E(R_3)]} \tag{10}$$

Proof. Let $P_{ji}(t) = \int_0^\infty P_{ji}(x,t)dx,\ j=1,2;\ i=1,2,\cdots,N$

$$\mu_1(t) = \int_0^\infty \mu_1(x)P_{1i}(x,t)dx / P_{1i}(t),\ i=1,2,\cdots,N$$

$$\mu_2(t) = \int_0^\infty \mu_2(x)P_{2i}(x,t)dx / P_{2i}(t),\ i=1,2,\cdots,N-1$$

$$\mu_3(t) = \int_0^\infty \mu_3(x)P_{2N}(x,t)dx / P_{2N}(t)$$

satisfying $\mu_j(t) \to \mu_j, t \to \infty (j=1,2,3)$ .Then the transition probability matrix can be obtained easily.

$$D = \begin{bmatrix} D_{11} & D_{12} & D_{13} \\ D_{21} & D_{22} & D_{23} \\ D_{31} & D_{32} & D_{33} \end{bmatrix} \tag{11}$$

where

$$D_{11} = \text{diag}\left(-(\lambda_h+\lambda_s), -(\lambda_h+a\lambda_s), \cdots, -(\lambda_h+a^{N-2}\lambda_s), -(\lambda_h+a^{N-1}\lambda_s)\right)_{n\times n}$$

$$D_{12} = \text{diag}\left(\mu_1(t), \mu_1(t), \cdots, \mu_1(t)\right)_{n\times n},$$

$$D_{13} = \begin{bmatrix} 0 & 0 & 0 & 0 & \mu_3(t) \\ \mu_2(t) & 0 & 0 & 0 & 0 \\ \vdots & & & & \\ 0 & 0 & \mu_2(t) & 0 & 0 \\ 0 & 0 & 0 & \mu_2(t) & 0 \end{bmatrix}_{n\times n},$$

$$D_{21} = \text{diag}\left(\lambda_h, \lambda_h, \cdots, \lambda_h\right)_{n\times n},$$

$$D_{22} = \text{diag}\left(-\mu_1(t), -\mu_1(t), \cdots, -\mu_1(t)\right)_{n\times n}, D_{23} = O_{n\times n},$$

$$D_{31} = \text{diag}\left(\lambda_s, a\lambda_s, \cdots, a^{N-2}\lambda_s, a^{N-1}\lambda_s\right)_{n\times n}, D_{32} = O_{n\times n},$$

$$D_{33} = \text{diag}\left(-\mu_2(t), -\mu_2(t), \cdots -\mu_2(t), -\mu_3(t)\right)_{n\times n}$$

Thus we can get the instantaneous failure frequency of the mathematical model with the method in Ref.[17] $M_N(t)=\sum_{i=1}^{N}(\lambda_h+a^{i-1}\lambda_s)P_{0i}^*(t)$.

Let $t\to\infty$, the steady-state failure frequency of the mathematical model is immediate

$$M_N=\lambda_h\sum_{i=1}^{N}a^{i-1}+\lambda_s Na^{N-1}\Big/\sum_{i=1}^{N}a^{i-1}[1+\lambda_h E(R_1)]+a^{N-1}\lambda_s[(N-1)E(R_2)+E(R_3)]$$

Certainly, we can obtain the formulations of the instantaneous reliability indices and their corresponding steady-state values as well, such as the reliability of the system, the probability of the system waiting to be repaired and so on.

4 Numerical Example

According to the following numerical analysis, the mathematical model availability decreases with the repair cycle (repair times) increasing, which is accord with theoretical analysis. It can be seen, after the system software failure, according to the needs of practical problems, in order to make the system availability at a certain interval range, we may overhaul the system after a number of minimal repair.

Table 1. Simulation Values

μ	μ_3	λ	a	A	N *(repair times)*
1	3	0.1	2	0.57	14
1	3	0.1	2	0.55	15
1	3	0.1	2	0.54	16
1	3	0.5	5	0.21	14
1	3	0.5	5	0.20	15
1	3	0.5	5	0.19	16
2	6	0.1	2	0.72	14
2	6	0.1	2	0.71	15
2	6	0.1	2	0.70	16
2	6	0.1	5	0.63	14
2	6	0.1	5	0.62	15
2	6	0.1	5	0.60	16
2	6	0.5	2	0.34	14
2	6	0.5	2	0.33	15
2	6	0.5	2	0.32	16
2	6	0.5	5	0.26	14
2	6	0.5	5	0.24	15
2	6	0.5	5	0.23	16

Acknowledgments. This work is supported Heilongjiang Province Education Department Project of China under grant 11553004 and is supported by Doctor Startup Fund of Daqing Normal University under grant 10ZR06.

References

1. Rao, T.S.S., Gupta, U.C.: Performance modelling of the M/G/1 machine repairman problem with cold-,warm- and hot- standbys. Computers & Industrial Engineering 38, 251–267 (2000)
2. Edmond, J.V., Stanislav, S.M.: On Gavers parellel system sustained by a cold standby unit and attended by two repairmen. Operations Research Letters 30, 43–48 (2002)
3. Wang, K.H., Ke, J.C.: Probabilistic analysis of a repairable system with warm standbys plus balking and reneging. Applied Mathematical Modelling 27, 327–336 (2003)
4. Zhang, Y.L., Wang, G.J.: A deteriorating cold standby repairable system with priority in use. European Journal of Operational Research 183, 278–295 (2007)
5. Michael, J.A.: Age repair policies for the machine repair problem. European Journal of Operational Research 138, 127–141 (2002)
6. Hu, H.D., Wang, J.M.: Asymptotic stability of software system with rejuvenation policy. In: Proceedings of the 26th Chinese Control Conference, pp. 646–650 (2007)
7. Barlow, R.E., Proschan, F.: Statistical theory of reliability and life testing. Holt, Reinehart and Winston, New York (1975)
8. Khalil, Z.S.: Availability of series system with various shut-off rules. IEEE Trans. Reliability R. 34, 187–189 (1985)
9. Chao, M.T., Fu, J.C.: The reliability of a large series system under Markov structure. Adv. Appl. Probab. 23, 894–908 (1991)
10. Zhang, Y.L., Wang, G.J.: A geometric process repair model for a series repairable system with k dissimilar components. Applied Mathematical Modelling 31, 1997–(2007)
11. Zhang, Y.L., Wang, G.J.: A deteriorating cold standby repairable system with priority in use. European Journal of Operational Research 183, 278–295 (2007)
12. Rao, F., Li, P.Q., Yao, Y.P., et al.: Hardware/software Reliability Growth Model. Acta Automation Sinica 22(1), 33–39 (1996) (in Chinese)
13. Mark, A.B., Christine, M.M.: Developing Integrated Hardware-Software Reliability Models: Difficulties And Issues. In: Proceedings of AIAA/IEEE Digital Avionics Systems Conference (1995)
14. Lam, Y.: Geometric Processes and Replacement Problem. Acta Math. Appl. 4, 366–377 (1998)
15. Shi, Z., He, X.G., Wu, Z.: Software Reliability and its evaluation. Computer Applications 20(11), 1–5 (2000)
16. Wu, X., Zhang, J., Tang, Y.H., Dong, B.: Based on the Software and Hardware Features the Reliability Analysis of the Computer System. Journal of Civil Aviation Flight University of China 17(1), 33–36 (2006) (in Chinese)
17. Cao, H., Cheng, K.: Introduction to Reliability Mathematics. Higher Education Press, Beijing (2006) (in Chinese)

Study on the Premises Distribution System in the Construction of Informationization

Hongwei Yang, Fang Zhang, and Xiaohui Li

College of Information and Electrical Engineering,
Sheyang Agricultural University, Shenyang, China
yhwsyau@163.com

Abstract. Premises distribution is an important and basic work of enterprise or institutions to realize networking, automation, digitization and intelligent construction. This paper introduces the design and application elements of network premises distribution system, especially introduces some modules included in the system.

Keywords: Premises Distribution System, Informationization construction, Topological structure, System structure.

1 Introduction

Premises distribution system is a practical technology and the inevitable outcome of the modern information society; it is the inevitable request to multi-function, intelligent building. Premises distribution system plays an important role to the overall function of the building based on all kinds of system resources and keeping each department high efficiency and long running. Premises distribution system, full name is building and complex premises distribution system, also called the open wiring system, is a kind of network system for comprehensive data transmission in the building and complex, it is the neural context of the whole intelligent building. Premises distribution system uses structure approach to connect voice switching inside building, intelligent processing equipment and other generalized data communication facilities. And uses necessary equipment to connect the outside building data network or telephone lines. The system includes all cables to connect above equipments and related wiring devices.

2 Characteristics of Premises Distribution System

Network Premises distribution system design should be considered some factors:

2.1 Using Mainstream Wiring Technology in Technology

The mainstream wiring technology in enterprise network construction is advanced technology and the popular technology, which is usually the engineering technology.

A. Xie & X. Huang (Eds.): Advances in Electrical Engineering and Automation, AISC 139, pp. 307–312.
springerlink.com

Currently, in most engineering practice, data communication uses multimode fiber primarily trunk line, 100 M super five-kind of UTP cable indoor. Voice communication uses three-kind of multi-cores cable route, three-kind or five-kind of UTP cable indoor.

2.2 Wiring System should Consider Enough Bandwidth

Enterprise should provide perfect information service for customers, suppliers and partners, including VOD, information navigation, information release etc. Realize these functions should consider enough bandwidth.

2.3 Products Selection should Notice Prices, Easy Maintenance and Use

Wiring system needs the low cost of products under meeting performance requirements and certain period development; achieve the highest wiring performance to price.

3 Type of Premises Distribution System

In order to make the engineering design specific of the intelligent building and intelligent Building Park, the integrated wiring system is divided into three design levels according to the actual need:

3.1 Basic Type

Apply to the low configuration standard occasions, use copper-core cable for network.
Basic integrated wiring system configuration:

(1) Each work area (station) has one information socket;
(2) The wiring cable of work area (station) is a four twisted-pair cable, which leads to the floor distribution frames;
(3) Completely using clip handover hardware;
(4) Each work area (station) trunk cable (total distribution frames wires of floor distribution frames to devices) at least two twisted-pair cable.

3.2 Enhanced Type

Apply to the medium configuration occasion, use copper-core cable for network.
Enhanced type integrated wiring system configuration:

(1) Each work area (station) has two or more information socket;
(2) The wiring cable of each work area (station) is a four twisted-pair cable, which leads to the floor distribution frames;

(3) Using clip handover hardware (110A) or Plug-in type hardware (110P);
(4) Each work area (station) trunk cable (total distribution frames wires of floor distribution frames to devices) at least three twisted-pair cable.

3.3 Comprehensive Type

Apply to the higher configuration occasion, use cable and copper-core cable for network.

Comprehensive type integrated wiring system configuration:

(1) Add cable system on the basis of basic type and enhanced type;
(2) At least equip two twisted-pair cable in trunk cable of each basic type work area;
(3) At least equip three twisted-pair cable in trunk cable of each enhanced type work area;

4 Design of Premises Distribution System

4.1 Basic Requirement

Premises distribution system should be open network topology structure, should support voice, data and images, multimedia business information transmission.

4.2 Engineering Design of Premises Distribution System should Appropriate to the Following Seven Parts

(1) Work area

An independent need to set up telecommunication equipment (TE) area should be divided into a work area. Working area should compose of connection cables that wiring system information telecommunication outlet (TO) extend to terminal equipment place and adapter.

(2) Wiring subsystem

Wiring subsystem compose of information socket module, wiring cable and optical cable that information socket module to fiber device(FD), equipment cable and jump line etc.

(3)Trunk subsystem

Trunk subsystem compose of trunk cables and cables that equipment to telecom room, building device (BD) installed in equipment room, equipment cable and jump line etc.

(4) Complex building subsystem

Complex building subsystem composes of trunk cables and cables that connect buildings, complex device (CD), equipment cable and jump line etc.

(5) Equipment room

To integrated wiring system engineering design, equipment room mainly installs building wiring equipment. Telephone exchanges, host computer equipment and entrance facilities also can install in-up with wiring equipment.

(6) Incoming line room

Incoming line room is entrance area between outside of the building and information pipeline, and can be used as installation site between entrance facilities and buildings wiring equipment.

(7) Management

Management should mark and record according to the mode of working area to work room, telecom room, equipment room, incoming line room, cables, information socket module and other facilities.

4.3 The Structure of Premises Distribution System should Meet the Following Requirements

(1) The basic Structure of Premises distribution system should meet the following requirements as figure 1.

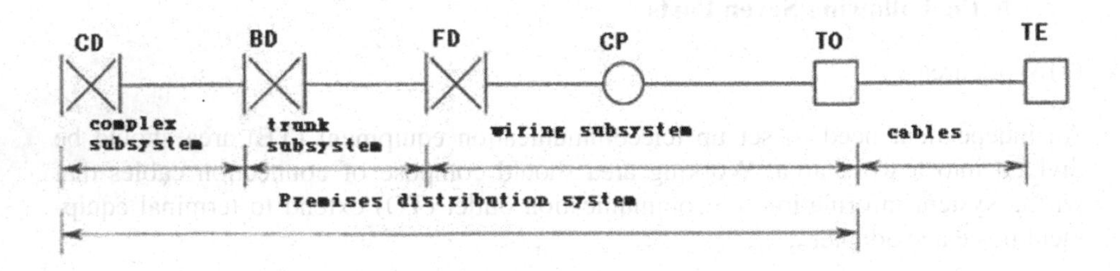

Fig. 1. The structure of Premises distribution system

Notice

Wiring subsystem can set concentrate point (CP), also not set concentrate point.

(2) The structure of Premises distribution subsystem should meet the following requirements as figure 2.

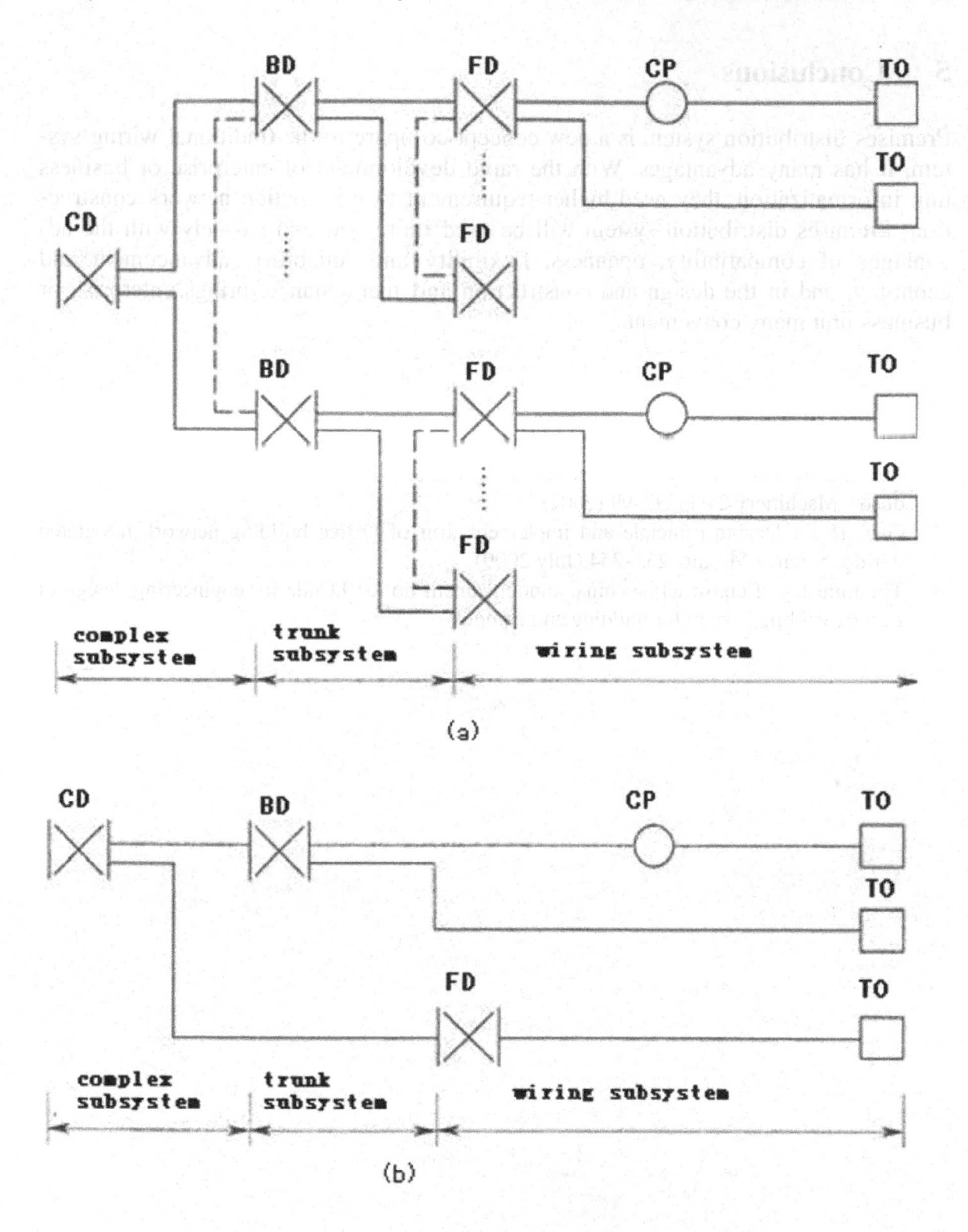

Fig. 2. The structure of Premises distribution subsystem

Notice

1 The dotted line in graph express that BD and BD, FD and FD can set trunk cable.
2 Building FD can connect CD straightly with trunk cables, TO can connect BD straightly through straight cables.

5 Conclusions

Premises distribution system is a new concept compare to the traditional wiring system, it has many advantages. With the rapid development of enterprise or business unit informatization, they need higher requirement to information network construction, Premises distribution system will be used more and more widely with the advantages of compatibility, openness, flexibility and reliability, advancement and economy, and in the design and construction and maintenance, brings enterprise or business unit many convenient.

References

1. Yao, R.-Q.: PDS design in the construction of enterprise of informationization. Light Industry Machinery 25(1), 97–99 (2007)
2. Gan, H.-F.: Design principle and implementation of Office building network integrated wiring. Science Mosaic, 253–254 (July 2009)
3. The ministry of construction,china, announcement no. 619,Code for engineering design of generic cabling system for building and campus

On Linear Network Coding and Matroid

Chunfeng Yang*, Chenxi Ming*, and Jiaqing Huang

Department of Electronics and Information Engineering,
Huazhong University of Science & Technology(HUST),
Hubei Key Lab of Smart Internet Tech.,Wuhan, 430074, P.R. China
yangchunfeng6731700@gmail.com

Abstract. The independence relation in a matroid implies the potential link between a matroid and network coding. Recently, Sun et al. proposed a algorithm of constructing generic linear network codes(GLNC) via representation of the *network matroid*. However, the construction algorithm was not explicitly given. Moreover, the discussion was merely on the case of GLNC. In this paper we can come to a connection between linear network coding(LNC). The main contributions of this paper are (1) to illustrate the representation algorithm of *network matroid* explicitly; (2) to apply the representation of three *induced matroid* to the construction of network code for its linear multicast (LM), linear broadcast (LB) and linear dispersion (LD) properties. Finally the complexity of the polynomial construction algorithm is analysed.

Keywords: linear network code, *network matroid*, *induced matroid*, representation.

1 Introduction

The pioneering work by Ahlswede et al. [1] laid the essential foundation of network coding which can be classified into acyclic and cyclic cases. Later Li et al.[2] proved that linear network coding can achieve the maximum capacity in a single-source finite acyclic network. To achieve this superiority, plenty of efforts were made to construct an effective network code. Jaggi-Sanders Algorithm in [3] was proposed to implemental polynomial time algorithm for linear multicast (LM).

Matroid theory is a branch of mathematics whose heart is the abstraction of "independence relation". There are various notions of independence such as linear independence in vector spaces or the acyclic property in graph theory.

A matroid M is an ordered pair (S, I), where S is a finite set and I is a set of subsets of S satisfying the following three conditions [4].

(a).$\Phi \in I$.
(b).If $i \in I$ and $J \subset i$, then $J \in I$.
(c).If i, $J \in I$ and $|J|<|i|$, then there is an element e of i-J such that $J \cup \{e\} \in I$.

The set S is called the ground set and the matroid M=(S, I) is called a matroid on S. The members of I are called independent sets and any subset of S not in I is called a

* These two authors contribute equally to this paper.

A. Xie & X. Huang (Eds.): Advances in Electrical Engineering and Automation, AISC 139, pp. 313–321.
springerlink.com

dependent set. A maximal independent set of a matroid is called a base of the matroid and a minimal dependent set is called a circuit.

Previous work has been done to deal with matroids on the edge set of a network. Through the structure of edge-disjoint paths, a single-source network is associated with a particular matroid, called the *network matroid* [5]. A matroid is said to be representable over a field F if it is isomorphic to the vector matroid of a matrix over F, and this matrix is called an F-representation for the matroid [4].

Sun et al.[5] illustrated the connection between generic linear network codes (GLNC) and the corresponding *network matroid*, that is, the representation of the *network matroid* induces GLNC for the network. With LM, linear broadcast (LB), linear dispersion (LD) and GLNC as a whole of the linear properties for acyclic networks as well as the independence property contained in matroids, we are explore the cases of LM, LB, LD via the mathematical method: matroid.

2 Related Work

Linear network coding as well as matroid theory is well investigated and also much attention has been paid to the connection between them, such as constructing matroidal networks from matroids [6]. But little was studied in the link between the construction of a linear generic code for a network and the representation of the *network matroid* of the network until Sun et al. proposed that every representation for the *network matroid* of an acyclic network induce a GLNC[5]. Also, they presented that via the association with an acyclic network case an algorithm of finding a representation in a cyclic case could be derived. However, these work only cover the case of generic code with corresponding *network matroid.* As the four properties of linear network coding are LM, LB, LD, GLNC, the other three cases, namely the constructions of LM, LB, LD, have not been associated with matroids. In addition, the construction algorithm of representation of matroid was not given an explicit presentation. Our contributions of this paper are: [1]We propose a way of representation of matroid. [2]We illustrate the construction algorithm of LM, LB, LD via three *induced matroids*.

3 Definitions

Definition 1): [3](local and global encoding kernel). An ω-dimensional linear network code on an acyclic network over a base field F consists of a scalar $k_{d,e}$ for every adjacent pair of channels (d,e) in the network as well as a column ω-vector f_e for every channel e such that:

(i) $f_e = \Sigma_{d\in In(t)} k_{d,e} f_d$ for $e\in Out(t)$.

(ii) The vectors f_e for the ω imaginary channels $e\in In(s)$ form the standard basis of the vector space F^{ω}.

The vector f_e is called the global encoding kernel for channel e. And $K_t = [k_{d,e}]_{d\in In(t), e\in Out(t)}$ is called the local encoding kernel at node t.

Definition 2): [3](linear multicast, linear broadcast and linear dispersion) Let vectors f_e denote the global encoding kernels in an ω-dimension F-valued linear network code on a single-source finite acyclic network. Let $V_t = <\{f_d : d \in In(t)\}>$ Then, the linear network code qualifies as a linear multicast, a linear broadcast, or a linear dispersion, respectively, if the following hold:

(a) $\dim(V_t) = \omega$ for every non-source node t with $\mathrm{maxflow}(t) \geqslant \omega$.
(b) $\dim(V_t) = \min\{\omega, \mathrm{maxflow}(t)\}$ for every non-source node t.
(c) $\dim(\cup_{t \in T} V_T) = \min\{\omega, \mathrm{maxflow}(T)\}$ for every collection T of non-source nodes.

Definition 3): [5](*network matroid*) Given the network topology, via the structure of edge-disjoint paths, a single-source network is associated with a *network matroid* which reflects the independence relation contained in the edge set.

Definition 4): [5](*induced matroid*) Given a linear network code L on a network, the matroid (E, I_L), in which E is the edge set and I_L is the family of subset of edges whose coding vectors are linearly independent, is called the *induced matroid* of the linear network code L.

4 Methods

4.1 Representation of *network matroid*

The method here is to induce the *network matroid* from network topology and then construct GLNC via this matroid. Our procedures of algorithm are as follows:

4.1.1 Induce *network matroid* from Network Topology

1).Label all the edges(except the imaginary edges) with integrals .
2).Find out all the paths from the source nodes to the sink nodes.
3).Select any two edges a and b. (since ω=2, the size of the base or the rank of the *network matroid* is 2). If there exist two different edge-disjointed paths m,n, satisfying: a).a is in path m, and b is in path n; b).the sub-path m0~a of path m does not include b, the sub-path n0~b of path n does not include a.
4). Similarly, we can obtain the other elements of the *network matroid.*

Remark 1. As for step 2), the paths are sorted by the output edges of the source nodes. In latter discussion of the construction, we all assume that these edges have been encoded because the coding vectors for these edges can be easily determined due to the practical application.

Remark 2. As for step 3), the above two sub-procedures a) and b) guarantee that all the possible independent edge pairs can be found since the above paths are integral ones from source nodes to sink nodes.

4.1.2 Construct Network Code from *network matroid*

Set the global encoding kernel F with size of w*n, the local encoding kernel K with size of n*n and the source message matrix U with size of w*n(Note: w is the number of the imaginary edges of the source node and n is the number of edges in this topology) .

Owing to the relation between global encoding kernel and local encoding kernel, we can obtain the formula: F=F*K+U, thcn F=U*[I-K]-1[7].

Our goal is to find out and assign proper non-zero values to K matrix components

1). Initialization for matrix K,U,F. (Refer to the latter example)

2). Put matrix K,U into the formula F=U*[1-K]-1.

3). Due to the constraint of mutual independence among the elements in *network matroid*, all matrices in the *network matroid* need to be of full rank.

The integrals in these matrices correspond to the columns in matrix F, so the corresponding matrices made up of the columns should be of full rank.

4). Put F into them and it comes to that some elements in matrix K are non-zero. Given any finite field to choose non-zero positive integrals, we can achieve the generic linear network coding.

4.2 Construction Algorithm for LM, LB, LD

We use the method that compute *induced matroids* from the network topology and then construct LM, LB, LD via them. Our procedures of algorithm are as follows:

i) LM

1). Label all the edges(except the imaginary edges) with integrals.

2). Find out all the paths from the source nodes to the terminal nodes. (Note: here in order to better show the process, the path should be combined with the nodes, which will be illustrated in latter example)

3). Search all the nodes in the paths, and find out the multicast nodes.

The sets of the input edges of a multicast nodes are bases of the *network matroid*.(Note: For nodes that have more than ω input edges, such as x edges, we can constitute a ω*x matrix whose rank is ω).

4). Get the matrix K as well as matrix U directly from the topology

5). Calculate the matrix F with matrices K and U

6). Take out vectors from matrix F in line with elements in sets got by step 3) as new matrices.

7). Assign values to the non-zero numbers in matrix K so that the new matrices in step 6) are of full rank.

ii) LB

All the steps are similar to the case of LM except for slight differences that in step 3) broadcast nodes rather than multicast nodes are the ones to be considered and in step7) the new matrices are not necessarily of full rank. (Note : For nodes that have less than ω input edges, such as x edges, we can constitute an ω*x matrix whose rank is x)

iii). LD

We can induce the construction method from the above algorithms for LM, LB with the distinction that in case of LD, sets of nodes in different number should be considered completely and respectively. Obviously, construction for LD is more complicated than for LM, LB.

5 Example

5.1 Representation of *network matroid*

Here we take the butterfly topology in Fig.1 as an example.

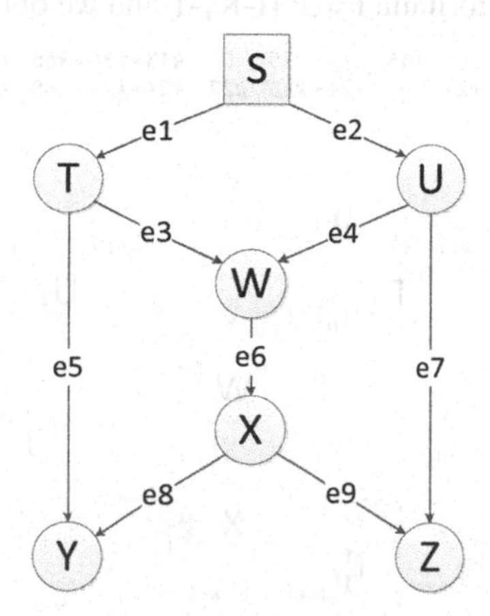

Fig. 1. Butterfly Topology with edges labeled

1). Label all the edges (except the imaginary edges) with the integrals 1~9.(Note: in latter discussion, we assume the integrals stand for the corresponding labeled edge)

2). Find all the paths from the source nodes to the sink nodes which in Table 1.

Table 1. All paths in Fig.1

ST	1-5
ST	1-3-6-8
ST	1-3-6-9
SU	2-7
SU	2-4-6-8
SU	2-4-6-9

3). In this case, for example, in edge pair {3,6}, 3 is in path 1-3-6-8 and 6 is in path 2-4-6-8, meanwhile, the sub-path 1-3 does not include 6, and sub-path 2-4-6 does not include 3. Therefore, {3,6} is an element of the *network matroid* due to the fact that the *network matroid* comprises all sets of possible independent edges.

4). Similarly, we can obtain the other elements in the *network matroid*. Thus the *network matroid* is:

{{1,2}{1,4}{1,6}{1,7}{1,8}{1,9}{2,3}{2,5}{2,6}{2,8}{2,9}{3,4}{3,6}{3,7}{3,8}{3,9}{4,5}{4,6}{4,8}{4,9}{5,6}{5,7}{5,8}{5,9}{6,7}{7,8}{7,9}{1}{2}{3}{4}{5}{6}{7}{8}{9}}

5). a). In matrix K, k13,k15,k24,k27,k36,k46,k68,k69 are unknown, so encoding is needed. That is:

$$K=\begin{bmatrix} 0 & 0 & k13 & 0 & k15 & 0 & 0 & 0 & 0 \\ 0 & 0 & 0 & k24 & 0 & 0 & k27 & 0 & 0 \\ 0 & 0 & 0 & 0 & 0 & k36 & 0 & 0 & 0 \\ 0 & 0 & 0 & 0 & 0 & k46 & 0 & 0 & 0 \\ 0 & 0 & 0 & 0 & 0 & 0 & 0 & 0 & 0 \\ 0 & 0 & 0 & 0 & 0 & 0 & 0 & k68 & k69 \\ 0 & 0 & 0 & 0 & 0 & 0 & 0 & 0 & 0 \\ 0 & 0 & 0 & 0 & 0 & 0 & 0 & 0 & 0 \\ 0 & 0 & 0 & 0 & 0 & 0 & 0 & 0 & 0 \end{bmatrix}$$

b). In matrix U, source node S transmit data to edge 1 and 2 directly. Thus:

$$U=\begin{bmatrix} 1 & 0 & 0 & 0 & 0 & 0 & 0 & 0 & 0 \\ 0 & 1 & 0 & 0 & 0 & 0 & 0 & 0 & 0 \end{bmatrix}$$

6). Put matrix K,U into the formula F=U*[I-K]-1 and we obtain:

$$F=\begin{bmatrix} 1 & 0 & k13 & 0 & k15 & k13*k36 & 0 & k13*k36*k68 & k13*k36*k69 \\ 0 & 1 & 0 & k24 & 0 & k24*k46 & k27 & k24*k46*k68 & k24*k46*k69 \end{bmatrix}$$

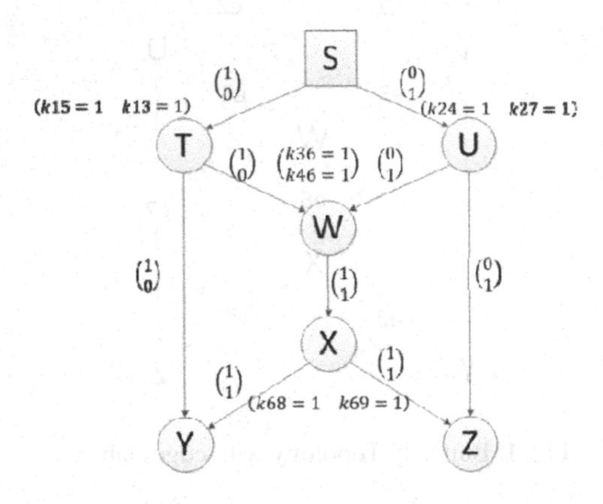

Fig. 2. Butterfly Topology with network code

7). All matrices corresponding to the elements of the *network matroid* need to be of full rank. (Note: In fact, according to the dependence between the edges, it can further reduce the number of the matrices that need to be calculated in the actual operation.)

8). Put F into them and it comes to that k13,k15,k24,k27,k36,k46,k68,k69 are non-zero. For example in the case of GF(2), let k13=k15=k24=k27=k36=k46=k68 =k69=1, we can obtain the final network code in Fig.2.

5.2 Construction for LM, LB, LD

We illustrate how to construct network codes for LM, LB, LD with the representation of corresponding *induced matroid* in a network in Fig.3 with assumption that global encoding vectors for the three output edges of the source node S have been set as $[0,0,1]^T$; $[0,1,0]^T$;$[1,0,0]^T$.

i). The case of LM.

1~3). Search all the nodes in the paths, and multicast node H can be got. The input edges of it comprises 4, 13, 14. Thus, the base of this *induced matroid* is {4, 13, 14}.

4). The K and U matrices can be calculated as the example in Section 5.1.

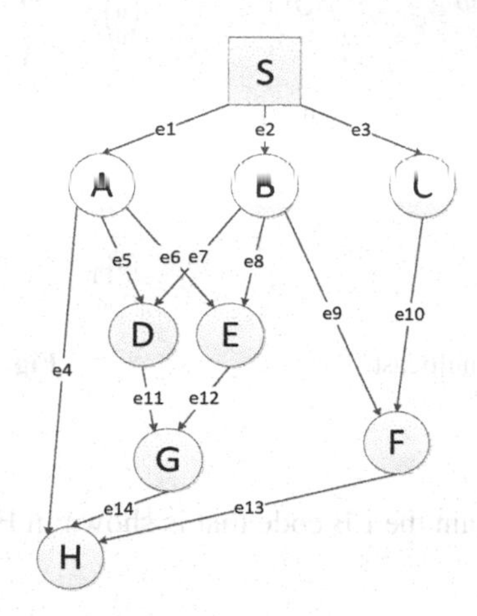

Fig. 3. Topology for code construction

5). Calculate with the above two matrices, and we can come to the F matrix:

$$F=\begin{bmatrix} 1 & 0 & 0 & k14 & k15 & k16 & 0 & 0 & 0 & 0 & k15*k511 & k16*k612 & 0 & k1114*k15*k511+k1214*k16*k612 \\ 0 & 1 & 0 & 0 & 0 & 0 & k27 & k28 & k29 & 0 & k27*k711 & k28*k812 & k29*k913 & k1114*k27*k711+k1214*k28*k812 \\ 0 & 0 & 1 & 0 & 0 & 0 & 0 & 0 & 0 & k310 & 0 & 0 & k1013*k310 & 0 \end{bmatrix}$$

6). Take out vectors from matrix F in line with elements in{4,13,14} as a new matrix:

$$\begin{bmatrix} k14 & 0 & k1114*k15*k511+k1214*k16*k612 \\ 0 & k29*k913 & k1114*k27*k711+k1214*k28*k812 \\ 0 & k1013*k310 & 0 \end{bmatrix}$$

7). It should be satisfied that this matrix is of full rank. That is: $K14*K1013*k310*(K1114*k27*k711+k1214*k28*k812)\neq 0$.We can set K14=K1013=k310=K1114 =k27 =k711=1 and k1214=k28=k812=0.Thus the final encoding outcome is shown in Fig.4:

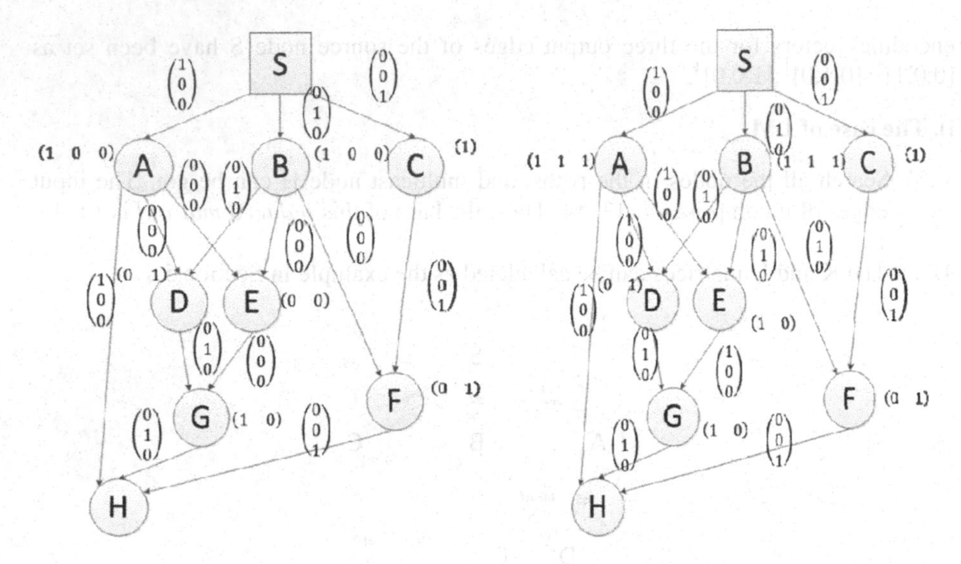

Fig. 4. Linear multicast **Fig. 5.** Linear broadcast

ii). The case of LB.

Similar to LM, we can obtain the LB code that is shown in Fig.5.

iii). The case of LD.

It is similar to the above two cases and we don't present it here due to limited space.

6 Complexity Analysis

Here, we mainly analyze the complexity of the construction algorithm for LM, and a similar result can be reached for that of LB and LD. Compared with Jaggi-Sanders Algorithm for LM in [3] which is achieved via the network topology itself, the construction algorithm in this paper mainly depends on the mathematical module: matroid, which leads the question into pure mathematical analysis. Consider the two parts of the construction respectively:

1). **Obtain *induced matroids* from the network topology.** We should traverse every node to search the input edges. If the maxflow of one node is no less than ω, then the node is a multicast node. The complexity of this procedure does not exceed $O(V*\omega)$ (V is the number of nodes);

2). **Construct LM via the *induced matroid*.** The first step is to find T multicast nodes, then find out the corresponding vector matrices with complexity of $O(\omega*\text{In}(T))$ and finally simplify these matrices to figure out the local encoding kernel in them with complexity of $O(T*\omega^3)$.

Thus, the construction algorithm for LM, LB, LD in this paper is a polynomial one.

7 Conclusion

In this paper, we explicitly propose the algorithm of representation of *network matroid* to construct the GLNC. Meantime, we illustrate the association between *induced matroids* and LM, LB, LD so that a connection between LNC and matroids as well as the code construction of LM, LB, LD is accomplished.

Acknowledgments. This research was supported by National Natural Science Foundation of China(No.60872005), National Foundation of Research Training Project for Undergraduates(No. 20112314).The authors thank anonymous reviewers for their comments.

References

1. Ahlswede, R., Cai, N., Li, S., Yeung, R.: Network information flow. IEEE Transactions on Information Theory 46(4), 1204–1216 (2000)
2. Li, S.-Y.R., Yeung, R.W., Cai, N.: Linear network coding. IEEE Trans. Inf. Theory 49(2), 371–381 (2003)
3. Yeung, R.W.: Information Theory and Network Coding. Springer (2008)
4. Dougherty, R., Freiling, C., Zeger, K.: Network Coding and Matroid Theory. Proceedings of the IEEE 99(3) (2011)
5. Sun, Q., Ho, S.T., Li, S.-Y.R.: On Network Matroids and Linear Network Codes. In: ISIT 2008, Canada (2008)
6. Dougherty, R., Freiling, C., Zeger, K.: Networks, matroids, and non-Shannon information inequalities. IEEE Trans. Inf. Theory 53(6), 1949–1969 (2007)
7. Li, S.-Y.R., Yeung, R.W.: On Convolutional Network Coding. In: ISIT (2006)

The Study and Implementation of Deep Integration Platform of Campus Information Based on SOA

Jiquan Shen[1] and Shui Ju[2]

[1] School of Computer Science and Technology, Henan Polytechnic University, Jiaozuo, Henan, 454000, China

[2] School of Mechanical and Power Engineering, Henan Polytechnic University, Jiaozuo, Henan, 454000, China

Abstract. Using gird technology, a distributed, shared, dynamic extended, service-oriented resource integrated architecture is established to achieve the basic grid service. Then, the register, publishing, discovering, denoting, and access mechanism of education information resources in the grid environment are implemented, which resolves some kernel problems about education information resources, such as dynamic extension, single system image, autonomous control, modeling and coordination of business process and so on. The architecture provides resource share, high-performance and high-quality data processing service for users and greatly improves the utilization Ratio of existing education resources accordingly.

Keywords: Grid, SOA, Information integration, Share.

1 Introduction

At present, it is the age of "information explosion", the feature of which is the means and mode of people's producing, achieving and using information are various. The main purpose of information systems based on computer technology is to resolve the problems in the process of people's producing, achieving and using information. In recent years, with the development of computer technology, the informationization degree of colleges and universities has a great improvement. Many management information systems (such as educational administration management system, financial management system, books management system, students management system, etc.) have appeared. The application of these systems has greatly enhanced the efficiency of relevant departments. However, these information systems are independent and can't share information. That is to say, there is a great deal of information islands. To obtain information, the users have to access every application system and gather the information. The whole process is cumbersome and complicated because of the restriction of various information schemas and access authority, which results in the existing information system can't be fully comprehensive using and the information in different systems can't be shared.

A. Xie & X. Huang (Eds.): Advances in Electrical Engineering and Automation, AISC 139, pp. 323–327.
springerlink.com

Information integration and sharing is a basic contradiction. How to do? There are many ways. At present, the basic approach in many enterprises and institutions is: when the demand bulk up to a certain extent, new plan and design is made and all previous n systems are integrated into a new platform. As time goes by, the requirements keep on changing, that may lead to the existence of m systems and a further integration. It is the most common condition in the process of informationizization. This condition make the systems have quickly changes, high costs and short service life. As a result, many enterprises and institutions even dare not carry out information construction.

2 Integration of Campus Information Based on SOA

2.1 Basic Idea

The basic idea of the deep integration platform of campus information based on SOA is adopting open technical standards and virtualization technology in the information integration. Open can solve the dynamic extended problem of information system, and meet the requirements of system extension and future seamless integration; Virtualization technology can solve the contradiction between the general pattern and special business, avoid the traditional way of case by case and reduce the cost of the system integration.

Now, the traditional depth integration approach of campus information is integrating the various types of information management systems by COM/DCOM, CORBA, RMI and so on. Because these technologies and methods have some limits more or less, the integrated system established by them seriously waste resources. For example:

- Repeated resources deployment.
- Tightly-coupled mechanism in the system integration.
- Update of the version and interfaces.
- Resources discovering.

2.2 Integration Model of Campus Information Based on SOA

OGSA extends the grid computing applications to the enterprise application field where the main feature is more widely distributed system service integration. OGSA establishes the basic concepts of grid service and abstracts everything as service. The service is a kind of entity provides certain ability for the customers through information exchange. The service can be defined as a specific information exchange sequence which leads to perform some operations. The service has the conceptual and application universality. The services includes from the low level of resources management function (e.g. storage service) to the high level of system monitor function to(e.g. fault disposing service). OGSA abstracting everything as service is helpful to manage and share various grid resources by the unified standard interface.

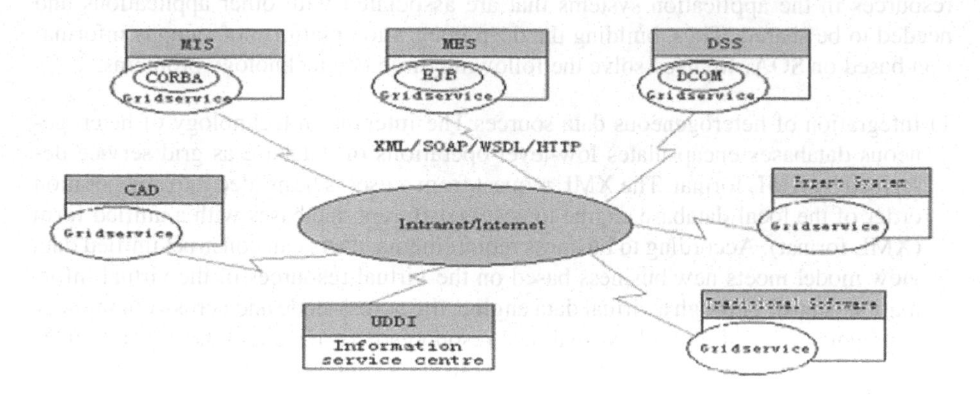

Fig. 1. The integration model of campus information based on SOA

Figure 1 shows the integration model of campus information based on SOA. In the model, we can see that all resources inner application systems use uniform object model of grid encapsulation in the external interfaces. All the service providers, service requesters and information service centers are interconnected by the Internet. The service providers publish the services provided by them to an information service center. The service requesters discover the needed services by the register mechanism of UDDI. Then, they find the service providers and gain the services through the remote calls of the Internet. It is needed to be explained that a service provider can also be a service requester and send service requests to other service providers.

The deep integration platform of campus information based on SOA can establish a dynamic, inter-organizational, overall and integrative information shared platform in the Internet environment, which can publish and manage information expediently. And the required information can be accessed timely, so we can improve the existing information resources utilization and avoid redundant construction. Simultaneously, solving the problems such as the integration of heterogeneous database, the data exchange among heterogeneous distributed systems, the dataflow process reengineering the problem and so on, the platform can exactly meet the foregoing integration and application functional requirements.

2.3 The Key Technologies

The aim of using the grid technology to implement the research of campus information integration is to achieve the information share and integration and keep the existing functions of application. Therefore, it must apply grid service encapsulation to existing application systems in order to provide a uniform standard interface. Applying grid service encapsulation to existing application systems, whether encapsulate all the functions of the application systems? The answer is negative. Because encapsulating all the functions of the application systems, we must thoroughly understand every detail of the application systems, and this is not possible. In fact, restructuring and integrating the heterogeneous distributed information resources just includes the information resources in the application systems that are associated with other applications and needed to be shared. Thus, building the deep integration platform of campus information based on SOA, we need solve the following three key technology problems.

1) Integration of heterogeneous data sources:The integration technology of heterogeneous databases encapsulates low-level operations of database as grid service described by XML format. The XML request from a user is translated into an operation order of the local database engine to access different databases with a unified form (XML format). According to business requirements, users can construct unified data view model meets new business based on the virtual resources of the virtual information centre. Through a virtual data engine, the across-node and across-community aggregate effect of multiple virtual data resources is realized and the dynamic integration requirements of multiple autonomous data sources are met. A query request from a user involves many heterogeneous distributed data sources. For the realization of efficient joint query from multiple many heterogeneous distributed autonomic data sources, the technology of virtual data engine can be used:

- Optimize the task decomposing technology.
- Load balancing.
- Service from the nearby.
- Buffer technology.
- Asynchronous access mechanism.
- Shared address space.
- Connection pool technology.
- Batch technology.
- Virtual database streaming query engine technology.

2) Professional-oriented process modeling: The professional-oriented business process modeling is a gird application programming model which is based on SOA and takes the business abstract service as the core element. It tackles the effective organization and real-time presentation of business service resources in a particular field, and the business process construction and the business process verification to solve the integration of grid service resources.

- Effective organization and real-time presentation of business service resources.
- Business process construction.

3) Business process coordination: Under the grid environment, the business process coordination means the interaction and data exchange among processes within the organization or across different organizations. The business process has the features of distributed autonomy, dynamic, asynchronous and business semantic variance. The solution of the question depends on the dynamic discovery of grid service, the asynchronous interaction and data exchange mechanism among services. The event-driven business process coordination can be realized by an event-driven service agent mechanism.

3 Conclusions

The integration model of campus information based on SOA is a distributed, shared, dynamic extended, service-oriented resource integrated architecture, which resolves some kernel problems such as dynamic system extension, single system image, autonomous control, resource share and so on. It provides basic guiding role and practice methods for information share construction of heterogeneous and distributed education resources, improves the existing education resources utilization and avoids redundant construction. Consequently, it can reduce the cost of the system integration and development and meet the requirements of education informationization.

Acknowledgments. This paper is supported by the Basic and Leading Research Project of Henan Province (No. 092300410216), the Foundation Project for Youth Key Teacher of Henan Province (No. 2009GGJS-035).

References

1. Foster, I., Kesselman, C., Jin, H., Yuan, P., Shi, K. (trans.): Grid computing, 2nd edn. Publishing house of electronics industry (2004)
2. Xu, Z., Feng, B., Li, W.: Grid computing technology. Publishing house of electronics industry (2004)
3. MCQuay, W.K.: The collaboration grid: Trend for next generation distributed collaborative environments. Enabling Technologies for Simulation Science (III), 359–366 (2162)
4. Chai, X., Liang, Y.: Web Services technology, structure and applications. Publishing house of electronics industry (2003)
5. Zhuang, X., Sun, B.: The research and implementation of data shared scheme of heterogeneous education resource databases based on Web Service. E-education Research (2), 47–50 (2003)

Discussing about Characteristics and Design Idea of Nokia Mobile Phone

Guobin Peng

Design and Art College
GuiLin University of Electronic Technology
Guilin, P.R. China
pengguobin@guet.edu.cn

Abstract. The daily industry products which people consume present their own personalities, just like people themselves. The most ideal objective of products is to provide each consumer the goods or service they need or meet other particular needs from customers, at the same time, we advocate the exist of personality and common development, but that is not to mean there will be no personality if commonality exists, a good design still can make products full of perfect quality with various personalities. Mobile phones are a kind of products with extremely individual personalities, which are very close to our daily life. A set of mobile phone is not merely a technical product, it also can provide good communication and other service functions, it can even reflect consumer sentiment and personality, the demonstration of such characteristics is guided by design concept, thus to achieve the harmony of shape and.

Keywords: Products personality, Products quality, Design concept, Design methods.

1 Introduction

In the psychology personality refers to the integrity of individual psychology characteristic which a person displays frequently in his life and practice, relatively stable, has certain tendentiousness. Why we discuss about personality is to discover each kind of characteristic and performance of personality, promulgate the essential characteristics of personality, find out its originality. Then take the object created by human being for example, it also has such personality characteristics of certain stability and tendentiousness, people entitle such personalities onto products through activities, facing multiplicity in real life, product existence form and performance characteristic also include diversification. There are more and more various kinds of daily products in the market, simultaneously product competition is also very intense and runs into white-hot stage, it is extremely critical how to find out a good design localization and meet market demand through the improvement of product design characteristic and the selling point, the design thread of Nokia mobile phones may provide some ponders and inspirations to us.Nokia mobile phones keep up with times forward tightly, by its invariable modeling design thread and good quality performance to face up changeable market, by its mature design concept and market strategy to win affection and trust from consumers,

A. Xie & X. Huang (Eds.): Advances in Electrical Engineering and Automation, AISC 139, pp. 329–334.
springerlink.com

finally creating its magnificent result in the mobile phone profession. All of these are worth us studying and analyzing so as to prepare for own brand handset design and market strategy.

2 Characteristic Analysis of Nokia Telephone Products

2.1 Qualitycharacteristic Analysis of Personality Modeling

In mobile phone industry, we can see thousands of design modeling, different factories develop different telephone models, even reaching hundred kinds, from multitudinous telephone brands Nokia cell-phones design brand give people deep impression. Nokia cell-phones continually keep its straight board cell-phone to excavate its potential and continuous innovation. Every time when Nokia promotes any exquisite products it always keeps personality quality of straight board cell-phone modeling. Introducing European's favorite straight board handset modeling into Asia too. Nokia always keeps the total image unchangeable, like horizontal design of Nokia 5510, spring type design of Nokia 8910. Nokia starts promoting straight slide cover design from 8110, as well as afterwards 7110, 8810, 8850, all continue this kind of style, on one hand is to make straight board handset more delicate and more practical, simultaneously increasing the screen to meet network function and fashion demand. this kind of design development thread that insisting on to excavate all kinds of new potential of straight board handset, leading the trend of times and keeping its own personality modeling characteristic obtains success, just like Nokia China area General Manager zhao kelin said： Nokia cell-phone design needs to keep consistent style, especially for the top grade cell-phones. Therefore this kind of keeping consistent style and full potential excavation should be the consistent characteristic performance of its personality, this kind of consistent style of personality characteristic maintenance and continuous innovation to differ from other telephone modeling, keeping its personality extension.Simultaneously Nokia telephone always regards high quality telephones at modeling and quality as its goal, design the man-machine interaction button more reasonable, pursue accurate localization, low error rate, the button arrangement is very precise and scientific, form reasonable layout together with the screen, the organization is more compact, the dialogue is clearer, good feeling to hold the complete machine, the structure is stable. All modeling design has its unique and scientific personality, design the modeling through scientific research of human, display its intrinsic quality by mutual union and expression of these two together. In random visits to the investigators, their view of quality characteristics of mobile phone Nokia is as follows in table 1.

Table 1. Investigation statistics of Nokia mobile phone quality characteristics

generous styling	solid and durable	good voice quality	feature-rich	multi-style	proper price
82%	89%	91%	67%	79%	82%

2.2 Function Characteristic Analysis

The market is the best standard to examine the superiority and inferiority of a product, we can see an obvious fact from the market, some handsets with unique contour not necessarily have good sales volume, just like the classic design of Motorola V70, but the market proved it is unable to become the type with big sales volume, such is an example to prove that good contour design not necessarily means mature function. The function of Nokia cell-phones is practical and convenient, the performance is stable, response speed is rapid, function design is professional, session layout operation is friendly, consider maturely, the localization is accurate, always famous by good quality, simultaneously pursue continuous new technology, enhance its function attraction, combing science technology and friendly design together.

We saw N73 which was newly promoted in year 2006 attracted quite many consumer's attention by its splendid hardware and software function performance. we list out the original function performance of N73 hardware and software. N73 has 2.4 Inch and 260,000 color TFT screen, resolution ratio high as 240 X 320 dpi, demonstration effect is extremely exquisite.N73 has an Carl Zeiss optics lens with 3,200,000 maximal pixel, f2.8 biggest diaphragms and 5.6mm focal distance, support automatic focusing, flash and has 1/1000~2m mechanical shutter exposure. two-way loudspeaker in the sound effect aspect, matched with 2G mini SD card, AD41 audio switcher and more convenient music broadcast function, support blue tooth, Pop-Port connection (support full speed connection of USB 2.0), faster and more conveniently to accomplish data exchange with other handsets or computers. The apron can well protect lens from outside destruction. Using Symbian9.1 system, the third S60 edition operation surface, broad software . N73 supports many kinds of office software, with Quick Office inside, can realize Word, Excel, powerpoint documents browsing, provide convenience for our work. In addition this machine also has 40MB fuselage memory and supports mini-SD memory card expansion, and supports hot inserts. This has only enumerated one section of Nokia classic handsets, we can see while Nokia handsets pursue new technology, it also applies mature technology to handset function design and enhances its function and technical charm, which keeps the quality performance that never falls behind in the respect of technology, this is only a part of its technology content, it also uses massive high-tech and friendly design at environmental protection, design and manufacture its fuselage and packing by using environment protected and recycled material, using isolated sound installment to protect human eardrum, hearing and so on, full concerns for human, realizing and meeting human demand, which also is the basis of design.

3 Idea and Method

Personalization quality performance is not leaves oneself open to rumor, but it comes from its core idea. Nokia design work and brand core idea is to design attractive appearance, build experience which people cherish. Not only need this kind of design to look attractive, but also need to let it work according to people's idea, fully manifest friendly design request.

Friendly design request must care about human and human behavior. Personalized characteristic design is one kind of experience with extreme personality color, therefore to understand how the handsets cater and strengthen people's life style is the key what lies in. We will think to understand its related design principle and method to enhance its value quality.

3.1 Design Principle

Design needs to be carried on under instructions of below three important principles:

1) Pursue the principle of using simple ways to create equipments which are designed in fashion, and easy to use. Make technology become simple and perfect experience through innovation way.
2) Practical, Nokia always pays great attention to learn specific popular tendency, life style and regional savors, manifest its maturity by its usual function and design.
3) Under the principle that all in order to get experience, most of the time the goal of using handsets is not to talk on the telephone, send short message, it is a necessary goods that people take along with themselves. This judges that executing design from people's experience is the key of exploration, also is the starting point of developing new products.

Above are several basic principles, we also may see it follows and pursues developing products by principle and method of science technology and the law of development.

3.2 Experience Way of Design

There are many design ways and methods, seeking an effective method is extremely important, in view of design demand of the handset telephone, different person has different understanding and opinions. To some people, the handset is an important accessory; but to other people, it is the conversation tool in daily work and life, some use it to photograph as camera, some use it as radio or mp3 entertainment tool, some use it as a way to store study material, many people hope it is better to have more handset function, or judge our need degree according to own interest and long-term function. Regarding above problems, there exist surveys show that different groups have different understanding of phone functional knowledge and need, as shown in table 2.

Table 2. The statistic of different mobile phones users' knowledge for mobile phones demand

university students	white-collar workers	farmers	ordinary workers
fashion, multi-function	quality,fashion	quality,cheap price	quality,proper price
86%	85%	92%	89%

As to these different situation Nokia does not only concentrates its attention on one kind of design or some specific market. On the contrary, the designers through detailed research in the world and street corner anthropology investigation, continually to track various kinds of demand and life style. Research home stay and people who rushes outside, observe how people use mobile phones in their daily life, or seek different objects to experience life, according to these crowd's life experience to carry on design. After such understanding, summarize and analyze these experiences, use technology to create the contour, the function and the concept which the people like and are willing to use, before product development going to the market we also can use same experience way to improve and observe the scientific nature of design. This is the commonly used experience development design.

3.3 Development Strategy

As a global brand it requests global strategy, Nokia team members come from 34 different countries and areas. In this team there are psychologist, researcher, designer, anthropologist, engineer and technical expert, distributed in main cities of the world. They pay attention to material, color and more widespread design wave, with new contour, color, material, function as well as interactive way between people and machine to carry on experiment, for example through broadening multimedia functions and content use scope, will appear new way which makes mobile internet experience more perfect, forecast ahead of time to conceive use way of handsets under future environment, cooperate with other outstanding designers, students, artists and manufacturers, expand design scope and experiment scope. We can collect the popular tendency information and inspiration in the world through this kind of network, describe the newest local fashion wave, and make this kind of wave obtain demonstration in the rousing product design. facing this fast changing world, it will continually experience new creativity, thus to create remarkable personality design, easy function use and outstanding quality to maintain good prestige.

4 Conclusion

Personality presentation and prominent quality characteristic are extremely important, Nokia handset just precisely seeks personality among commonality in design development, manifest its entire idea fully, keep its quality and personality characteristic. What we can learn from this paper is that it is extremely necessary for daily high-tech domestic commercial products to strengthen this research, we frequently see some factories design and develop serial products, but how many generations can they develop continually, one generation, two generations... only by continually deepening such good quality and personality can a company retain an good integrity image, this will be a powerful help to enterprise brand setting. Meanwhile we may see design development cannot be separated from true grasp of realistic environment, know the time, know the environment, know about ourselves and use all kinds of possible design method to develop this kind of potential and make products more mature, more competitive, prepare for brand strategy promotion.

References

1. Jian, Z.: Industry design methodology (revised edition). Beijing institute of technology publishing house, Beijing (2000)
2. Ulrich, K.T., Eppinger, S.D.: Product design and development. Higher education publishing house, Beijing (2004)

Discussion about Art and "Contradiction" and "Practice" Based on Information Technology

Guobin Peng

Design and Art College,
GuiLin University of Electronic Technology,
Guilin, P.R. China
pengguobin@guet.edu.cn

Abstract. Chairman Mao Zedong teach Marxist philosophy in Yan'an Anti-Japanese Military and Political University, the author of "Practice" and "Contradiction." "Contradictions" is talking about dialectics, but it is mainly about how to analyze social conflicts and solve social conflicts, to promote the transformation of materials, to achieve the purpose of changing the social structure, human freedom and liberation, we can also benefit from it when engage in artistic creation. "Practice" is the study of epistemology, but it is first about the characteristics of people and human nature, talk about the social and human transformation, focusing on transformation. " Contradiction" and "Practice" interrelate, unified in a whole system, it is very necessary for the artists to research this two points .

Keywords: contradictions, practice, artistic creation and innovation.

1 Contradictions and Contradictions Art

1.1 Conflict Is the "Key" of Science and Art Thinking

In "Contradiction," contradiction exists in all things, contradiction covers throughout the development process of every thing, contradiction has a variety of features, both universal and special; both the identity and fighting spirit. Also it is divided into major conflicts and the main aspects of contradictions. To grasp these features of the contradictions, we will master the "key"of scientific thinking.

Therefore, adherence to use comprehensive, connected and developing point of view, that is to say, with conflicting views to observe things, analyze things, it is helpful to be close to objective, far from subjective, truely reflection true face of things, will help deepenly understand the internal law of things, to achieve scientific thinking, three-dimensional thinking. One works in art creation also exists in a contradiction, it should reflect the core ideas and values of society and culture, and it must be reflected in the objective contradictions world, or content of the work will be very superficial, contradictory thinking and attitude requires us to use rational contradictory concept to think and create.

A. Xie & X. Huang (Eds.): Advances in Electrical Engineering and Automation, AISC 139, pp. 335–338.
springerlink.com

1.2 Contradiction Is the "Power" to Promote Artistic Creation and Innovation

Contradictory things and its each side have its own characteristics. Contradictions of different quality, only with different qualitative methods can we resolve contradiction. This is which "contradiction" theory tells us. In real work, it is because we are sticking to make specific analysis for the specific things that can we promote innovation of every task. Similarly for art creation, must ultimately implement on the specific work and practice, for a specific problem, to analyze the characteristics from all its aspects, make a profound experience and understanding, understanding the greatest contradiction characteristic, find solutions way, and finally make it up to another level, also pay attention to solve small problems of other aspects, weaken it, strengthen and weaken some of the characteristics of things is the surpass and progress on the front road, at least this way of thinking understanding and spirit is necessary.

Therefore, as long as we are good at capturing the contradictions at work, good at grasping, analysis and making good use of contradiction, we will be able to promote practice innovation, theoretical innovation and innovation of other aspects, and constantly promote forward of the artistic thinking and creation. In this sense, conflict is the driving force of creation.

1.3 The Art of Contradiction Can Add Strength to Build a Harmonious Society

Since the contradiction is ubiquitous, since the root cause of the development of things is not something external but internal things and the internal contradictions of things. So in dealing with interpersonal relationships, building a harmonious society should insist on finding a cause more from the inside of things. Art is the spirit and power, the charm of art is to wake up people's love for life, and infinite moving, currently material culture is wealthy, but really to a large extent, we lack of spiritual civilization, in a highly paced and competitive age, if you can not keep up on spiritual power, it is very terrible and great harmful for the individual and society, and this damage is infinite, in fact, people now need the power of love and spiritual power, this power source is supplied by the society as a whole, improve the common and healthy spirit sense of the whole society, to a large extent by the mold of the true art air, not to stay in a small number of artists making money and fame, as well as all entertainment and a small self-playing, of course, the artist's way of lifebeing is not discussed, how far can the art truly become spiritual food that people pursuit and love. Contradictions art can enrich and solve some problems, to promote from the hot and difficult of harmonious society.

Some survey indicates that two different artistic approaches lead to different results for domestic business and dispute resolution, as shown in table 1:

Table 1. Different results statistic of artistic approaches to contradictions

outside for business	outside for business	resolve conflicts and disputes	resolve conflicts and disputes
expression of comic art	direct expression	direct intervention	indirect mediation
80%	25%	30%	75%

At the same time conflicts also has the identity, can be transformed into each other, that is to say, conflicts not only has a positive kernel, but also has a negative kernel, the result is distinctive, the key is how to treat and deal with. So long as proper control and treat, there will be unexpected results. As what told in "Contradiction," the opposition and struggle of different ideas inside the party is often the case, it is what social class contradictions and the contradictions of old and new things reflected in the party. If there is no contradiction and ideological struggles to resolve the conflicts between the party, the Party's life would stop. The case for a party is also same for a unit, espccially for a country. As long as we persist in using conflicting views and artisc ways to deal with interpersonal relationships, we will reduce conflicts, resolve conflicts, transform conflict and promote harmony.

2 Practice and Practicc Art

In "Practice", about people awareness, both for nature and for society, is a development process step by step from junior to senior, it is progressively, from one-sided to more areas, indicating the level of art is also constantly deepening from junior to senior, art creation needs to follow the times, and even beyond and guide the community spirit, encouraging people to greater spiritual power, practice theory refers to our creative approach,which is a good magic weapon.

What practice told the understanding process, the first step, belong to feeling stage; the second step, belong to concepts, judgments and reasoning stage, art is intangible and belongs to feeling stage, next is to judge and reason for this feeling. From a number of the material that is very rich (not piecemeal) and realistic (not delusion) sense, obtained good and correct results. If you want a correct understanding, you have to fully understand the reality things, and feel deeply and rich, in order to ultimately form a correct conclusion.

According to the survey, most of people engaged in artistic activities view that life is the source of art, and can be re-created, as shown in table 2.

Table 2. Agreement survey statistic of life and practice is the source is art

painter	actor	movie director	animation creators
creation inspiration comes from life	performing arts need to experience life	creation is established in life	role actions and stories come from the nature
75%	85%	89%	92%

What " Practice" told and said: When an objective process moves and changed from one development stage to another development stage over time, we should be good at making their own subjective understanding followed over time, that is to make a new work program, suitable for the new circumstances change. We oppose to the persistence of conservative ideology, but also against empty talk recklessness, our party leaders as an art industry managers do have to study practice theory. We should learn in practice, and constantly sum up experience, improve our cognitive abilities, we will understand from not understand to few understand. Artistic development is ongoing, much of our current stay in the traditional culture, traditional is certainly very important, if only to stay and love of the traditional culture, how our future culture development make progress, historical development is unstoppable,arts and culture is also unstoppable, and we should on the basis of retaining the traditional development, actively encourage the development of the fine healthy art civilization.

Practice theory and knowledge theory requires us to pay attention to the accumulation of knowledge and experience of art creation, on the other hand, a person's knowledge, nothing more than two parts of direct experience and indirect experience. We need pay attention to learn from the experience of others, take doctrine, take people's strongness to make up our own weaknesses, to help us complete artistic innovation, which provides theoretical approach for the creation of artists, how to understand the plagiarism problem, how to understand innovation, how is is the pursuit of originality. Take a very simple example, if the other side of the desert is the truth, if the person seeking the truth is only 1 km away from the other side of the desert, at this time he has not get the truth. Then the descendants follow his path would obtain the truth.

What "Practice" said: In the long course of absolute truth, people's understanding of a certain specific process in various development stages has relative truth. The sum of innumerable relative truths is the absolute truth. Occurrence, development and eradication process in social practice is endless. The occurrence, development, and the elimination processof people's understanding is endless. Changes and movement in the objective real world will never end, people's knowledge of the truth in practice will never end.Therefore, we must continue to understand, learn, improve, and ultimately achieving an accurate grasp of things, to better carry out creation.

3 Conclusion

In summary, in the future process, I will furtherly strengthen the learning of "Contradiction theory" and "practice theory", with contradictory views and practical methods to research and solve problems, adhere to use lofty ideals to inspire ourselves, with firm belief to spur ourselves, to constantly improve ourselves, continue to exceed ourselves and realize the greatest value.

References

1. Yongtao: Practice, contradictory theory and Marxism in China. Marxist Philosophy Research, 44–47
2. Cao, W.: Civilization because of culture and culture because of civilization - a new interpretation on the original meaning of civilization. 4th Culture and Arts Studies (2010)

The Control System Design of Intelligent Robot

Yi Gao[1], Xing Pan[2], and Yong Pan[2]

[1] College of Information Technical Science,
NanKai University,
TianJin, China
Gaoyi@mail.nankai.edu.cn
[2] Research and Development Department,
Qicheng Science and Technology Co., Ltd.,
TianJin, China

Abstract. The control system of intelligent robot uses 32bit high performance ARM CPU as the control kernel of the whole intelligent robot. The robot is wheel driven, model founded and computer programmed. ARM is adopted as the core controller, and it has low consumption and outstanding performance. The embedded Linux is used as the operating system of the bottom controlling software, and its performance is stable and its real-performance is good. The module extension can be done, which is convenient for the secondary development. The graphic programming interface accelerates the secondary development for the users.

Keywords: intelligent robot, control system, ARM11, motor control, embedded system.

1 Introduction

As the development of the sensor technology, computer science, and artificial intelligence, the robot is developing toward the intelligent. Intelligent robots have to have the ability of sensing their operating environment, mission planning and decision-making. At present, the research on intelligent robots are mainly focused on embedded hardware platform designs[1-3] and the application of intelligent robots[4-5].

The control system design of intelligent robot uses 32bit high performance ARM CPU as the control kernel of the testing robot. Taking advantage the features of the embedded processor including high performance, multiple access interface and convenience in implant and that of the embedded operating system including high real-time, high reliability and abundant resources, we use embedded technology as the development platform of the testing robot, changing the application of the technology concerning robots into re-development on the embedded platform, which shortens the time consuming in development and saves resources. The robots are wheel driven, model founded and computer programmed. The embedded Linux is used as the operating system of the bottom controlling software, and its performance is stable and its real-performance is good.

A. Xie & X. Huang (Eds.): Advances in Electrical Engineering and Automation, AISC 139, pp. 339–344.
springerlink.com

2 Hardware System Design

The hardware used in this embedded intelligent robot platform mainly comprises embedded central processor and other peripheral equipments. Hereby, this paper will make a detailed introduction on the principal function module of the control system. Figure 1 shows the typical structured flowchart of the control system.

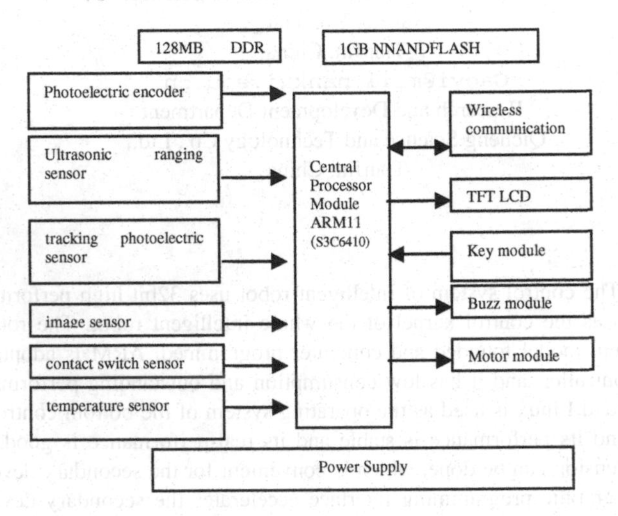

Fig. 1. Hardware System Design

2.1 Central Processor Module

The core processor of the central processor module is S3C6410 which is a low-cost, low-power, high performance, 16/32-bit RISC ARM1176JZF-S core microprocessor. The CPU has a complete MMU to process virtual memory management and separate 16 KB instruction and 16 KB Data Cache as well as 16 KB instruction and data TCM. The CPU with great performance and low energy consumption ensures the powerful data processing abilities of the terminal and provides the hardware for the complicated applications. Besides, the CPU module also includes 128MB DDR RAM and 1GB NAND FLSH.

2.2 LCD Display Module

Intelligent robot requires sophisticated human-machine interaction system to display the image acquisition information and the condition of the robot. This design use the 4.3 inches TFT LCD Innolux AT043TN24.

2.3 Motor Module

The design is the differential motion drive. Differential motion refers to integrating two different motions or two independent motions into one motion. In the robot, there are two DC motors which control two drive wheels respectively. When the robot is in operation, the rotating speed of the two motors can be set respectively which can be combined to realize various kinds of motion, for example, going straightly, making a turn, moving in curveted. This is the differential drive.

The L298 which is constant current constant voltage bridge driver chip is used in this robot to drive two motors simultaneously, control clockwise veer and anti-clockwise veer of the robot, and modulate pulse width in the enable pin so that clockwise and anticlockwise turn of the wheel can be realized and the speed of the wheel can be controlled. The detailed motor drive module is shown in the following figure 2:

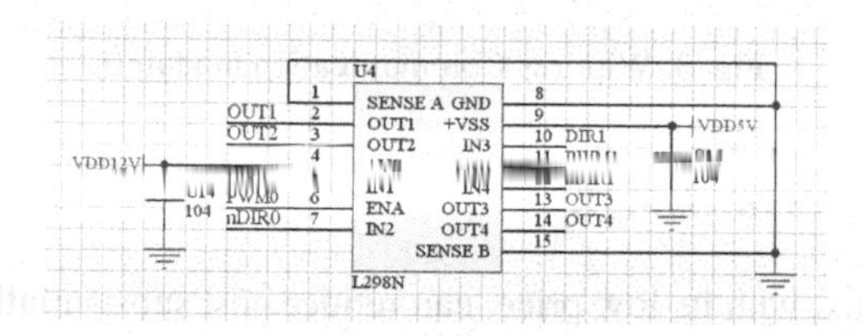

Fig.2. Motor Module Design

2.4 Sensor Module

Sensor module, an important part that enables the robot to detect information from the environment, involves many peripheral detecting types of equipment, for example, infrared photoelectric sensor, ultrasonic ranging sensor, tracking photoelectric sensor, image acquisition sensor, contact switch sensor and temperature sensor. This paper takes the temperature sensor and infrared encoder as an example.

Temperature module uses digital temperature sensor with low power dissipation DS18B20。 The DS18B20 digital thermometer provides 9-bit to 12-bit Celsius temperature measurements and has an alarm function with nonvolatile user-programmable upper and lower trigger points. The DS18B20 communicates over a 1-Wire® bus that by definition requires only one data line (and ground) for communication with a central microprocessor.

2.5 Wireless Communication Module

nRF24L01 is a single radio frequency send-receive chip, which has low energy consumption. Under the same working mode, it can save about 1/3 more energy than CC2420 chip made by Chipcon Company. Its working frequency range is 2.4 GHz ~ 2.5GHz., its operating voltage is 1.9~3.6 V and it can select as many as 125 channels. The chip has built-in functional modules such as frequency synthesizers, power amplifier, crystal oscillator and modulation. It also integrates enhanced Shock Burst and the output frequency and communication channel can be deployed by programs. The chip energy consumption is very low. When signals are sent at -6dBm, its working current is only 9 mA. When receiving signals, the working current is only 12.3 mA. Many kinds of low frequency working modes make the energy-saving design more convenient. The design is shown in Figure 3 :

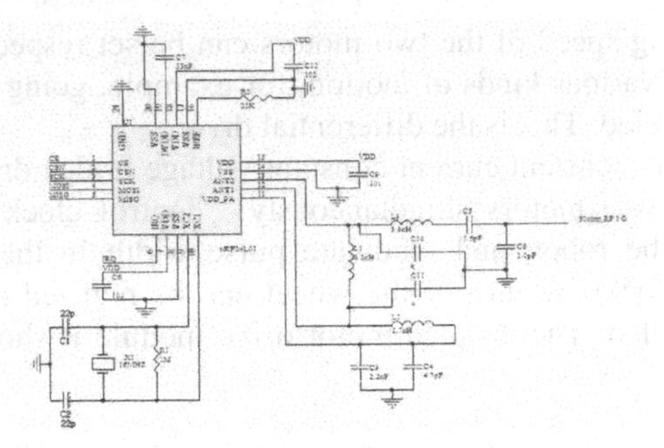

Fig. 3. Wireless Communication module

3 Software System Design

The embedded Linux, with its low price, can reduce cost substantially, thus becoming the first choice of the embedded operating system. However, applied in general operating system, it has certain technical barrier in real-time applications, as a result, the kernel of the Linux shall adjusted when the Linux is applied in embed system. Because the ability of the hard real-time interaction with the external environment is demanded by many embedded equipments, it is important to transform the Linux which primarily targets at the time-sharing system into a system that can support real-time operating systems of software. [6-7]

The software system of the intelligent robot uses the embedded Linux which has went through real-time optimization as the operating system of the bottom-layer controlling software. It has stable performance as well as good real time quality. What's more, the modular structure can be expanded and further developed.

3.1 Real-Time Optimization of Embedded LINUX Operating System

In the design of the kernel of mainstream Linux, interrupt can preempt tasks with highest priority, making the maximum time of being blocked unsure. Besides, the interrupt of the kernel with a view to protecting the critical section further increase the blocking time of the tasks with the highest priority. The rather high precision scheduling requirement of the real-time system can not be fulfilled by the Linux's kernel time management through million second level cyclic interrupt produced by hard ware clock programming. The precision of kernel timer is limited by the clock interrupt either which makes it fail to fulfill the demand of high precision of the real-time system.

Increase the granularity of the clock. With interrupt cycle being 10ms, the clock precision of the standard Linux system is not accurate enough to fulfill the demand on precision of the response time in certain embedded application area. Therefore, in the application of the real-time Linux, its clock granularity should be thinned. There are two ways to resolve the problem of the clock granularity: one is to directly modify the initial value of the kernel timing parameter HZ to thin the clock granularity.

Take advantage of simulation technology of soft interrupt. The problem of interrupt can be solved by adding an interrupt simulation software layer between the Linux kernel and the hardware interrupt controller. When hardware interrupt occurs, the system will set a variable of the interrupt simulator or just record this event in an event list. Instead of processing the interrupt, the interrupt simulator will return the right to control the CPU to the system.

3.2 Design of Embedded Application Program

At present, design of the embedded application program in the embedded system is mainly the development of the real-time multi-task software based on the embedded real-time operating system. This paper is therefore based on RTOS and conforms to the process of developing real-time multi-task software. Currently, the design tasks are: (1) seek tasks(photoelectric detection tasks)automatically (2) infrared detection tasks (3)ultrasonic detection tasks (4)image accumulation tasks (5)temperature detection tasks (6)key board scan tasks (7) liquid crystal display tasks (8) tasks of transmitting and receiving through ultrasonic (9) electric motor controlling tasks. The flow-process diagram is shown below Figure 4.

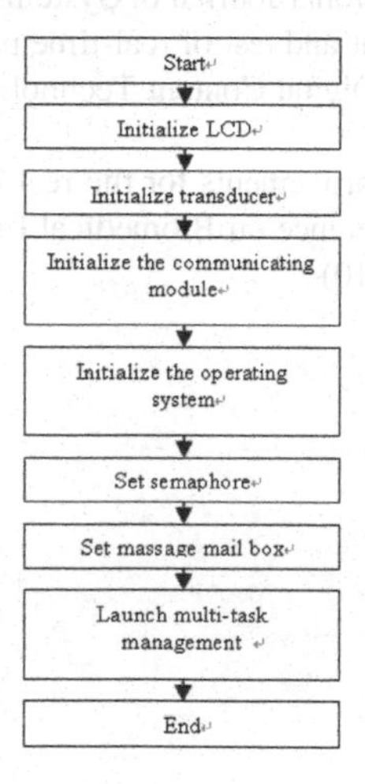

Fig. 4. The Initial flow-process diagram

4 Conclusion

The low cost, high performance and many peripheral equipment connections of ARM microprocessor provides stable and reliable hardware architecture for the mobile robot controller, which makes it possible to install operating system in the robot control system. The development of the operating system in the robot control system makes it easier to develop and more convenient to maintain the robot control system. At the same time, the intelligence, stability and reliability of the system are also enhanced. The experiments have proved this intelligent robot system can accurately accomplish various actions and it has good dynamic performance and controlling precision. Besides, it possess other advantages of modularity, easiness in enlargement, translatability, small hardware, low power consumption, high real-time and high reliability, etc.

References

1. Fang, Z., Tong, G., Hu, Y., Xu, X.: An embedded platform for intelligent mobile robot. In: Proceedings of the World Congress on Intelligent Control and Automation (WCICA), vol. 2, pp. 9104–9108 (2006)
2. Matson, E.T.: Embedding intelligent agents to enable physical robotic and sensor organizations. In: IEEE International Symposium on Computational Intelligence in Robotics and Automation, CIRA 2009, pp. 309–315 (2009)
3. Kim, D.-S., Lee, S.-S., Choi, B.-H.: A real-time stereo depth extraction hardware for intelligent home assistant robot. IEEE Transactions on Consumer Electronics 56(3), 1782–1788 (2010)
4. Sanders, D.A., Lambert, G., Graham-Jones, J., Tewkesbury, G.E., Onuh, S., Ndzi, D., Ross, C.: A robotic welding system using image processing techniques and a CAD model to provide information to a multi-intelligent decision module. Assembly Automation 30(4), 323–332 (2010)
5. Chang, Y.-C., Yen, H.-M., Wu, M.-F.: An intelligent robust tracking control for electrically-driven robot systems. International Journal of Systems Science 39(5), 497–511 (2008)
6. Liping, Y., Kai, S.: Improvement and test of real-time performance of embedded linux 2.6 kernel International. Journal of Digital Content Technology and its Applications 5(4), 247–253 (2011)
7. Xu, H., Tang, R.: Study and improvements for the real-time performance of Linux kernel. In: 2010 3rd International Conference on Biomedical Engineering and Informatics, BMEI 2010, vol. 7, pp. 2766–2769 (2010)

The Development of E-Learning System for Fractal Fern Graphics Based on Iterated Function System

Fucheng You and Yingjie Liu

Information & Mechanical Engineering School,
Beijing Institute of Graphic Communication, Beijing, China
youfucheng@yahoo.com.cn, liuyingjiea@hotmail.com

Abstract. Traditional methods of designing fractal fern graphics only provide graphics with single color, as well as the programs are rather complex. In this paper, C# language is used to develop an E-learning environment of fractal fern graphics and briefer source codes are provided. Students can get colorful fractal graphics by iterative operation in this E-Learning environment, also can get different and random colors for fractal graphics too, which can inspire the interests and attract the attention of students during learning process. Fixed points and limits can be demonstrated in E-learning environment, which is favorable for students to study the conception of fractal fern graphics. It is easy for students to master program design and understand the structures of fractal fern graphics.

Keywords: Fractal Fern Graphics, E-Learning System, Iterated Function System.

1 Introduction

E-learning is teaching and learning activities through digital contents of computer network and multi-media. E-learning takes full advantages of technologies of modern communication, computer, multi-media and virtual reality, which provide learning environment with innovative communication mechanism and abundant education resource and result in an innovative learning style [1].

Research on learning environment is main content in E-learning system research. According to learning theory of constructivism, research contents of learning environment include learning scene design, human-computer interaction technology and learning behavior analysis in E-learning system [2]. The purpose of learning environment research is to construct more favorable E-learning environment for digital learning.

E-learning implementation usually needs an E-learning system or environment through elaborate design network course, learning resource and tools. As a computer application, fields of E-learning system are learning. Therefore, design of E-learning system must accord with advanced learning theory, reflect human's cognitive law and adapt to personalized learning demand.

A. Xie & X. Huang (Eds.): Advances in Electrical Engineering and Automation, AISC 139, pp. 345–351.
springerlink.com

Maslow Rogers makes more research on learning, who divides learning into meaningless learning and meaning learning [3]. Meaningless learning only involves knowledge accumulation and has no relationship with learner. Meaning learning refers to individual behavior [4].

Fractal graphics is an important branch of computer graphics. The traditional computer graphics has no consider with color change and only draws final graphics with default color. The graphics are fixed with no change. It's also difficult for students to understand programs of fractal graphics.

According to requirements of E-learning and learning theory, meaningless learning should be changed to meaning learning. Therefore, program codes of color display are added firstly in the paper. Fractal fern graphics can be drawn in different colors, which can attract students' interests because human eyes are sensitive to color changing. Secondly, changing codes of iterated function system, fractal fern graphics can be formed with different shapes, which can guide students' attention and consideration. Rules should be found during changing process of colors and structures, which can improve students' desires on knowledge research.

2 Algorithm of Iterated Function System

2.1 Similar Transformation and Affine Transformation

Theory of iterated function system is an important branch of fractal theory. Graphics are regarded as collaged small pieces which are similar with the whole after transformation by iterated function system. Self-similarity is implemented through similar transformation. Self-affinity is implemented through affine transformation.

Similar transformation is a kind of scaling transformation with same scale in different directions. While affine transformation is a kind of scaling transformation with different scales in different directions [5].

Affine transformation is given by Eq. 1, in which ω denotes affine transformation, x and y denote the coordinates before transformation, x' and y' denote the coordinates after transformation, a,b,c,d,e,f denote the coefficients of affine transformation.

$$\omega:\begin{cases} x' = ax + by + e \\ y' = cx + dy + f \end{cases} \tag{1}$$

For more complex fractal graphics, more different affine transformations are need. Affine transformation group $\{\omega_n\}$ determine the structure and shape of fractal graphics. In addition, call probability of every affine transformation in group $\{\omega_n\}$ are not always same. Let P denote the call probability. Then a,b,c,d,e,f and P form the codes of iterated function system.

2.2 Iterated Function System Generation of Fractal Fern Graphics

Fractal fern graphics can be denoted by follow affine transformations from Eq. 2 to Eq. 5. Let probabilities of $\omega_1, \omega_2, \omega_3, \omega_4$ be $P_1 = 0.02, P_2 = 0.84, P_3 = 0.07, P_4 = 0.07$ and $P_1 + P_2 + P_3 + P_4 = 1$.

$$\omega_1 : \begin{Bmatrix} x' = 0x + 0y + 0 \\ y' = 0x + 0.25y - 0.14 \end{Bmatrix} \tag{2}$$

$$\omega_2 : \begin{Bmatrix} x' = 0.85x + 0.02y + 0 \\ y' = -0.02x + 0.83y + 1 \end{Bmatrix} \tag{3}$$

$$\omega_3 : \begin{Bmatrix} x' = 0.09x - 0.28y + 0 \\ y' = 0.3x + 0.11y + 0.6 \end{Bmatrix} \tag{4}$$

$$\omega_4 : \begin{Bmatrix} x' = -0.09x + 0.28y + 0 \\ y' = 0.3x + 0.09y + 0.7 \end{Bmatrix} \tag{5}$$

3 C# Program Codes of Fractal Fern Graphics

E-learning system can be divided into single color drawing and random color drawing. Single color drawing is fixed black. Random color drawing can be divided into close drawing and far drawing. Interface of E-learning system for fractal fern graphics is shown as Fig. 1.

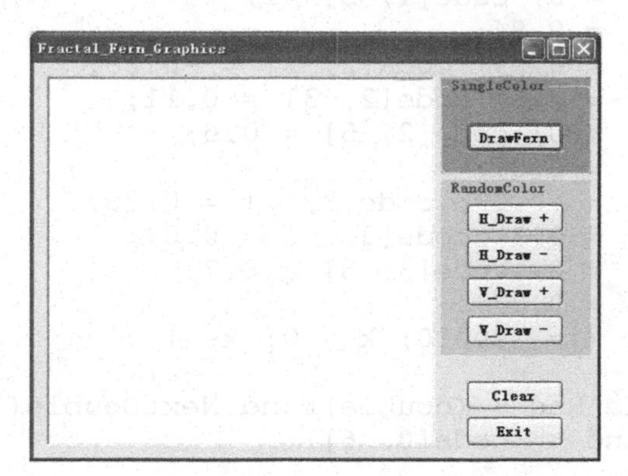

Fig. 1. Interface of E-Learning system for fractal fern graphics

Generation algorithm of fractal fern graphics is below.

Step 1: Generate random number *Rnd* and limit the range of *Rnd* between 0 and 1.

Step 2: Define the probability range of ω_1 and ω_2 as $[0,0.888128]$ and $[0.888128,1]$.

Step 3: Call codes of iterated function system for respective affine transformation according to *Rnd*, then assign the respective coefficients a,b,c,d,e,f.

Step 4: Calculate the coordinates x' and y' after affine transformation according to Eq. 1.

Step 5: Draw a point on x' and y'.

Step 6: Repeat from step 1 to step 5. Let x' and y' be x and y for the next calculation.

Firstly, define and package generation function *DrawFern()* for fractal fern graphics generation. C# program codes of *DrawFern()* are as follows.

```
private void DrawFern()
{
  g = pictureBox1.CreateGraphics();
  Pen pen1 = new Pen(this.BackColor, 1);
  pen1.Color   =   Color.FromArgb(rand.Next(0,   255),
                rand.Next(0, 255), rand.Next(0, 255));
  x = 10; y = 10;
 code[0, 0] = 0; code[0,1] = 0;
 code[0, 2] = 0; code[0, 3] = 0.25;
 code[0, 4] = 0; code[0, 5] =-0.14;
 code[0, 6] = 0.02;
 code[1, 0] = 0.85; code[1, 1] = 0.02;
 code[1, 2] = -0.02; code[1, 3] = 0.83;
 code[1, 4] = 0; code[1, 5] =1;
 code[1, 6] = 0.84;
 code[2, 0] = 0.09; code[2, 1] = -0.28;
 code[2, 2] = 0.3; code[2, 3] = 0.11;
 code[2, 4] = 0; code[2, 5] = 0.6;
 code[2, 6] = 0.07;
 code[3, 0] = -0.09; code[3, 1] = 0.28;
 code[3, 2] = 0.3; code[3, 3] = 0.09;
 code[3, 4] = 0; code[3, 5] = 0.7;
 code[3, 6] = 0.07;
   for (int k = 100000; k > 0; k--)
    {
      double Rnd = (double)rand.NextDouble();
      if (Rnd <= code[0, 6])
      {
       a = code[0, 0]; b = code[0, 1]; c = code[0, 2];
      d = code[0, 3]; e = code[0, 4]; f = code[0, 5];
      }
      else if(Rnd <= code[0, 6]+[1,6])
        {
```

```
        a   =   code[1,   0];  b   =  code[1,   1];   c   =
code[1, 2];
       d   =   code[1,   3];   e   =   code[1,   4];   f   =
code[1, 5];
       }
      else          if(Rnd          <=          code[0,
6]+code[1,6]+code[2,6])
       {
        a   =   code[2,   0];  b   =  code[2,   1];   c   =
code[2, 2];
       d   =   code[2,   3];   e   =   code[2,   4];   f   =
code[2, 5];
       }
       else
       {
        a   =   code[3,   0];  b   =  code[3,   1];   c   =
code[3, 2];
       d   =   code[3,   3];   e   =   code[3,   4];   f   =
code[3, 5];
        }
    x1 = (a * x) + (b * y) + e;
    y1 = (c * x) + (d * y) + f;
    x = x1; y = y1;
    g.DrawEllipse(pen1,   (int)(Xscal   *   x   +   150),
      (int)(pictureBox1.Bottom - Yscal * y - 20), 1, 1);
     }
 }
```

Run above C# program codes. Click button "*Drawfern*", fractal fern graphics can be achieved. Control scaling coefficients, i.e., Xscal=Xscal +5, Yscal=Yscal+5 or Xscal=Xscal-5, Yscal=Yscal-5. Then fractal fern graphics with random color can be zoomed in or out.

Click button "*H_Draw+*" continually, enlarging fractal fern graphic horizontally is shown in Fig. 2.

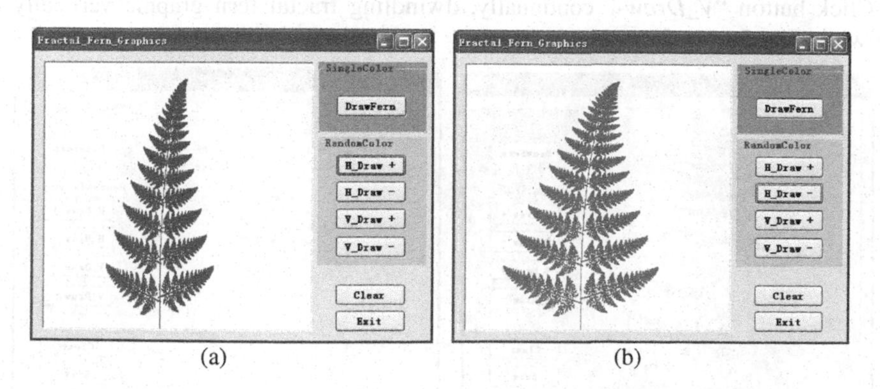

(a) (b)

Fig. 2. Colorful fractal fern graphics enlarges horizontally from (a) to (b)

Click button "*H_Draw-*" continually, dwindling fractal fern graphic horizontally is shown in Fig. 3.

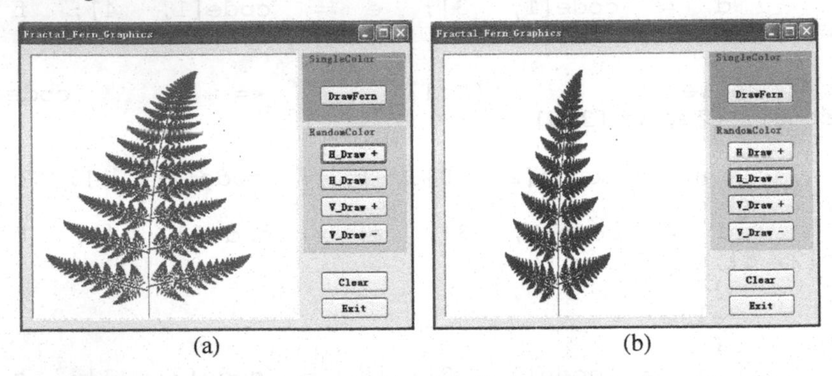

(a) (b)

Fig. 3. Colorful fractal fern graphics dwindles horizontally from (a) to (b)

Click button "*V_Draw+*" continually, enlarging fractal fern graphic vertically is shown in Fig. 4.

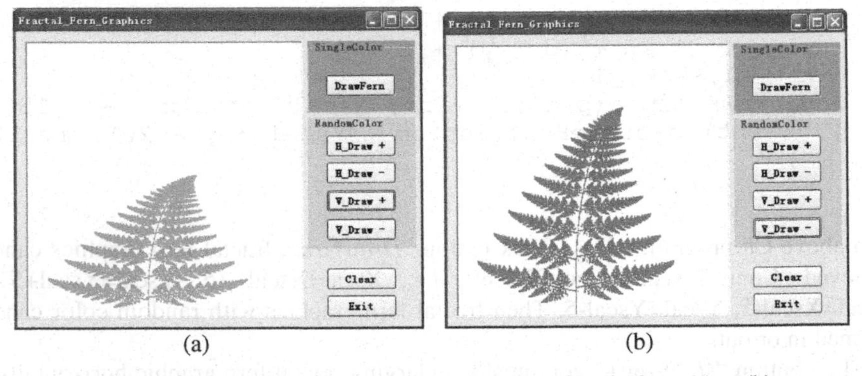

(a) (b)

Fig. 4. Colorful fractal fern graphics enlarges vertically from (a) to (b)

Click button "*V_Draw-*" continually, dwindling fractal fern graphic vertically is shown in Fig. 5.

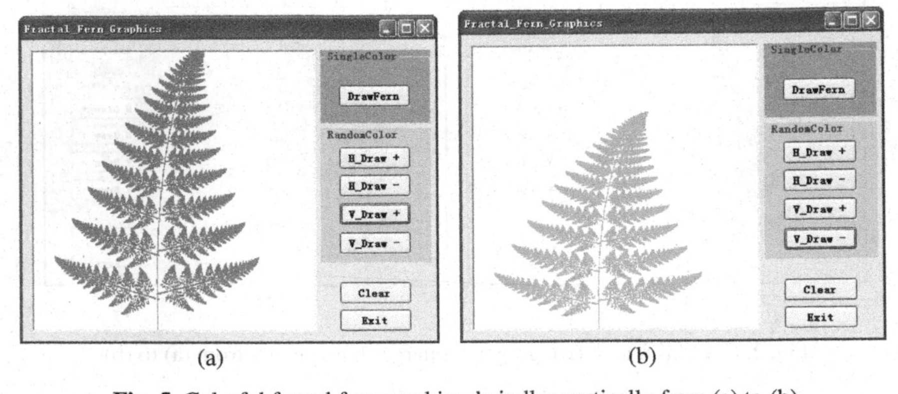

(a) (b)

Fig. 5. Colorful fractal fern graphics dwindles vertically from (a) to (b)

4 Conclusion

From above analysis, color display codes of fractal graphics are added in the program firstly according to E-learning theory. Then different colors are displayed for fractal fern graphics. Click button "*Drawfern*", fractal fern graphics can be achieved. Click button "*H Draw+*" and "*V Draw+*", colorful fractal fern graphics can be displayed and zoomed in. Click button "*H Draw-*" and "*V Draw-*", fractal fern graphics can be zoomed out. It will inspire students' learning interests. Structure change rules of fractal fern graphics can be mastered by students during the dynamic change of graphic color and shape. Then meaningless learning can be changed to meaning learning and learning efficiency will be improved.

Acknowledgments. This paper is supported by Project for Beijing Municipal Party Committee Organization Department (No. 10000200118) and the Funding Project for Academic Human Resources Development in Institutions of Higher Learning under the Jurisdiction of Beijing Municipality.

References

1. Kienle, A., Wessner, M.: The CSCL community in its first decade: development, continuity, connectivity. Computer-Supported Collaborative Learning, 9–13 (2006)
2. Dieheva, D., Diehev, C.: Authoring educational topic maps-can we make it easier. In: Fifth IEEE International Conference on Advanced Learning Technologies 2005, Taiwan, China, pp. 216–218 (2005)
3. Shi, L.F.: Theory of Learning, pp. 57–62. People Education Press, Beijing (2001)
4. Hao, X.W.: Research on ontology Based E-learning System. Doctor Degree Paper. Shandong University, pp.41–46 (2007)
5. Sun, B.: Fractal algorithms and programming: realized by Java, pp. 47–49. Science Press, Beijing (2004)

The Development of E-Learning System for Fractal Pine Perspective Graphics Based on Iterated Function System

Fucheng You and Yingjie Liu

Information & Mechanical Engineering School,
Beijing Institute of Graphic Communication, Beijing, China
youfucheng@yahoo.com.cn, liuyingjiea@hotmail.com

Abstract. Traditional methods of designing fractal pine perspective graphics only provide graphics with single color, as well as the programs are rather complex. In this paper, C# language is used to develop an E-learning environment of fractal pine perspective graphics and briefer source codes are provided. Students can get colorful fractal graphics by iterative operation in this E-Learning environment, also can get different and random colors for fractal graphics too, which can inspire the interests and attract the attention of students during learning process. Fixed points and limits can be demonstrated in E-learning environment, which is favorable for students to study the conception of fractal pine perspective graphics. It is easy for students to master program design and understand the structures of fractal pine perspective graphics.

Keywords: Fractal Pine Perspective Graphics, E-Learning System, Iterated Function System.

1 Introduction

E-learning is teaching and learning activities through digital contents of computer network and multi-media. E-learning takes full advantages of technologies of modern communication, computer, multi-media and virtual reality, which provide learning environment with innovative communication mechanism and abundant education resource and result in an innovative learning style [1].

Research on learning environment is main content in E-learning system research. According to learning theory of constructivism, research contents of learning environment include learning scene design, human-computer interaction technology and learning behavior analysis in E-learning system [2]. The purpose of learning environment research is to construct more favorable E-learning environment for digital learning.

E-learning implementation usually needs an E-learning system or environment through elaborate design network course, learning resource and tools. As a computer application, fields of E-learning system are learning. Therefore, design of E-learning system must accord with advanced learning theory, reflect human's cognitive law and adapt to personalized learning demand.

A. Xie & X. Huang (Eds.): Advances in Electrical Engineering and Automation, AISC 139, pp. 353–359.
springerlink.com

Maslow Rogers makes more research on learning, who divides learning into meaningless learning and meaning learning [3]. Meaningless learning only involves knowledge accumulation and has no relationship with learner. Meaning learning refers to individual behavior [4].

Fractal graphics is an important branch of computer graphics. The traditional computer graphics has no consider with color change and only draws final graphics with default color. The graphics are fixed with no change. It's also difficult for students to understand programs of fractal graphics.

According to requirements of E-learning and learning theory, meaningless learning should be changed to meaning learning. Therefore, program codes of color display are added firstly in the paper. Fractal pine perspective graphics can be drawn in different colors, which can attract students' interests because human eyes are sensitive to color changing. Secondly, changing codes of iterated function system, fractal pine perspective graphics can be formed with different shapes, which can guide students' attention and consideration. Rules should be found during changing process of colors and structures, which can improve students' desires on knowledge research.

2 Algorithm of Iterated Function System

2.1 Similar Transformation and Affine Transformation

Theory of iterated function system is an important branch of fractal theory. Graphics are regarded as collaged small pieces which are similar with the whole after transformation by iterated function system. Self-similarity is implemented through similar transformation. Self-affinity is implemented through affine transformation.

Similar transformation is a kind of scaling transformation with same scale in different directions. While affine transformation is a kind of scaling transformation with different scales in different directions [5].

Affine transformation is given by Eq. 1, in which ω denotes affine transformation, x and y denote the coordinates before transformation, x' and y' denote the coordinates after transformation, a,b,c,d,e,f denote the coefficients of affine transformation.

$$\omega : \begin{Bmatrix} x' = ax + by + e \\ y' = cx + dy + f \end{Bmatrix} \tag{1}$$

For more complex fractal graphics, more different affine transformations are need. Affine transformation group $\{\omega_n\}$ determine the structure and shape of fractal graphics. In addition, call probability of every affine transformation in group $\{\omega_n\}$ are not always same. Let P denote the call probability. Then a,b,c,d,e,f and P form the codes of iterated function system.

2.2 Iterated Function System Generation of Fractal Pine Perspective Graphics

Fractal pine perspective graphics can be denoted by follow affine transformations from Eq. 2 to Eq. 3. Let probabilities of ω_1, ω_2 be $P_1 = 0.888128, P_2 = 0.111872$ and $P_1 + P_2 = 1$.

$$\omega_1 : \begin{cases} x' = -0.632407x - 0.14815y + 3.840822 \\ y' = -0.54537x + 0.659259y + 1.282321 \end{cases} \quad (2)$$

$$\omega_2 : \begin{cases} x' = -0.0.361111x + 0.444444y + 2.071081 \\ y' = 0.210185x + 0.037037y + 8.330552 \end{cases} \quad (3)$$

3 C# Program Codes of Fractal Pine Perspective Graphics

E-learning system can be divided into single color drawing and random color drawing. Single color drawing is fixed black. Random color drawing can be divided into close drawing and far drawing. Interface of E-learning system for fractal pine perspective graphics is shown as Fig. 1.

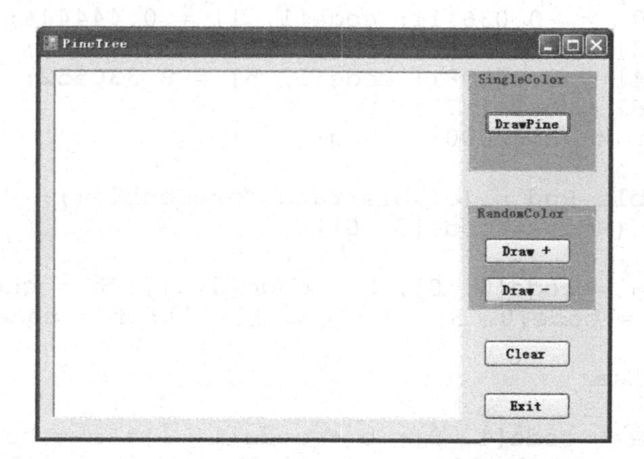

Fig. 1. Interface of E-Learning system for fractal pine perspective graphics

Generation algorithm of fractal pine perspective graphics is below.

Step 1: Generate random number Rnd and limit the range of Rnd between 0 and 1.

Step 2: Define the probability range of ω_1 and ω_2 as $[0, 0.888128]$ and $[0.888128, 1]$.

Step 3: Call codes of iterated function system for respective affine transformation according to Rnd, then assign the respective coefficients a, b, c, d, e, f.

Step 4: Calculate the coordinates x' and y' after affine transformation according to Eq. 1.

Step 5: Draw a point on x' and y'.

Step 6: Repeat from step 1 to step 5. Let x' and y' be x and y for the next calculation.

Firstly, define and package generation function *DrawPine()* for fractal pine perspective graphics generation. C# program codes of *DrawPine()* are as follows.

```
  private void DrawPine()
{
  g = pictureBox1.CreateGraphics();
  Pen pen1 = new Pen(this.BackColor, 1);
  pen1.Color   =   Color.FromArgb(rand.Next(0,    255),
                      rand.Next(0, 255), rand.Next(0, 255));
  x = 10; y = 10;
  code[0, 0] = -0.632407; code[0, 1] = -0.614815;
  code[0, 2] = -0.54537; code[0, 3] = 0.659259;
  code[0, 4] = 3.840822; code[0, 5] =1.282321;
  code[0, 6] = 0.888128;
  code[1, 0] = -0.036111; code[1, 1] = 0.444444;
  code[1, 2] = 0.210185; code[1, 3] = 0.037037;
  code[1, 4] = 2.071081; code[1, 5] = 8.330552;
  code[1, 6] = 0.111872;
   for (int k = 100000; k > 0; k--)
      {
        double Rnd = (double)rand.NextDouble();
         if (Rnd <= code[0, 6])
          {
           a = code[0, 0]; b = code[0, 1]; c = code[0, 2];
          d = code[0, 3]; e = code[0, 4]; f = code[0, 5];
          }
          else
          {
           a = code[1, 0]; b = code[1, 1]; c = code[1, 2];
          d = code[1, 3]; e = code[1, 4]; f = code[1, 5];
           }
     x1 = (a * x) + (b * y) + e;
     y1 = (c * x) + (d * y) + f;
     x = x1; y = y1;
     g.DrawEllipse(pen1,    (int)(Xscal    *    x    +    200),
        (int)(pictureBox1.Bottom  -  Yscal  *  y  -  20),  1,
        1);
     }
  }
```

Run above C# program codes. Click button "*DrawPine*", fractal pine perspective graphics can be achieved. Control scaling coefficients, i.e., Xscal=Xscal+2,

Yscal=Yscal+2 or Xscal=Xscal-2, Yscal=Yscal-2. Then fractal pine perspective graphics with random color can be zoomed in or out.

Click button "*Draw+*" continually, enlarging fractal pine perspective graphic is shown in Fig. 2.

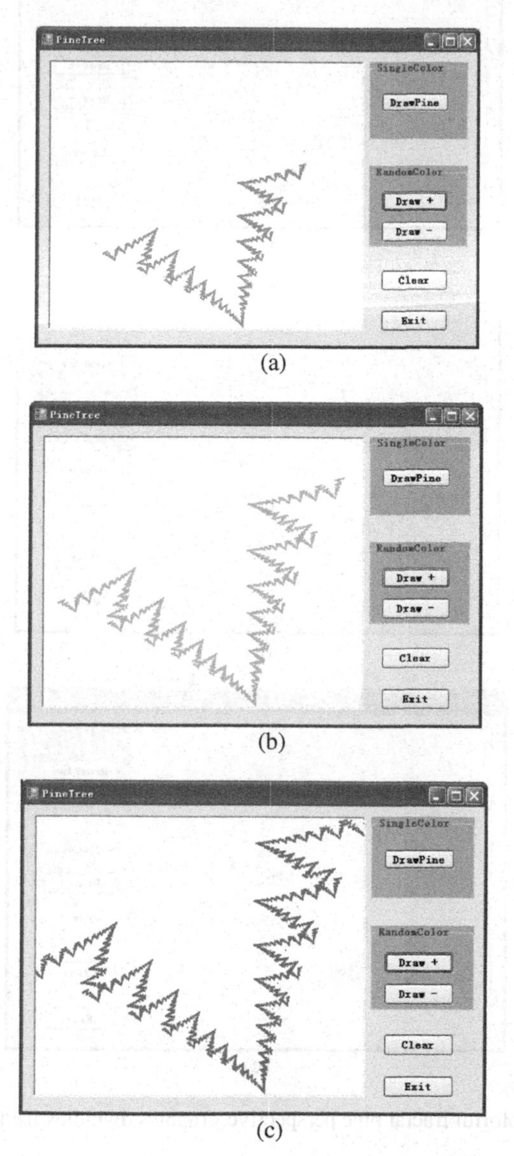

Fig. 2. Colorful fractal pine perspective graphics enlarges from (a) to (c)

Click button "*Draw-*" continually, dwindling fractal pine perspective graphic is shown in Fig. 3.

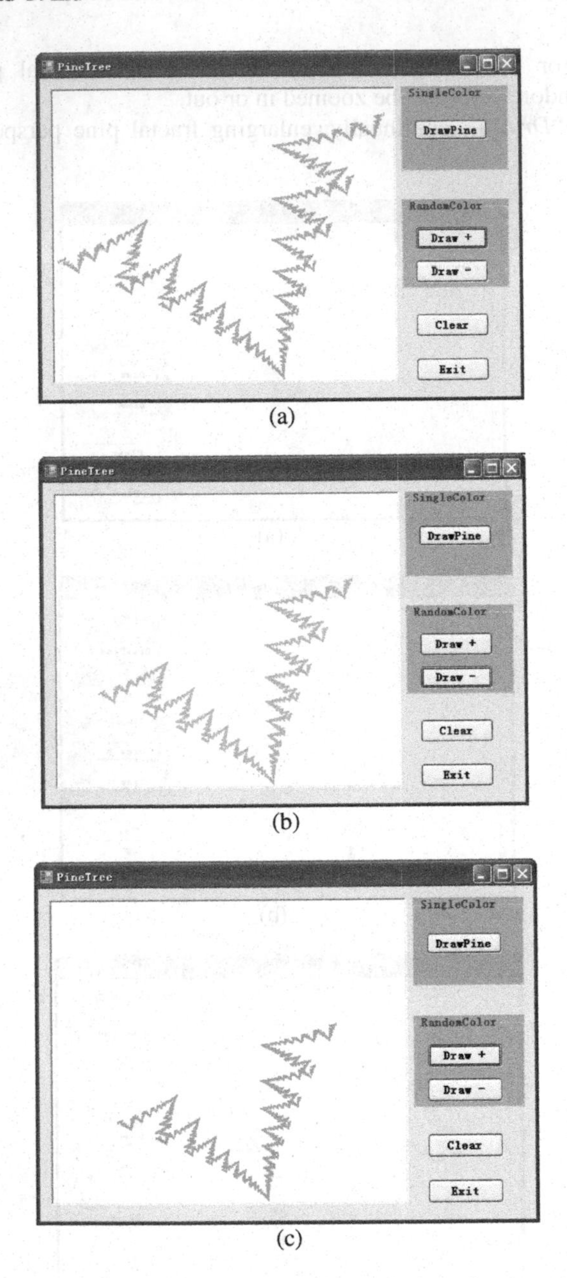

Fig. 3. Colorful fractal pine perspective graphics dwindles from (a) to (c)

4 Conclusion

From above analysis, color display codes of fractal graphics are added in the program firstly according to E-learning theory. Then different colors are displayed for fractal pine perspective graphics. Click button "*DrawPine*", fractal pine perspective graphics can be achieved. Click button "*Draw+*", colorful fractal pine perspective graphics can be displayed and zoomed in. Click button "*Draw-*", fractal pine perspective graphics can be zoomed out. It will inspire students' learning interests. Structure change rules of fractal pine perspective graphics can be mastered by students during the dynamic change of graphic color and shape. Then meaningless learning can be changed to meaning learning and learning efficiency will be improved.

Acknowledgments. This paper is supported by Project for Beijing Municipal Party Committee Organization Department (No. 10000200118) and the Funding Project for Academic Human Resources Development in Institutions of Higher Learning under the Jurisdiction of Beijing Municipality.

References

1. Kienle, A., Wessner, M.: The CSCL community in its first decade: development, continuity, connectivity. Computer-Supported Collaborative Learning, 9–13 (2006)
2. Dieheva, D., Diehev, C.: Authoring educational topic maps-can we make it easier. In: Fifth IEEE International Conference on Advanced Learning Technologies 2005, Taiwan, China, pp. 216–218 (2005)
3. Shi, L.F.: Theory of Learning, pp. 57–62. People Education Press, Beijing (2001)
4. Hao, X.W.: Research on ontology Based E-learning System. Doctor Degree Paper, Shandong University, pp.41–46 (2007)
5. Sun, B.: Fractal algorithms and programming: realized by Java, pp. 47–49. Science Press, Beijing (2004)

A Developmental Approach to Robotic 3D Hand-Eye Coordination

Lin Hu, Fei Chao*, Min Jiang, MingHui Shi, and Pan Wang

Cognitive Science Department,
Fujian Key Laboratory of Brain-like Intelligent System,
Xiamen University, Xiamen, Fujian, P.R. China
fchao@xmu.edu.cn

Abstract. Inspired by infant development and brain research, this paper, combining the two research hotspot: developmental robotics and robotic hand-eye coordination, proposes to employ incremental process to implement robotic hand-eye coordination using double-neural networks under the experimental condition of two cameras and a robotic arm. This setup is to imitate the developmental progress of human infant to capture objects; and the networks are implemented by the constructive network which can simulate infant's brain development. Furthermore, this paper delicately describes the details of the experiment plan and steps to show the combining course of the two hotspot.

Keywords: Developmental Robotics, Robotic Hand-eye Coordination, Constructive Network.

1 Introduction

Hand-eye coordination is a kind of basic skills for the intelligent robot to survive and work in the human world, it is also the essential part and the significant research topic in the field of intelligent robot [1]. The core technique of hand-eye coordination of robot is the mapping from the visual sensory to the arm operation of robot, including hand-eye coordination with calibration and without calibration [2]. Mostly, the traditional robot hand-eye coordination is based on the system with calibration, but regretfully, there are many limitations, including (1) it is difficult to build the desired model in reality. (2) it needs re-calculation once there is a tiny change of system. (3) Usually, it is restricted in an incommodious area of the camera calibration [3].

Considering the above limitations, the hand-eye coordination without calibration can be classified as follows: (1) Hand-eye coordination with one camera on hand; (2) 3D hand-eye coordination with one static camera; (3) 3D hand-eye coordination with two static cameras; (4) 2D hand-eye coordination with one immobile camera and one camera on hand; (5) hand-eye coordination with two cameras on hand [3]. Where, a robotic system of 3D hand-eye coordination with two static cameras is composed of

* Corresponding author.

A. Xie & X. Huang (Eds.): Advances in Electrical Engineering and Automation, AISC 139, pp. 361–367.
springerlink.com

one robotic arm which can move in 3D circumstance, and the two static cameras are used to support object tracking. Another research hotspot of intelligent robot is "developmental robot". The distinctions between developmental robot and traditional robot are: the developmental robot can not only finish the scheduled tasks, but also perceive the unknown world to take a certain action purposefully and intentionally [4]. Besides it can do online learning imitating human infant, without interruption of human, guarantee to organize and store the learnt knowledge reasonably as well.

There are also several limitations on home and international researches, for example, most of the hand-eye coordination systems only use one camera, and they carry out 2D experiment which would lead orientation inaccurately [5][6][7][8], without imitating the developmental progress of human infants capturing the objects [9][10].

Combining the two hotspot mentioned above, this paper proposes an incremental process model, employs duo-neural network to carry out hand-eye coordination with two static cameras, with the goal to train the robot to imitate human infant developmental process and to achieve 3D hand-eye coordination to capture object accurately.

2 The Related Work and Background

In order to build a development-driven approach, it is necessary to understand human infant development procedure, and to abstract the significant developmental features from the procedure. Through simulating those features, a computational learning system is created to support our robot to develop hand-eye coordination ability.

Infant Hand-eye Coordination Learning. At birth, human infants are capable of visually orienting to laterally placed sounds and of making directed hand movements toward visual targets. These “pre-reaching” movements are not successful in making contact with targets, but are visually directed in that infants show more reaches to the hemifield in which a target appears. Researches demonstrate that it is clear that infants focus their visual attention on the target and do not view the hand during reaching. Early reaching is based on vision of the target and on proprioceptive signals from the arm. Later, towards the end of the first year of life, infants acquire the ability to independently move their fingers, and vision of the hand gains importance in the configuring and orienting of the hand in anticipation of contact with target objects. The jerky reaching movements of infants reflects a sequence of corrected submovements [11].

By researching and learning the development of human infant reaching process, we can see that human infant move her arm around the objects at first, then repeat this time and time again to capture the objects. We apply the infant reaching developmental process upon robotic arms to capture objects. This paper proposes to design two neural networks by using the experience of infant hand-eye coordination. One neural network is trained to do large amplitude arm movements around the objects, imitating the early infant pre-reaching. The other neural network is designed to do tiny amplitude arm movements to calibrate, imitating the later movement of infant reaching. The design of double-neural networks perfectly reflects the two phases of the developmental process of infant reaching.

Constructive Network. Our robotic system employs constructive learning system to build its developmental learning/control structure. The current researches applied static networks to build their systems, most static neural networks need to predefine the network structure and learning can only affect the connection weights. Those situations are not consistent with developmental psychology. Because scientists from developmental psychology emphasize that two kinds of developments in the brain: quantitative adjustments as well as qualitative growth [12]. Quantitative adjustments mean the synapse connection weights' adjustments of the neural network, and qualitative growth refers to the changes of the neural network's topology. Therefore, we focus on looking for a type of artificial neural network that is able to correspond to such two properties.We applied a radial basis function network (RBF) with resource allocating algorithm in this paper. We refer to plasticity in the mapping network as the result of two forms of change: an increase or decrease in the number of neurons or nodes in the network; or a change in an existing nodes parameters, either as a shift of location of the covering field or a change in size of that coverage. These two kinds of change in the network have different mechanisms but both represent plasticity for growth and development. There, the RBF network is selected for good function approximation ability, explicitencoding of the learning process, and biological plausibility.

Developmental Perspective. In this paper, we raise two parameters to restrict the maturation of the whole developmental robot hand-eye coordination system, one is the visual definition, and another is the movement amplitude of robotic arm. Through the introduction of the constraints, we gradually release the constraints to train the neural networks. The whole procedure can be described as follows: at the beginning, the vision is obscure, the arm can only wave in large amplitude. At this time, the system is not matured. As the release of both constraint parameters, the vision becomes increasing distinct and the robotic arm start to do fined calibration until the arm could capture the objects accurately. Then, the whole system becomes stable and matured.

3 An Experimental System for Hand-Eye Coordination

This section illustrates our robotic system, and its control system. The robotic system consists of two cameras and one robotic arm. The control system includes the developmental learning algorithms, and the computation model to support those algorithms.

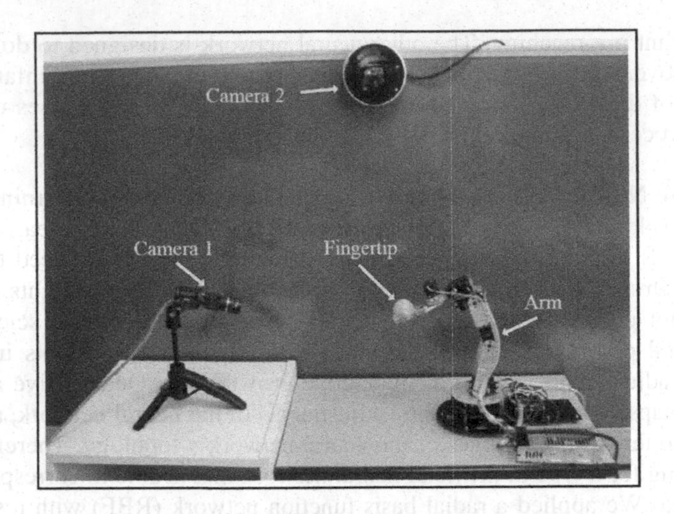

Fig. 1. The experiment system

Experimental System. Fig. 1 illustrates the experimental robotic system: we use an "AS-6 DOF" aluminium alloy robotic arm including 6 Degree-of-Freedoms(DOF), the arm is mounted on a workspace. We lock 2 DOFs of the hand, use the rest 4 DOFs to finish reaching movements, the used 4 DOFs are labeled in Fig. 2. This setup can support the robotic arm to move and capture objects in 3D environment. Each rotational joint of the robot arms has a motor drive and also an encoder which senses the joint angle, thus providing a proprioceptive sensor reading. The joint values of the 4 motors are labeled as θ_1, θ_2, θ_3 and θ_4. 2 cameras are applied to build the robotic vision system in this work, since one camera cannot give enough information for 3D reaching movement. A camera is mounted on a frame placed next to the arm (Camera 1 in Fig. 1); another camera is mounted above and looks down on the work space installed above the workspace (Camera 2 in Fig. 1).

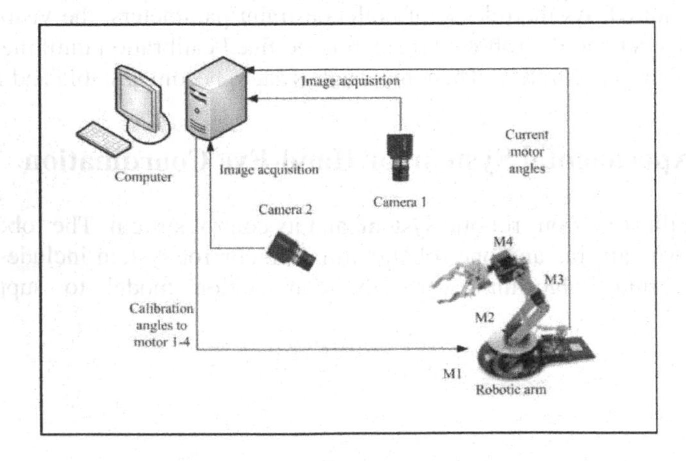

Fig. 2. The control system to the robot

Fig. 2 demonstrates the control system of the robot. The arm is controlled by an AVR board, which can receive the desired position of each motor from the control system, and feedback each motor's current position. The board can not only drive the motors to absolute positions, but also it can drive the motors to relative positions. The vision system receives images from the two cameras, it can detect whether objects or the arm appear within the workspace, and it also can give the arm or object's position within the image.

Learning Algorithms. Taking into account the training of the two neural networks, we set up a threshold δ to control choosing which one to train. There, we calculate the distance d between the coordinates captured by the two cameras the first time and the second time. If d is greater or equal to δ, we train the neural network N_1. Vice versa, we train network N_2. N_1 is trained to output the motor angles to calibrate the amplitude of the robotic arms, and N_2 is employed to calibrate in tiny amplitude of arm movements to finish seizing objects.

As the times of arm movements increasing, the error of the neural network becomes smaller and smaller, and the system tends to become more mature. When the error is stabilized less than a fixed value e, we could regard the neural network is mature. Therefore, we change vision definition to acquire new coordinates to retrain the neural network, until the error of the whole system is stabilized less than e. We train the neural network time and time again until the largest vision definition, at that moment, the robotic arm can do improved calibration and the whole system is in a stable situation. After training the two neural networks, we use them to control the arm to capture objects.

The robotic arm behaves spontaneous movements which stand for the arm moves randomly. Before each movement, the both cameras can calculate the fingertip position, x_1, y_1 indicates the fingertip position in Camera 1, x_2, y_2 gives the position in Camera 2; meanwhile, the joint values (θ_1, θ_2, θ_3, θ_4) of the robotic motors are acquired for the robotic controller. After one movement, both the fingertip position and the joint values are varied, and we use x_1', y_1', x_2', y_2' and θ_1', θ_2', θ_3', θ_4' to identify the new values. Here, the Euclidean distance d is the distance between x_1, y_1, x_2, y_2 and x_1', y_1', x_2', y_2'. $\Delta\theta_{1-4}$ are the different values between θ_{1-4} and θ_{1-4}'. A threshold δ which can be used to determine which network is trained or used to control the arm. If d is greater or is equal to δ, N_1 network is selected. x_1', y_1', x_2', y_2' are the input of N_1, and θ_{1-4}' are the network's expected output. If d is less than δ, N_2 is trained, its input contains x_1, y_1, x_2, y_2 and Δx_1, Δy_1, Δx_2, Δy_2, its expected output of N_2 is $\Delta\theta_{1-4}$.

The Constructive Network. The RBF network starts with no hidden units, and with each learning step, i.e. after the system observes the consequence after an action, the network grows or shrinks when necessary or adjusts the network parameters accordingly. The network growth criteria are based on the novelty of the observations, which are: whether the current network prediction error for the current learning observation is bigger than a threshold, and whether the node to be added is far enough from the existing nodes in the network, as shown in equations 1 and 3. The criterion in equation 3 is to check the prediction error within a sliding window to ensure that growth is smooth. m is the length of the sliding window.

$$\|e(t)\| = \|y(t) - f(x(t))\| > e_1 \tag{1}$$

$$\sqrt{\sum_{j=t-(m-1)}^{t} \frac{\|e(j)\|^2}{m}} > e_2 \tag{2}$$

$$\|x(t) - \mu_r(t)\| > e_3 \tag{3}$$

4 Experimental Results

This section shows the experimental procedure and the expected experimental results. The learning system has been implemented, however, this work's experiment part is still on going. In terms of the learning system, the experimental procedure can be designed as follows: Mark the hand of the robotic arm and the target by using special colour. Firstly, no target is put in the workspace, the learning system only generates random movements, according to the changing of two constraints, the robotic arm behaves no small movements at the beginning, but will generate several small movements in the end of the experiment. We can also observe that N_1 will has more chances to be trained when the system starts to run, N_2 has very few chances to be trained, however, during the middle phase of the experiment, N_1 still can be trained, but the weights of N_1 will not have significant change; meanwhile, N_2 starts to train its weight more frequently. All this situations are caused by the changing of the two constraints. Therefore, the entire robotic system's developmental procedure is driven by the constraints.

5 Conclusion

This paper has shown how ideas and data from psychology and primate brain science can inspire effective learning algorithms that could have considerable impact in robotics applications. Our experimental system is incremental and cumulative in its learning. Moreover, two distinct stages of behaviour were produced from a single method; this shows how competence at one level may support the emergence of further levels. There exists a large gap between our psychological understanding of development and our ability to implement working developmental algorithms in autonomous agents.

This paper describes one study on the path towards closing that gap and indicates the potential benefits for future robotic systems. In the future work, we could use motorized 2 DOFs cameras to replace the two static cameras in the system, this can avoid the flaws of immobile camera like invisibility.

Acknowledgments. This work is funded by the Natural Science Foundation of China (No. 61003014), and the Natural Science Foundations of Fujian Province of China (No. 2010J01346 and No. 2010J05142).

References

1. Connolly, C.: Artificial intelligence and robotic hand-eye coordination. Industrial Robot: An International Journal 35(6), 496–503 (2008)
2. Xie, M.: A developmental principle for robotic hand-eye coordination skill. In: Proceedings of the 2nd International Conference on Development and Learning, pp. 108–113 (2002)
3. Su, J.: Robotic Uncalibrated Hand-eye Coordination. Publishing House of Electronics Industry, Beijing (2004)
4. Lungarella, M., Metta, G., Pfeifer, R., Sandini, G.: Developmental robotics: a survey. Connection Science 15(4), 151–190 (2003)
5. Metta, G., Sandini, G., Konczak, J.: A developmental approach to visually-guided reaching in artificial systems. Neural Networks 12(10), 1413–1427 (1999)
6. Meng, Q., Lee, M.H.: Automated cross-modal mapping in robotic eye/hand systems using plastic radial basis function networks. Connection Science 19(1), 25–52 (2007)
7. Chao, F., Lee, M.H.: An autonomous developmental learning approach for robotic eye-hand coordination. In: Artificial Intelligence and Applications, Innsbruck, Austria (2009)
8. Huelse, M., McBride, S., Law, J., Lee, M.: Integration of active vision and reaching from a developmental robotics perspective. IEEE Transactions on Autonomous and Mental Development 2(4), 355–367 (2010)
9. Marjanovic, M.J., Scassellati, B., Williamson, M.M.: Self-taught visually-guided pointing for a humanoid robot. In: Proceedings of the 4th Int. Conf. on Simulation of Adaptive Behavior, pp. 35–44 (1996)
10. Andry, P., Gaussier, P., Nadel, J., Hirsbrunner, B.: Learning invariant sensorimotor behaviors: A developmental approach to imitation mechanisms. Adaptive Behavior 12(2), 117–140 (2004)
11. Berthier, N.E.: The syntax of human infant reaching. In: Unifying Themes in Complex Systems: Proceedings of the Eighth International Conference on Complex Systems, vol. VIII, pp. 1477–1487. NECSI Knowledge Press (2011)
12. Shultz, T.R.: Computational Developmental Psychology. MIT Press, MA (2003)

Design and Implementation of Induction Motor Control

Genghuang Yang, Hinsermu Alemayehu, Feifei Wang, Shigang Cui, and Li Zhao

Tianjin Key Laboratory of Information Sensing and Intelligent Control,
Tianjin University of Technology and Education, Tianjin, China, 300222
ygenghuang@126.com

Abstract. This thesis details design and implementation of three phase induction motor control system. The system is Atmega128 based 3-phase induction motor control. The controller drives the motor through a pulse-width modulated inverter which utilizes space vector modulation for the generation of its waveforms. The controller is implemented using Atmega128 microcontroller which connected to PC through HyperTerminal for monitoring purposes. It is used to drive a 3-hp motor with load. The experimental results presented favorably compare speed transients between data taken from the 3-hp motor and data from a MATLAB simulation based on an analysis of the entire system. The motor fault status and normal status is also displayed. At the last we derived three phase induction motor successfully and the speed is controlled well.

Keywords: 3-phase induction motor, MATLAB, PWM, Atmega128, feedback control.

1 Introduction

Since the invention of the induction motor by Nikola Tesla in 1886, the use of three phase induction motors has grown tremendously. The advantage of using an induction motor comes with its rugged and economic design and reliable performance even in adverse conditions. Induction motors are regularly installed outdoors, exposed to rain and sandstorms, and have even been found at the bottom of oil wells. As a result, the induction motor has found widespread use throughout the world. Approximately 60% of electrical energy world-wide passes through the windings induction motors [1].

The main reason for the slow speed AC motor speed control is to determine the frequency of the AC main factors and changes in motor torque control is difficult, so that the stability of AC variable speed, reliability, economy and efficiency cannot be met production requirements. Purpose of this research is to design, implementation and control of three phase Induction motors using Atmega128 microcontroller [2].

2 The Structure of 3- Phase Induction Motor Control System

The structure of research mainly contains: 3-phase power supply, 3 phase rectifier, C Filter, 3 phase PWM inverter, DC supply, gate drivers, PC, RS-232, and Atmega128 module, optoisolators, induction motor and so forth [3]. The structure of three phase induction motor control system is shown in figure 1.

A. Xie & X. Huang (Eds.): Advances in Electrical Engineering and Automation, AISC 139, pp. 369–373.
springerlink.com

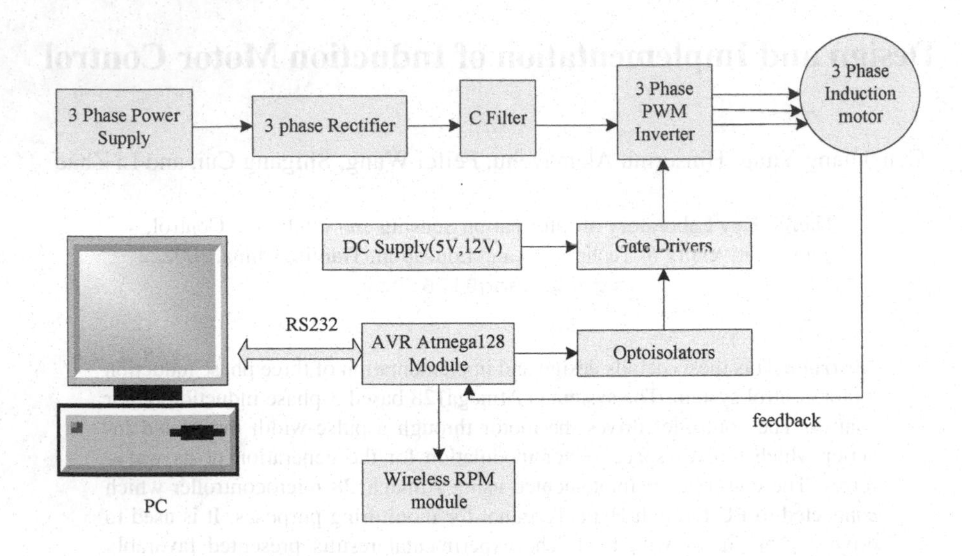

Fig. 1. The structure of three phase induction motor control system

The whole system block diagram contains: power supply, power circuit, control circuit, motor circuit and communications.

The power circuit consists of three phase rectifier, C filter and three phase PWM inverter (Mosfet).The phase rectifier is used to covert AC to DC voltage, while C filter is used as dc link and three phase PWM inverter is used to control the speed of induction motor depending signal generated from microcontroller or oscilloscope [4].

The control unit consists of gate drivers which are used to drive three phase PWM inverter, Optoisolators which is used to connect AVR Atmega128 and gate drivers to generate three phase signal from or oscilloscope, feedback which include sensor and voltage detection circuit. The microcontroller controls the whole systems such as speed control, under/over voltage control. I used Atmega128 for this purpose [5].

The motor circuit consists of three phase induction only. Induction motors are the most widely used motors for appliances, industrial control, and automation; hence, they are often called the workhorse of the motion industry.

In this project I used RS-232 communication. RS-232 is an asynchronous serial communication protocol widely used in computers and digital systems. It is called asynchronous because there is no separate synchronizing clock signal as there are in other serial protocols like SPI and I2C. The protocol is such that it automatically synchronizes itself. We can use RS232 to easily create a data link between our MCU based projects and standard PC) [6].

3 System Design for Three Phase Induction Motor Control System

3.1 Hardware System Design

The hardware system design section contains: power converter module, control module, feedback module and communications.

This describes the design of a Three Phase Power Converter (AC/DC/AC) as a The system implemented converts a three phase input voltage of 400 Vrms at a power rating of 500 W to DC by using a three phase diode bridge rectifier, and a Capacitive filter which assists in stabilizing the output DC voltage produced. After that, the system provides the gate drive signal to a three phase pulse width modulated (PWM) inverter driving an induction motor. The pulse width modulation signal is generated by MATLAB [7].

3.2 Control Module Design

As shown below the control module is the main part of the project and used to control the whole project. Each pin used for specific and general purpose. So I decided to connect pins for many functions. Pin PA0-PA7 (portA) is connected to LCD of pin DA0-DA7 which is connected to EEPROM and used to save data temporarily and display result. Pin PB1, PG0, PG1 are connected to LCD pin of RS, RW and E respectively. Port F pins PF0-PF4 is connected feedback module which is used to send analog to digital microcontroller. Pin PF0, PF1, PF2, PF3 and PF4 are connected to fault, overload, vout1, vout2 and short circuit feedback control respectively. Port E Pin PE0 (PD0) and PE1 (PD1) are connected to serial of RS232 of RXD0 and TXD0 respectively. Port E pins PE3-PE5 is connected to PWM1, PWM2, and PWM3 respectively. Port C Pins PC0, PC1, PC2, PC3 are connected to potentiometer (pot) which is used to vary the frequency, forward/reverse and run /stop and S1 respectively. S1 has two options. When it pressed it gives 5V and when it pressed again it will be ground. The LED0 and LED1 are connected to PC4 and PC5 respectively.

3.3 Feedback Module Design

When fault is occurred the HCPL-788J will send signal to microcontroller and the fault is displayed by led (light emitting diode. The module I designed here is isolation amplifier with short circuit and overload detection. Three phases phase A, phase B and phase C are directly connected three phase PWM inverter output phases phase A, phase B and phase C respectively. When fault is occurred the motor stopped automatically. It is connected to port F pins of PF0-PF4 are connected the pins of HCPL-788J.

3.4 Software System Design

The software system design consists of MATLAB, ICC AVR and AVR studio. The MATLAB is used for modeling of asynchronous motor control such as speed control of the motor. The ICC AVR is used for C programming and AVR studio is used for downloading HEX file to AVR microcontroller.

3.5 Microcontroller Based Software System Design

The main motor control flowchart is as figure 2 shows.

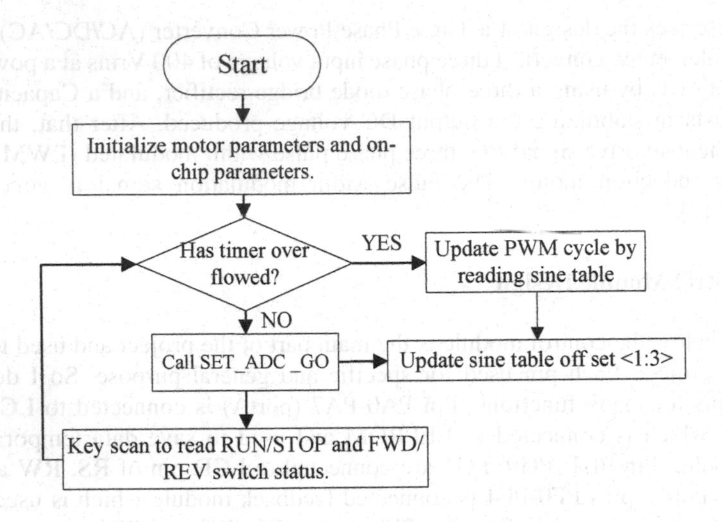

Fig. 2. Main motor control flowchart

4 Conclusion

The device will automatically stop the motor when fault occurred. The circuit will take full control of the motor and it will protect the motor from several faults such as over voltage and under voltage, over current and the circuit will switch on the motor under safety conditions. This also protects Induction motor from single phasing which is also a major fault. To drive 3-phase induction motor using microcontroller this can be used for industrial purposes. To reduce cost for driving induction motor which controlled by plc and to replace plc function by microcontroller. To make microcontroller more applicable for industrial purposes.

Acknowledgement. This work is supported by China Postdoctoral Science Fund (20090460501) , Tianjin Higher School Science & Technology Development Fund (20090704) and Fund of Tianjin University of Technology and Education (YJS10-03, YJS10-05, KJY11-08).

References

1. Mi, C., Shen, J., Natarajar, N.: Field-oriented control of induction motorsbased on a MC68322 microcontroller. In: The 7th IEEE Workshop on Power Electronics in Transportation (WPET 2002), Michigan, USA, October 24-25, pp. 69–73.
2. Rashid, M.H.: Power electronics: circuits, devices and applications, 2nd edn. Prentice-Hall, Englewood Cliffs (1993)

3. Yiliang, T., Wenjin, C., Xiaorong, X., Yingduo, H., Wong, M.: 80C196MC microcontroller-based inverter motor control and IR2130 six-output IGBT driver. In: International Conference on Electric Machines and Drives (IEMD 1999), Seattle, USA, May 9-12, pp. 655–657 (1999)
4. Minas, G., Martins, J.S., Couto, C.: A microcontroller based voltage space vector modulator suitable for induction motor drives. In: International Symposium on Industrial Electronics (ISIE 1999), Bled, Slovenia, July 12-16, vol. 2, pp. 469–473 (1999)
5. ATMEL Corporation, AVR Enhanced RISC Microcontroller Datebook (2000)
6. Information, RS-232 communication, http://www.extremeelectronics.com
7. Houldsworth, J.A., Grant, D.A.: The Use of Harmonic Distortion to Increase the Output of a Three-Phase PWM Inverter. IEEE Trans. on Industry Applications IA-20(5) (1984)

Using Operational Processes Model to Improve Software Development

Aziguli Wulamu[1], Jian Wang[2], and Chun Zuo[3]

[1] School of Information Engineering,
University of Science & Technology Beijing, Beijing, China
[2] School of Computer and Information Technology,
Beijing Jiaotong University, Beijing, China
[3] Technology Center of Software Engineering,
Institute of Software of Chinese Academy of Sciences, Beijing, China
ali@bsw.gov.cn, wangjian@bjtu.edu.cn, zuochun@sinosoft.com.cn

Abstract. The development of key processes software systems in special domain, such as medication or insurance, becomes complexity. The biggest problem is how to share knowledge among end users, developers and managers. In order to upgrade quality of software development, this paper aims to present a novel operational process model that assist to share or reuse the knowledge through different stage of software engineering. The operational process model is the knowledge transformation and computing model and it will prompt the automation of software design; also enhance the standardization of knowledge representation.

Keywords: Operational Process Model, Knowledge Representation, Ontology, Software Engineering.

1 Introduction

With the evolution of communication and computer technologies, the traditional process system has been changed to information-based networks computer operation system (IB-NCOS). The development of IB-NCOS becomes more and more difficulty and complexity.

In generally, the software development process can be divided into several steps, such as shown in Fig. 1. Every step has the inputs and outputs such as A, B, C and D. For example, A is a common management file (CMF) which includes operation processes, restriction characterization and so on. The software analysts make reference to this CMF that the report of software requirement is given. After software designing, C is given to use in software development. Subsequently D is the computer system which is offered to end users.

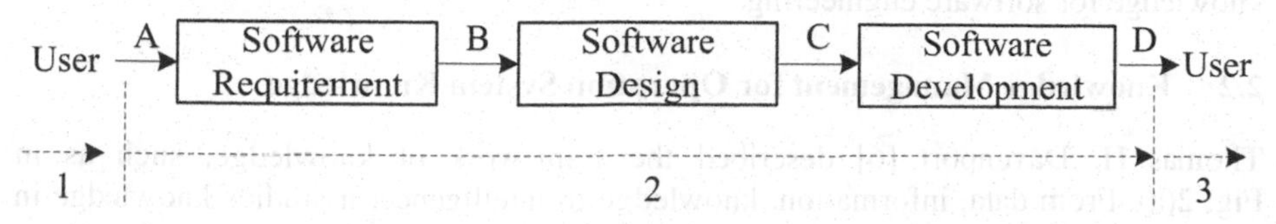

Fig. 1. Steps of Software Development

A. Xie & X. Huang (Eds.): Advances in Electrical Engineering and Automation, AISC 139, pp. 375–380.
springerlink.com

However, the problems of sharing software engineering knowledge among end users, developers and managers are obvious. In despite of most time is used to analysis user's requirements and design software, but there often have ambiguities and difference between users and designers. This problem would influence the development of IB-NCAS, how to prompt the quality of service becomes more urgency. To solve this question that could consider many different aspects, such as programming technique, or high-speed hardware environment. Many researchers have presented some solution schemes [1-4], which use knowledge management, semantic technique, and software architecture and so on.

In this paper, it focuses on knowledge represent, represent technique. On the sight of knowledge, it could describe how the knowledge to flow from A, B, C and D. At the position of A that logics of operation processes are given. Otherwise, at the position of C those physical design logics for developing are achieved. Each step has its own knowledge represent modality, at the same time the knowledge of each step has relationship.

This paper would present one operational model which is an operational process model in section 3. The knowledge represents discuses in section 4, and the example would be given.

2 Background Knowledge

2.1 Localization of Common Management File

IB-NCOS is the management information system in essence. It follows the CMF which is written by users. On the sight of software engineering, the frame of process production is divided into three phases:

1) In the management of operation phase, the production is the CMF.
2) In the software requirement / design phase, the production is the requirement and design document.
3) In the computer system phase, the production is the management information system.

During these phases from CMF to the management information system, the CMF is the fundamentality of software development.

The CMF is written for process management that it is used to manage in some domains [5]. It is important to analysis the software requirements. However, because the different domains knowledge is used, during analyzing the requirement it is difficult to clear up ambiguities in concepts and process. The CMF doesn't have the concept of object, and it has no defined file format, that why it could not be used to represent knowledge for software engineering.

2.2 Knowledge Management for Operation System Knowledge

Thomas H. Davenport [6] described the framework of knowledge, such as in Fig. 2(a). From data, information, knowledge to intelligence, it studies knowledge in the abstract.

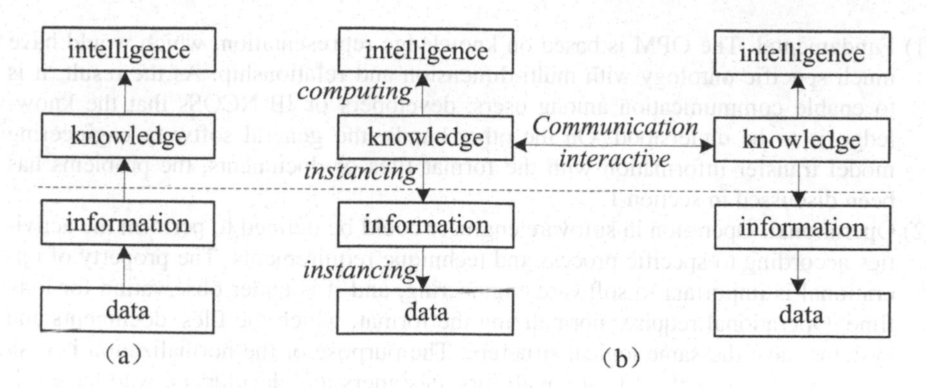

Fig. 2. The Framework of Knowledge

In Fig. 2(b), that knowledge is in the central position, so the new relationships can be given. Next the meanings of data, information and knowledge would be listed in this paper. Data is consisted of dimensionality tables, code tables, and activity descriptions and so on. Information contains log datasheets, relation charts and so on in the data environment. Knowledge is abstracted from information, which could be drawn form man's experience and methods of re resolving. Thus knowledge includes rules and restrictions on the specific environment.

Traditionally, it follows the flow of software engineering:

1) Users write the CMF for process operations management and transfer the data to software analyzers and designers.
2) Software analyzers and designers do vast working in order to understand, digest and abstract the real requirements of user in some specific domain. It has many restrictions for analyzers and designers who need to master different specialty. In generally, lots of time is used to do repeated working for data abstraction, which is always on information layer. In spit of some knowledge databases (KDB) are built for some systems, but these knowledge represents are only isolation and samples and it needs to add relations, rules, restrictions and so on.

In knowledge represent opinion, one method is presented to solve the problem of how to share software engineering knowledge among end users, developers and managers. Firstly, knowledge representation is given by using ontology. Then based on knowledge it could computes, with rules and restriction, to achieve intelligence behavior. On the other hand, knowledge representation is one format for using to descript the information and data. The method would be discussed in section 4.

3 The Operational Process Model

The operational processes model (OPM) will be defined in this section. To compare OPM with the general software engineering model, there are many differences. The major differences are shown on:

1) Fundamental. The OPM is based on knowledge representation, which would have much specific ontology with multi-dimension and relationship. As the result, it is to enable communication among users, developers or IB-NCOSs that the knowledge is to be understood. On the other hands, the general software engineering model transfer information with the format files or documents, the problems has been discussed in section 1.
2) Operational. Operation in software engineer could be defined to perform the activities according to specific process and technique requirements. The property of operational is important to software engineering, and it is under observation for long time. Operational requires normalizing the format, which the files, documents and systems have the same logical structure. The purpose of the normalization is easy to communicate with software analyzers, designers and developers, who have different domain background. Ontology is used to represent knowledge of software engineering in OPM. Because ontology is used to refer to what exists in a system model, and in the area of computer science which could represent concepts, properties, relations and restrictions. It supports logical operation, so it fits to do computing by intelligence systems. However the format files or documents don't have the ability to perform logical operation.

As shown in Fig. 2(b), that different domains communicate between the knowledge layers to interactive knowledge. In this paper, the OPM is committed, which has a property of knowledge exchanging from software requirement analyzing to computer system development.

Definition 1. Operational process model (OPM) is used to assist the implementation of software engineering, which prompts understanding of software engineering knowledge. The input of OPM is the CMF ontology, an intermediate result is the requirement analyzing and design document ontology, and the output result is the product system ontology. The processing includes mapping, derivation and normalization.

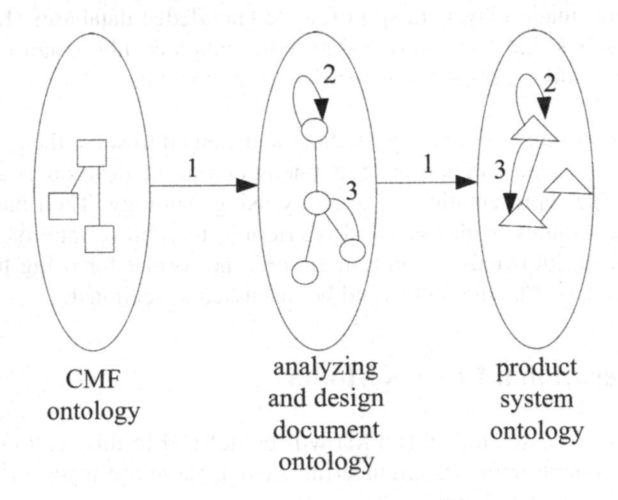

Fig. 3. Processing of OPM

It is same with the general software engineering's development procedure that users should describe their requirements of their process system. As shown in Fig. 3, there have three stages which are the same with shown in Fig. 1. In OPM, it gives different ontology instance, which are used to respective representation the knowledge of stages. Knowledge is transferred among different ontology. This definition is in a novel viewpoint, which distinguishes with Wongthongtham [3] that it presents a conception of software engineering ontology; the software engineering ontology is the same with document ontology in OPM. The major goal of the software engineering ontology is to assist the communication between analyzers and developers.

However, the major working of this paper is to present the OPM, which has multi-ontology, and it defines the processing of ontology. Fig. 3 gives a gradation model. There has three different domain ontology on the top level and each of ontology has a hierarchical knowledge structure. The processing of mapping is used to link up the top level ontology, such as CMF ontology and document ontology. Derivation and normalization are working in separate ontology. Each of stage knowledge has specific concepts, relations and restrictions, so after mapping processing, it needs to instance ontology, sub-ontology based on its own hierarchical knowledge structure.

The OPM is the knowledge transformation and computing model, which presents a novel method to unified operating all stages of software engineering on the viewpoint of knowledge.

4 The Knowledge Representation

In recent years, ontology is not only to be used in artificial intelligence, but also more and more to be used in computer science, especially in semantics processing, knowledge engineering, intelligent information integration and so on [7-9]. In definition of OPM, it includes ontology to assist knowledge processing.

In [3], it has defined that software engineering ontology includes instances, properties and classes. In essence, software engineering ontology represents data, relations and specific restrictions (includes the environment).

In section 3, it describes the operational processes model, in which presents multi-ontology for knowledge transformation and computing. It has a hierarchal structure, which descript CMF ontology, requirement analyzing and design document ontology and product system ontology, also includes sub ontology and processing of mapping, derivation and normalization. The different ontology, in this paper, is same with [3], which consists of instances, properties and classes.

The main purpose of hierarchal ontology model is for the domain knowledge sharing and reuse during the life cycle of IB-NCOS. It fills the gap between CMF and software requirement document, and also it includes specific domain knowledge, such as computer software technique, so it establishes the logical knowledge structure among user domain, software analyzing and developing domain and online systems domain. This knowledge representation's instance would involve amount of data description, properties and conceptions, at the same it should follow restrictions of domain.

Using ontology to eliminate misunderstandings and miscommunication would enhance the development quality and productivity of software.

For example, objects' designing is important for orient-object development. So following the same structure of ontology, they could instance ontology in specific domain. When it instances ontology, that main ingredient of ontology, such as data, properties and classes, is filled by data, relations and restrictions. Consequently, objects' designing is an isomorphism between CMF and software requirement document.

5 Summary

Operational process model is the knowledge transformation and computing model, that user's process management, software engineering and other knowledge is represented by the OPM's multi-ontology. This work, in this paper, has a main purpose to solve problem of how to share software engineering knowledge among end users, developers and managers. To compare with others work, OPM is a novel model for software development, it fits all stages of software engineering life cycle, which could upgrade quality of software design. Also it will prompt the automation of software design; also enhance the standardization of knowledge representation.

Acknowledgements. This paper is supported by the projects "Researching Cognitive model actuation of acquiring of knowledge technique based on magnanimous Chinese medicine medical case (61070101)" from the National Natural Science Foundation China and "Knowledge Base of Key industry information and service system building (D0106006040291)" from the Beijing Science and Technology Committee.

References

1. Mei, H., Shen, J.R.: Progress of Research on Software Architecture. J. Soft., 1257–1275 (2006) (in Chinese)
2. Zhang, S., Zheng, G., Chang, Q., Wang, Y., Li, P.P.: The Architecture and Implementation of the New Generation Business System in a Commercial Bank. In: International Conference on Business Intelligence and Financial Engineering, pp. 501–504 (2009)
3. Wongthongtham, P., Kasisopha, N., Chang, E., Dillon, T.: A Software Engineering Ontology as Software Engineering Knowledge Representation. In: 3rd International Conference on Convergence and Hybrid Information Technology, pp. 668–675 (2008)
4. Medvidovic, N., Grünbacher, P., Egyed, A., Boehm, B.W.: Bridging Models across the Software Lifecycle. J. Sys. Soft., 199–215 (2003)
5. Zuo, C.: Key Process System. Technical report, Software Development Forum (2005) (in Chinese)
6. Davenport, T.H., Prusak, L.: Working Knowledge, 2nd edn. Harvard Business School Press, USA (2000)
7. Wand, Y., Storey, V.C., Weber, R.: An Ontological Analysis of the Relationship Construct in Conceptual Modeling. ACM Transaction on Database System, 495–528 (1999)
8. Lu, R.Q., Shi, C.Y., Zhang, S.M., Mao, X.P., Xu, J.H., Yang, P., Fan, L.: Agent-oriented Commonsense Knowledge Base. Science in China (Series E), 453–463 (2000) (in Chinese)
9. Li, S.P., Yin, Q., Hu, Y.J., Guo, M., Fu, X.J.: Overview of Researches on Ontology. J. Com. Res. Dev., 1041–1052 (2004) (in Chinese)
10. Technical blog for writing operational files, http://blog.csdn.net/zuochun/article/details/6738815 (in Chinese)

Statistical Modeling for Issue of Higher Education Tuition Standards in China

Chun Cai and Ning Yu

Dept. of Information and Science,
College of Arts and Science, Beijing Union University,
Beijing, 100191, China

Abstract. Through statistic analysis on the issue of higher education tuition standards in China, we found three major factors of making education tuition standards by modeling based on those factors, we drew a conclusion that the tuitions of higher schools were correlated to disposable family income of urban resident, net family income of rural resident, per capita expenditure of common higher schools. By experiments on relevant data from 2001-2006, we found the data of higher education expenditure in Beijing area in 2004 were unreasonable in some way in comparison with other years.

Keywords: higher education tuition, linear regression, disposable income, per capita expenditure.

1 Introduction

The tuition standard has direct influence on the future of the China. With transition of China's higher education from elite education to mass education, more and more people have chance to enter colleges. So how to make the reasonable higher education tuition standard has become an emerging sensitive complicate issue before us. Too high tuition is unaffordable for a lot of families and too low tuition is so unfair for higher schools that they cannot assure teaching quality for financial problem. Higher education is non -compulsory education with budgets mainly from government appropriation, fundraising of schools, social donation and tuition in all countries in the world. The eligible impoverished students may get support through approaches like loan or reduction, exemption and subsidiary of tuition and some of them are able to get scholarships from government, schools, enterprises and others for excellence both in morality and academy. All of specific appropriation of local government in education, local living level and total annual family income in different provinces and cities will have influence on local tuition standard. Considering that Chinese has a vast territory, totally different economic strengths in different provinces and cities and a large gap of incomes between urban and rural areas, it is unfair to make higher education tuition of different provinces and cities at the same standard. So we calculated the higher education tuitions by area first, and then further classified the tuition standards on the basis of the family of the undergraduate –urban or rural family-in terms of specialty of the schools in different areas -science departments and arts departments in key higher school and science departments and arts departments in common school.

A. Xie & X. Huang (Eds.): Advances in Electrical Engineering and Automation, AISC 139, pp. 381–387.
springerlink.com

By these models we calculated the tuition standards in different areas and finally established an evaluation model by area to calculate tuition standards in different areas.

2 Data Processing

We reviewed the data of statistical yearbooks of Beijing from 2001 to 2006 on website of Beijing Municipal Bureau of Statistics, and selected four sets of data required respectively: total annual family income of urban resident(TAFIUR), disposable annual family income of urban resident(DAFIUR), net annual family income of rural resident(NAFIRR) and total annual expenditure of common higher school(TAECHS). See Tab 1 for details.

Table 1. Relevant data of Beijing area from 2001 to 2006

Year	TAFIUR (Yuan)	DAFIUR (Yuan)	NAFIRR (Yuan)	TAECHS (Qian Yuan)
2001	13768.8	11577.8	5098.8	17388720
2002	14052.5	12464.3	5880	20235164
2003	14959.3	13882.6	6496.3	21393401
2004	17116.5	15637.8	7172.1	24767669
2005	19533	17653	7860	26458894
2006	22417	19978	8620	29244117

During research, we calculated relevant coefficients of each row in Tab. 1 and weeded data of total family income of urban resident by linear correlation theory correlation coefficient approximates 1. And then we calculated per capita input of higher education with data of total expenditures of common higher school. We also chose the number of undergraduate students under the paragraph "conditions of students in higher schools in Beijing" from sections "Education" and "Culture" in Beijing Statistical Yearbook from 2001 to 2006. The number of undergraduate students of common higher school is 340284, 398573, 450789, 387628, 409613, 425396 respectively from 2001 to 2006.

From 2001-2006, Annual tuition standard of higher schools in Beijing area: 5,500 Yuan, 5,000 Yuan, 4,600 Yuan and 4,200 Yuan for students of science departments and arts departments in key school and in common school respectively. During research, we subtracted seven items of expenditures including employee's welfare fees and social insurance premiums, grants, expenses of business affairs, equipment purchase costs, repair costs, other expenses and expenditures for capital construction from total expenditure of common higher schools representatively and just reserved salaries because other expenses, believed to not belong to student cultivation costs, should be provided by government as other education expenditures. Thus, we can obtain the per capita expenditure of common higher schools (PCECHS). We can find, except data of 2004, per capital expenditures in other five years basically show a linear relation. Through investigation and verification, we confirmed the per capita expenditure 8823 in 2004 was unreasonable in some way. So we have to weed 2004 data as shown in Tab 2.

Table 2. Relevant calculation data of Beijing area

Year	DAFIUR (Yuan)	NAFIRR (Yuan)	PCECHS (Yuan)
2001	11577.8	5098.8	8587
2002	12464.3	5880	9432
2003	13882.6	6496.3	11971
2005	17653	7860	13292.03
2006	19978	8620	14031.14

3 Modeling

3.1 Multiple Linear Regression Based Model

Based on the joint of influence of disposable family income of urban resident, disposable family income of rural resident and government input in higher education per student on the higher education tuition, we chose a multiple linear regression. Actually there is a huge gap of disposable family income between urban area and rural area. If we chose per capita disposable income, the tuition afforded by rural population would be enhanced intangibly and the tuition afforded by urban population would be reduced correspondingly, so we established two models for different conditions in urban areas and in rural areas.

The higher education tuition model taking disposable family income of urban resident and government input in higher education per student as correlation factors:

$$Y = \beta_0 + \beta_1 x_1 + \beta_3 x_3 + \varepsilon \tag{1}$$

Among equation (1), Y explains tuitions in different areas, x_1 explains disposable income of urban resident, x_3 explains per capita input in higher education.

The higher education tuition model taking disposable family income of rural resident and government input in higher education per student as correlation factors:

$$Y = \beta_0 + \beta_2 x_2 + \beta_3 x_3 + \varepsilon \tag{2}$$

Among equation (2), x_2 explains disposable income of rural resident.

In order carry out further analysis, we classified the higher schools into key schools and common schools according to provisions of current standards and the tuition standards of both classes are also different. So we further classified above two models into arts departments in key schools, science departments in key schools, arts departments in common schools and science departments in common schools for further discussion.

3.2 Model Solution

Model solution of average annual tuition of Category-1 science departments' students from urban areas. In terms of standard of science departments of key schools in Beijing, we obtained following result through data regression analysis with Excel.The result shows, the regression intercept β0 is 0.49841, the correlation coefficient of disposable income of urban resident β1 is -0.0266; the correlation coefficient of input in higher education per student β3 is 0.0433; from the residual, 2, we also find the fitting degree of equation is very high.

$$Y = 0.4984 - 0.0266x_1 + 0.0433x_3 \tag{3}$$

Model solution of average annual tuition of Category-2 science departments' students from urban areas:

In terms of science departments standard of common higher schools in Beijing, through data regression analysis with excel, we obtain the equation:

$$Y = 0.4399 - 0.0050x_1 + 0.0241x_3 \tag{4}$$

Model solution of average annual tuition of Category-1 arts departments' students from urban areas. In terms of arts departments standard of common higher schools in Beijing, through data regression analysis with excel, we obtain the equation:

$$Y = 0.4985 - 0.0266x_1 + 0.0433x_3 \tag{5}$$

Model solution of average annual tuition of Category-2 arts departments' students from urban areas. In terms of arts departments standard of common higher schools in Beijing, through data regression analysis with excel, we obtain the equation:

$$Y = 0.4114 - 0.0606x_1 + 0.0856x_3 \tag{6}$$

Model solution of average annual tuition of Category-1 science departments' students from rural areas:

$$Y = 0.5690 + 0.0521x_2 - 0.0474x_3 \tag{7}$$

Model solution of average annual tuition of Category-2 science departments' students from rural areas:

$$Y = 0.4410 - 0.0521x_2 + 0.0474x_3 \tag{8}$$

Model solution of average annual tuition of Category-1 arts departments' students from rural areas:

$$Y = 0.5026 - 0.1128x_2 + 0.0715x_3 \tag{9}$$

Model solution of average annual tuition of Category-2 arts departments' students from rural areas:

$$Y = 0.4204 - 0.2252x_2 + 0.1313x_3 \quad (10)$$

We established 8 multiple linear regression models with data collected in this paper.

4 Test and Analysis on Models

4.1 Tuitions Prediction in Beijing Area in 2007

With models foresaid we predict the data of tuitions of higher schools in Beijing were basically consistent with the fact in 2007. See Tab.3 for the result:

Table 3. Tuitions Prediction of Beijing area

Class of student	Category -1 science (Yuan)	Category -1 arts (Yuan)	Category -2 science (Yuan)	Category -2 arts (Yuan)
Urban	5481	4993	4618	3952
Rural	5540	4925	4560	3846

The prediction result of tuition of higher schools in Beijing area with the models is relatively consistent with the fact, so we may make further test on the models.

4.2 Tuitions Prediction in Other Cities from Urban Areas

The goal of calculation of average annual tuition of students from urban areas in higher schools in other provinces and cities with the models is to validate whether or not the higher education tuition models established are applicable to the whole country. See Tab.4 for the calculation result of average annual tuition of students from urban areas in higher schools in other provinces and cities (10 thousand Yuan).

Table 4. Tuitions prediction of area from urban areas

City	DAFIUR	PCECHS	Category-1 science	Category -1 arts
Shanghai	1.6683	0.6181	0.5635	0.4809
Zhejiang	1.5882	0.5849	0.5639	0.4816
Fujian	1.1175	0.5034	0.5635	0.4906
Sichuan	0.7710	0.2845	0.5671	0.4904

4.3 Tuitions Prediction in other Cities from Rural Areas

See Tab.5 for the calculation result of average annual tuition of students from rural areas in higher schools in other provinces and cities (10 thousand Yuan).

Table 5. Tuitions prediction of area from rural areas

City	NAFIRR	PCECHS	Category-1 science	Category -1 arts
Shanghai	0.7337	0.6181	0.5779	0.4641
Zhejiang	0.7584	0.5849	0.5808	0.4589
Fujian	0.2206	0.5034	0.5566	0.5137
Sichuan	0.3805	0.2845	0.5753	0.4800

After prediction on the charge standard of higher schools in other provinces and cities, we found, between developed areas like Shanghai and Zhejiang and developing areas like Sichuan, there was not only a huge gap of urban per capita disposable incomes and rural per capita net incomes, but also a similarity in tuition standards. We think that is unreasonable. The coastal cities like Shanghai have very high education level and obtain more education expenditures from the state in comparison with other areas so that the appropriation per student is relatively high in such areas, while the expenditures of the state for cultivating a university student is relatively steady. So we may draw a conclusion that, in the areas with high education level, the education cost sharing of the state is more than that in other areas and thus the cost sharing of the individual is less than that in other areas; in this way, in the areas with low education level, the cost sharing of the state is less and thus the cost sharing of the individual is more because education cost of a university student is relatively steady.

Besides, from Tab. V we found, in some rural areas, with sum of net incomes in two years, a rural family cannot afford a university student in one year. We think such phenomenon is correlated to following factors: 1. too many university students in higher schools in the year make education expenditure cannot be allocated reasonably; 2. the appropriation from central government was limited; 3. the central government provided impoverished students with state-subsidized student loan and relevant subsidies, but which are not considered in our modeling.

5 Summary

We found the data calculated were basically consistent with the fact through validation on the models, so we believe the modeling was reasonable. But in real life the factors influencing the making of tuition standard are not as simple as imagined. With the mathematical models built up based on Beijing area, the paper made predictions on different areas in China like Shanghai and Sichuan and found, for students from urban areas and from rural areas, there were different high education tuition standards implemented in above areas. Such standards were basically consistent with the fact. More importantly the mathematical models obtained in the research provide a scientific reference for state to make tuition standards more scientific, reasonable and suitable for current conditions in China.

In the view of above analysis, the models obtained in the research will be more complete, more scientific if we had more true original data and more detailed data materials. We compared some factors between coastal cities like Shanghai and some hinterland cities in Sichuan with the models we made and found there was no difference of tuitions paid by students between Sichuan and coastal cities like Shanghai basically, though per capita GDP in Shanghai was far higher than that in Sichuan. So we believe less higher education cost sharing of government in Sichuan is the reason of the unfair tuition foresaid. Based on these models we bring out some suggestions as below:

The state should make the standard for appropriation in higher education per student and determine tuitions in different areas in consideration of disposable family incomes in different cities.

The state should make appropriation in higher education per student in consideration of disposable famous incomes of urban and rural residents in the same areas and enhance appropriation for students from rural areas so as to lighten overburden of their families and enable more rural families to send their children to university easily.

The state should enhance limit of grant and loan of university students to reduce pressure on them and their families, especially the impoverished families. More importantly the state should give any eligible person a chance to accept higher education.

Acknowledgment. The paper is supported by Academic Human Resources (PHR201 008292) and Beijing Education Committee Research Project (KM201111717004).

References

1. Editorial Committee, Beijing Municipal Bureau of Statistics. Beijing Statistical Yearbook 2002. China Statistics Press, Beijing (2002)
2. Editorial Committee, Beijing Municipal Bureau of Statistics. Beijing Statistical Yearbook 2003. China Statistics Press, Beijing (2003)
3. Editorial Committee, Beijing Municipal Bureau of Statistics. Beijing Statistical Yearbook 2004. China Statistics Press, Beijing (2004)
4. Editorial Committee, Beijing Municipal Bureau of Statistics. Beijing Statistical Yearbook 2005. China Statistics Press, Beijing (2005)
5. Editorial Committee, Beijing Municipal Bureau of Statistics. Beijing Statistical Yearbook 2006. China Statistics Press, Beijing (2006)
6. Editorial Committee, Beijing Municipal Bureau of Statistics. Beijing Statistical Yearbook 2007. China Statistics Press, Beijing (2007)
7. Editorial Committee, Shanghai Municipal Bureau of Statistics. Shanghai Statistical Yearbook 2005. China Statistics Press, Shanghai (2005)
8. Editorial Committee, Zhejiang Provincial Bureau of Statistics. Zhejiang Statistical Yearbook 2005. China Statistics Press, Shanghai (2005)
9. Editorial Committee, Fujian Provincial Bureau of Statistics. Fujian Statistical Yearbook 2005. China Statistics Press, Fujian (2005)
10. Editorial Committee, Sichuan Provincial Bureau of Statistics. Sichuan Statistical Yearbook 2005. China Statistics Press, Sichuan (2005)
11. Jiang, Q., Xie, J., Ye, J.: Mathematical Models, 3rd edn. China Higher Education Press, Beijing (2003)

Chaos Synchronization between Two Different Hyperchaotic Systems with Uncertain Parameters

Hong Zhang

Information Engineering College, Minzu university of China,
100081, Beijing, China
Zhang_hong__@163.com

Abstract. This work investigates chaos synchronization between two different hyperchaotic systems (HS). The analytical conditions for the synchronization of these pairs of different HSs with uncertain Parameters are derived by utilizing Lyapunov function method. Furthermore, synchronization between two different HSs is achieved by utilizing sliding mode control method in a quite short period and both remain in hyperchaotic states. Numerical simulations are used to verify the theoretical analysis using different values of parameter.

Keywords: Hyperchaotic systems, synchronization, sliding mode control.

1 Introduction

Since Pecora and Carroll established a chaos synchronization scheme for two identical chaotic systems with different initial conditions [1], Chaos synchronization has been extensively studied due to its potential application in technological applications [2–6]. A wide variety of approaches have been proposed to achieve synchronization such as active control [7,8], adaptive control [9], linear and nonlinear feedback control [10,11], among many others [12–18]. However, most of them are valid only for the chaotic systems which system parameters are accurately known. But in practical situation, some system's parameters can not be exactly known in advance. These uncertainties will destroy the synchronization and even break it. So, synchronization of hyperchaotic systems in the presence of unknown parameters is very essential.

The adoption of hyperchaotic systems has been proposed for secure communication and the presence of more than one positive Lyapunov exponent clearly improves security of the communication. Spectral analysis also shows that the systems in the hyperchaotic mode has an extremely broad frequency bandwidth of high magnitudes, verifying its unusual random nature and indicating its great potential for some relevant engineering applications (see, e.g. [19,20]).

In this paper, sliding control method is applied to two different hyperchaotic systems. The detailed arrangement is as follows. In Section 2, the synchronization of uncertain hyperchaotic systems is studied by using sliding control method. Numerical simulations are included to verify the results in Section 3. Section 4 draws some conclusions.

A. Xie & X. Huang (Eds.): Advances in Electrical Engineering and Automation, AISC 139, pp. 389–394.
springerlink.com

2 The Design of the Sliding Mode Controller

Sliding mode controller design. Now, consider the nonlinear system described by the dynamics as the master system

$$D(X) = A(X,\theta)X \tag{3}$$

The controller U is added into the slave system, which is given by

$$D(Y) = B(Y,\varphi)Y + U \tag{4}$$

Where $X \in R^n$, $Y \in R^n$, $A,B \in R^{n \times n}$, θ denotes the master system's parameters, φ denotes the slave system's parameters. Thus, the control problem considered is to design an appropriate sliding mode controller U, such that two systems become synchronized, in the sense that $\lim_{t\to\infty}\|Y - X\| = 0$. To achieve the goal, defining the synchronization error as $e=Y-X=(e_1,e_2,\ldots e_n)^T \in R^n$ and subtracting the system (3) from the system (4), the error dynamics is determined by

$$\begin{aligned} D(e) &= B(Y,\varphi)Y + U - A(X,\theta)X \\ &= B(Y,\varphi)Y - B(X,\varphi)X + B(X,\varphi)X - A(X,\theta)X + U \\ &= B(Y,\varphi)Y - B(X,\varphi)X + B(X,\varphi)X - A(X,\tilde{\theta})X + A(X,\tilde{\theta})X - A(X,\theta)X + U \end{aligned} \tag{5}$$

We define the errors between the master and the slave systems as fellow

$$\begin{aligned} &C(X,Y)e = B(Y,\varphi)Y - B(X,\varphi)X, e_\theta = \theta - \tilde{\theta} \\ &E(X)e_\theta = A(X,\tilde{\theta})X - A(X,\theta)X, F(X) = B(X,\varphi)X - A(X,\tilde{\theta})X \end{aligned} \tag{6}$$

$\tilde{\theta}$ denotes the estimated master system's parameters. Using Equations (6), the error dynamics can be rewritten as

$$D(e) = C(X,Y)e + E(X)e_\theta + F(X) + U \tag{7}$$

Sliding surface design. Once a proper switching surface has been chosen, it is followed by choosing a sliding mode controller to drive all system trajectories onto the sliding surface. Here the sliding surface can be defined as follows

$$S(e,e_\theta) = [e,e_\theta] \tag{8}$$

Using (7) and (8), the control input is determined as

$$U = -C(X,Y)e - E(X)e_\theta - F(X) - K_1 e - K_2 e_\theta \tag{9}$$

Control Stability analysis. The error dynamics can be obtained using (9) in (7)

$$D(e,e_\theta) = [-K_1, -K_2][e,e_\theta]^T \tag{10}$$

Choose the following Lyapunov function candidate as follows:

$$V = \frac{1}{2}(e^2 + e_\theta^2) \tag{11}$$

Therefore, the differential of the Lyapunov function candidate along the trajectories of synchronization error system (10) is

$$D(V) = -(e\dot{e} + e_\theta \dot{e}_\theta) \tag{12}$$

Obviously the matrix [K_1,K_2] is positive semi-definite, D(V) is semi-negative definite. On the basis of Lyapunov stability theory, synchronization error system (10) asymptotically stable, hyperchaotic system (3) and the hyperchaotic system (4) implementation of the synchronization. When in sliding mode, the controlled error system satisfies the following conditions

$$S(e, e_\theta) = 0, D(e, e_\theta) = 0 \tag{13}$$

Once reaching the sliding surface, the system trajectories will remain on the surface.

3 Using Sliding Mode to Synchronize Two Different Hyperchaotic Systems with Uncertain Parameter

In 1976, Rössler put forward the concept of super-chaos, and construct a hyperchaotic system. On the example of Rössler hyperchaotic system, in 2009 Zhou Ping has established a simple hyperchaotic system with only one nonlinear term [21]. We will use sliding mode to synchronize the two hyperchaotic systems with uncertain Parameter.

$$\begin{cases} \dot{x}_1 = \alpha x_1 - 1.2x_2 \\ \dot{x}_2 = x_1 - 0.1x_2 x_3^2 \\ \dot{x}_3 = -x_2 - 1.2x_3 - 5x_4 \\ \dot{x}_4 = x_3 + 0.8x_4 \end{cases} \tag{14}$$

Zhou Ping studied the system and gave out when $0.45<\alpha\leq 0.59$ the system is chaotic, when $0.59<\alpha\leq 0.69$ the system is hyperchaotic(which has two positive and one negative Lyapunov exponents. Fig. 1(a) shows an example). To the system as the drive system, response system is Rossler hyperchaotic system (Fig. 1(b) shows an example):

$$\begin{cases} \dot{y}_1 = -y_2 - y_3 + u_1 \\ \dot{y}_2 = y_1 + ay_2 + y_4 + u_2 \\ \dot{y}_3 = b + y_1 y_3 + u_3 \\ \dot{y}_4 = -cy_3 + dy_4 + u_4 \end{cases} \tag{15}$$

The synchronization error is

$$\begin{cases} \dot{e}_1 = -e_2 - e_3 - x_2 - x_3 - \tilde{\alpha}x_1 + 1.2x_2 + (\tilde{\alpha} - \alpha)x_1 + u_1 \\ \dot{e}_2 = e_1 + ae_3 + e_4 + ax_2 + x_4 + 0.1x_2x_3^2 + u_2 \\ \dot{e}_3 = y_1y_3 + b + x_2 + 1.2x_3 + 5x_4 + u_3 \\ \dot{e}_4 = -ce_3 + de_4 - (c+1)x_3 + (d-0.8)x_4 + u_4 \\ \dot{e}_\theta = -k_\theta(\tilde{\alpha} - \alpha) \end{cases} \tag{16}$$

Select

$$U = \begin{bmatrix} x_2 + x_3 + \tilde{\alpha}x_1 - 1.2x_2 - (\tilde{\alpha} - \alpha)x_1 + k_1e_1 \\ -e_1 - ae_3 - e_4 - ax_2 - x_4 - 0.1x_2x_3^2 + k_2e_2 \\ -y_1y_3 - b - x_2 - 1.2x_3 - 5x_4 + k_3e_3 \\ ce_3 - de_4 + (c+1)x_3 - (d-0.8)x_4 + k_4e_4 \end{bmatrix} \tag{17}$$

So, chooses $[k_1,k_2,k_3,k_4,k_\theta]$ positive semi-definite, D(V) is negative semi-definite. Therefore, on the basis of Lyapunov stability theory, the response and drive systems are asymptotically synchronized.

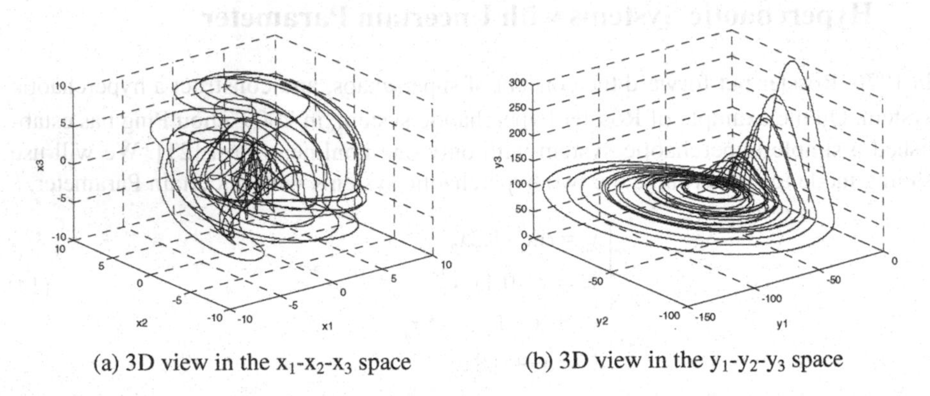

(a) 3D view in the x_1-x_2-x_3 space (b) 3D view in the y_1-y_2-y_3 space

Fig. 1. (a) The orbit of system (14) with parameter α=0.67. (b) The orbit of system (15) with parameter a=0.25,b=3,c=0.5,d=0.05. The initial conditions are set to be $x_1(0)=-6$, $x_2(0)=-2$, $x_3(0)=3$, $x_4(0)=2$ and $y_1(0)=-6$, $y_2(0)=-2$, $y_3(0)=3$, $y_4(0)=2$.

3.1 Numerical Results

In order to verify the performances of the proposed method, some numerical simulations are presented in this section. Based on the fourth-order Runge-Kutta method, the system of differential equations is solved with time step size equal to 0.001 in all numerical simulations. The results of the numerical simulations show the trajectories of drive system and response system diverge from each other at the first stage. However, as soon as the controllers are applied at time t=60, the synchronization of all variables is achieved, respectively.

We choose the value of the feedback gains to be $k_1=0$, $k_2=-6$, $k_3=-1$, k4=-1,$k_\theta=-1$.

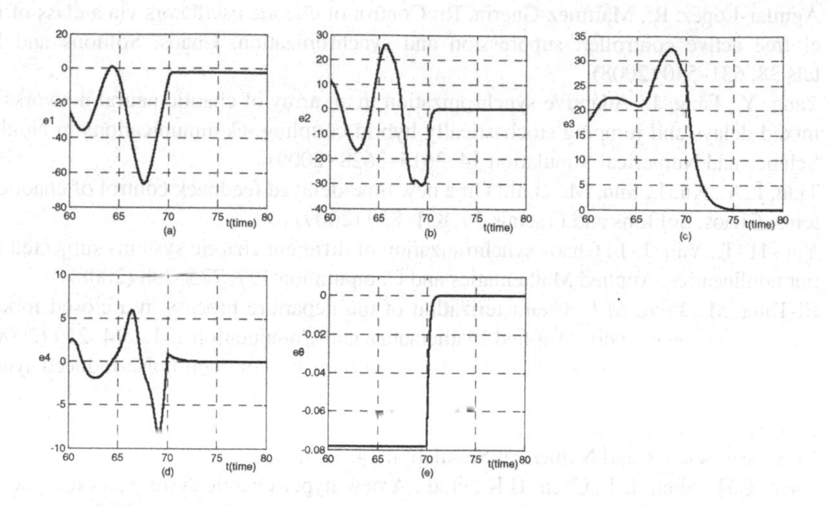

Fig. 2. The time histories of synchronization errors with sliding mode control. The orbit of system (16) with parameter α=0.67, $\tilde{a}$=0.598, so the time response of e_θ as (e).

4 Conclusions

In this paper, chaos synchronization between two different HSs has been studied using control sliding mode technique in a quite short period. The new HS has been used to drive the Rossler HS. Numerical simulations have been carried out using different initial values to show the effectiveness of the proposed synchronization techniques.

References

1. Pecora, L.M., Carroll, T.L.: Synchronization in chaotic systems. Physical Review Letters 64, 821–824 (1990)
2. Nijmeijer, H., Mareels, I.M.Y.: An observer looks at synchronization. IEEE Transactions on Circuits and Systems I 44, 882–890 (1997)
3. Sira-Ramírez, H., Cruz-Hernández, C.: Synchronization of chaotic systems: a generalized Hamiltonian systems approach. International Journal of Bifurcation and Chaos 11, 1381–1395 (2001)
4. Posadas-Castillo, C., Cruz-Hernández, C., López-Gutiérrez, R.M.: Synchronization of chaotic neural networks with delay in irregular networks. Applied Mathematics and Computation. 205, 487–496 (2008)
5. Li, R.-H.: A special full-state hybrid projective synchronization in symmetrical chaotic systems. Applied Mathematics and Computation 200, 321–329 (2008)
6. Boukabou, A.: On the control and synchronization design for autonomous chaotic systems. Nonlinear Dynamics and Systems Theory 8, 151–167 (2008)
7. Naseh, M.R., Haeri, M.: Robustness and robust stability of the active sliding mode synchronization. Chaos, Solitons and Fractals 39, 196–203 (2009)

8. Aguilar-López, R., Martínez-Guerra, R.: Control of chaotic oscillators via a class of model free active controller: suppression and synchronization. Chaos, Solitons and Fractals 38, 531–540 (2008)
9. Tang, Y., Fang, J.: Adaptive synchronization in an array of chaotic neural networks with mixed delays and jumping stochastically hybrid coupling. Communications in Nonlinear Science and Numerical Simulation 14, 3615–3628 (2009)
10. Tian, L.X., Xu, J., Sun, M., et al.: On a new time-delayed feedback control of chaotic systems. Chaos, Solitons and Fractals 39, 831–839 (2009)
11. Yau, H.-T., Yan, J.-J.: Chaos synchronization of different chaotic systems subjected to input nonlinearity. Applied Mathematics and Computation 197, 775–788 (2008)
12. El-Taha, M., Jafar, M.J.: Characterization of the departure process in a closed fork–join synchronization network. Applied Mathematics and Computation 181, 214–219 (2006)
13. Yoshimura, K., Muramatsu, J., Davis, P.: Conditions for common-noise-induced synchronization in time-delay systems. Physica D: Nonlinear Phenomena 237, 3146–3152 (2008)
14. Lou Albert, C.J.: A theory for synchronization of dynamical systems. Communications in Nonlinear Science and Numerical Simulation 14, 1901–1951 (2009)
15. Chen, C.H., Sheu, L.J., Chen, H.K., et al.: A new hyper-chaotic system and its synchronization. Nonlinear Analysis: Real World Applications 10, 2088–2096 (2009)
16. El-Dessoky, M.M.: Synchronization and anti-synchronization of a hyperchaotic Chen system. Chaos, Solitons and Fractals 39, 1790–1797 (2009)
17. Pototsky, A., Janson, N.: Synchronization of a large number of continuous one-dimensional stochastic elements with time-delayed mean-field coupling. Physica D: Nonlinear Phenomena 238, 175–183 (2009)
18. Laoye, J.A., Vincent, U.E., Kareem, S.O.: Chaos control of 4D chaotic systems using recursive backstepping nonlinear controller. Chaos, Solitons and Fractals 39, 356–362 (2009)
19. Grassi, G.: Observer-based hyperchaos synchronization in cascaded discrete-time systems. Chaos, Solitons and Fractals 40, 1029–1039 (2009)
20. Aguilar-Bustos, A.Y., Cruz-Hernández, C.: Synchronization of discrete-time hyperchaotic systems: an application in communications. Chaos, Solitons and Fractals 41, 1301–1310 (2009)
21. Zhou, P., Wei, L., Cheng, X.: A hyperchao s system with only one nonlinear term. Acta Physica Sinica 58, 5201–5208 (2009)

MESA Finger: A Multisensory Electronic Self-Adaptive Unit for Humanoid Robotic Hands

Bowen Li[1], Jiangxia Shi[2], and Wenzeng Zhang[3]

[1] Dept. of Automation, Tsinghua University, Beijing, 100084, China
[2] Dept. of Automotive Engineering, Beihang University, Beijing, 100191, China
[3] Dept. of Mechanical Engineering, Tsinghua University, Beijing, 100084, China
lbw09@mails.tsinghua.edu.cn, shijiangxia@ae.buaa.edu.cn, wenzeng@tsinghua.edu.cn

Abstract. The current under-actuated hands with self-adaptation are complex, cumbersome and their grasping forces on objects cannot be adjusted. This paper proposed a novel idea of robotic finger, multisensory electronic self-adaptive (MESA) finger. The MESA finger can self-adaptively grasp objects by multisensory feedback control, instead of under-actuated mechanisms. The MESA finger can also be called electronic under-actuated (EUA) finger. A typical MESA finger with one joint was designed in detail, where there are three sensors and a control module. The finger makes use of signals from sensors about the relative location between the finger and the object to fulfill self-adaptive grasp automatically, and the grasping force can be adjusted easily. The MESA finger could be used to construct multi-fingered robotic hands with high degrees of freedom, high self-adaptation, and low control requirement.

Keywords: Humanoid robot, Dexterous hand, Multisensory electronic self-adaptive grasp, Electronic under-actuated finger, Multisensory feedback control.

1 Introduction

Humanoid robot is a hot spot of robot research, and the research of robotic hands is one of the most important subjects. The research of robotic hands can be mainly divided into two branches: dexterous hands and under-actuated hands. A dexterous hand has more than 3 fingers and more than 9 DOF, which has lots of sensors and a controller to receive signals from sensors and thus drive multiple motors (active joints) according to a complex control algorithm. The control program, however, is always difficult to design and the cost is quite high. For example Utah/MIT Dexterous Hand [1], Gifu Hand II [2], High-speed hand [3,4] belong to dexterous hands.

Under-actuated hands are carefully designed in mechanical structures and they can fulfill self-adaptive grasping and releasing without sensors but they are always complex, cumbersome and their grasping forces to objects cannot be adjusted. For example Thierry Laliberte's under-actuated mechanical hands [5], the prosthetic hand by HIT [6], the anthropomorphic hand [7] and TH series hands of Tsinghua Univ. [8-10] are among this branch. By reviewing the above-mentioned hands, it is clear that a new

A. Xie & X. Huang (Eds.): Advances in Electrical Engineering and Automation, AISC 139, pp. 395–400.
springerlink.com

type of robotic hand is needed which has low control difficulty and the grasping force can be adjusted.

This paper puts forward a novel idea of robotic finger, multisensory electronic self-adaptive (MESA) finger. The MESA finger can self-adaptively grasp objects by multisensory feedback control, instead of under-actuated mechanisms. The MESA finger can also be called electronic under-actuated (EUA) finger.

2 Design of the MESA Finger

2.1 Mechanical Design

A MESA finger with one MESA joint is designed, shown in Fig. 1. The finger contains two segments: the first and second segments. The motor is embedded in the first segment and connected with the joint shaft with two gears to drive the joint. There are three kinds of sensors placed in the finger.

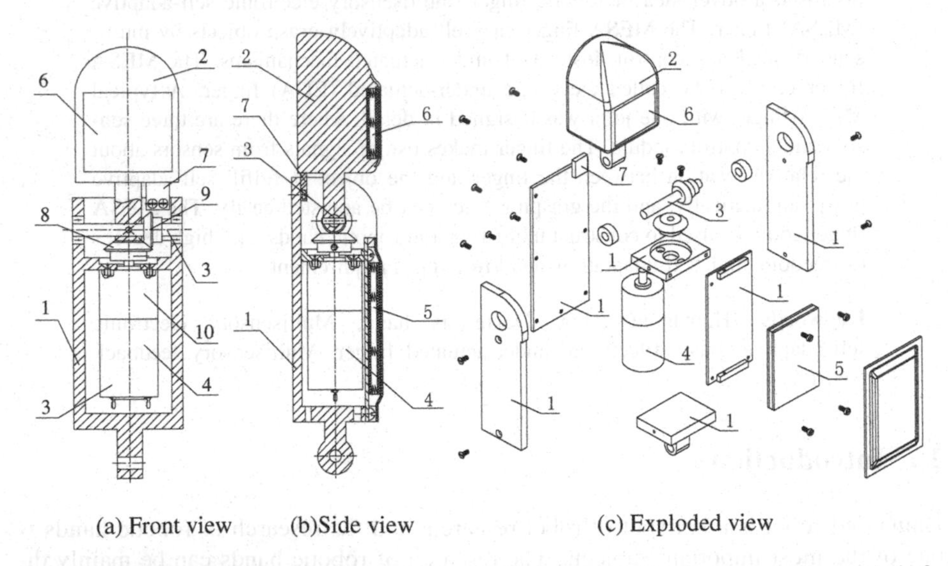

Fig. 1. Mechanical design of a MESA finger. 1-The 1st seg, 2-The 2nd seg, 3-The joint shaft, 4-Motor, 5-TG sensor, 6-SR sensor, 7-SG sensor, 8-The 1st Gear, 9-The 2nd Gear, 10-Reducer.

The first sensor is the Trigger (TG) Sensor, which is placed on the grasping surface of the first segment. TG sensor detects if the grasped object has touched the first segment, and produce a Trigger signal to start the bending of the finger.

The second sensor is the Stop Grasping (SG) Sensor, which is place on the grasping surface of the second segment. SG sensor detects if the grasped object has touched the second segment, if any object touched the second segment, the finger has bended by a appropriate angle to grasp the object, and a SG signal is produced.

The third sensor is the Stop Releasing (SR) Sensor, which is placed on the back board of the first segment, on the joint or on the circuit board. SR sensor detects if the finger has been extended fully and then produces a SR signal.

The MESA finger can be connected serially to multi-joint MESA fingers. A robotic hand can be constructed with multiple MESA fingers.

2.2 Control Algorithm Design

The electronic components of MESA finger are shown in Fig. 2. The finger contains three kinds of sensors, a control module and a motor driving circuit. TG sensors and SG sensors should be more than one and be placed in array in order to reduce noises which can cause wrong actions of the finger. Displacement sensors, pressure sensors, and torque sensors can be used as the TG, SG and SR sensors. In addition, current probes can also be used as SR sensors if we set it on board. In Fig. 2, adjustable capacitance pressure transducers are used as SG and TG sensors, and a switch is used as a SR sensor. The control module connects with all sensors, a motor driving circuit and an active input through which we can send reset signal to unbend the finger.

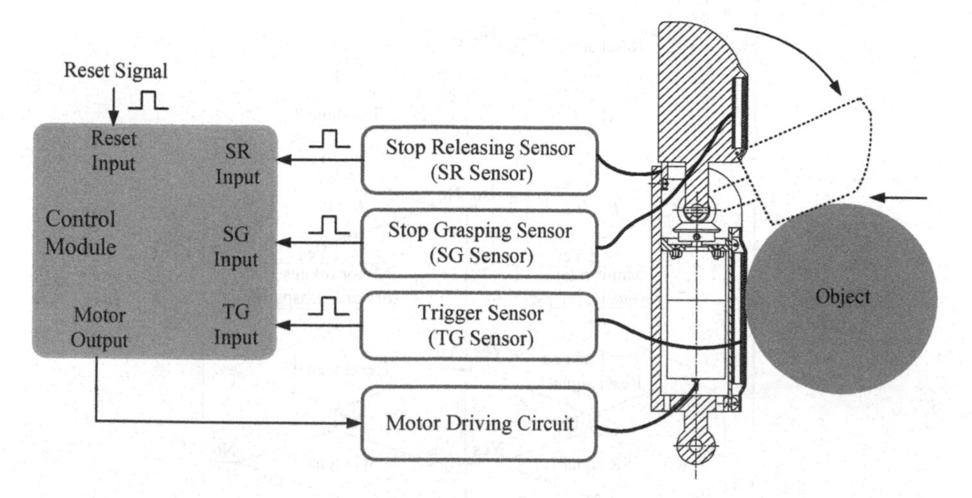

Fig. 2. Electronic components of the MESA finger

The control algorithm of MESA finger is shown in Fig. 3. The control algorithm includes following steps:

(a) Set R as the reset mark bit, set B as the stop releasing mark bit, set F as the stop grasping mark bit; when the algorithm starts, set R=0;
(b) If a SR signal is received then set B=1; otherwise set B=0;
(c) If a SG signal is received then set F=1; otherwise set F=0;
(d) If a Reset signal is received then set R=1 and go to (f); otherwise go to (e);
(e) If a TG signal is received then go to (j); otherwise go to (f);
(f) If B=1 then go to (n); otherwise go to (g);
(g) Drive the motor to rotate backward in a fixed short period of time Δt, go to (h);
(h) If a Reset signal is received then go to (i); otherwise set R=1 and go to (n);
(i) If a SR signal is received then go to (n); otherwise go to (g);
(j) If F=1 then go to (n); otherwise go to (k);
(k) Drive the motor to rotate forward in a fixed short period of time Δt, go to (l);

(l) If a Reset signal is received then go to (n); otherwise go to (m);
(m) If a SG signal is received then go to (n); otherwise go to (k);
(n) Stop the motor, go to (a).

The grasping process of the MESA finger is shown in Fig. 4.

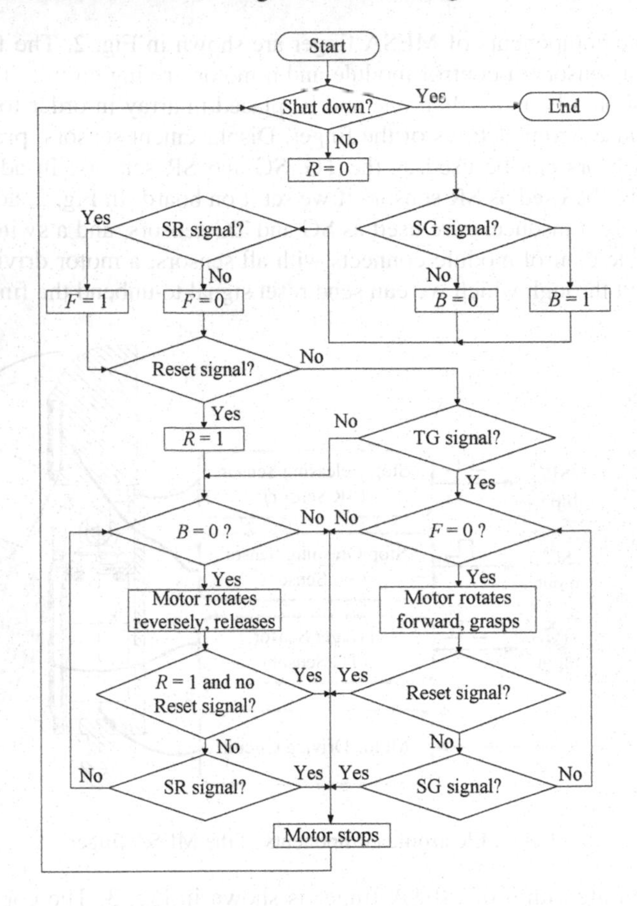

Fig. 3. Flow chart of control algorithm of the MESA finger

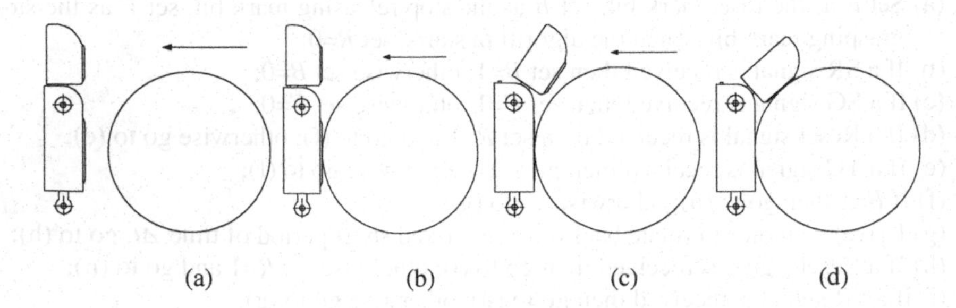

Fig. 4. Grasping process of the MESA finger

3 Grasp Analysis of the MESA Finger

MESA fingers can grasp objects self-adaptively. The grasp space of MESA fingers is the key whether this new grasp pattern will work. The analysis is deduced based on the situation shown in Fig. 5 with relating geometric parameters. In order to simplify the question, grasping a spherical object is considered. In Fig. 5,

H: the distance between the joint shaft and the bottom of the 1st segment, mm
D: the distance between the joint shaft and the grasping surface of the finger, mm
Y: the distance between the object's center and the bottom of the 1st segment, mm
R: the radius of the spherical object, mm
α: the tilting angle of the 2nd segment, °

One has

$$\alpha = 180^\circ - 2\arctan\frac{D+R}{H-Y}\,. \tag{1}$$

Because of the limit of the mechanical structure, α must be not greater than 90°, one can conclude:

$$D - H + R + Y \geq 0\,. \tag{2}$$

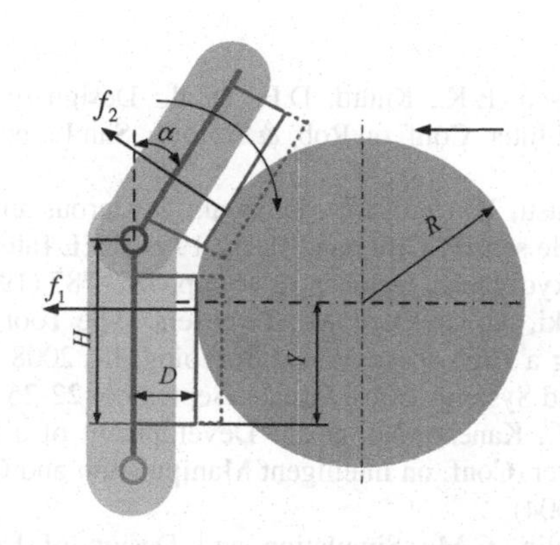

Fig. 5. Grasp space analysis sketch of the MESA fingers

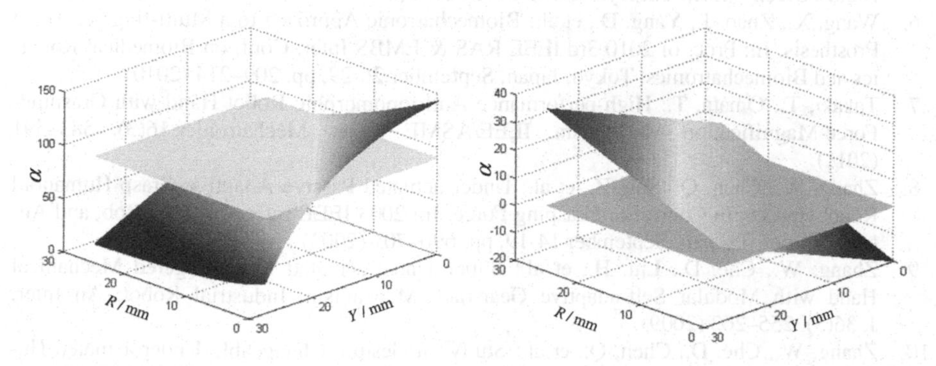

Fig. 6. The relationship among α, R and Y **Fig. 7.** The relationship among α, R and Y

The surface underneath the yellow plane in Fig. 6, and the surface above the blue plane in Fig. 7 represent situations that the finger can grasp the object self-adaptively. This analysis demonstrates that if the object grasped is proper, the object touches the finger in the middle of the 1st segment, the finger can self-adaptively grasp successfully.

4 Conclusions

This paper proposed a novel idea of robotic finger, multisensory electronic self-adaptive (MESA) finger. Theory and simulation analysis show that the finger collects information with the sensors and can fulfill self-adaptive grasping and holding according to the information collected automatically; its structure is simple, the transmission chain is short, the controlling algorithm is simply constructed, and these advantages make it a robotic finger with high reliability and wide applications. The MESA finger could be used to construct multi-fingered robotic hands with high degrees of freedom, high self-adaptation, and low control requirement.

Acknowledgments. The work was funded by NSF of China (No.50905093).

References

1. Jacobsen, S.C., Iversen, E.K., Knutti, D.F., et al.: Design of the Utah/MIT Dexterous Hand. In: 2001 IEEE Inter. Conf. on Rob. & Autom., San Francisco, CA, USA, pp. 1520–1532 (1986)
2. Kawasaki, H., Komatsu, T., Uchiyama, K., et al.: Dexterous anthropomorphic robot hand with distributed tactile sensor: Gifu hand II. In: 1999 IEEE Inter. Conf. on Systems, Man, and Cybernetics, Tokyo, Japan, October 12-15, pp. 782–787 (1999)
3. Mizusawa, S., Namiki, A., Ishikawa, M.: Tweezers Type Tool Manipulation by a Multi-fingered Hand Using a High-speed Visual Servoing. In: 2008 IEEE/RSJ Inter. Conf. on Intelligent Robots and Systems, Nice, France, September 22-26, pp. 2709–2714 (2008)
4. Namiki, A., Imai, Y., Kaneko, M., et al.: Development of a High-speed Multifingered Hand System. In: Inter. Conf. on Intelligent Manipulation and Grasping, Genoa, Italy, July 1-2, pp. 85–90 (2004)
5. Laliberte, T., Gosselin, C.M.: Simulation and Design of Under-actuated Mechanical Hands. Mech. Mach. Theory 33(1/2), 39–57 (1998)
6. Wang, X., Zhao, J., Yang, D., et al.: Biomechatronic Approach to a Multi-fingered Hand Prosthesis. In: Proc. of 2010 3rd IEEE RAS & EMBS Inter. Conf. on Biomedical Robotics and Biomechatronics, Tokyo, Japan, September 26-29, pp. 209–214 (2010)
7. Takaki, T., Omata, T.: High-performance Anthropomorphic Robot Hand with Grasping-Force-Magnification Mechanism. IEEE/ASME Trans. Mechatronics 16(3), 583–591 (2011)
8. Zhang, W., Chen, Q., Sun, Z., et al.: Under-actuated Passive Adaptive Grasp Humanoid Robot Hand with Control of Grasping Force. In: 2003 IEEE Inter. Conf. on Rob. and Autom., Taipei, Taiwan, September 14-19, pp. 696–701 (2003)
9. Zhang, W., Che, D., Liu, H., et al.: Super Under-actuated Multi-fingered Mechanical Hand with Modular Self-adaptive Gear-rack Mechanism. Industrial Robot: An Inter. J. 36(3), 255–262 (2009)
10. Zhang, W., Che, D., Chen, Q., et al.: Study on Gesture-Changeable Under-actuated Humanoid Robotic Finger. Chinese J. of Mechanical Engineering 23(2), 142–148 (2010)

The Applications in Channel Assignment Based on Cooperative Hybrid Artificial Bee Colony Algorithm

JunXia Liu[1], ZhenHong Jia[1], XiZhong Qin[1], Chun Chang[2], GuoJun Xu[2], and XiaoYan Xia[2]

[1] College of Information Science and Engineering, Xin jiang University, Urumuqi, 830046, P.R. China
[2] Xinjiang Mobile Communication Company, Urumqi, Xinjiang 830091, P.R. China

Abstract. The frequency resources of mobile communication network are limited. In order to improve utilization of wireless resource, we used the cooperative hybrid artificial bee colony algorithm to solve the wireless channel assignment problem. The proposed algorithm used the dynamic step to balance local and global search capability, the dynamic step was cosine rule with the increasing times of iteration; The artificial bee colony were divided into two subgroups with different evolutionary strategy co-evolution, the algorithm was easy to jump out of local optimal solution in this way; The introduction of the single-individual selective mutation increased diversity of population and the speed of convergence. Simulation results show that: the proposed algorithm can be better to solve the wireless channel assignment.

Keywords: mobile communication network, channel assignment, double artificial bee colony algorithm, selective mutation.

1 Introduction

Mobile communication system is facing the contradiction between rapid growth of mobile users and limited frequency resources, in order to improve utilization of limited frequency resource , usually we use different algorithm for optimal channel assignment scheme. The neural network algorithm, simulated annealing (SA), genetic algorithm (GA) are common channel allocation algorithm[1]-[2].But there is still a long time of convergence and it is easy to fall into local optimal solution .

In order to solve those problems, we propose the cooperative hybrid artificial bee colony (CHABC) algorithm for channel assignment based on artificial bee colony (ABC) algorithm. The ABC algorithm is an optimization algorithm based on a particular intelligent behavior of bee colony, it has the advantage of simple algorithm, easy implementation, fast convergence, adaptability, etc [3].

A. Xie & X. Huang (Eds.): Advances in Electrical Engineering and Automation, AISC 139, pp. 401–406.
springerlink.com

2 Channel Allocation Model

Channel assignment problem is frequency assignment problem, the goal is under no violating the interference constraints that all cells can be assigned to the required number of frequency. Usually consider only the three main interference Constraints: co-channel interference (CCC), adjacent channel interference (ACC), co-site interference (CSC). we use an $N \times N$ dimensional compatible matrix C to represent the main interference constraints (N is the number of the cell of cellular systems), the diagonal elements C_{ii} of matrix C represent minimal interval between channels assigned to the cell i, the non-diagonal elements C_{ij} ($i \neq j$) of matrix C represent minimal channel interval between the channels assigned to the cell i and the cell j. Vector $R=[r_i]$ $(i=1,2,\ldots,N)$ represents the number of frequency requirement of each cell, the element r_i of the vector R is the number of frequency requirement of the cell i. The set of available frequency is $FN=\{1,2,\ldots,FNum\}, FNum=max\{f_{ik}\}$, let f_{ik} represent that the frequency point of the k-th position is assigned to the cell i, f_{ik} is a positive integer, f_{jl} is similar. They should meet $|f_{ik}-f_{jl}| \geqslant C_{ij}$, $(1 \leqslant i, j \leqslant N, 1 \leqslant k, 1 \leqslant r_i.)$ The fitness $S(F)$ of the objective function is defined as the total number of constraints violated, F_{ikjl} describes whether f_{ik} and f_{jl} meet the constraints or not, the mathematical model as follows[4]:

$$S(F) = \sum_{i=1}^{N} \sum_{k=1}^{ri} \sum_{j=1}^{N} \sum_{l=1}^{rj} F_{ikjl} \tag{1}$$

$$F_{ikjl} = \begin{cases} 0, others \\ 1, | f_{ik} - f_{jl} | < C_{ij} \end{cases} \tag{2}$$

known C, R and available frequency set FN to find the minimal value of the objective function, that is channel assignment scheme F of $S(F)=0$.

3 Cooperative Hybrid Artificial Bee Colony Algorithm

3.1 Adjustment Neighborhood Search Step

In ABC algorithm, Assume that the number of bee colony is M, at the t-th iteration the existing set of nectar source$\{\theta i(t)\}, i=1, 2, \cdots M$, the preference of a food source by an onlooker bee depends on the nectar amount $H(\theta i(t))$ of that food source. the probability with the food source located at $\theta i(t)$ will be chosen by a onlooker bee can be expressed as formula (3):

$$Pi = H(\theta i(t)) / \sum_{k=1}^{M} H(\theta_k(t)) \tag{3}$$

After that,,employed bees search for new nectar source by formula (4) in the neighborhood of selected area[5].

$$\theta i(t+1) = \theta i(t) \pm \psi i \tag{4}$$

this paper search by formula (5), dynamic step with Cosine law replace randomly generated dynamic step ψi:

$$\theta i(t+1) = \theta i(t) + \cos(t/t_{max})(\theta_{g1} - \theta i(t)) \tag{5}$$

where t_{max} is the maximal times of iteration, θ_{g1} is the best nectar location of the subgroup N_1 in the t-th iteration.

3.2 Double-Artificial Bee Colony Algorithm

According to the literature [6], the artificial bees colony are divided into two subgroups N_1、 N_2 . they have different optimized strategy.

The bees of subgroup N_1 use formula (5) for neighbor search. Bees choose the larger the nectar amount of the food source located $\theta i(t)$ and $\theta i(t+1)$ to seek honey.

The bees of subgroup N_2 adopt updating the local worst solution to update subgroup, the specific method as followed: randomly selected *nn* nectar sources by probability P_j, such as formula (6) as shown [7]:

$$P_j = 2(N_2+1-j)/[N_2(N_2+1)],\ 1 \leqslant j \leqslant N_2 \tag{6}$$

where P_j represents the selected probability of the j-th nectar source. According to nectar amount of the food source we calculate local optimal solution θ_{mg}. and local worst solution W of the *nn* nectar sources. In the channel assignment scheme, the each row of solution matrix C represents a cell, it can be randomly selected i cell (i is a random number, $1 \leqslant i \leqslant N$) to replace corresponding part of the worst individual W by local optimal solution θ_{mg}.If the nectar amount of *NEWw* is little than ones of W ,then give up individual *NEWw* updating nectar scource. After then, we calculate global optimal solution θ_{g2} of subgroup N_2, randomly select i cell in the global optimal solution θ_{g2} of subgroup N_2 to replace corresponding part of W.

3.3 The New Selective Mutation Technique

In order to improve the diversity of population and speed of convergence, the selective mutation technique is introduced.

Selective mutation technique is a single individual evolutionary approach, firstly we calculate the value of selected individual fitness, that is, the value of formula (1). If $S(F)>0$, then walk through each frequency point f_{ij} of the solution F, to examine whether the frequency point f_{ij} and other frequency point allocated meet the interference constraints by formula (2) or not. If the frequency point meets the constraints, then do nothing; if it is not satisfied with constraints, then the frequency point replaced the frequency point f_{ij} must satisfy CSC, that is C_{ii}, it must also meet ACC and CCC with the frist *pcell* cells. They are difficult to assign frequency point for these cells. The formula for the degree deg_i of assignment difficulty and *pcell* are formula (7) and formula (8) respectively [8].

$$\deg_i = \left(\sum_{j=1}^{N} r_i c_{ij} \right) - c_{ii}, 1 \le i \le N \tag{7}$$

$$PCell = \begin{cases} [N * \mu], S(F) > MinV * \varepsilon \\ [N * \lambda], S(F) \le MinV * \varepsilon \\ 3, 2 < S(F) < MinV * \varepsilon, \Omega \\ 1, S(F) = 2, \Omega \\ 1, S(F) = 1 \end{cases} \tag{8}$$

Pcell is a dynamic variable controlled by fitness function ,Where '[ ]'denotes rounding, *N* is the total number of cell , $\mu \in (0, 1)$, $\varepsilon \in (0, 1)$, $\lambda \in (0, 1)$, *MinV* is the minimal fitness value of this iteration, Ω Indicates that when the algorithm loops to *m* multiple generations ,it find that the value of fitness function does not change with continuous *l*-generation.

4 Channel Assignment Algorithm and the Steps

4.1 The Relationships between Behavior of Honey Bees and the Corresponding Channel Assignment Problem

Nectar source located θi represents the feasible frequency assignment scheme *Fi*; the nectar amount $H(\theta i(t))$ of food source corresponds to the quality of feasible frequency assignment scheme, specifically the relationship shown in Equation (9); the largest nectar amount of food source corresponds to the best channel assignment scheme; the speed of search best nectar source represents the speed of finding the best channel assignment scheme, that is algorithmic Convergence speed.

$$H(\theta i(t)=1/(1+S(Fi)) \tag{9}$$

4.2 The Steps of Channel Assignment Algorithm

Step 1: The parameters needed by our approach are determined: Population numbe *M*, t_{max}, N_1, N_2, *FNum*,*m* , *l*, μ, ε,λ,*limit*, *nn*.

Step 2: Initialize *M* feasible solutions, calculate *S(F)*, $H(\theta i(t))$, θ_{max}. If *S(F)=0*, then the algorithm ends. Otherwise , go to next step.

Step 3: The two subgroups adopt co-evolution of different ways to update the nectar source. The specific way see section 4.2.

Step 4: corresponding nectar source of all feasible solutions adopt selective mutation of section 4.3. Calculate the nectar amount $H(\theta i(t)^{\sim}$ of mutated the nectar source $\theta i(t)^{\sim}$. If $H(\theta i(t))^{\sim}$ is larger than previous ones, then accept this mutation. Otherwise follow the nectar source of no mutation.

Step 5: If some nectar source do not improve after the *limit* cycle, then give up this nectar source, meanwhile the corresponding employed bee become into a scout to randomly search for a new nectar source.
Step 6: if $H(\theta i(t))$ >H(θ_{max}),then update θ_{max}, that is, $\theta_{max} = \theta i(t)$.
Step 7: Calculate $S(F)$ of the all nectar source corresponding feasible solution, if $S(F)=0$, then the algorithm ends, otherwise jump to Step 3.
Step 8: If the algorithm has reached t_{max} iteration, it can't find $S(F)=0$, then the algorithm ends ,and we think it does not converge.

5 Algorithm Simulation

Algorithm with MATLAB7.0 simulating, population size M=40,employed bees number=onlookers number $=M, N_1=N_2=20$, $t_{max}=500$, $limit=10$, $m=3$, $l=6$, $\mu=0.6$, $\varepsilon=0.8$, $\lambda=0.75$, $nn=12$. In this paper, a group of classical benchmark problem selected from literature [1] are done 50 times simulation respectively, and we calculated the average generation of convergence (AGC)and rate of convergence (RC), the simulation results and comparing results show in Table 1.

Table 1. Comparison of simulation results

	Literature [9]		literature[10]		CHABC algorithm	
Problem	RC (%)	AGC	RC (%)	AGC	RC (%)	AGC
1	100	0.0	100	0.0	100	0.0
2	100	2284.6	100	0.0	100	4.0
3	100	34.7	100	42.4	100	2.8
4	100	12.0	100	0.0	100	1.6
5	100	136.5	100	28.1	100	5.4
6	100	20.2	100	0.0	100	3.2
7	98	1977.0	85	85.5	100	14.0

Literature [9] combines genetic algorithm with the minimal cost function , then the literature [10] adopts the methods with genetic algorithm of adaptive crossover probability combining the minimal cost function. As can be seen from Table 1: proposed algorithm to solve the problem 2,4,6 has bigger generations of convergence comparing with literature [10], while among the other remaining problem the proposed algorithm not only guarantees 100 % rate of convergence, but also much litter average generation of convergence than literature [10].On the rate of convergence and average generation of convergence compare with the literature [9], superiority of the algorithm are obvious.

6 Conclusion

ABC algorithm is improved in this paper. We proposed CHABC algorithm based on it and apply the proposed algorithm for fixed channel assignment. Simulation results show that the proposed algorithm can be better to solve the wireless channel assignment problem, the rate of convergence and the speed of convergence of the algorithm is excellent comparing with GA algorithm.

Acknowledgement. The authors would like to acknowledge the financial support from the Xinjiang Mobile Communication Company.

References

1. Funabiki, N., Takefuji, Y.: A neural network parallel algorithm for channel assignment problems in cellular radio networks. IEEE Trans. Veh. Technol. 41(4), 430–437 (1992)
2. Ngo, C.Y., Li, V.O.K.: Fixed channel assignment in cellular radio networks using a modified genetic algorithm. IEEE Trans. Veh. Technol. 47(1), 163–172 (1998)
3. Karaboga, D.: An idea based on honey bee swarm for numerical optimization. Technical Repert-TR06. Erciyes University, Turkey (2005)
4. Aardal, K.I., Hoesel, S.P.M.V., Koster, A.M.C.A., et al.: Models and solution for Frequency assignment problems. Springer, Berlin (2001)
5. Karaboga, D., Basturk, B.: On the performance of artificial bee colony (ABC) algorithm. Applied Soft Computing 8(1), 687–697 (2008)
6. Sierra, M.R., Coello, C.A.: Improving PSO-based multi-objective optimization using crowding, mutation and e-dominance. In: Proc. of the 3rd Int'l Conf. on Evolutionary Multi-criterion, Optimization, Mexico, pp. 505–519 (2005)
7. Han, Y., Cai, J.-H., Zhou, G.-G., et al.: Advances in Shuffled Frog Leaping Algorithm. Computer Science 37(7), 16–18 (2010)
8. Seyed, A.G.S., Hamidreza, A.: A hybrid method for channel assignment problems in cellular radio networks. In: Proceeding of IEEE WCNC, USA (2006)
9. Li, M.-L., Wang, Y.-N., Du, L., et al.: Research on Fixed Channel Assignment Method in Cellular Systems. Mini-micro Systems 25(8), 1420–1421 (2004)
10. Zhong, X.-Y., Jin, M., Zhong, X.-Q., et al.: Channel Assignment in Cellular Network Based on Self-adaptive Genetic Algorithm. Computer Engineering 36(17), 189–191 (2010)

Design and Implementation of the Improved Control Algorithm for Switching Capacitor Banks

Tao Yuan[1], Qinggaung Yu[2], and Zhengping Wu[1]

[1] Electrical Engineering & Renewable Energy School, China Three Gorges University, Yichang, 443002, China
[2] Department of Electrical Engineering, Tsinghua University, Beijing, 100084, China
YT1986YT@163.com

Abstract. This paper proposed three kinds improved control algorithms for switching capacitors banks, it can solve the traditional algorithms' repeated switching and other problems effectively. Algorithm.1 is sequence and cycle Var compensating of queue model; Algorithm.2 is a way of binary data's addition and subtraction for Var compensation. Algorithm.3 is according to the way of combining the 8421 BCD code to switch the capacitors. This paper presented the principle and method of the switch control algorithm and designed a automatic var compensation device based on the PLC controller. The experimental results indicate that the programs of PLC according to the three algorithms running more accurate and reliable than the traditional algorithms.these algorithms can be used in many kind of var compensation to compensate the reactive power or power factor.

Keywords: Sequence and cycle of queue model, the way of combination the 8421 BCD code for compensation, binary data's addition and subtraction for compensation.

1 Introduction

At the present, with the development of the reactive power compensation technology that many methods of var compensation had appeared in our country. The mostly used way of var compensation is shunt compensative capacitor banks which had be more full developed, this way has several advantages such as simply constricted, fewer wires, low cost, easy maintenance and take small place, it was used in coal industry and other heavy industry. To sum up the above arguments, this paper has designed three Improved control algorithms of Var compensation which can be used more effective and feasible in Shunt compensative capaacitor banks than the traditional ways.

The designed algorithm of shunt power capacitor banks by this paper is mainly used in reactive power compensation and Power Factor compensation, so during the process design should fellow those principles : (1)Minimize the switching times of capacitor banks to lengthen the switchs and the capacitors' service life.(2) Zero-crossing Compensation to reduce the electric shock to the electrical net. (3)If the charge- discharge of the capacitors was unfinished ,it can not to be compensated.

A. Xie & X. Huang (Eds.): Advances in Electrical Engineering and Automation, AISC 139, pp. 407–413.
springerlink.com

2 Design and Implementation of Sequence-Cycle Var Compensation Algorithm

2.1 The Principles and Characteristics of Sequence-Cycle Var Compensating

on/off	The number of the Capacitor Bank						
	1	2	3	4	5	···	N
on	+						
on	+	+					
on	+	+	+				
on	+	+	+	+			
off		+	+	+			
off			+	+			
on			+	+	+		
on			+	+	+	···	+
off				+	+	···	+
off					+	···	+
on	+				+	···	+
on	+	+			+	···	+
off	+	+				···	+

Fig. 1. Switching sequence of the cycle-sequence way. In the figure,"+" means the capacitors were compensated. The figure is just a simple process of this switching way, the fellow will give you more information about it.

The sequence and cycle var compensating is in the form of queue model,FIFO(First In First Out), it connected the capacitor banks from beginning to the end, so we can cycling switch the capacitor banks and it can connect more capacitor banks. In the actual cases, the capacitors' switching will be not frequent,so the capacitors first switched will have more time to discharge.

If the controller connected N groups capacitors which can have different values, every capacitor banks have a number from 1 to N. In this way the controller should compensate the capacitor banks form 1 to N, if beyond the number N it came to the beginning number. If want to switch off the capacitor banks, it also from the number 1 to N(the number be compensated). As shown in figure 1 is the switching sequence of the sequence and cycle var compensating way.

2.2 The Implementation of Sequence-Cycle Model for Var Compensation

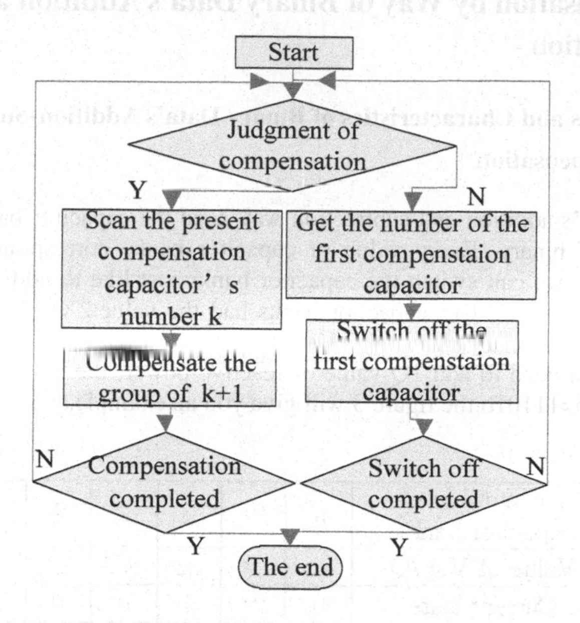

Fig. 2. Flow chart of sequence-cycle. This shows, at the beginning of the program,it need to measure the reactive power and power factor to judgment when to switch the capacitor banks.

As shown in Figure 2, it is flow chart of sequence and cycle Var compensating. This paper implement the program by using the PLC programming language ladder diagram in the compensation of reactive power or power factor, The experimental evidence that it run very stable.If one capacitor banks had been compensated, the program will record the number of this capacitor banks(N=k), if keep on compensating the new record number is k+1.When it wants to switch off the capacitors banks, it need to know the first compensated capacitor banks' number. If there were two adjacent capacitors, the first one of them is switched off and the second one is switched on, it can be sure that the second one is the earliest one be compensated, if it need to switch off the capacitors, it should switch off this one.

3 Design and Implementation of the Algorithm of Var Compensation by Way of Binary Data's Addition and Subtraction

3.1 Principles and Characteristics of Binary Data's Addition-Subtraction for Var Compensation

The binary data's addition and subtraction way need the capacitor banks arranged as the bits(1/0) of binary. Every value of capacitor banks corresponds to the value 2^n(n=1to n),then we can switch the capacitor banks just like to add and subtract the binary data. If there are five capacitor banks had the value:2^4Q, 2^3Q,2^2Q,2^1Q,2^0Q(a unit),it can form an binary data like 11100b(the capacitor banks 1,2,3 on,4,5 off),if the electrical system need to add 1Q value of reactive power, the calculation formula is 11100b+00001b=11101b,the figure 3 will give you an example.

The number of capacitor banks	n	···	3	2	1	0
Value of Var /Q	2n	···	23	22	21	20
Current state	0	···	1	0	0	1
+Addtion△Q0	0	···	0	0	0	1
State one	0	···	1	0	1	0
+Addtion△Q1	0	···	0	0	1	0
State two	0	···	1	1	0	0
subtraction△Q2	0	···	0	1	0	0
State three	0	···	1	0	0	0
···	···					

Fig. 3. Binary data's addition-subtraction way. In the figure3 ,the capacitor banks from 0 to n corresponds to the value from Q to 2^nQ,"0" means off,"1" means on. The ΔQ is the value of the Var that the system need to be compensated.

3.2 Implementation of the Binary Data's Addition-Subtraction Way for Var Compensation

As shown in figure 4, If the system need more reactive power,it adds the binary data Q_2 to Q_1,so it got the new state $Q_3=Q_2+Q_1$. At the same time if the system want to switch off some capacitor banks. We subtract the binary data Q_4 and the new state $Q_5=Q_1-Q_4$.When it got the new state data Q_3 or Q_5, the controller output on/off Signal to the capacitor banks corresponding the binary bits of the data Q_3 or Q_5.

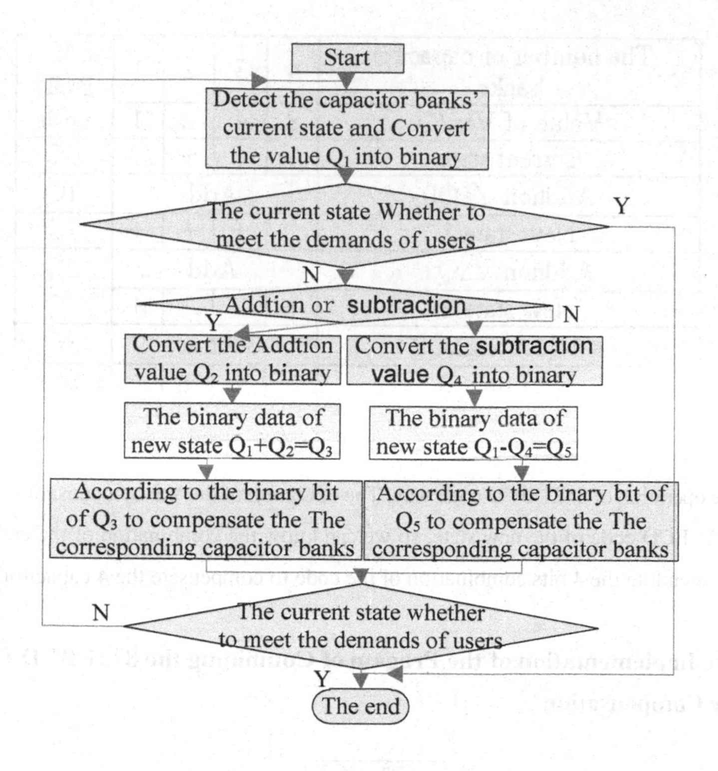

Fig. 4 . Flow chart of binary data's way

4 Design and Implementation of Combining 8421 Code Way

4.1 The Principles and Characteristics of Combining the 8421 BCD Code for Var Compensation

The way of combining the 8421 BCD should arrange the capacitor banks' value as 8,4,2,1 multiply by Q from left to right. If combine these value that can get such value:1,2,…,15 multiply by Q. In this way we can get form 0 to 15 come to 16 kinds of capacitor banks' value, it was similar to a kind of stepless way of Var compensation.

The number of capacitor banks	4	3	2	1	8421 BCD
Value of Var /Q	8	4	2	1	code
Current state		+	+		6Q
Addtion $\Delta Q0$	Add				1Q
New state 1		+	+	+	7Q
Addtion $\Delta Q1$	Add				2Q
New state 2	+			+	9Q
subtraction$\Delta Q2$	Subtract				3Q
New state 3		+	+		6Q
···	···				

Fig. 5. The operation of 8421 BCD code way. The operation of the Var compensation should first get the 8421 BCD code of the new state, so we can know the combination of the code, the program corresponding the 4 bits combination of the code to compensate the 4 capacitor banks.

4.2 The Implementation of the Pragam of Combining the 8421 BCD Code for Var Compensation

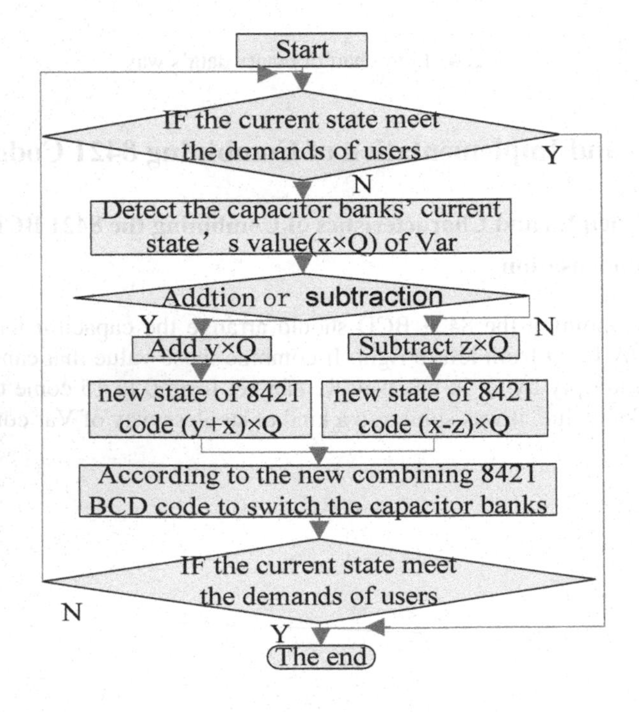

Fig. 6. Flow chart of combining 8421 BCD code

According to the algorithm of combining the 8421 BCD code, the 4 capacitor banks' value are Q, 2Q,4Q and 8Q.The program need to get the value of Q(unit of Var) to know the other capacitor banks' value and distribution. We know that combine the 8421 BCD code can get the value from 0 to 15 in total 16 codes, and every code can change into the other 15 codes. Every kind of change is knowable and can be inferred, so those 240 kinds of changes need to be written in program as the subroutine to be called.

5 Conclusion

The paper had completed the above 3 algorithms' program design and design a reactive power compensation controller based on PLC. The actual operation of this controller is very accurate and stable. According to the above principles, the improved algorithms can avoid repeated switching and protect the capacitor banks and the switches. It is a Very flexible and save way, the influence on the system is very small.

References

1. Xie, X., Jiang, Q.: Flexible AC transmission system principle and Application, pp. 148–149. Tsinghua University press, Beijing (2006)
2. Wang, Y.: Harmonic and reactive compensation device in coal mine application. Coal Mine Safety (3), 63–64 (2009)
3. Philip, M., Ashmole, P.: Flexible AC Transmisson Systerms. Power Engineering Review 20(3), 4–9 (2000)
4. Zhang, W.: 10kv substation bus reactive power and voltage controller of an integrated, p. 5. Tsinghua University Library, Beijing (2009)

Study on Improved LANDMARC Node Localization Algorithm

Tao Jiang, Yourui Huang, and Yanlin Wang

School of Electronical and Information Engineering,
Anhui University of Science and Technology,
Huainan, 232001, Anhui province, China
505925321@qq.com

Abstract. The node location technology is the key technology of Internet of Thing(IOT). Location Identification based on Dynamic Active radio frequency identification Calibration(LANDMARC) is an algorithm to locate the object based on the radio frequency identification technology. The beacon nodes contained in the LANDMARC network reduce costs effectively and improve the positioning accuracy. However, the limited node positioning range of it and the positioning errors caused prevent it from large-scale application. In this paper, it is found that using improved BP neural network algorithm to optimize the traditional LANDMARC algorithm can expand the effective range of the node-positioning, reduce costs and improve the node positioning accuracy. Meanwhile a way to promote the network's robustness is also proposed here.

Keywords: LANDMARC, Node Positioning, BP Neural Network, Robustness.

1 Introduction

With the development of the wireless technology, mobile devices and micro system technology, IOT is more and more attractive, while many applications of IOT need to know the specific physical location of objects. In recent years, people have put forward many schemes about the node positioning, such as triangulation, scene analysis, proximity and LANDMARC. But in practice, all these methods have different kinds of disadvantages. Therefore, the thought how to optimize the methods has become one way to solve the problem of node positioning.

Since the cost of the card reader is higher than that of reference tag, the LANDMARC algorithm, which replaces the card reader by the reference tag reduces the cost effectively. It can also improve the precision of positioning through increasing the number of reference tags appropriately. But in the experiment, it is found that the original LANDMARC algorithm can't get the coordinates of unknown nodes accurately. As the effective positioning range is narrow, the network has to set more reference tags to achieve better location performance, which lead to higher costs. And the traditional LANDMARC algorithm is apt to make mistakes when obstacles exist in the network. For these disadvantages, LANDMARC algorithm needs to be optimized.

A. Xie & X. Huang (Eds.): Advances in Electrical Engineering and Automation, AISC 139, pp. 415–421.
springerlink.com

By optimizing LANDMARC localization algorithm with the improved BP neural network and a new proposed method, the performances of LANDMARC such as range, cost, precision and the robustness of the network are optimized.

The structure of the paper is as follows. In the second part, the related works are introduced briefly. The third part introduces the LANDMARC localization algorithm in detail. The fourth part finds a way to optimize the LANDMARC location algorithm based on analyzing the characteristics of the improved BP neural network, and proposes a method to improve the robustness of networks. The fifth part shows the experimental process and gives the results. And the full paper is summarized in the sixth part.

2 Related Works

As the practical value of LANDMARC localization algorithm is very high, a lot of researches have been done around LANDMARC localization algorithm. The latest researches include: By introducing frequency parameters, Zhang Ying and Miao Quanli put forward a kind of location algorithm based on Radio Frequency Identification technology(RFID), which reduces interference with the Multi-channel affection in radio frequency [3]; By using the strategy that refining region can eliminate the influence of too many unconcerned reference tags and card readers on positioning calculation, Chen Congchuan proposed a kind of detailed regional indoor locating algorithm [4]; Zhu Juan and Zhou Shangwei found a kind of sub-regional method to calculate the coordinate reference tag which is based on Lagrange interpolation formula [5]; Gu Jia, Qian Yubo etceteras came up with two improved LANDMARC localization algorithms. One considers the point of equal potential and another is based on path loss model, which improve the localization algorithm's positioning precision and stability [6]. But the result of these algorithms is often not accurate as the relationship between signal strength and the network signal position is nonlinear relationship which is difficult to be described by the general formula. In addition, when obstacles exist in the network, the algorithms can't adapt to the changes, which lead to positioning error.

The paper optimizes several performances of the LANDMARC algorithm by introducing BP neural network and finds a way to shield the network from obstacles.

3 LANDMARC Approach

LANDMARC is based on active RFID. An RFID system has several basic components including RFID readers, RFID tags, and the communication between them. The most distinguishing feature of LANDMARC algorithm is that replace part of RFID readers by RFID tags with known positions as beacon nodes in the network to help position the unknown nodes.

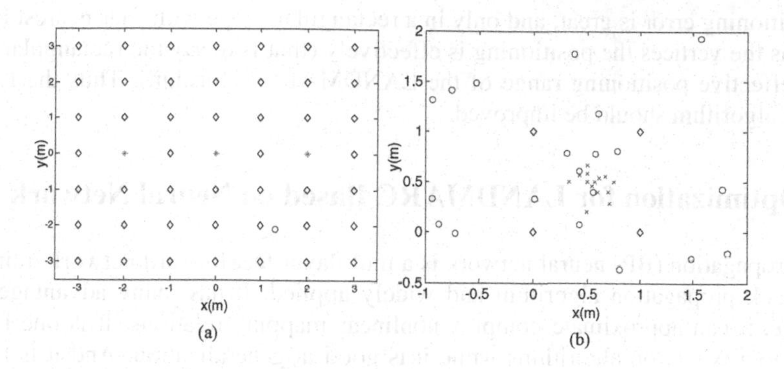

Fig. 1. (a) The Network topology of LANDMARC, (b) The coordinates of unknown nodes positioned by LANDMARC algorithm

As shown in figure 1(a), stars denote RFID readers (receiving nodes). Diamond nodes denote RFID reference tags (beacon nodes). And Circles denote tracking RFID tags (unknown nodes). It is supposed that there are n receiving nodes and m beacon nodes. The signal strength vector of an unknown node is defined as $S=(S_1,S_2,\ldots,S_n)$ where S_i indicates the signal strength perceived on the first i reader ($i\in(1,n)$) from the unknown node. The beacon nodes Signal Strength vector is $\theta=(\theta_1, \theta_2,\ldots,\theta_n)$ where θ_i denotes signal strength. Each of the unknown nodes has its E vector as $E=(E_1, E_2,\ldots,E_n)$ where $E_j=\sqrt{\sum_{i=1}^{n}(\theta_i-S_i)^2}$ ($j\in(1,m)$). E_j represents the location relationship between the beacon nodes and the unknown node that the nearer beacon node to the unknown node is supposed to have a smaller value of E_j. The unknown node's coordinate is obtained by the formula (1):

$$(x,y)=\sum_{i=1}^{k} w_i(x_i,y_i) \tag{1}$$

where k denotes the number of recently beacon nodes while the coordinates calculated are most appropriate when k=4 which will be used as the default of k below. (x_i,y_i) denotes the coordinate of the first i beacon node. w_i is the weighting factor to the first i recently beacon node which is obtained by the formula (2):

$$w_i=\frac{1/E_i^2}{\sum_{i=1}^{k} 1/E_i^2} \tag{2}$$

But it is found that these formulas above will make a great error in practice. For example, there is a RFID system with transmitting power of 20 dB, transmitting antenna gain of 17 dB, receiving antenna gain of 24 dB, cable and cable head loss of 1 dB. Its receiving signal strength is RSS=20+17+24-1-20lgD=60-20lgD, where D means distance (the unit is kilometer). The positioning result as shown in figure 1(b) is worked out from formula (1) and (2). Circles denote 20 unknown nodes generated randomly. The crossings covered each over in the center of graph denote the calculated coordinates of unknown nodes. And diamond nodes denote beacon nodes. It is found that the positioning error is great, and only in a rectangular scope with four nearest beacon nodes as the vertices the positioning is effectively (that is to say the rectangular scope is the effective positioning range of the LANDMARC algorithm). Thus the LANDMARC algorithm should be improved.

4 Optimization for LANDMARC Based on Neural Network

Back Propagation (BP) neural network is a multilayer feedforward network trained by error back propagation algorithm and widely applied. It has many advantages. For example, it can approximate complex nonlinear mapping relations; It is one kind of global approximation algorithms while it is good at generalization; And it is less affected by the damage of the individual neurons on the relationship of input and output through using the parallel distributed processing method.

But there are still several shortcomings in the traditional BP neural network algorithm such as its slow speed of convergence, being easy to fall into the local minimum value and being hard to determine the hidden layer and the number of the hidden layer node, etc. That is because that when the traditional BP neural network algorithm fixes the first i step of weight w(i) it can only make the adjustment in accordance with the output in the first i step, not taking into account the accumulation of experience. This paper uses an improved BP neural network algorithm, adding part of last adjustment quantity of weight to which of this time gained from the error calculation as the actual adjustment quantity of weight:

$$\Delta w(i) = -\eta \Delta E(i) + \alpha \Delta w(i-1) \tag{3}$$

where α is the momentum coefficient ($0<\alpha<0.9$) and η is the learning rate ($0.001<\eta<10$). The α reduces the oscillation tendency of learning process, improves the convergence and suppresses the trend of the network into local minimum.

The reason why it runs out is that the formula (1) and (2) are summed up by the experience which can't map out a good fit of the relation from input to output. As the improved BP neural network can map out a good fit of the relation from input to output, an improved BP neural network is designed here to describe the relation from the vector E to the coordinates of the unknown nodes. And it will be found that using BP neural network can not only improve the positioning precision, but also expand the effective range of node positioning in the experiment.

5 Analysis of Experiment

In this study, it's assumed that the four neighboring beacon nodes have been found in a RFID system as the example in the section 3, and select 80 nodes for training. There are two receiving nodes with the coordinates of (0,0) and (2,0) while each receiving node need to read four groups of data (θ_i-S_i, $i \in (1,4)$). As the result the BP neural network's input layer contains eight nodes. The Output layer contains 2 nodes for the coordinate (x, y) in two dimensions. And with repeated tests, it is concluded that the training effect with 10 nodes in the hidden layer is the best. The momentum coefficient takes α=0.008, and the learning rate takes η=0.006.

As shown in figure 2(a), the horizontal axis indicates the number of training, and the vertical axis indicates the sum of all nodes positioning errors, it is obvious that the error is reducing in terms of more training. When the precision is up to a certain value at the end of training, the trained network will be recorded.

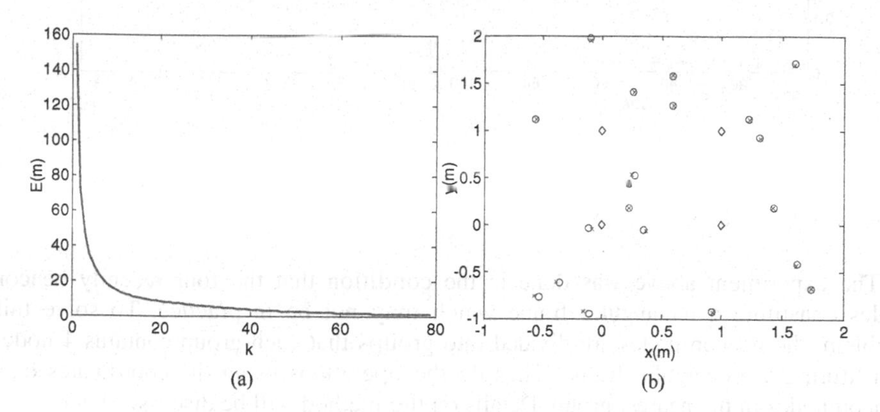

Fig. 2. (a) The change of error in terms of more training, (b) The coordinates of unknown nodes positioned by the improved LANDMARC algorithm

In order to test the result, 20 nodes are generated randomly whose coordinates are calculated by the improved LANDMARC algorithm as shown in figure 2(b). Diamond nodes say the beacon nodes, circles say the coordinates of the unknown nodes and red crossings say the operated coordinates.

It can be found that the LANDMARC algorithm optimized by the improved BP neural network is optimized in three aspects:

First of all, comparing figure 2(b) with figure 1(b), it is known that the effective range of node positioning is expanded 9 times from a rectangular scope with the vertices of four nearest beacon nodes to the whole graph.

Next, the cost of node positioning is reduced for the density of nodes can be reduced as the effective range of node positioning is expanded.

Finally, the node positioning precision is enhanced. As shown in Figure 3(a), the horizontal axis indicates the number of the unknown nodes while the vertical axis indicates the average location error. The blue line denotes the change of average error in terms of more unknown nodes derived from the original method. The red line denotes the change of average error in terms of more unknown nodes derived from the improved LANDMARC algorithm. It is found that the node location precision is increased 40 times with the optimization of improved BP network.

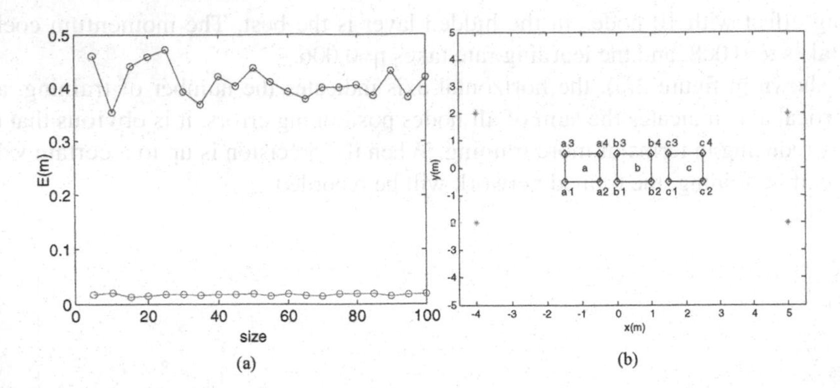

Fig. 3. (a) The change of average error in terms of more unknown nodes, (b) The network topological graph with groups of nodes

The experiment above was done in the condition that the four recently beacon nodes constitute a rectangular frame which may not be in practice. To solve this problem, the beacon nodes are divided into groups that each group contains 4 nodes constituting a rectangular frame. Then do the operations using the coordinates of 4 beacon nodes in the nearest group. Details on the method will be discussed later.

It will cause positioning errors that the original algorithm can't adapt to the changes that the connectivity of network will be reduced with the obstacles existing in the network. The method that takes the effective group by sampling multiple beacon node groups can ensure the algorithm effectiveness. Details are as follows:

Firstly, the beacon nodes will be divided into groups. Figure 3(b) is a network topological graph with diamond nodes indicating beacon nodes and stars indicating receiving nodes. Each of rectangular frames indicates a group of nodes. And the 12 beacon nodes are divided into 3 groups named a, b, c.

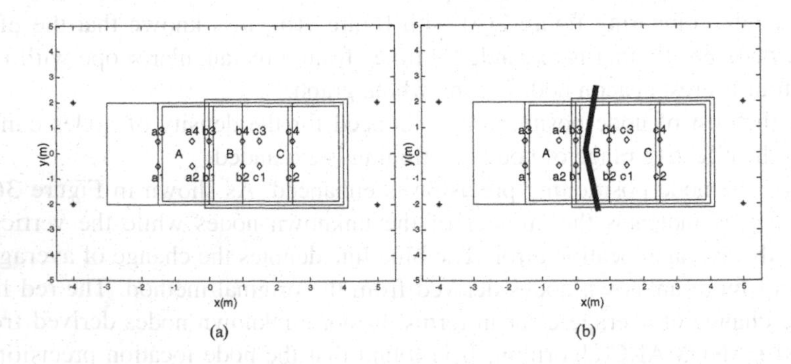

Fig. 4. (a) The positioning ranges of 3 groups, (b) The network with an obstacle

Secondly, beacon nodes will be distributed into the network obeying the rule that each point in the network can be located by at least two groups of beacon nodes. The frames in figure 4(a) indicate the 3 effective positioning ranges named A, B and C

(The frame of single lines denotes A. The frame of double lines denotes B. And the frame of three lines denotes C). It can be found that each point in the area B can be covered by at least two pieces of region (the area B is the focus).

Thirdly, when an obstacle exists in the area B (the thick brush in the graph as shown in figure 4(b)), the location of the obstacle can be positioned in the range among b1, b2, b3 and b4 generally through comparing the signal strengths received by the receiving nodes from the 12 beacon nodes.

Finally, comparing the signal strengths received by the receiving nodes from the unknown nodes and the beacon nodes, it can be determined whether the unknown node is on the right or on the left of obstacles. If the unknown node is on the left, the data from group a will be used to calculate, and if the unknown node is on the right, the data received from group c will be used to calculate.

The interference of obstacles can be shielded by using the method above.

6 Conclusions

This paper puts forward an idea that using the improved BP neural network can optimize LANDMARC node localization algorithm and verify the feasibility of this idea in the experiment. It shows that the improved LANDMARC node localization algorithm is optimized at the precision, expanding positioning range, and reducing the cost. By the way, a method is also proposed to shield obstacles in this paper.

Acknowledgements. The work was supported by Program for New Century Excellent Talents in University (NCET-10-0002) and Anhui Provincial Natural Science Foundation (1108085J03).

References

1. Liu, J.: Intelligent Control, pp. 117–139. Publishing House of Electronics Industry, Beijing (2010)
2. Liu, D.: The plane target recognition based on an improved BP neural network. China Hi-Tech Enterprises (16), 12–13 (2011)
3. Zhang, Y., Miao, Q.: Implementation of a kind of location algorithm based on RIFD technology. Information Technology (11), 117–118 (2011)
4. Chen, C., Cheng, L.: RFID Indoor Location Algorithm Based on Range Diving. Computer Applications and Software (1), 50–52 (2011)
5. Zhu, J., Zhou, S., Ma, Q.: An indoor location algorithm based RFID. Microcomputer Information (8-2), 160–162 (2009)
6. Gu, J., Qian, Y., Sun, H., Wang, J.: Research on Tool Indoor Positioning Technology. Transaction of Beijing Institute of Technology 9(9), 1056–1058 (2010)
7. Ni, L.M., Liu, Y.H., Lau, Y.C., et al.: LAND-MARC: Indoor Location Sensing Using Active RFID. Wireless Networks 10(6), 701–720 (2004)

Application of Log-Normal Distribution and Data Mining Method in Component Repair Time Calculation of Power System Operation Risk Assessment

Zelei Zhu, Jingyang Zhou, and Lijie Chen

China Electric Power Research Institute,
100192, Beijing, China
zhuzelei@epri.sgcc.com.cn

Abstract. Power system operation risk assessment can comprehensively take into account the occurrence possibility and severity of disturbances, and it is an effective complement to traditional deterministic security analysis. Calculating transient state probability of components in future short time duration is one of the key issues in power system operation risk assessment. In this paper, a novel probability distribution named'log-normal distribution'was presented for component fault repair time. The shape of probability density of log-normal distribution can describe the distribution characteristics of component fault repair times well. When the history data is limited, it can be used to calculate the instantaneous state probability, the method is simple and easy to use. But, if the system has a large amount of historical data, it will be suitable to establish the discrete model of repair time with data mining methods. Based on the information entropy weight allocation method, a real-time discrete model was established in this paper. The model can relatively accurate predict component repair time under various operation conditions. The results of example show that, if historical data is limited, a log-normal distribution can better meet characteristics of probability distribution of the actual sample, the transient state probability will be got accurately. When there is a large amount of historical data, discrete model established by data mining method is accurate and effective, can be very good to meet the field.

Keywords: Risk Assessment, Power System Operation, Repair Time, Log-normal, Data Mining.

1 Introduction

In 1997, CIGRE has given the definition of power system operation risk assessment based on probability analysis method [1], which is mainly adopted to assess the impact of the contingency based on occurrence possibility and severity. The traditional EMS has been using the deterministic model and analysis method. Due to grid scale and complex operation mechanism, operation process is full of uncertainty and the deterministic analysis methods cannot account the randomness of power system disturbances so that operation decisions are often conservative based on the most serious

A. Xie & X. Huang (Eds.): Advances in Electrical Engineering and Automation, AISC 139, pp. 423–431.
springerlink.com

cases. So only the uncertainties model is more scientific to describe power system disturbance's characterization [2-15].

It is crucial to establish a scientific and reasonable component outage random process model so that component transient state probability valves in investigated period can be correct calculated for operation risk assessment. The fault repair time of power system equipment is a random variable. It is an important work to choose an appropriate probability distribution to describe the repair time for establishing a scientific component outage random process model. It dirctly affects the transient state probability calculation and brings a new topic to operation risk assessment.

Through statistical analysis about massive historical data, the distribution of component repair time under statistical significance can be got. But these are generally qualitative conclusions. Tests show that the probability distribution of component repair time obtained by the data of different segments is lack of consistency when it is described by a simple exponential distribution.

In this paper, the log-normal distribution is proposed to describe fault repair time. When the history data is limited, it can be well characterized the distribution characteristics of fault repair time; When the system has a large amount of historical data, it will be suitable to establish the discrete model of repair time with data mining methods. Based on the information entropy weight allocation method, a real-time discrete model was established in this paper. The model can relatively accurate predict component repair time under various operation conditions.

2 Log-Normal Probability Distribution of Component Repair Time

Determination of component transient state probability in future short time duration is one of the key issues in power system operation risk assessment. It is crucial to establish a scientific and reasonable component outage random process model so that component transient state probability valves in investigated period can be correct calculated for operation risk assessment. The fault repair time of power system equipment is a random variable. It is an important work to choose an appropriate probability distribution to describe the repair time for establishing a scientific component outage random process model. In the traditional reliability assessment, component fault repair time described by exponential distribution does not affect the correctness of reliability assessment results. But this distribution type doesn't suitable for operation risk assessment calculation.

In fact, fault repair time of power system equipment includes the logistic delay time, the technical delay tome, the fault location time, the fault recovery time, the verification time and other several stages. Namely the repair time will generally focused on a time scale, is not too short but most not too long. To describe the repair time with the exponential probability density function is not fully consistent with the actual power system repair.

Reference [16] presented a probability distribution named'splice exponential distribution'to describe component fault repair time. The shape of probability density of splice exponential distribution likes a bell.

2.1 Fault Repair Time Model

Referenced the ideas in [16], presenting a novel probability distribution named 'log-normal distribution' to describe the component repair time.

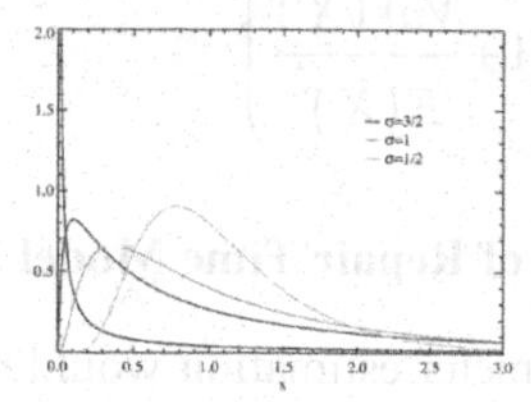

Fig. 1. Density function curve of log-normal distribution

In probability theory and statistics, the log-normal distribution is a type of probability distribution that logarithmic is any random variable of the normal distribution. As can be seen from fig. 1, the probability density function curve for the 'bell-shaped', its shape and size were determined by the parameters μ and σ^2. The log-normal distribution is suitable for establishing mathematical model of equipment repair time and reliability growth process model of electronic products. The log-normal probability distribution density function is:

$$f(x;\mu,\sigma)=\begin{cases}\dfrac{1}{x\sigma\sqrt{2\pi}}e^{-(\ln x-\mu)^2/2\sigma^2} & ,x>0\\ 0 & ,x\le 0\end{cases} \tag{1}$$

Its distribution function is:

$$F(x;\mu,\sigma)=\int_0^x f(x;\mu,\sigma)dx=\Phi\left(\frac{\ln x-\mu}{\sigma}\right) \tag{2}$$

Here, $\Phi(\cdot)$ is the standard normal distribution function, μ and σ are mean value and standard deviation of variable logarithmic respectively. The expectation and variance are:

$$E(X)=e^{\mu+\sigma^2/2} \tag{3}$$

$$Var(X)=\left(e^{\sigma^2}-1\right)e^{2\mu+\sigma^2} \tag{4}$$

According to statistical results based on historical data, the expectation and variance of repair time can be solved. The parameters μ and σ^2 can be solved through (3) and (4).

$$\mu = \ln\left(E(X)\right) - \frac{1}{2}\ln\left(1 + \frac{Var(X)}{E(X)^2}\right) \tag{5}$$

$$\sigma^2 = \ln\left(1 + \frac{Var(X)}{E(X)^2}\right) \tag{6}$$

2.2 Parameters Estimation of Repair Time Model

When less historical data, parameter estimation would be done by the linear unbiased estimation (LUE); When the sample size is large, the MLE of μ and σ is estimator having good properties. And MLE μ and σ is:

$$\mu = \frac{1}{n}\sum_{i=1}^{n} \ln x_i \tag{7}$$

$$\sigma^2 = \frac{1}{n}\sum_{i=1}^{n}\left(\ln x_i - \frac{1}{n}\sum_{i=1}^{n} \ln x_i\right)^2 \tag{8}$$

These can be obtained by historical data directly.

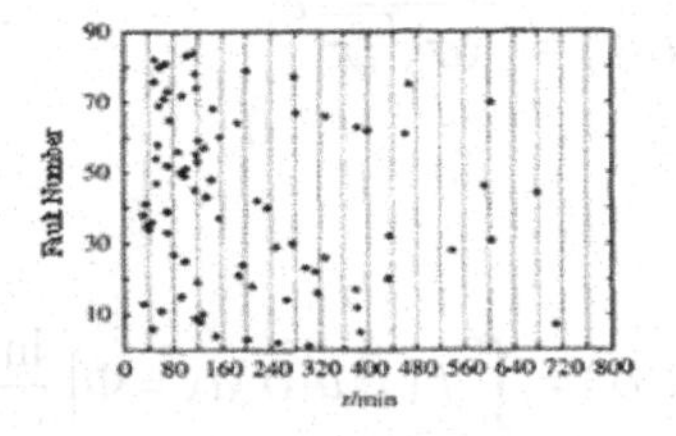

Fig. 2. Statistic result of real faults' repair time

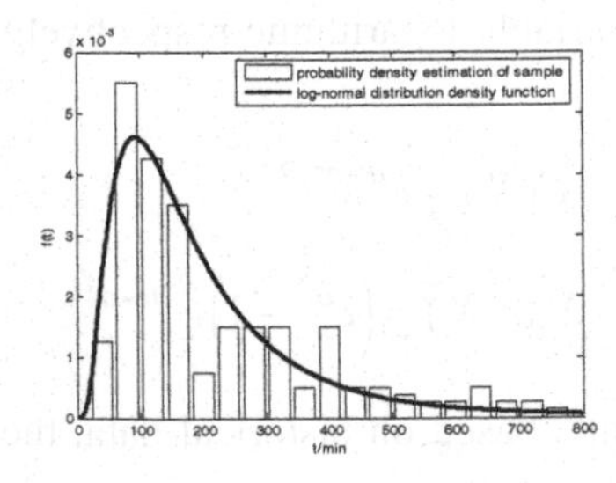

Fig. 3. Effect diagram of log-normal distribution's density function

Fig. 3 is repair time statistical distribution diagram of permanent fault occurred in the 10 kV overhead line of a power company between September and October in 2007. Each point represents the occurrence of one fault and there is a total of 84 fault instances. Can be seen from the density distribution of star points in vertical interval in figure 3, fault repair time mostly concentrated in [40,160] interval range, in line with experience. The expectation and variance of 84 fault samples are 202.13 min and 27435 min respectively.

With the log-normal probability describes to repair time, and through the MLE parameter estimation to determine distribution parameters(μ =5.0521, σ =0.7167). Relative to the exponential distribution which has only one parameter, the proposed log-normal distribution with two parameters can meet requirement of sample expectation and variance simultaneously. Fig. 4 gives the log-normal distribution probability density function with probability density estimation the sample data in comparison. It can be seen that log-normal distribution is very consistent with the sample.

3 Mathematical Model of Component Repair Time Based on Data Mining Method

System clustering method is a fuzzy clustering analysis approach based on fuzzy equivalence relationship. In classical clustering analysis method, it is usually utilized classical equivalence relationship to cluster the sample set X, which would not loss any case information. With information entropy theory, when data processing, the importance of each factor can be supplied. In this paper, both methods are combined to study the problems of multiple factors[17-20].The weight distribution method based on information entropy is as following.

1). Determining the sample object needed to deal with and extracting factor data. Assuming the set composed of n samples to be processed is:

$$X = \{X_1, X_2, X_3 \cdots X_n\} \tag{9}$$

Each sample can be expressed by m index feature vectors:

$$X_i = \{X_{1i}, X_{2i}, X_{3i} \cdots X_{ni}\} \tag{10}$$

Then the set can be expressed with the $m \times n$ order eigenvalue matrix, which completes clustering of the sample object.

$$X_i = \begin{Bmatrix} x_{11} & x_{12} & \cdots & x_{1n} \\ x_{21} & x_{22} & \cdots & x_{2n} \\ \vdots & \vdots & \ddots & \vdots \\ x_{m1} & x_{m2} & \cdots & x_{mn} \end{Bmatrix} \tag{11}$$

2). Establishing fuzzy similarity relation. First, each eigenvalue of the matrix is normalized, then the fuzzy similarity matrix $\boldsymbol{R}_{n \times n}$ can be established with max-min method, which is as following:

$$R_{ij} = \sum_{k=1}^{m} x_{ik} \wedge x_{jk} \Big/ \sum_{k=1}^{m} x_{ik} \vee x_{jk} \tag{12}$$

3). Using the fuzzy equivalence closure method, to get the fuzzy equivalence matrix, the classification number will be determined by the fuzzy equivalence matrix.

4). Determine the system's initial information entropy, then according to the definition of conditional entropy to confirm mutual information quantity I_{a_i} when each factor is deleted in each confidence levels. On a different level of confidence, the mutual information weighted factor is used to indicate the amount of information, and it can be expressed as:

$$M_j = \sum_{i=1}^{p} \alpha_i \times \frac{1}{I_{\alpha_i}} \tag{13}$$

5). According to the relative size of information which the factors contained, the weight can be normalized to solve, weight allocation formula is:

$$w_j = M_j \Big/ \sum_{j=1}^{m} M_j \tag{14}$$

3.1 Data Warehouse of Repair Time

There are many factors which affect repair time of power system component. For different fault types, the repair time distribution is not the same. The main factors impact on fault repair time contained wear resistance, corrosion resistance, fracture resistance, aging resistance, human economy influence and natural condition effect. At random failure period and early failure period, respectively, the data of different fault were collected from the historical data. After data preprocessing, save the data to the database and create a simple data warehouse [20].

3.2 To Establish Discrete Model of Component Repair Time

When there is a fault, the factors which influenced repair time: logistic delays, technical delays, fault location, fault repair, verification time and operational personnel.

First of all, from the data warehouse to extract a group of data constituted by the N latest repair time data and its six factors under a fault:

The fault repair time $T: t_1, t_2, \cdots t_n$

The logistic delays $A: a_1, a_2, \cdots a_n$

The technical delays $B: b_1, b_2, \cdots b_n$

The fault location $C: c_1, c_2, \cdots c_n$

The fault repair $D: d_1, d_2, \cdots d_n$

The verification time $E: e_1, e_2, \cdots e_n$

The operational personnel $F: f_1, f_2, \cdots f_n$

Then apply the information entropy weight distribution method to calculate weight of six factors which effects fault repair time, these were respectively: $w_1, w_2, \ldots, w_n$, wherein $w_1, w_2, w_3, w_4, w_5, w_6$ respectively indicated effect weight six factors on components repair time. Predicts that T is components repair time:

$$T = \begin{bmatrix} 1-(w_1+w_2+\cdots w_n) \\ \vdots \\ w_i \\ \vdots \\ 1-(w_1+w_2+\cdots w_n) \end{bmatrix}^T \cdot \begin{bmatrix} A \\ B \\ C \\ D \\ E \\ F \end{bmatrix} \cdot \begin{bmatrix} t_1 \\ \vdots \\ t_i \\ \vdots \\ t_n \end{bmatrix}^T \cdot \left\{ \begin{bmatrix} 1-(w_1+w_2+\cdots w_n) \\ \vdots \\ w_i \\ \vdots \\ 1-(w_1+w_2+\cdots w_n) \end{bmatrix}^T \cdot \begin{bmatrix} A \\ B \\ C \\ D \\ E \\ F \end{bmatrix} \cdot \begin{bmatrix} 1 \\ 1 \\ \vdots \\ 1 \end{bmatrix}^T \right\}^{-1} \tag{15}$$

On the elements of each fault, according to the above method to predict the component repair time$T$, resulting in a group of data which associated with a fault and its repair time:

$$(s_1, T_1), (s_2, T_2), \cdots (s_m, T_m) \tag{16}$$

Here, m is the fault type number; s_i is the fault. These data constituted the discrete model of real-time component repair time:

$$T = f(s_i), i = 1, 2, \cdots, m \tag{17}$$

4 Simulation Example Analysis

Table 1. Accuracy rate of two methods predicts the real repair time

component	Log-normal Distribution	Data mining method
transformer	92.233%	96.144%
overhead line	90.020%	93.909%
breaker	94.891%	96.319%

From the real-time / historical database in power system of the real-time and historical data of component repair time had been obtained and the calculation results back into the database. Choose three different devices in power system, statistical results of the repair time distribution are shown in Table 1.

As can be seen from Table 1, when there is huge amounts of data, data mining method can more accurately describe the component repair time distribution regularity. So, the validity of the method is verified.

5 Conclusion

This paper puts forward with a log-normal distribution which describes repair time after the fault, can be well characterized fault repair time's distribution characteristics when history data is limited; Based on data mining method such as fuzzy clustering and rough set theory, discrete model of component repair time was established by the massive data under operation conditions. The discrete model can be corrected in real-time, can accurately describe the component repair time distribution characteristics. This simulation result shows that two methods are effective. And, it is significance for real power system operation.

References

1. CIGRE Task Force 38.03.12.: Power system security assessment. Electra 175, 49–77 (1997)
2. Feng, Y., Wu, W., Sun, H., Zhang, B., He, Y.: A preliminary investigation on power system operation risk evaluation in the modern energy control center. In: Proceedings of the CSEE, vol. 25, pp. 73–79 (2005)
3. Xue, Y.-S.: The Way from a Simple Contingency to System-wide Disaster-Lessons from the Eastern Interconnection Blackout in 2003. Automation of Electric Power Systems 27, 1–5 (2003)
4. Feng, Y., Zhang, B., Wu, W., Sun, H., He, Y.: Power system operation risk evaluation based on credibility theory: Part one a survey of operation risk assessment. Automation of Electric Power Systems 30, 17–23 (2006)
5. Feng, Y., Zhang, B., Wu, W., Sun, H., He, Y.: Power system operation risk evaluation based on credibility theory: Part two theory fundament. Automation of Electric Power Systems 30, 11–15 (2006)
6. Feng, Y., Zhang, B., Wu, W., Sun, H., He, Y.: Power system operation risk evaluation based on credibility theory: Part three engineering application. Automation of Electric Power Systems 30, 11–16 (2006)
7. Wan, H., Mccalley, J.D., Vittal, V.: Increasing Thermal Rating by Risk Analysis. IEEE Trans on Power Systems 14, 815–828 (1999)
8. Zhanng, J., Pu, J., Mccalley, J.D.: A Bayesian Approach for Short-term Transmission Line Thermal Overloads risk Assessment. IEEE Trans on Power Delivery 17, 770–778 (2002)
9. Shi, H.-J., Ge, F., Ding, M.: Research on on-line assessment of transmission network operation risk. Power System Technology 29, 43–48 (2005)
10. Lu, B., Tang, G.: Application of risk-based security assessment in power system, vol. 24, pp. 61–64 (2000)
11. Zhao, S., Zhou, Z., Zhang, D.: Risk assessment index of dynamic stability for large-scale interconnected grids and its application. Power System Technology 33, 68–72 (2009)
12. Ding, M., Li, S., Wu, H.: Integrated evaluation of power system adequacy and stability. In: Proceedings of the CSEE, vol. 23, pp. 20–25 (2003)
13. Wei, Y.-H., Liu, S.-G., Su, J.: Risk Assessment of Urban Power Network Based on Enumerative Sampling Method. Power System Technology 32, 62–66 (2008)
14. Chen, W., Jiang, Q., Cao, Y.: Risk-based vulnerability assessment in complex power systems. Power System Technology 29, 12–17 (2005)

15. Ming, N., McCalley, J.D., Vittal, V.: Online risk-based security assessment. IEEE Trans. on Power Systems 18, 258–265 (2003)
16. Ning, L.-Y., Wu, W.-C., Zhang, B.-M.: A Novel Probability Distribution of Component Repair Time for Operation Risk Assessment. In: Proceedings of the CSEE, vol. 29, pp. 15–20 (August 2009)
17. Huang, D.: Means of weights allocation with multi-factors based on impersonal message entrogy. Systems Engineering Theory Methodology Applications 12, 322–324 (2002)
18. Hu, Q., Yu, D., Xie, Z.: Fuzzy probabilistic approximation spaces and their information measures. IEEE Transactions on Fuzzy Systems 14, 191–201 (2006)
19. Hu, Q., Yu, D., Xie, Z.: Information-preserving hybrid data reduction based on fuzzy-rough techniques. Pattern Recognition Letters 27, 414–423 (2006)
20. Zeng, D., Yang, T., Chen, X., Liu, J.: Application of Data Mining Method in Real-time Optimal Load Dispatching of Power Plant. In: Proceedings of the CSEE, vol. 30, pp. 109–114 (2010)

Application of BP Neural Network PID Control to Coordinated Control System Based on Balance Thought

Zelei Zhu, Cuihui Yan, and Zhongxu Han

China Electric Power Research Institute,
100192, Beijing, China
zhuzelei@epri.sgcc.com.cn

Abstract. Coordinated control system of thermal power unit is a time-varying, uncertainties, strong-coupling and multi-variable control system. Its control quality is a key to improve the thermal processes automation and achieve power system automatic generation control. In this paper, the non-linear mathematical models of drum-boiler turbine unit and once-through boiler unit are given. Analysis of the conventional boiler feed-forward control system and based on the thought that dynamic energy and static energy separation, the boiler BP neural network PID control feed-forward controller is structured and applied. Simulation results show that not only load response and anti-interference ability are improved, but also dynamic overshoot greatly reduce. It is conducive to the safety, economy and stability of generator unit.

Keywords: coordinated control system, state feedback, neural network, mathematical model.

1 Introduction

As a result of power grid capacity increased, thermal power units need to respond to AGC peak-shaving. It is a higher demand for control system, the control system performance, reliability are becoming important factors that affected development of unit. Coordinated control system is a strong coupling, multiple-input and multiple-output system. Load and the main steam pressure control are interdependence, mutual constraints[1-7]; it is non-linear dynamic characteristics of unit; boiler side emerge large delay. Therefore it is very difficulty of obtaining a satisfactory control effect for conventional control strategy. The controller designed by A variety of advanced control methods such as optimal control, predictive control, intelligent control and neural network PID control was studied and applied to coordinated control system and achieve a certain effect[8-12].

Concept of generalized intelligent control [13-14] which combination of mechanical, experience, intuitive and logical thinking way, take classical control, modern control, special measurement techniques [15] and narrow intelligent control in unified study platform, the better solution to control problems the large capacity of supercritical unit [16], and further expand the breadth and depth that control theory research

A. Xie & X. Huang (Eds.): Advances in Electrical Engineering and Automation, AISC 139, pp. 433–440.
springerlink.com

and explore. In order to improve the coordination control system performance, this paper cites the concept of BP neural network PID control [16] and applies the concept to coordinate control system. The algorithm is simple to achieve and can be achieved in computer control system through various existing algorithm directly. It does not rely on proprietary neural network PID control algorithm module and it is propitious to engineering applications. The application of this algorithm can overcome the big lag and the big inertia of boiler. It can effectively improve rate of coordinate control system responding to AGC.

2 Non-linear Mathematical Model of Coordinated Control System

Using a block diagram to express a unit of a practical nonlinear mathematical model of boiler-turbine shown in Fig 1[17].

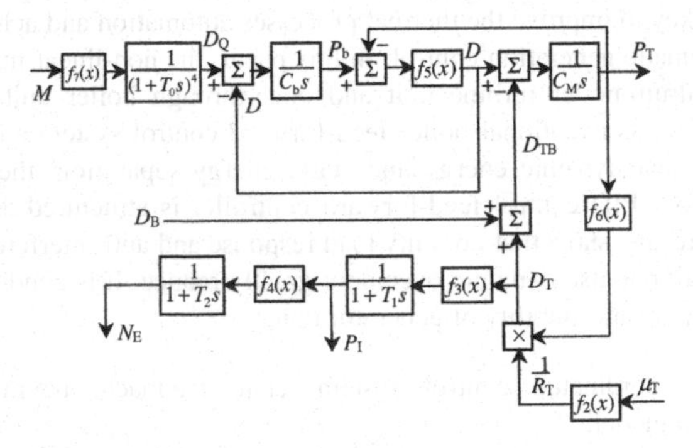

Fig. 1. The practical non-linear mathematical model of boiler-turbine coordinated control system

Where, M: the total fuel flow; K_M: the proportional coefficient of the changed values between fuel and load; τ_M: the lag time that the load change caused by the fuel changed; D_Q: the heat flow of entering into boiler expressed by dimension of boiler steam flow; D: the boiler steam flow; C_b: the preserving coefficient, corresponding to the boiler steam flow changing value when the drum pressure varies 1 MPa; P_b: the drum pressure or vessel pressure; $R_g r$: the dynamic resistance of supper heater; D_{TB}: the steam flow in boiler; D_B: the steam flow in turbine by-pass; D_T : the steam flow in turbine; C_M: the capacity factor of the steam header; P_T: the turbine throttle pressure; $f_6(x)$: function of P_T ; R_T: the dynamic resistance of the steam header; μ_T: the turbine regulated valve scale; p_1 : the turbine first stage pressure; N_E: the active power.

In Fig. 1:

$$f_2(u_T) = 0.0001u_T^{\ 3} - 0.0061u_T^{\ 2} + 0.1926u_T \quad (1)$$

$$f_3(D_T) = 0.0091D_T + 0.0002 \quad (2)$$

$$f_4(p_1) = 33.0204p_1 + 0.0077 \quad (3)$$

$$f_5(\Delta p) = -80.9\Delta p^6 + 860.4\Delta p^5 - 3505.2\Delta p^4 + 6828.5\Delta p^3 - 6536\Delta p^2 + 3483.1\Delta p + 3.9 \tag{4}$$

$$f_6(p_T) = -8p_T^5 + 192p_T^4 - 2416p_T^3 + 15846p_T^2 - 42175p_T \tag{5}$$

$$f_7(w) = 0.8w^3 - 59.3w^2 + 1662.1w \tag{6}$$

$$f_8(M) = -0.1M^4 + 5.8M^3 - 256.5M^2 + 4789.1M \tag{7}$$

$$f_9(M) = 1888.8M \tag{8}$$

$$f_{10}(p) = 0.0007p^6 - 0.0279p^5 + 0.4576p^4 - 2.01p^3 - 32.6532p^2 + 428.3708p + 100 \tag{9}$$

Reference [18] gave an experience once-through boiler model. This paper considers the dynamic characteristics which a greater impact on the control system in running process of supercritical unit and must pay full attention to when design and debugging coordination control system comprehensively, and given the structure of the mathematical model as shown in Fig 2.

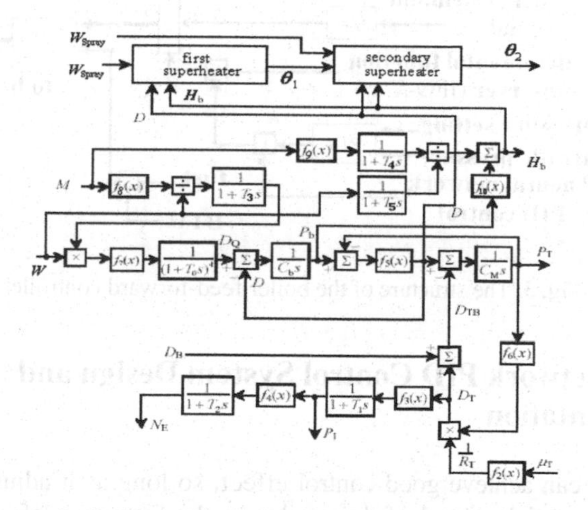

Fig. 2. Mathematical model of controlled object of coordinated control system for generator unit with supercritical unit

Where, W: boiler feedwater flow; H_b: micro-superheated steam enthalpy value; W_{Spray}: I or II superheater spray water flow; θ_1 : I superheater outlet temperature; θ_2 : II superheater outlet temperature.

The main difference between once-through boiler dynamic characteristic and drum boiler's is that the controlled objects have the characteristics of MIMO strong coupling, nonlinear and variable parameters. Drum boiler mass balance mainly is reflected by the drum water level, so its steady-state energy balance is directly reflected by the fuel and steam flow in the case of neglecting steam temperature change; once-through boiler mass balance is directly reflected by the feedwater flow and steam flow from baric flow channel. The change value of separator outlet enthalpy value reflects the ratio of fuel and water in the inlet, simultaneously the change of steam temperature in the outlet is considered, then comprehensive assessment the situation of energy balance.

There are great delay and inertia in boiler side. It is difficult to solve this problem for control systems used the conventional PID controller to control, Particularly integral role of the system often produce excessive accumulation and lack of energy which would result parameters fluctuation in adjustment process. The introduction of feed-forward control will made energy imbalance to be limited in a small scope when energy balance will be lost or just an imbalance occurs between turbine and boiler. The energy imbalance will be eliminated as soon as possible in dynamic process. Conventional load instruction is main feed-forward way and it is easy to make overshoot in the adjusting process and result in instability.

The structure of the boiler feed-forward controller as follow in Fig 3, including five main parts.

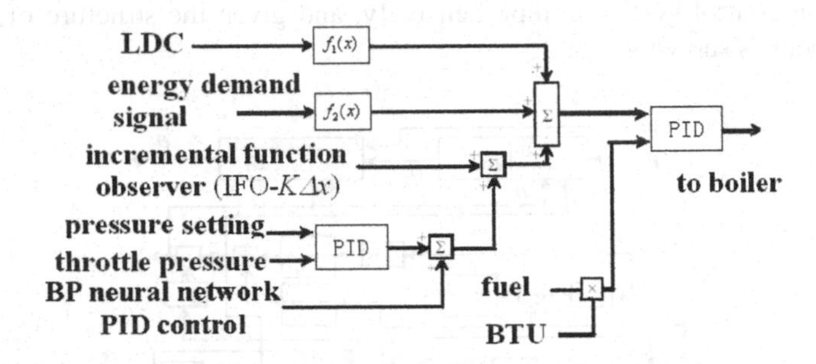

Fig. 3. The structure of the boiler feed-forward controller

3 Neural Network PID Control System Design and Implementation

The PID control can achieve good control effect, so long as it adjust the proportion, integral and differential control of three roles in the formation of control amount of interaction and mutual restriction relationship between them, the relationship is not a simple linear combination, so it must find out the optimal relationship with countless changes from nonlinear combination. Neural network with arbitrary nonlinear representation capability can pass on the system performance of the learning to achieve the best combination of the PID control. BP neural network can approach any nonlinear function, the structure and learning algorithm is simple and clear. Through the neural network self learning can find an optimal control law of P, I, D parameters. Based on the BP neural network PID control system structure is shown in Fig 4[16].

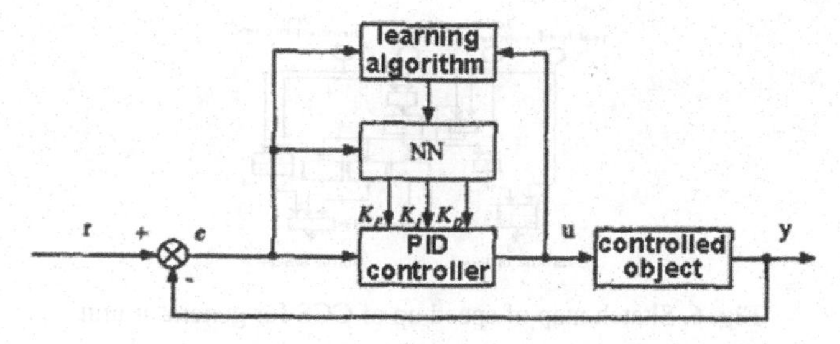

Fig. 4. The system structure based on the BP neural network PID control

The controller is composed of two parts, one is the classical PID controller, which is directly on the object process closed-loop control. And three parameters K_P K_J K_D are adjusted on line, the other is neural network (NN), which is according to the operational status of the system to adjust the PID controller parameters, aimed at achieving a performance optimization. Even the output layer neurons to the output state corresponds to the three adjustable parameters K_P K_J K_D of PID controller, through neural network self learning, weighted coefficient adjust, so that the stable state corresponds to the PID controller parameters under an optimal control law. Classical incremental digital PID control algorithm is as follows:

$$u(k) = u(k-1) + K_p[e(k-1)] + K_L e(k) + K_D[e(k) - 2e(k-1) + e(k-2)] \quad (10)$$

Considered K_P K_J K_D as variable coefficient depended on the system state, it can be described as:

$$u(k) = f[u(k-1), K_p, K_L, e(k), e(k-1), e(k-2)] \quad (11)$$

In type $f(\bullet)$ is nonlinear function concerned with K_P K_J K_D, $u(k-1), y(k)$, which can use BP neural network(NN) as shown in Figure 5, through the training learn to find an optimal control law.

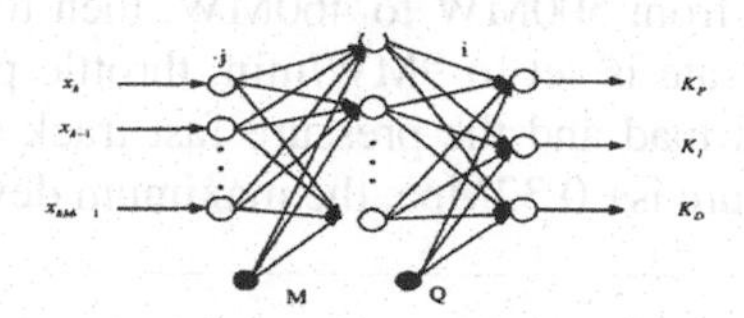

Fig. 5. The network structure of NN-BP

When AGC target load is in AGC way, the unit receives load instructions from the dispatching centre. When control transfer from the "AGC" to the "CCS", the target load instructions of the unit is decided by operator's manual setting. LDC is different from target load, the latter can be immediately set to run within a reasonable value, while LDC needs to achieve the value of targets load according to "LDC changing rate" over time.

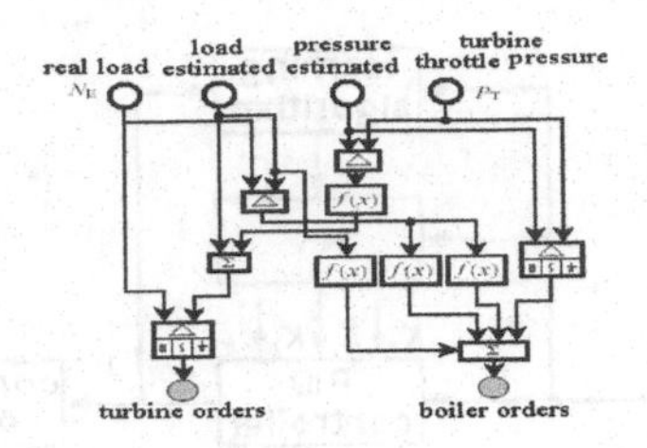

Fig. 6. Skatch map of speed up of CCS for generator unit

In original design, the boiler's main control loop contained the "speed up" circuit, the differential signal of load estimated and real load was formed by of function generator, that is a nonlinear proportional regulator, when the differential between load estimated and real load signal in a large margin, means the boiler can not keep up the change of turbine, the output order would send to the main controller of boiler, that would made the system increase or decrease the amount of coal, thus is the "accelerator" action, according to the real demand of coordinated control with once-through boiler added a second-class "speed up" circuit in debugging process and considered the press bias and effect of some other working condition, and is shown in fig 6.

4 Simulation Test

Fig 7, 8 and 9, respectively, are simulation curve that control system is applied in mathematical model in sliding pressure mode and constant pressure mode.

Fig 7 is 60 minutes simulation trend of Load from 350MW to 300MW, then return to 350Mw in constant pressure mode. Load change rate is set to 9MW/min and throttle pressure is set to 20Mpa. The actual load fast track the target load and there is no overshoot phenomenon. The maximum deviation of throttle pressure is ± 0.28Mpa, the maximum deviation of load is ± 3MW. Fig 8 is 60 minutes simulation trend of Load from 440MW to 400MW, then to 360MW in sliding pressure mode. Load change rate is set to 9MW/min, throttle pressure slide to 19.3MPa from 22.7MPa, the actual load and the pressure fast track the target. The maximum deviation of throttle pressure is± 0.40Mpa, the maximum deviation of load is ± 3MW. Fig 9 is 60 minutes simulation trend of Load from 500MW to 460MW, then to 420MW in sliding pressure mode. Load change rate is set to 9MW/min, throttle pressure slide to 21.6MPa from 24.0MPa, the actual load and the pressure fast track the target. The maximum deviation of throttle pressure is± 0.32Mpa, the maximum deviation of load is ± 3MW.

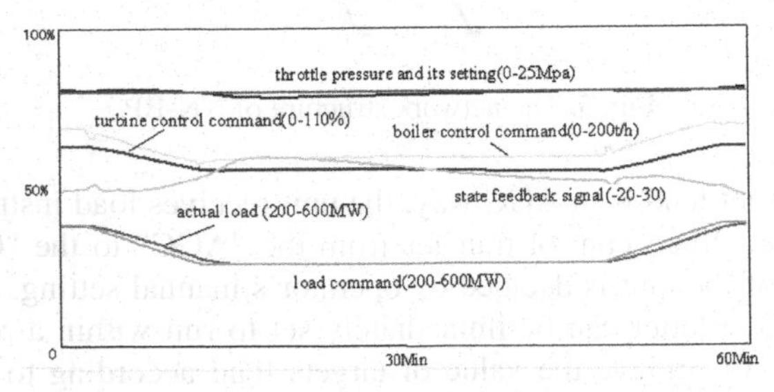

Fig. 7. The trend when load drop in constant pressure mode

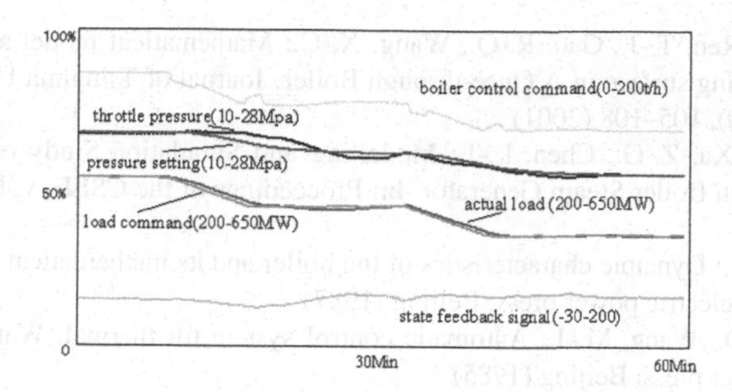

Fig. 8. The trend when load drop in sliding pressure mode

Can be seen from Fig 7, 8 and 9, the feed forward controller of BP neural network PID control has a good effect, can effectively improve the system's response rate and reduce the system overshoot.

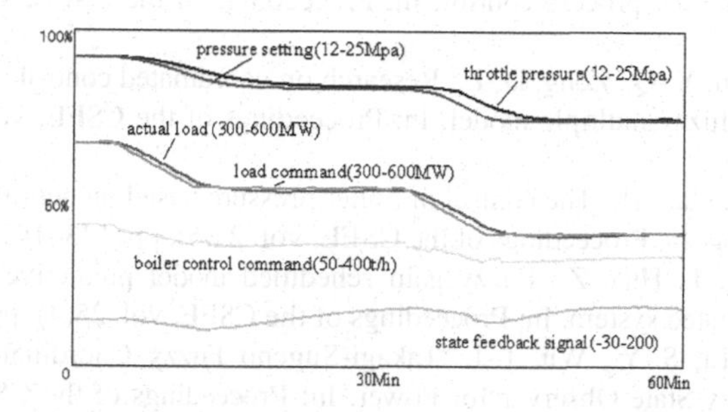

Fig. 9. The trend when load drop in sliding pressure mode

5 Conclusion

Boiler-turbine coordinated control system of thermal power unit is a time-varying, uncertainties, strong-coupling and multi-variable control system. Its control quality is a key to improve the thermal processes automation and achieve power system automatic generation control. BP neural network PID control is applied to coordinated control system. Through system parameters are reasonably set, dynamic energy and static energy can maintain a balance. Coordination control system design method is simple, but the control effect is very good. Neural network PID control system design method is simple, with the traditional advantages of PID, on-line intelligent tuning of PID controller parameters, to adapt the parameters of controlled object and the structure change and the input reference signal changes, and against external disturbance.

References

1. Wang, G.-J., Zhang, C.-Y.: A Universal Mathematical Model for Both Performance Analysis and Real-Time Simulation of Evaporation Zone in Once-through Boiler. In: Proceedings of the CSEE, vol. 1(1), pp. 61–66 (1995)

2. Wang, W., Ren, T.-J., Gao, R.-Q., Wang, X.-C.: Mathematical model and simulation of the evaporating surface in A Once-through Boiler. Journal of Tsinghua University (Sci. & Tech.) 41(10), 105–108 (2001)
3. Fan, Y.-S., Xu, Z.-G., Chen, L.-J.: Modelling and Simulation Study on a Supercritical Once-through Boiler Steam Generator. In: Proceedings of the CSEE, vol. 18(4), pp. 246–253 (1998)
4. Zhang, C.-Y.: Dynamic characteristics of the boiler and its mathematical model. Water resources and electric power press, Beijing (1987)
5. Zhang, Y.-D., Wang, M.-J.: Automatic control system for thermal. Water resources and electric power press, Beijing (1985)
6. Chai, T., Liu, H., Zhang, J.: Novel design method for the coordinated control system based on fuzzy reasoning and adaptive control and its application. In: Proceedings of the CSEE, vol. 20(4), pp. 14–18 (2000)
7. Yu, D.-R., Xu, Z.-Q.: Supercritical Unit Control Technology and Development. Thermal Power Engineering 16(2), 115–121 (2001)
8. Liu, Z.-Y., Lu, J.-H., Chen, L.-J.: Prospects of application of intelligent PID controller in power plant thermal process control. In: Proceedings of the CSEE, vol. 22(8), pp. 128–134 (2002)
9. Liu, J.-Z., Chen, Y.-Q., Zeng, D.-L.: Research on coordinated control system of a 500MW unit based on fuzzy multiple model. In: Proceedings of the CSEE, vol. 23(6), pp. 2790–2794 (2003)
10. Yu, D., Xu, Z., Guo, H.: The control of boiler pressure based on neural networks decoupling linearization. In: Proceedings of the CSEE, vol. 22(5), pp. 143–147 (2002)
11. Zhang, T., Lu, J., Hua, Z.: Fuzzy gain scheduled model predictive control for boiler-turbine coordinated system. In: Proceedings of the CSEE, vol. 25(4), pp. 158–165 (2005)
12. Luan, X.-C., Li, S.-Y., Wu, J.-J.: Takagi-Sugeno Fuzzy Coordinated Control System Based on Fuzzy State Observer for Power. In: Proceedings of the CSEE, vol. 26(4), pp. 76–81 (2006)
13. Han, Z.-X., Huang, H.-P., Li, D.: Limitations of the Narrow Fuzzy-Control and the Concept of Generalized Fuzzy-Control. In: 3rd IEEE International Conference on Cybernetics and Intelligent Systems (CIS 2008), China Chengdu, p. 1186 (2008)
14. Han, Z.-X.: Concept of Generalized Intelligent Control as well as Research and Application on Thermal Control Field. In: Proceedings of the CSEE, vol. 28(suppl.), pp. 84–90 (2008)
15. Han, Z.-X., Zhou, C.-X., Li, D., Zhang, W., Ma, H.-J., Zhang, C.-J.: Soft Measurement Technology of Coal-fired Heat and Its Application on Supercritical Generating Unit Control System. In: Proceedings of the CSEE, vol. 28(35), pp. 90–95 (2009)
16. Li, J., Yang, J.: Design of Intelligent Control Algorithm Based on NN PID Control. Automatic Measurement and Control 24(8), 300–302 (2008)
17. Han, Z.-X., Qi, X.-H., Sun, Y.: The Method Heightening Response Rate to AGC by Analog Flexible Fuzzy Pre-feed Coal Control and Its Application. Power System Technology of China 30(suppl.), 1–6 (2006)
18. Han, Z.-X., Qi, X.-H., Pan, G., Zhou, G., Xiang, T.-C., Liu, D.: A new design method of coordinate control system for once-through boiler and its engineering application. In: Proceedings of the CSEE, vol. 25(21), pp. 121–127 (2005)

Control Virtual Human with Speech Recognition and Gesture Recognition Technology

Wei Zhao[1,*], XiaoFang Xie[1], and XiangHong Yang[2]

[1] Department of Ordnance Science and Technology of NAAU, China
plazhaowei@tom.com
[2] New Equipment Training Center of NAAU, China

Abstract. In accordance with the few kinds of virtual human control device, the application of speech recognition and gesture recognition in virtual human control was researched. The development status of limited collection isolated word recognition technology was introduced, the basic principle of data glove was analyzed, its correcting and data process technology were discussed. The principle of gesture recognition based on vision and its research status was introduced in detail. The realization of control virtual human with speech recognition and gesture recognition in engineering were introduced, and got anticipated result.

Keywords: virtual human, speech recognition, gesture recognition, virtual reality, data glove.

1 Introduction

Virtual Human is an essential part of virtual training system and game, it can enhance the reality of the virtual environment, allow users access to more immersion and realism. The current research on virtual human focus on modeling and rendering, motion generation and control, autonomous navigation, expression systems, etc., have made great progress, but the virtual human control equipments is still very tiny, seriously affecting the user's operation experience.

The mainly human-computer-device used to control virtual human are keyboard, mouse, Joystick, motion capture equipment and various types of movement sensors. The game roles is often controlled with keyboard and mouse, such as "Counter Strike", "Battlefield", usually use mouse to control the direction, and use keyboard to control move forward or back. Joystick generally containing 2 to 4 axes and more than 6 buttons, two of the axes provide X, Y coordinates, which used to control the positions of the virtual human; The rotation handle often used to control the forward direction of the virtual human; an axis used as a throttle, which used to control the virtual human's movement speed; the buttons can used to control the action switch. The advanced Joystick such as Microsoft ForceFeedBack2 and WingMan Strike Force3D, have force feedback function, make the control more realistic[1] [2].

* Corresponding author.

A. Xie & X. Huang (Eds.): Advances in Electrical Engineering and Automation, AISC 139, pp. 441–446.
springerlink.com

Motion capture equipments also increasingly used for virtual human control, but they are very complexity and expensive, generally only applied in the laboratory or large-scale film studio. Several types of motion sensors appeared in recent years, such as shooting simulator for football game, etc., the operator can use a more natural way to control the virtual football players. Our laboratory has designed an immersive roaming device, operators can control the virtual human movement with natural freely walking, it's also a new type of virtual human control equipment. This paper introduces the development of speech recognition and gesture recognition technology, researches their implementation and application in virtual human controlment [3] [4].

2 Development Status

Back in the early 80s, Japan developed the Voice-Q voice answering system for Shinkansen train, which recognizes the inquiries, and use synthesis speech to response. Microsoft and IBM, provide a large number of investments every year in the speech recognition system, they already have commercial speech recognition engine; Oki, Samsung and Philips, etc., invested heavily on develop speech recognition ASIC.

Current speech recognition technology is more mature, especially small and medium vocabulary speaker-independent speech recognition system, have reached the practical requirements. With the development of the large scale integrated circuit technology, this complex speech recognition system can make special chip for mass production. There are a large number of speech recognition products in people's lives now, such as most of mobile phones already have voice dialing function.

The current gesture recognition technology is divided into data glove-based gesture recognition technology and vision-based gesture recognition technology. Limited to the current technology of data gloves, there are still many technical issues need to be resolved, such as must be calibrated before use, expensive and easily damaged. Vision-based gesture recognition technology is the focus of current research, but its theory is still not perfect, especially for dynamic gesture recognition, there are still many issues need to be resolved, such as identification accuracy, the use conditions, and so on, limited its practical application.

3 Speech Recognition Technology

Voice control has more advantages than other interactive methods, but the signal of natural language is very complex, it's difficult for the computer to recognize natural language. In recent years speech recognition technology has reached a practical stage, especially for isolated word recognition has made great progress. Isolated word recognition is the simplest and most mature speech recognition technology. The current study for the isolated word recognition, both in small or large vocabulary words, specific or non-specific persons all have reached a very high level.

According to our experience, the voice commands to control virtual human are generally not more than 100, so take a limited set of isolated word recognition technology will fullfill the requirements. We use Microsoft's Speech SDK for the development of voice recognition. Speech SDK is a COM-based development kit for voice recognition. Using the Speech SDK for development of a limited set of isolated word recognition follow these steps:①Initialize the COM library.②Create a speech recognition engine object.③Create a speech recognition context object. ④Through the message map mechanism associated the speech recognition context object with message processing function.⑤Defined a set of voice commands and grammar rules.⑥ load voice commands set and grammar rules, and active the grammar rules.

Use the Aircraft Maintenance Virtual Training System(AMVTS) project as an example, to introduce the voice control technology in virtual human controlment. This project has four control units, connected together to form a distributed simulation training system, can give training to a missile-hang-truck driver, two ground crew, and a commander, everyone has a agent in the virtual world, they work together to hanging the missiles under the aircraft wings. The commander use voice commands to control his agent, the system can deal the voice commands list in Table 1, when recognize a voice instruction, the corresponding code would sent to all federal members, and control the commander's agent to complete the corresponding action. After several tests, in the case of small noise (<40db), the recognition rate can reach 98%.If the operator accomplish voice training before use the training system, will get higher recognition rate [5] [6].

Table 1. Voice command set of supplementary training module

Voice command	code	State change
Turn left	4001	Keep turn left
Turn right	4002	Keep turn right
Forwards	4003	Move forward along the current direction
Backwards	4004	Move backwards along the current direction
Stop	4005	Stop move
Go straight	4006	Control the truck go straight
…	…	……

4 Gesture Recognition Technology

4.1 Data Glove-Based Gesture Recognition Technology

Data glove is an input device which measure the finger bending angle in real-time, some kinds of data glove can measure the hand position and attitude data. There are a variety of commercial data gloves, such as Immersion's Cyber Glove System, 5DT's Data Glove, etc., 5DT's data glove use fiber optic sensors to sense the bending of the fingers, the luminous flux will change when finger bending. The raw data of the data glove is a measure of luminous flux. However, the size of each user's fingers, bending limit are so different, different users of the same finger bending angle does not guarantee the same luminous flux, so the raw data of the data glove can not be directly used to describe the degree of finger bending, the data glove must be calibrated before use. When calibration the data gloves, the user wear the data gloves and do the following three gestures: ①fingers close together and keep straight.② clenched fist.③ Fingers open as possible. The three gestures determine the maximum, minimum bending angle and maximum ringent amplitude of data glove sensor. If you need to get the position and attitude data, you need to use a position tracking devices, you can get more details about the use of position tracking devices in [1]. System must have position reset function, when an error occurred in the position tracking device, and seriously affected the data, the operator can reset the initial position.Using the following formula to processes the raw data:

$$\varphi = \frac{RAW - RAW_{min}}{RAW_{max} - RAW_{min}}(\varphi_{max} - \varphi_{min}) + \varphi_{min}$$

φ is the fingers bending angle, φ_{max} and φ_{min} is the fingers maximum and minimum bending angle, RAW express the measured raw data of data glove, RAW_{max} and RAW_{min} is the maximum and minimum raw data [7] [8] [9] [10] [11] [12].

4.2 Vision-Based Gesture Recognition Technology

Vision-based gesture recognition system obtain video data stream from one or more cameras, the system test whether gestures appear in data stream based on interaction model, if find then split out the gesture from the video signal, then select the gesture model to analysis. Vision-based gesture recognition has the following steps:

(1) Hand gesture segmentation. Gesture segmentation is split the interested area from the gesture images. Common methods are: ①Increase restrictions, such as use homophony background color, or wear special gloves etc. ②Large gesture database methods. That is creating a hand image database that has a variety of gestures in different positions, proportion and time. ③Method based on skin color model, quickly find the hand candidate regions and narrow the scope of the follow-up testing by the skin color. ④Subtraction method and its improved algorithms, the object image subtract background image, this method has obvious effect on elimination of the background image.⑤Contour tracking method, typically a gesture segmentation model based on Snake model.

(2) Gesture modeling. Model selection depending on the application, if you want to achieve natural human-computer interaction, it must establish a precise gesture model, making the identification system make the right response to the vast majority of gestures made by the user.

(3) Gesture analysis. The mission of gesture analysis is to estimate the selected gestures model parameters, generally composed of feature detection and parameter estimation. In feature detection process, we must first locate the main body. When feature detection end, you can estimate the parameters.

(4) Gesture recognition. Gesture recognition is to classify the trajectory of the model parameter space to a subset of this space. The mainly gesture recognition technology are[13]:①Template matching technology. It matches the gesture parameters with the pre-stored template parameters, by measuring the similarity between the two to complete the recognition task[14].②Neural network technology. This technology has the self-organizing and self-learning ability, can effectively deal with noise and incomplete mode[15].③ Hidden Markov Models (HMM). HMM been

successfully used for continuous speech recognition, handwriting recognition and other fields, is the mainstream method of dynamic recognition currently, the HMM makes handling random gestures possible. ④Gesture recognition based on geometric features. Geometric feature-based gesture recognition technology use the gesture edge feature and gesture regional features as identifying characteristics, there are a variety of specific implementation approaches [14] [15].

In AMVTS project, we use vision-based gesture recognition technology for the ground crew to control they agents, as show in Figure 1.

Fig. 1. The application of vision-based gesture recognition technology in project

5 Conclusions

This article introduced the research status, basic theory of isolated word speech recognition, data glove-based gesture recognition and vision-based gesture recognition technology, researched their applications in virtual human control, verified them in the actual project. The methods introduced have useful reference and application value.

References

1. Zhao, W., Xie, X.-F.: Development of Virtual Human Technology and Its Engineering Application. Journal of System Simulation 21(17), 5473–5476 (2009) (in Chinese)
2. Adams, E., Rollings, A.: Fundamentals of game design, pp. 90–102. Pearson Education, Inc., USA (2007)
3. Zhao, W., Xie, X.-F.: Immersing Ramble Device Based on Freedom-walk. Computer Engineer. 15(095), 265–266 (2009) (in Chinese)
4. Zhao, W., Xie, X.-F., Li, D.-D.: An improved design of ramble device based on freedom-walk. In: ICIS 2010, vol. (3), pp. 73–75 (2010)
5. He, H.-Y.: Speech recognition technology and its application. Popular Science 6(80), 70–72 (2005) (in Chinese)
6. Zhao, W., Xie, X.-F., Tu, S.: Virtual soldier battle efficiency evaluation. In: ICCASM 2010, vol. (6), pp. V6-577–V6-581 (2010)
7. Zhao, W., Xie, X.-F.: Study on comfortableness of immersing virtual reality system. Computer Engineer and Design 18(033), 4251–4253 (2009)

8. Xie, X.-F., Ou, Y.-Z., Mi, Y.-L.: Virtual reality technology and its applications, pp. 153–165. Hai Chao Press, BeiJing (2002) (in Chinese)
9. Ascension Technology Corporation. Flock of Birds Installation and Operation Guide. Ascension Technology Corporation, USA (2002)
10. Fifth Dimension Technologies, 5DT Data Glove 14 User Manual (2006)
11. Liu, J.-S., Yao, Y.-X., Li, J.-G.: Virtual Manipulation Based on Data Glove in Virtual Assembly. Journal of System Simulation 16(8), 1744–1747 (2004) (in Chinese)
12. Liu, Y.-Z., Wan, G., You, X.: Virtual arm model based on data gloves and position tracker. Journal of Institute of Surveying and Mapping 22(4), 272–274 (2005) (in Chinese)
13. Sun, L.-J., Zhang, L.-C., Guo, C.-L.: Technologies of Hand Gesture Recognition Based on Vision. Computer Technology and Development 18(10), 214–216 (2008) (in Chinese)
14. Zhang, L.-G., Wu, J.-Q., Gao, W.: Gesture Recognition Based on Hausdorff Distance. Journal of Image and Graphics 7(7), 1144–1149 (2002) (in Chinese)
15. Yuan, X.-R., Chen, T.X.: Gesture recognition based on neural network. Journal of Shandong Jiaotong University 14(2), 63–66 (2006) (in Chinese)

Research of Vector Quantization Algorithm in Face Recognition System

JingJiao Li, Haipeng Li, Aixia Wang, Aiyun Yan, and Dong An

College of Information Science and Engineering,
Northeastern University,
ShenYang, China

Abstract. This thesis proposes a face recognition algorithm based on vector quantization. It precisely introduces the algorithm of vector quantization and implements the face recognition system on PC based on vector quantization algorithm. The face eigenvalue is extracted and vector quantified, and the results are regarded as the training data for Hidden Markov Model. The training target is the Hidden Markov Model. Experiments prove that this algorithm increases the speed of multiple images recognition.

Keywords: Hidden Markov Model, vector quantization, codebook, code vector.

1 Introduction

Face recognition has a great potential of application in all quarters of present-day society. This paper proposes a vector quantization algorithm which used in face recognition system. This algorithm applies image vector quantization algorithm to the image data compression which can reach the purpose of reducing processing time and improving system efficiency.[1].

VQ (Vector Quantization) is a kind of information source coding techniques which developed in the 1980s and it proposed base on information theory of Shannon. Vector quantization is divided the sample face image into a number of sub-image and each sub-image data compos a vector quantization ,And then quantified for each vector which means quantify each vector as a whole[2]. Vector quantization is an effective method of data compression which is using the vector correlation between all the components of the VQ and inhibits the signal redundancy between the signal quantization processes.

2 The Process of Vector Quantization

Suppose a vector in the n-dimensional vector space is:

$$x = (x_1, x_2, \ldots, x_N) \tag{1}$$

The element $x_i (1 \le i \le N)$ of this vector is a time-continuous real extraneous variable. To quantify the vector x is to quantify the elements it contains as a unity.

A. Xie & X. Huang (Eds.): Advances in Electrical Engineering and Automation, AISC 139, pp. 447–453.
springerlink.com

That is to say, vector quantization is a process of which the n-dimensional extraneous vector x mapping to a new n-dimensional vector y . It can be written as follows:

$$q[x] = y \tag{2}$$

In this formula, y is the vector in the vector space Y which is called a code book. Y is made up of a number of vectors whose quantity is L[3]. That is:

$$Y = \{y_i, 1 \le i \le L\} \tag{3}$$

L is the size of the code book and it is named the number of levels. The vector y_i is called the code vector. The code vector is a n-dimensional vector which can be written as follows:

$$y_i = (y_{i1}, y_{i2}, \dots, y_{iN}) \quad 1 \le i \le L \tag{4}$$

Therefore, the process of vector quantization for vector x is a process of finding the nearest code vector to x in the code book Y. There is a unique distortion measure standard to distinguish which code vector is nearest.

Before the vector quantization, the code book should be designed. To design a L-level code book Y, we need to divide the n-dimensional vector space X into L different regions. These regions are called cells which represented by C_i,

At the same time, make each cell chamber Ci and a N-dimensional vector yi linked, and look the yi as the representative of cell Ci. So vector yi which as many as L constitutes the codebook Y . The processes of designing codebook are dividing of random vector space X and selection of the representative vectors of each cell. The process of designing codebook is also known as training process[4].

When the design of codebook is completed, the random vector x can be quantized. If the vector x is in the cell Ci, vector quantizer will represent the cell's vector yi as the result of vector x's quantization, as

$$q(x) = y_i \quad x \in C_i \tag{5}$$

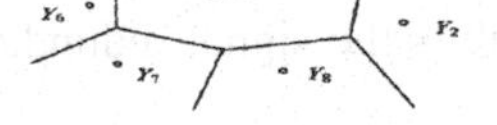

Fig. 1. Divide two dimensional vector space into eight cells

3 Codebook Design

3.1 Standard of Codebook Design

There is significant relationship among the design of codebook , the divination of the N-dimensional space and the determination of each cell's code vector. In all L level vector quantizers, the vector quantizers whose average distortion is the least is called best vector quantizer. The best vector quantizer should satisfy two requirements:

The best vector quantizer use the least distortion rule or the nearest neighbor selection rule to achicve. The nearest neighbor selection rule is expressed as follows: $q(x) = y_i$ if and only if:

$$d(x, y_i) \le d(x, y_j) \tag{6}$$

In this $j \ne i, 1 \le j \le N$. This means, only if the distortion of x and yi is smaller or equal to the distortion between any other two code vectors yi , the vector x is belong Ci and quantized as yi .

Use the the least average distortion rule choose code vector , it's the same means the cell Ci's code vector yi should make the follow formula which means average distortion smallest.

$$D_i = E[d(x, y) \mid x \in C_i] = \int_{x \in C_i} d(x, y) p(x) dx \tag{7}$$

Vector y is called the centroid of cell Ci.

In practice, there are M vectors, assume that there are Mi vectors among these M vectors are belong Ci , so this cell's average distortion is :

$$D_i = \frac{1}{M_i} \sum_{x \in C_i} d(x, y_i) \tag{8}$$

y_i is the code vector to be determined in cell C_i. According to the second principal, we should choose the vector y_i whose D_i is smallest as the codebook vector. Calculation of the centrid of the cell is determined by the definition of distortion, when distortion is defined by mean square error or weighted mean square error, the centriod of cell is:

$$y_i = \frac{1}{M_i} \sum_{x \in C_i} x \tag{9}$$

That is to say, regard the typical value of M_i training vectors in C_i as the code vector of C_i.

If wc usc thc dcfinition of Bancang-Zhaiteng distortion factor, the design procedure of the centre of figure is as follows:

First calculate the typical value of normalized autocorrelation functions of all the training vectors in cells. That is:

$$\Phi_{y_i}(k) = \frac{1}{M_i} \sum_{x \in C_i} \Phi_x(k) \quad 0 \le k \le N \tag{10}$$

In Formula (11), $\Phi_x(k)$ is the normalized autocorrelation function of all vectors, it is defined as:

$$\Phi_x(k) = \frac{R_x(k)}{R_x(0)} \tag{11}$$

In Formula (12), $R_x(k)$ is the short-time autocorrelation function of all vectors. Put $\Phi_x(k)$ into Formula (13) as a modulus and then solve it. The solution is the centre of cell C_i.

$$\sum_{l=0}^{p} a_{pl} R(k-l) = \begin{cases} \varepsilon_{\min} & k=0 \\ 0 & 1 \le k \le p \end{cases} \tag{12}$$

In the formula $a_{p0} = 1$。

3.2 Codebook Design Algorithm—LBG Algorithm Abstract the Eigenvector

Codebook design algorithm is commonly used LBG algorithm, which referred to as K means algorithm in Pattern Recognition. Where K = L, which is an iterative clustering algorithm. LBG algorithm can be used for codebook design of both consistent probability distribution of source and given training vectors[5].

The main ideas of LBG algorithm as follows: First, We should set up an initial codebook. Second, According to the principle of nearest to classification the training vector and then to compute the centroid and the smallest types of distortion of all the class. If the average of the distortion is not small enough, then using centroid to instead of the initial codebook and at the same times re-classify the training data, computing the centroid and average of distortion. After the above iterative, we continued to improve the code book until it meet the requirement or can not have been a marked improvement in. LBG algorithm is to divide the training vector into L class and make them to meet the two criteria of the optimal vector quantizer.

Take m as the number of iterations, $C_i(m)$ as the *i-th* class of the *m-th* iteration, and $y_i(m)$ as the centroid of $C_i(m)$.

Then LBG algorithm is described as followed:

Initialization: Set m = 0, and initialize original codebook $\{ y_i(0), 1 \le i \le L \}$.

Classification: According to the principal of nearest theory, divide vector X to be trained into L classes, and express with Ci(m).

Updating codebook vector: m=m+1, calculate centroids of every class of trained vector X, and replace original vectors by these centroids,

$$y_i(m) = cent[C_i(m)] \quad 1 \le i \le L \tag{13}$$

Terminative condition: if the decrease of the average distortion D(m) of m-th iteration is less than the decrease of the average distortion D(m-1) of (m-1)-th iteration, and the is more the threshold, then stop the iteration calculation; or else, return 2).

The algorithm above can always gain the local best codebook. If choose some different original codebook, and take the iteration for several times, we can get the approximative global best codebook.

3.3 Issues about Codebook Designing

Selection of original codebook

The Selection of original codebook has a lot of affection on the best codebook designing. There is many method to constructing original codebook, such as random codebook algorithm, product codebook algorithm, division algorithm and so on, here will introduce two kinds of codebook designing algorithm.

Random codebook algorithm

Random codebook algorithm is select L vectors from trained vector set X randomly to construct the original codebook. But this method is not very good, because these selected vectors are not distributed equality in the vector set, so it doesn't have any representation. This method will cause slow constringency even no constringency in the training course. But we can overcome this bug with the following method, firstly set a distortion threshold ε, and select one vector from training vector set X randomly as the first codebook vector; and then judge vectors rest in the set one by one, if the current judging vector and the first vector is less than ε, this vector can not be a codebook vector, or else, this vector can be the second codebook vector; and then judge rest vectors in the training vector set X, if the distortion of any vector and two before codebook vectors is large than ε, it can be the third codebook vector; and the like, until all the vectors is judged in the vector set X. If the number of the vectors is just L, the algorithm is over; but if the number is less than L, it should reduce ε, and calculate again; or else, add ε, and calculate again. Check the L codebook vectors gained again. If there is too few vectors in the cell where i-th vector in, delete the original codebook vector in the class set, and replace it by other vectors.

Splitting method

First of all find the training vector set X of centroid, Then find the maximum value of this centroid distortion vector Xi among the X, after that find the distortion value of the largest vector Xi which is called Xj; Next to Xi and Xj as the basis for classification, are two sub-classes Ci and Cj; These two subsets handled by the same approach, can be 4 sub-set; And so on, if $L = 2^b$ (b s an integer), to b times as long as the split sub-class of L can be obtained. This sub-class of L center of mass as the initial code words can be.

There will be a better performance original codebook which is initialed by splitting method. So there also will be a better performance optimal vector quantizer by LBG algorithm which using original codebook initialed by splitting method. But its disadvantage is that large amount of calculation.

Distortion Measure

Quantization error or distortion with d (x, y) to represent is arising by the process of random vector x to code vector y by vector quantization. To measure the distortion, one hand, taking into account the ease of calculation, on the other hand have to take into account their subjective assessment of image quality and consistent.

mean square error

MSE is the most widely used distortion measure method, defined as

$$d_2(x,y)=\frac{1}{N}\sum_{i=1}^{N}(x_i-y_i)^2 \tag{14}$$

The method is simple, which is widely used mean square error of one of the reasons.

mean r-th power error

Distortion can also be used to define the Lr norm, that is:

$$d_r(x,y)=\frac{1}{N}\sum_{i=1}^{N}|x_i-y_i|^r \tag{15}$$

When r = 2, the calculation error is the mean square error. Thus formula (17) is the Formula (16) promotion. When $r \neq 2$,There are two situations are commonly used:
The average absolute error
When r = 1, the formula (17) simplify to

$$d_1(x,y)=\frac{1}{N}\sum_{i=1}^{N}|x_i-y_i| \tag{16}$$

The formula to calculate the average absolute error is the error.

Maximum average error

When r approaches infinity, the maximum average error is defined as:

$$\lim_{r\to\infty}[d_1(x,y)]^{1/r}=\max\{|x_i-y_i|,1\le k\le N\} \tag{17}$$

Empty cell treatment

Empty cell, Appearing In the formation process of codebook, will also affect the quality of codebook. The general approach is to remove the empty cell in the yard loss, then most of the cell that contains the training vectors into two cavities, get new results.

4 Realization of Face Recognition System

The system features extracted face image vector quantization parameters, the results obtained as a hidden Markov model training data.

- Human face image feature extraction.
- On the extracted feature value vector quantization, the data obtained as a sequence of observations.

- Establish a common HMM model$\lambda = (A, B, \Pi)$ and determine the model number of states, state transition allowed and observation sequence vector size.
- Split the training data evenly,corresponding to N states, caculate the initial parameters of the model.
- Substitute Viterbi Partition for Uniform Partion, caculate the parameters's initial estimate again.
- Use Baum-Welch Algorithm to reevaluate the parameters, ajust the model parameters to maximize the observation probability.

Experiment was done on ORL face database, and each person has 10 face images . Take each 5, a total of 200 images for training, with another 200 images for recognition. Take L = 5, P = 4, for each sample window, we take 5×1 as the size of the singular value vector, and for each image, we take 108 as the length T of the observation sequence. We use 200 images that training didn't use and achieved 86.5% recognition rate, the average time to recognize an image is 1.1 s.

This paper proposes a multiple images complex background HMM face recognition algorithm based on vector quantization, the model derived feature parameters using the vector quantization process, the algorithm solves a problem which is recognizing multi pieces of images take long executive time, however, from the ideal situation needs further improvement , and remains to be carried out in the follow-up perfect.

References

1. Nefian, A.V., Hayes, M.H.: Face recognition using an Embedded HMM. In: Proeeedings of IEEE International Conferenee on Audio and Video based Biometrie Person Authentieation, vol. 12(3), pp. 19–21 (1999)
2. Fan, C.M., Namazi, N.M., Penafiel, P.B.: A new image Motion estimation algo rithm based on the EM technique. IEEE Trans. Patt. Anal. Machine Intell. 18(3), 3482352 (1996)
3. Yang, M.H., Kriegman, D., Ahuja, N.: Detecting faces in images a survey. IEEE Trans. PAMI 24(1), 34258 (2002)
4. Widjojo, W., Yow, K.C.: A color and feature-based approach to human face detection. In: 7th International Conference Control, Automation, Robotics and Vision, ICARCV 2002, vol. 1, p. 5082513 (2002)
5. Mao, X., Hu, G.: A Gradient Based Estimation Method for HMMs. Journal of Shanghai Jiaotong University 36(5), 72–74 (2002)

Design of High-Speed Electric Power Network Data Acquisition System Based on ARM and Embedded Ethernet

Anan Fang and Yu Zhang

Electronic Department of Information Engineering School,
NanChang University, NanChang, P.R. China
fanganan@ncu.edu.cn, zhangyu3722693@yahoo.com.cn

Abstract. Electric Power Network is the main artery of national economy. It plays an important role in national development.With the current rapid development of the Internet, regulating power quality relying on the network has always been an important research subject for various countries. Coverage of Internet is wide, full use of its resources make it to be able to control electricity better, and electricity get reasonable regulation during transmission according to the actual situation of power units. As the power network current, voltage, frequency, power factor can get the best match and delivery. Presently, traditional RTU is still widely used in data acquisition, whose communication adapt the way of CAN bus. However, it has been constrained by RS-232, RS-485, which results in the lower communication rates significantly. In order to completely surmount this bottleneck, it is necessary to reform the means of communication, and now through analyzing a large number of network communication, a network communication method is proposed in this paper based on embedded TCP/IP protocols. An intelligent measurement module with microprocessor runing independently is adopted to construct the core telecontrol acquisition part of SCADA system, by which the purpose of high-speed remote data acquisition in electric power system definitely.

Keywords: Electric Power Network, TCP/IP, ARM, Data Acquisition.

1 Introduction

Internet has been widely applied in the people's daily life, more and more people like to query and manage the information dependent on the Internet. However, the means of monitoring are mostly to use RS485 bus currently, the transmission speed is 10Mb/s in the 12m, and when the distance is 1200m or more, the speed only will be 100kb/s. In this case, it has been unable to support the current huge communication network data's requirements, and RS485 bus network is in the master-slave structure, so only one network terminal can send or receive data at the same time.

With the Ethernet's rapid development and widely applied, the power system data acquisition has a new way. It has strong real-time, distributed widely, simple topology, high communication speed, good compatibility.With these advantages, the prospects of

A. Xie & X. Huang (Eds.): Advances in Electrical Engineering and Automation, AISC 139, pp. 455–459.
springerlink.com

the Ethernet communication applying to power network's data acquisition is bright. Thus, High-speed Electric Power Network Data Acquisition System Based on ARM and Embedded Ethernet is designded for monitoring the voltage, current, frequency, power factor of monitoring point real-time in this article, and design the man-machine interface system through Visual C++ to display the real-time data.

2 System Design

2.1 Hardware Design

According to system requirements, achieve network for data acquisition by the ARM-based and embedded TCP/IP technology. In this paper, the collected data is concluded in ARM chip whose name is ARM920T. Its Main frequencye is 180MHz, we use H9200E-AT91RM9200 development board to complete our expriment, whose core is ARM920T, and it has 32/64 SDRAM, 4M NOR Flash.

In this paper, we use the chip RTL8169 to realize network communications part of Ethernet, this module is connected to the PC via Ethernet as data communication process, shown in Fig. 1.

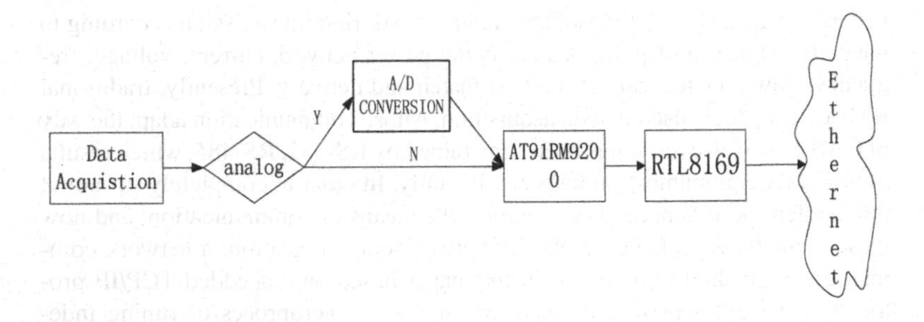

Fig. 1. System Topology Diagram

2.2 System Software Design

a. The data acquisition module

The software design is divided into data acquisition, data transfer, data packing, and communication parts. When the value of the flag for data is Y, collect the data. This is the data of the voltage acquisition module circuit. The measure of this paper is to transform voltage into frequency value and then transfer it down to the ARM processor according to the frequency's equivalent voltage value ratio. So the processor calculated it to get the instantaneous voltage.

b. TCP/IP protocol stack's realization

First of all, reduce the traditional TCP/IP protocol families, and keep the common agreement for the server, to provide the reliable network service. Then, we simplified the function and implement to improve the transmission speed and processing efficiency. We defines the Web service layer and various monitoring function interface on it, so various monitoring function definition can call these interface directly to package monitor options into Web page, no longer need to consider the realization of lower strata.

c. Communication modules

Through RJ45 interface of the RTL8169 Ethernet card chip directly connected to internet, the upper computer can be directly receiving data packets from the upper-bit computer SOKET. The simplified protocol stack is shown in Fig. 2. It call Rec_Data_Pac and Udp_Data_Process function to realize the driver and the UDP data receiving, with infinite loop calling the receive function in order to receive the waiting data.

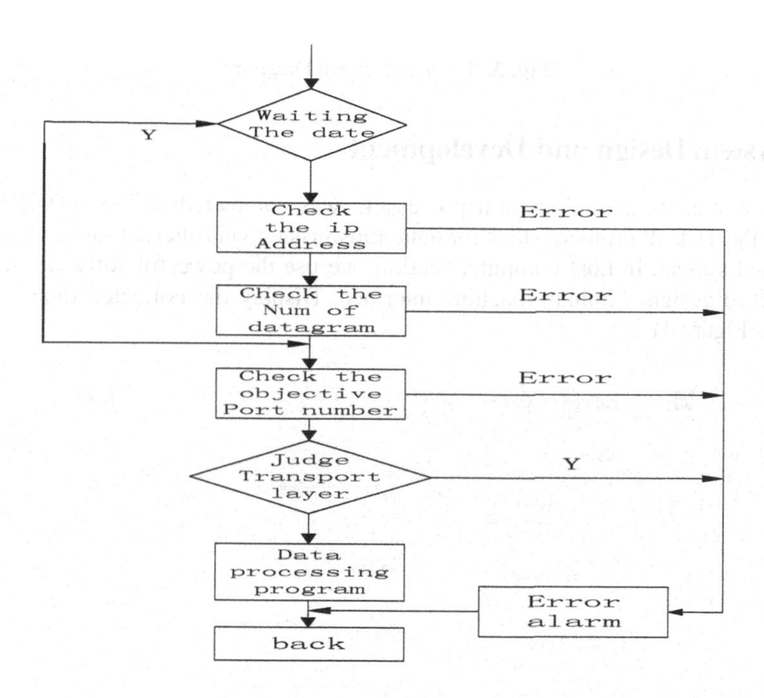

Fig. 2. The Simplified Protocol Stack System Flow

d. Man-machine interface system

In this system, man-machine interface system was designed by Visual C++, it would display the Voltage, frequency and power parameters. We just need use the function recvfrom( ) from Soket to receive the data. And then put the data into the the cache Data_Buf. The controls of interface can call the date from the cache, and display it. (software system flow chart is shown in Fig. 3)

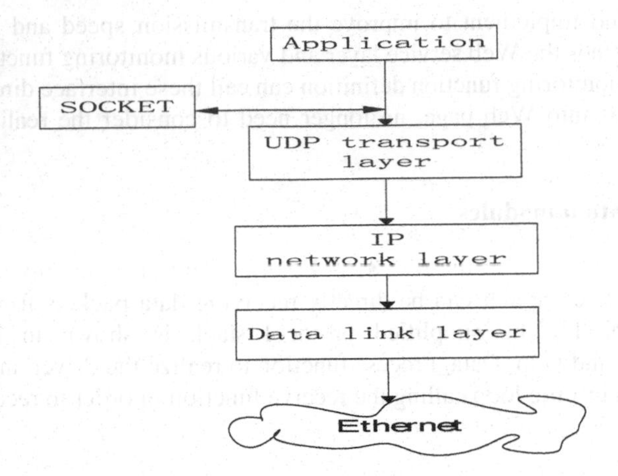

Fig. 3. Software System Diagram

3 System Design and Development

In this system, the part operation processing was completed in the AT91RM9200, realizes the TCP/IP protocol stack for data transmission via Ethernet through using the embedded system. In host computer section, we use the powerful software of Visual C++ 6.0 to design a human-machine interface. Display the collected data real-time (Such as Figure 4).

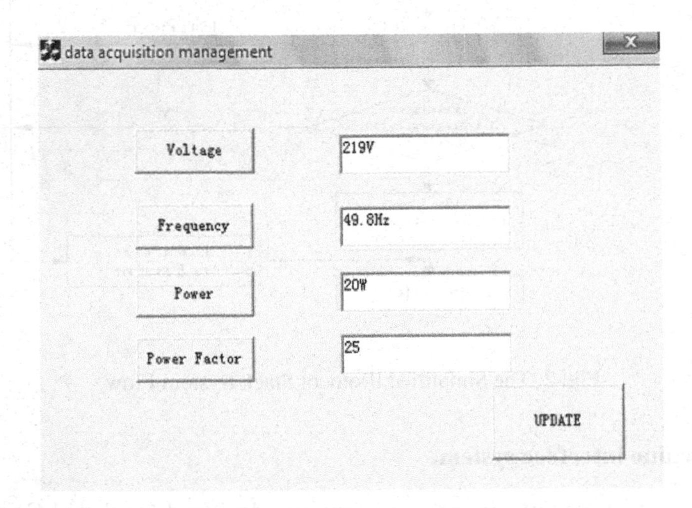

Fig. 4. Data Acquisition Management System

4 Conclusions

This paper realizes data acquisition based on TCP/IP protocol in power network, the transfer rate of network data have improved greatly for completing the TCP/IP protocol stack through embedded system. Due to network structure of Ethernet is many-to-many, it solved the problem of embarrassment from a network terminal can only receive data. It is easier for us to monitor the data, As Internet used more and more widely, we believe that the system designed in this article will be used wide, and development of it will be deeper and farther.

References

1. Hong, Y.: The Application of Embedded System Based on ARM. Microcomputer Information 32(12), 23–26 (2007)
2. Atmel Corporation, Embedded web server. AVR 460 (January 2010), http://www.atmel.com
3. Xi, B., Fang, Y.J.: Application of embedding technology in networking of serial-port devices. Electric Power Automation Equipment 27(8), 99–101 (2007)
4. Stevens, W.R.: TCP/IP Illustrated, vol. 1. Addison-Wesley, Reading (1994)

Research of On-Line Monitoring System for Malfunction in Distribution Network

Anan Fang and Khamsian Sisenglath

Electronic Department of Information Engineering School,
NanChang University, NanChang, P.R. China
fanganan@ncu.edu.cn, ssl_sian@hotmail.com

Abstract. Energy is the basis of survival and development for human society. AS the second form of energy, there are several advantages of it, such as simple, convenient, and reliabity. And it is the major energy in our modern society. Adequate and reliable electric power is the basic premise for rapid economic development and normal operation of the community. With the accelerated development of social economy, microelectronics, computers, power electronics and a variety of large-scale electrical equipment are put into use, which cause voltage fluctuations, flicker and negative sequence components. It makes power quality deterioration seriously and endangers the safety and economic operation of power system. We achieved power-line monitoring to establish communication links. And then we realized intelligence system for data acquisition, transmission, and comprehensive analysis for statistics. It prompted some countries to begin to explore the research of remote monitoring based network for sparing and making full use of network resources. And it is considered that the various parts of the power system are linked. Therefore, the authors conducted a lot of research and practice for the data collection and comprehensive analysis of LAN network. We use power line carrier technology to analyses statistical data in the network coverage area. Wire communication is successful to rate up to 100kb/s with debugging constantly. Its characteristics are not need to structure the network again. As long as wires exist, the data can be transmitted through LAN. So it will reduce the pressure of internet without reduce the rate of transmission. In long-distance transmission, we will still use internet to transmit the data.

Keywords: distribution network, power line carrier, transmission rate.

1 Introduction

With the rapid development of microelectonics, computers, power electronics technology, the wide application in industrial automation and the investment of large-scale electrical equipment, which take the serious deterioration of power quality ,the increasing impact load and grid voltage and current waveform distortion. microelectronics, computers, power electronics technology. We abuse the semiconductor rectifier technology to take the increasing non-linear load. It take serious harm for our power system operation and safety. Because of the parts of power system is linked,

A. Xie & X. Huang (Eds.): Advances in Electrical Engineering and Automation, AISC 139, pp. 461–465.
springerlink.com

every stoppage will bring huge loses. The search of power line carrier is urgent, it is a technology of automatic collection, transmission, and comprehensive analysis statistics which use microelectonics, computers, power electronics. Achieve real-time intelligent control and improve the quality of supply through intelligent control system who establish the communitcation link by network.

2 System Design

2.1 LAN Network Monitoring System Design

The system is constituted by monitoring center and fault detection terminal.It is shown in Fig 1.

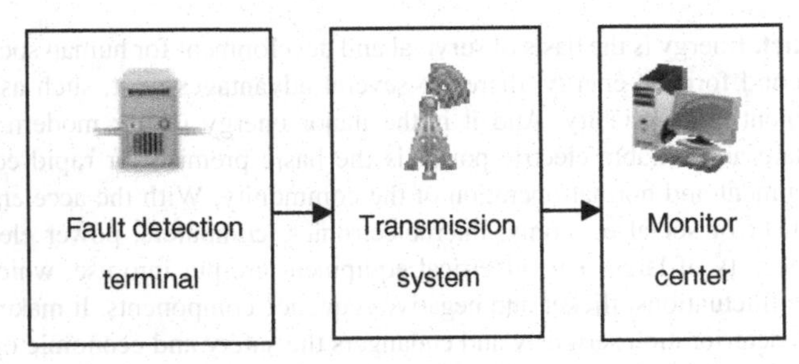

Fig. 1. Local area network system components

a. Monitor center

Monitor center is constituted monitoring software, a communications swich and a computer server. It collected and take notes of the date which passed back from the monitoring point. And then we will execute the processing according to the malfuntion. It can record the fault, release the data and auto alarm. What's more, it can display the online data on PC,and send the message to the manager to slove the problem. We can use this to tell manager the detailed content, managers can send commands to get the imformation they want.

b. The monitoring terminal distribute in distribution network

Each terminal has a monitoring PLC (Power Line Carrier) master, PLC master station transfer the collected date to GSM network area through the signal relay transmission. And it will upload all date to monitor center, it is shown in Fig 2.

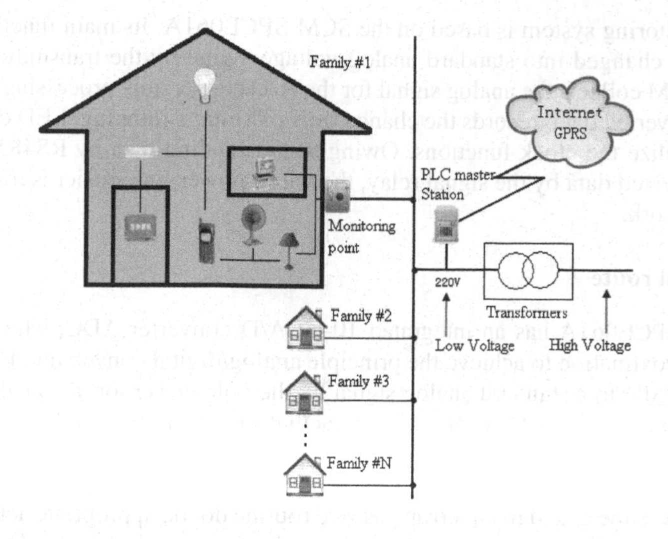

Fig. 2. The monitoring terminal

2.2 The Structure and Basic Function of Distribution Network Malfuction On-Line Monitoring System

a. System construction

Distribution network malfuction on-line monitoring system is mainly constituted by monitor, carrier communication module and SPCE061A chip.The figure of the structure is shown in Fig 3.

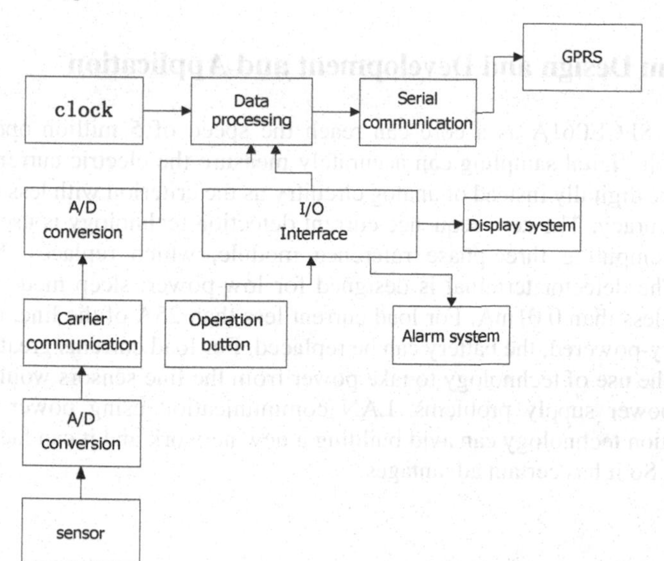

Fig. 3. System construction diagram

The monitoring system is based on the SCM SPCE061A. Its main functions are the field date is changed into standard analog voltage signal by the transmitter or sensor data; the SCM collects the analog signal for the A/D conversion, processing and storing the A/D converted data; records the change time of voltage jumping. LED displays and software realize the clock functions. Owing to the sending data by RS485 serial port and the collected data by the signal relay, the wired power line carrier is transmitted to a GSM network.

b. Technical route

The SCM SPCE061A has an integrated 10-bit A/D converter ADC, which uses successive approximation to achieve the principle analog/digital conversion. Field voltage is transformed into a standard analog signal by the voltage sensor, the analog signal is inputted from LINE_IN channel IOA0. In actual program, through 8-site continuous sampling and the average filter, the average of A/D conversion will be taken as the result., A interrupt will be requested once 0.25seconds through the overflow signal generated by Time A and the interrupt service routine do the appropriate action, such as software clock, A/D conversion, serial communication, etc. to create a low cost, high stability and strong anti-interference of the user network monitoring system. Using power line carrier communication through A / D converter as the transport system can improve the accuracy of data collection grid. Using either I2C or SPI serial interface, memory, clock calendar chip can reduce hardware connection, simplifying PCB routing, saving board space and lower system cost. Using the keyboard of management chips can reducie the burden on the software queries. Using VC ++6.0 to write monitoring software to avoid the server port is too busy and monitor the server port effectively to receive data packets, In that way, the basic problems in the network fault will be solved effectively.

3 System Design and Development and Application

Using SCM SPCE061A as a core can reach the speed of 5 million operations per second; 10-bit digital sampling can accurately measure the electric current. All measurements are digitally instead of analog circuitry as the criterion with less interference and high accuracy. The zero sequence current detection technology is used in ground detection, computing three-phase reference module, which replaces the old data detection. The detector terminal is designed for low-power, sleep mode power consumption is less than 0.01mA. For load current less than 25A of the line, using a dedicated battery-powered, the battery can be replaced. For load currents greater than 25A of the line, the use of technology to take power from the line sensors would not worry about the power supply problems. LAN communication using power line carrier communication technology can avid building a new network and it can facilitate home networking. So it has certain advantages.

4 Conclusions

Most important features of the PLC network are no need to re-erected the wired network , no need to increase the cost of setting up a special wired network. data can be passed if there is the electric wires. So it will undoubtedly become one of the best solution in data transmission of a smart home. At the same time because the data transmission is only in the family plot , applications 5 big problem bounding PLC will cease to exist. We use the traditional network to connect to the PC and then control the home appliances to realize the remote control of home appliances .PLC modem module is much lower than the cost of the wireless module. Thc binary bus between the devices transmits information through the two lines of SDA and SCL to connect with the bus. Each node on the bus has a fixed node address, according to the address to identify each device, you can easily constitute a multi-machine system and peripheral devices expanding system, the transmission rate can reach 100kb/s.

References

1. Sew, A.E., Ivan, S.: A wireless remote monitoring system: Application in the Northeast Corridor railtrack. In: 7th International Symposium on Field Measurements in Geomechanics, Boston, MA, United states, p. 93 (2007)
2. Alyosha, M., Rahul, M.: A two-layer wireless sensor network for remote sediment monitoring. In: American Society of Agricultural and Biological Engineers Annual International Meeting 2008, Providence, RI, United states, pp. 5401–5418 (2008)
3. Hu, H.: Application of Information Fusion Technology in the Remote State On-line Monitoring and Fault Diagnosing System for Power Transformer. Nanjing University of Science and Technology (2007)

Research on Improved Method in Kinematics for Humanoid Robot

QingYao Han, JianHao Di, and XiaoGuang Zhu

Department of Mechanical Engineering, North China Electric Power University, 071003, BaoDing, HeBei Province, China
jianhao777@sohu.com

Abstract. Because the kinematic equations are nonlinear, their solution is not always easy or even possible in a closed form. Also, problems about the existence of a solution and about multiple solutions arise. Based on specific structural features of humanoid robot, this paper improves the formula of inverse kinematics of the robot leg. The solution needs only once multiplication by the inverse matrix. Compared with the general numerical solutions, the solution greatly reduces the number of matrix inversion operation and the number of the excess solutions. And this method is a simple and effective, convenient and quick calculation.

Keywords: Humanoid Robot, Forward Kinematics, Inverse Kinematics.

1 Introduction

Kinematics is the science of motion that treats motion without regard to the forces which cause it. Within the science of kinematics, one studies position, velocity, acceleration, and all higher order derivatives of the position variables (with respect to time or any other variables). Hence, the study of the kinematics of robots refers to all the geometrical and time-based properties of the motion. A very basic problem in the study of mechanical manipulation is called forward kinematics (FK). This is the static geometrical problem of computing the position and orientation of the end-effector of the robot. Specifically, given a set of joint angles, the FK problem is to compute the position and orientation of the tool frame relative to the base frame. Sometimes, we think of this as changing the representation of robot position from a joint space description into a Cartesian space description. The inverse kinematics (IK) problem is that given the position and orientation of the end-effector, calculate all possible sets of joint angles that could be used to attain this given position and orientation. In some ways, IK problem is the most important element in a robot system [1]. Because the kinematic equations are nonlinear, their solution is not always easy or even possible in a closed form. Also, problems about the existence of a solution and about multiple solutions arise. This is due to the joint variables cannot be solved directly only through other joint variables, therefore, there are some issues about whether the scope of these indirect variables are appropriateness or not.

This paper focuses on the IK problem of the lower limbs of humanoid robot. In this paper, referenced the solutions that had been proposed in the literature [2] and [3], the

A. Xie & X. Huang (Eds.): Advances in Electrical Engineering and Automation, AISC 139, pp. 467–472.
springerlink.com

improved IK method involved the characteristics of analytic and numerical solutions is proposed. During the process of calculating, it only has once operation of multiplying inverse matrixes. Compared with traditional solutions, this solution greatly reduces the computation time, has explicit solution formula, and reduces the number of excess solutions. Finally, the solution is entirely reliable and is able to provide reference about data and motion by checking calculation that verified the solutions of forward and inverse in MATLAB.

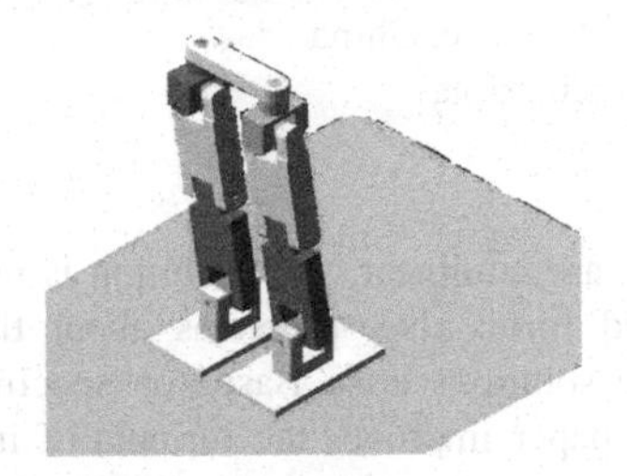

Fig. 1. Lower limb of humanoid robot

Fig. 2. Humanoid robot's kinematic diagram

2 Kinematic Modeling and Forward Kinematics

Each leg of the biped robot has six degrees of freedom, and there are 12 degrees of freedom in total. All articulated joints are axes of rotation, shown in Fig.1. In order to analysis the kinematics problem, each link is fixed a local frame by D&H convention, as demonstrated in Fig.2. Establishing the frame in this paper adopts the D&H convention that modified by John J. Craig, the professor of Stanford University. Because of the symmetry of the lower limb of humanoid robot, this paper illustrate with the right leg of humanoid robot as an example. And the link parameters are listed in the Table 1. Joint parameters, such as d_i、 a_{i-1} and α_{i-1}, are determined by the specific structure of the biped robot, and θ_i is joint variables.

Table 1. Font sizes of headings. Table captions should always be positioned *above* the tables.

	α_{i-1} [°]	a_{i-1} [mm]	d_i [mm]	θ_i [°]	Extent of joint variables[°]
1	0	0	0	θ_1	-60～60
2	-90	0	0	θ_2	-30～90
3	0	H	0	θ_3	-90～0
4	0	H	0	θ_4	-25～115
5	90	0	0	θ_5-90	-60～45
6	-90	0	0	θ_6	-60～45

According to the link parameters given in Table 1, the general form of ${}^{i-1}T_i$ (i = 1, 2~6) can be obtained as follows:

$$ {}^{i-1}T_i = \begin{bmatrix} C_i & -S_i & 0 & a_{i-1} \\ S_i C\alpha_{i-1} & C_i C\alpha_{i-1} & -S\alpha_{i-1} & -S\alpha_{i-1} d_i \\ S_i S\alpha_{i-1} & C_i S\alpha_{i-1} & C\alpha_{i-1} & C\alpha_{i-1} d_i \\ 0 & 0 & 0 & 1 \end{bmatrix}. \quad (1) $$

This transformation indicate that transforms vectors defined in {i} to their description in {i-1}. Where $\cos\theta_i$ can be abbreviated to C_i, and $\sin\theta_i$ can be abbreviated to S_i, etc (the same below). Then, the link transformations can be multiplied together to find the single transformation that relates frame {6} to frame {0}:

$$ {}^0T_6 = {}^0T_1\,{}^1T_2\,{}^2T_3\,{}^3T_4\,{}^4T_5\,{}^5T_6 = \begin{bmatrix} Nx & Ox & Ax & Px \\ Ny & Xy & Ay & Py \\ Nz & Xz & Az & Pz \\ 0 & 0 & 0 & 1 \end{bmatrix}. \quad (2) $$

This transformation is FK equation. If the robot's joint variables are known, the Cartesian position and orientation of the last link can be computed by Eq.2. It constitutes the kinematics of the robot's right leg. These are the basic equations for all kinematic analysis of the robot.

3 Kinematic Modeling and Forward Kinematics

The problem of solving the kinematic equations of a robot is a nonlinear one. Given the numerical value of ${}^0T_6$, we attempt to find values of $\theta_1 \sim \theta_6$. Consider the equations given in Eq.2. The precise statement of IK problem is as follows: given ${}^0T_6$ as twelve numeric values, to solve Eq.2 for six joint angles θ_1 through θ_6. For the case of a leg with six degrees of freedom, there are 12 equations and 6 unknowns. However, among the 9 equations arising from the rotation-matrix portion of ${}^0T_6$, only 3 are independent. These, added to the 3 equations from the position-vector portion of ${}^0T_6$, give 6 equations with six unknowns. These equations are nonlinear, transcendental equations, which can be quite difficult to solve. As with any nonlinear set of equations, we must concern ourselves with the existence of solutions, with multiple solutions, and with the method of solution.

The IK solution can be split into two broad classes: closed-form solutions and numerical solutions. Because of their iterative nature, numerical solutions generally are much slower than the corresponding closed-form solution. A sufficient condition that the robot with six revolute joints has a closed-form solution is that the last three neighboring joint axes intersect at a point. At present the relatively mature numerical method for IK problem of robotics is that both sides of the Eq.2 is premultiplied by the inverse matrix of the link transform matrix, then 5 matrix equations can be obtained. This solution is introduced by Cai Zixing in the literature [4]. Subsequently, the value of θ_1 can be solved through the equations, which only includes θ_1 in its left side. Then the value of $\theta_2, \theta_3, \theta_4, \theta_5$ and θ_6 can be solved through θ_1 from the last four equations in turn. For this IK problem of humanoid robot that contained six rotated joints, there may be 8 results. Because of the structure of humanoid robot's lower limb, the range of each joint variable is limited, therefore the value of joint variables outside the range of solutions can be directly ruled out. And then a group of satisfied solution that selected from the remaining solutions to meet the practical requirements for the robot.

This solution requires a lot of multiplications of inverse matrixes, in which have large quality of calculation. Especially after the multiplication, there are a large number of trigonometric functions with multiplications about each joint variable on the both side of the equation. The equations must be simplified at first, and then through observation both side of the equation to find right counterparts, finally the joint variables can be solved.

4 Improved Solution of Robotics Inverse Kinematics

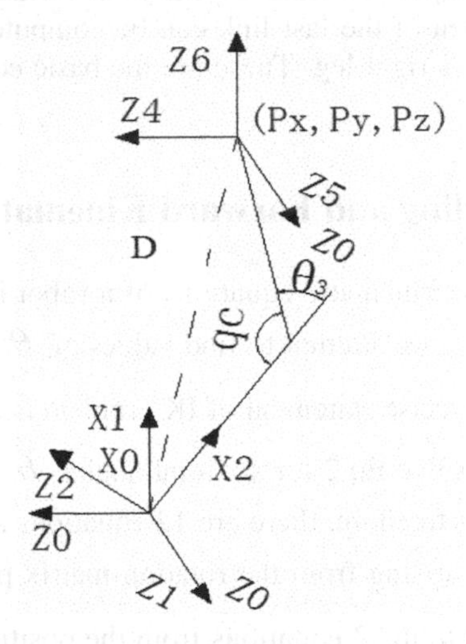

Fig. 3. Geometry associated with a robot's leg

It's been showed that the triangle formed by the links of the right leg, and the line defined as D joining the origin of frame {0} with the origin of frame {6}, and the corresponding angle of the triangle is defined by q_c, this angle is complement with the absolute value of knee's joint variable, in the Fig.3. Because the leg is planar, we can apply plane geometry directly to find a solution. The solution for θ_1 and θ_3 may be written as:

$$\theta_1 = a\tan 2(Py, Px). \tag{3}$$

$$\theta_3 = \pi - \arccos\left(\frac{2H^2 - D^2}{2H^2}\right). \tag{4}$$

Then Eq.2 is premultiplied by ${}^0T_2^{-1}(\theta_1,\theta_2)$, and postmultiplied by ${}^5T_6^{-1}(\theta_6)$. Eq.5 is obtained as follows:

$${}^0T_2^{-1}(\theta_1,\theta_2){}^0T_6\,{}^5T_6^{-1}(\theta_6) = {}^3T_4(\theta_4){}^4T_5(\theta_5). \tag{5}$$

Equating the (1, 4) elements both sides of Eq.5, θ_2 can be solved. Applying the same method one more time, θ_4, θ_5 and θ_6 can be obtained as follows:

$$\theta_2 = 2\arctan(u) = 2\arctan\left(\frac{-P_z \pm \sqrt{P_z^2 + m^2 - n^2}}{m+n}\right). \tag{6}$$

$$\theta_4 = a\tan 2(S_{34}, C_{34}) - \theta_3. \tag{7}$$

$$\theta_5 = \begin{cases} a\tan 2\left(y, \pm\sqrt{1-y^2}\right) \\ a\tan 2\left(\pm\sqrt{1-x^2}, -x\right) \end{cases}. \tag{8}$$

$$\theta_6 = 2\arctan\left(\frac{x \pm C_5}{y - S_5}\right). \tag{9}$$

Above the equations, it is showed that θ_2, θ_5 and θ_6 don't have only one solution. And the solutions need to be checked which is more in keeping with requirements.

Checking calculation of IK, firstly, assuming that 6 joint variables θ_i and link length H have been known, ${}^0T_6$ can be obtained through the FK equations. Secondly, using the IK equations can solve the joint variables reversely, and then to find the position parameters that correspond to the joint angles. Lastly, comparing the solutions with the assumed variables, the results can be obtained as follows:

Table 2. Comparison of angles and positions of FK and IK of Robot's right leg

	Assumed/ Result	Assumed/ Result	Assumed/ Result
θ_1 [°]	10/10.0	-10/-9.9981	5/5.0019
θ_2 [°]	-10/-9.9999	-10/-9.9981	-5/-5.0019
θ_3 [°]	-30/-29.9999	-30/-30.0001	-15/-15.0
θ_4 [°]	-10/-9.9992	-10/-9.9981	-10/-10.0038
θ_5 [°]	10/10.0	-10/-10.0038	10/10.0038
θ_6 [°]	0/0.0	0/-0.0001	5/4.9905
P_x [mm]	129.3190/129.3191	129.3190/129.3190	144.6391/144.6391
P_y [mm]	22.8024/22.8025	-22.8024/-22.8024	12.6543/12.6543
P_z [mm]	61.2327/61.2324	61.2327/61.2327	32.1882/32.1882

5 Conclusions

This paper focuses on the problem of humanoid robot kinematics, using the D&H convention to establish its kinematic model and calculating the FK solution. After summarizing variety solutions of the IK problem, this paper improves the conventional numerical solutions so that it can avoid much inverse matrix multiplication. Compared with the traditional geometric solutions and numerical solutions, this algorithm not only simplifies the solving process of the IK equations with multiple degrees of freedom, but also reduces the number of multiple solutions as much as possible, that can provide effectively help for learning and researching robot.

References

1. Craig, J.J.: Introduction to Robotics: Mechanics and Control, 3rd edn. Pearson Education International, London (2005)
2. Chen, P., Liu, G.: Research on Improved Algorithm of MOTOMAN-UPJ Manipulator Kinematics. Jixi Chuandong 30(4), 23–27 (2006)
3. Wang, X., Hao, J.: Derivation and solving Inverse Kinematics of MOTOMAN Manipulator. Journal of China University of Minering & Technology 30(1), 73–86 (2001)
4. Cai, Z.: Robotics, 2nd edn. Tsinghua University Press, Beijing (2009)

Performance of Creative Knowledge Work Team in Enterprises and Its Influencing Factors

Yi Liao[1,2]

[1] College of Mechanical Engineering, Chongqing University, Chongqing, P.R. China
[2] Department of Business Administration, Chongqing Education College, Chongqing, P.R. China
liaoyi2996@sina.com

Abstract. The concept of creative knowledge work team in enterprises has been put forward and characteristics of creative knowledge work team in internet environment have been analyzed. The performance of creative knowledge work team has also been defined. An input-operation-output mode has been established to investigate influencing factors of the performance of creative knowledge work team in enterprises in internet environment.

Keywords: Enterprise, Creative knowledge work team, Performance, Influencing factors.

1 Introduction

With fast development of high-techs and the consequent changes in the requirements for more diverse and personalized products, enterprises have to accelerate the process of new product development, build in more added-value to the new products and deliver improved service. Today, enterprises are exposed to a complicated, dynamic and competitive environment, which is characterised by the globalization, informatization, knowledge work and networking. The ability to deliver knowledge innovation becomes the core competitiveness of enterprises as they respond quickly to the changing marketing. As a consequence, the main types of work in the enterprises have changed from manual work to mental work and, correspondingly, knowledge workers become the main force of enterprises in the 21st century. Since most of the knowledge work and tasks could only be fulfilled by a team of knowledge workers instead of an individual, the team of knowledge workers is the true work force of the enterprises. However, to evaluate the work performance of a group of knowledge workers is more challenging than that of an individual knowledge worker and calls for systematic investigation.

There are already some researches regarding the performance of team work. In the aspect of how a team is operated, Guzzo and Shea put forward an input-process-output (IPO) model [1]. Salas et al. (1992) [2] put forward a heuristic model, which considered all possible factors influencing the performance of a team and tried to establish an integrated model. Concerning the measurement of the performance of a team, Cohen and Bailey (1997) pointed out that the performance of a team includes efficiency, productivity and responding speed of the team; quality of output; customer satisfaction; creativity, attitude and behaviour of team members [3]. However, it is rare to see any

A. Xie & X. Huang (Eds.): Advances in Electrical Engineering and Automation, AISC 139, pp. 473–477.
springerlink.com

researches on creative knowledge work team in networking environment. In this paper, the characteristics of creative knowledge work team in network environment have been analyzed, with theoretical investigation on the performance evaluation of creative knowledge work team.

2 Analysis of the Characteristics of Creative Knowledge Work Team

2.1 Definition of Creative Knowledge Work Team

With the reduction of common tasks in enterprises, the required abilities of workers become more and more dynamic and diverse. Therefore, it becomes very difficult for individual employees to complete a complicated task relying on their own knowledge. Instead, enterprises rely on team work, in which the team members are flexible, creative and from different fields of profession. Team members communicate and share knowledge through information technologies such as computers and internet/intranet. This is a knowledge work team. It is a critical issue for the enterprises to manage the knowledge work team. Unfortunately, there has not been a general definition for the knowledge work team. Here, the knowledge work team is defined as a group of knowledge workers who are from different professional backgrounds and work cooperatively to fulfil a mutual task such as design a new product or service. The R&D team in enterprises is a typical creative knowledge work team.

2.2 Characteristics of the Creative Knowledge Work Team

A creative knowledge work team is composed of knowledge workers and possesses general characteristics of a knowledge work team, namely the team members are knowledge workers and the target of the team is a creative output. However, with the development of computer science, information and network technologies, many new changes have occurred in the work process of the creative knowledge work team in enterprises. (1) The obtaining of knowledge resources has changed from a static environment (the obtaining, storing, transferring and processing of knowledge are limited) to a dynamic and open environment (there is a knowledge network with knowledge transferring between multichannel; the knowledge resources are updated continuously). (2) The communication methods has changed from the traditional face-to-face communication to high-tech-supported communications, which enable the team members to exchange information and get feed back in different place and time all over the world. (3) Team members are not limited to local regions but can be from different countries and regions in the world.

3 Performance Evaluation and Influencing Factors of Knowledge Work Team in Enterprises

3.1 Definition of the Performance of Creative Knowledge Work Team

Performance is a combination of achievement and efficiency, i.e., it describes how important the results are and how fast this process has taken. The performance of the creative knowledge work team in enterprises refers to the work efficiency and resultant achievement of a knowledge work team. The two aspects of performance support and affect each other. The efficiency of the team affects thc potential of each team member and the achievement of the whole team; correspondingly, the achievement of the team feedbacks on the control and improvement of the process.

3.2 A Model for Creative Knowledge Work Team in Enterprises

In order to study the performance of the creative knowledge work team in internet environment, the author carried out a field-research in a design-type enterprise, taking a chair design process as an example. A model was established according to the characteristic of such creative knowledge work team and theoretical analysis, as shown in Figure 1. Knowledge work team is an operation system, which follows a cyclic process of input, operation and output.

3.3 Influencing Factors

3.3.1 Influencing Factors during the Input Stage

In this stage the influencing factors include team member characteristics, team characteristics, task characteristics, supporting resources and the matching of them.

Since the fulfillment of a task relies on creative knowledge work of each team members and their cooperation between each other. Therefore, the personalities of each team member, their professions and competitiveness have effects on the performance of the knowledge work team.

Team characteristics include the difference between team members, the scale of the team and the roles each team member play in the team work. The diversity of personalities of team members determines that different arrangement of team members will have significant influence on the performance of the team. Magjuka & Baldwin (1991) [4] found that the relationship between the team characteristics and the performance of the team is more significant when the scale of the team and difference between each members are larger, and the communication is more smooth.

Task characteristics include target, type and complexity of the task. Task is the premise and foundation for the existence of a team. It is also the guideline for the arrangement of resources during the input stage. It was suggested that a diverse team will likely have best achievement when the task demands for high creativity and innovation. In addition, when the task of the team is consistent with the target of individual team members, there will be enhanced cohesion and therefore enhanced output.

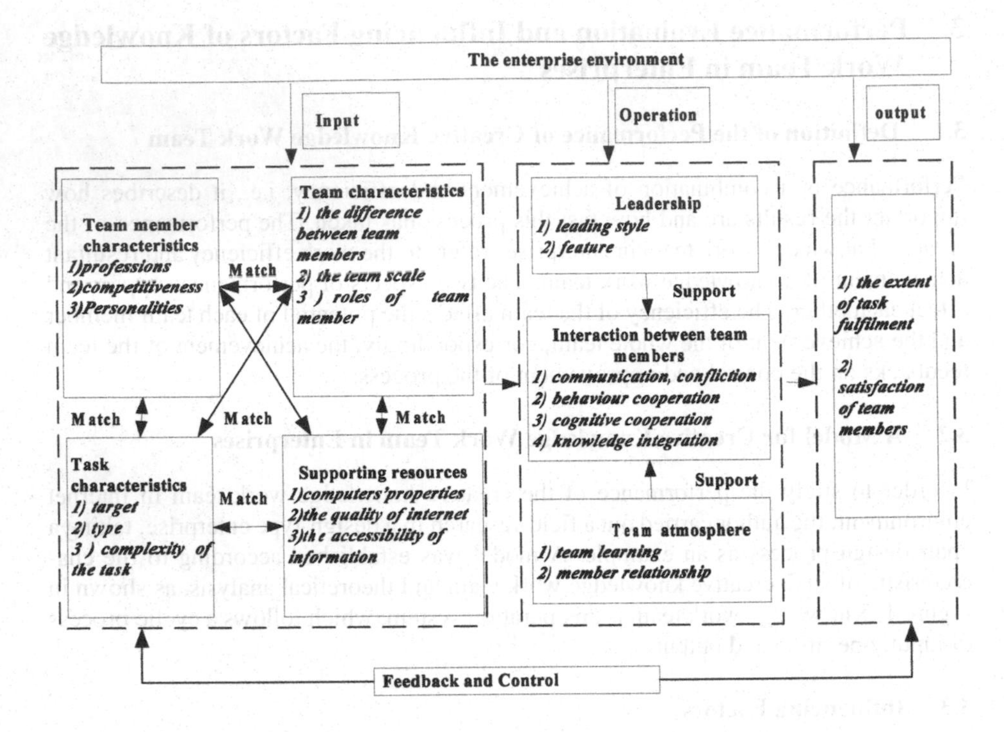

Fig. 1. Influential factors of group knowledge work efficiency

Supporting resources include the properties of computers, the quality of internet and the accessibility of information. Supporting resources directly influence the comfortableness and the efficiency of human-information system interaction. For example, well-designed hardware, user-friendly software and fast internet will highly improve the performance of the team.

3.3.2 Influencing Factors during the Operation Stage

(1) Interaction between team members. It includes communication, confliction, behaviour cooperation, cognitive cooperation and knowledge integration. The innovative nature of the knowledge work determines that team members have to communicate intensively to solve problems and to learn from each other. It was suggested that appropriate confliction between team members, especially the knowledge confliction, is beneficial to the achievement of the team [5]. Cooperation is a process that more than two components in the system work together to achieve the global properties. The essence of cooperation is to break the barriers and boundaries between components so that they can operate together to accomplish a mutual task. Behaviour cooperation refers to a process that team members can work synchronously or asynchronously, in terms of time and space, supported by cooperative tools. Behaviour cooperation enables timeliness of communication between team members and smoothness of the operation process and, therefore,

improves the efficiency of team. Cognitive cooperation means that team members are clearly aware of the target of the team. Cognitive deviation among team members can exist due to difference in mind-set, experience, knowledge background and cognitive ability of each team member. Consequently, communication and knowledge transferring are blocked and team members may have completely different understanding of the target.

(2) Team atmosphere. It includes team learning and member relationship. Learning is crucial for sharing knowledge, eliminating cognitive deviation and achieving cognitive cooperation. The achievement of a team is largely determined by the learning ability of the team. By learning, a harmonic and dynamically-balanced relationship within the team and between the team and external environment can be maintained. Jehn & Shah (1997) claimed that there is a positive correlation between relationship of team members and the performance of the team [6].

(3) Leadership. The leadership of a team is positively correlated to the cohesion and the performance of the team. The leadership is the medium connecting the team and the external environment, responsible for cooperation, integration, guidance and motivation of team members, guaranteeing the smooth operation of the team. The effectiveness of leadership mainly relies on the style and personality of the leader.

3.3.3 Influencing Factors during Output Stage

The output of a knowledge work team includes the extent of task fulfilment and satisfaction of team members. The fulfilment of the task can motivate team members since knowledge workers usually have strong sense of achievement. The satisfaction of team members is critical to the development of a team. High satisfaction encourages team learning and cooperation while low satisfaction disappoints team members and consequently leads to less achievement and low efficiency.

Acknowledgement. The author would like to gratefully acknowledge the Chongqing Education College for the support of the project under the grant number KY201125B.

References

1. Wang, F.H.: Knowledge Management. Shanxi Economy Publishing House, Shanxi (1999)
2. Salas, E., Dickison, T.L., Converse, S.A., Tannenbaum, S.I.: Toward an Understanding of Team Performance and Training. In: Swezey, R.W., Sanas, E., Ablex, N.J. (eds.) Teams: Their Training and Performance (1992)
3. Cohen, S.G., Bailey, D.E.: What Makes Teams Work: Group Effectiveness Research From the Shop Floor to The Executive Suit. Journal of management 3, 239–290 (1997)
4. The Research Process on Group Performance and Team Effectiveness, `http://cc.manaren.com/tuanduiguanli/200809/688_2.html`
5. Zhang, G., Fang, L.: Knowledge Conflict and Group Effectivenes:An Empirical Study. Science Research Management 6, 12–21 (2007)
6. Jehn, K.A., Shah, P.P.: Interpersonal Relationships and Task Performance: An Examination of Mediating Processes in Friendship and Acquaintance Groups. Journal of Personality and Social Psychology 4, 775–790 (1997)

Maneuver Target Tracking Based on Kalman IMM Algorithm

Wei Wu, Peng Cao, and Zhen Dong Pan

School of Electrical Engineering, Shenyang University of Technology,
{wuweiedu,beyondabcd,zhengdong176}@163.com

Abstract. In order to avoid tracking a maneuver detection algorithm. An interacting multiple model algorithm based on the Kalman was proposed to improve the accuracy of interacting multiple model(IMM). The target motion is divided into two parts, namely global motion and local motion, which are separately predicted according to their own dynamic behavior. Builds CA(Constant Accelerator) and CT(Constant Turn) target model of interaction to achieve the state of the target maneuver adaptive estimation. The simulation results show IMM algorithm for CV, CT maneuvering target tracking can achieve good results.

Keywords: IMM, Motive Target, CV, CT.

1 Introduction

In modern target tracking system, In the modern target tracking system, for the basic motion model doesn't change the moving object, the system state transition equation and observation equation is linear[1]. Then the linear minimum mean square error (LMMSE) , α-β filter with, Kalman filtering etc linear filtering algorithm can get good filtering effect. To achieve accurate tracking maneuvering targets, the essential problem is to make the established target motion model matches the actual target motion model[2]. The currently used how model (MM), interactive multiple model (IMM), switching model and so on. Taking into account the complexity of the algorithm, storage capacity and the demand for engineering applications, IMM algorithm has been applied widely[3].

2 IMM Filter

IMM is the principle of motion of the system as a model set of schema mapping, residuals using the output of each filter the model information and a priori information[4], rules based on a hypothesis testing, obtained corresponding to each filter model for the current moment matching model of the probability of the system (called model probability), system state estimation is the estimated probability of each model weighted fusion filter.

A. Xie & X. Huang (Eds.): Advances in Electrical Engineering and Automation, AISC 139, pp. 479–483.
springerlink.com

3 Algorithm Principle

$$\mathbf{X}(k+1)=\mathbf{\Phi}_j\mathbf{X}(k)+\mathbf{\Gamma}_j\mathbf{W}_j(k),\quad j=1,\ldots,r \tag{1}$$

Which, $\mathbf{W}_j(k)$ is mean zero, Covariance matrix for $\mathbf{Q}_j$ of the white noise sequence. With a Markov chain to control the transition between these models, Markov chain transition probability matrix is[5]

$$\mathbf{P}=\begin{bmatrix} p_{11} & \cdots & p_{1r} \\ \vdots & \ddots & \vdots \\ p_{r1} & \cdots & p_{rr} \end{bmatrix}$$

Measurement model for the:

$$\mathbf{Z}(k)=\mathbf{C}_j(k)\mathbf{X}_j(k)+\mathbf{V}_j(k) \tag{2}$$

IMM algorithm steps can be summarized as follows:

A. Corresponding to the model $M_j(k)$ $\hat{\mathbf{X}}^{oj}(k-1/k-1)$, $\mathbf{P}^{oj}(k-1/k-1)$ and $\mathbf{Z}(k)$ as input for Kalman filtering.

Prediction

$$\hat{\mathbf{X}}^j(k/k-1)=\mathbf{\Phi}_j(k,k-1)\hat{\mathbf{X}}^{0j}(k-1/k-1) \tag{3}$$

Predicting error variance array

$$\mathbf{P}^j_{\tilde{X}}(k/k-1)=\mathbf{\Phi}_j(k,k-1)\mathbf{P}^{0j}_{\tilde{X}}(k-1/k-1)\mathbf{\Phi}^T_j(k,k-1)+\mathbf{\Gamma}_j(k-1)\mathbf{Q}(k-1)\mathbf{\Gamma}^T_j(k-1) \tag{4}$$

Kalman gain

$$\mathbf{K}_j(k)=\mathbf{P}^j_{\tilde{X}}(k/k-1)\mathbf{C}^T(k)\left[\mathbf{C}(k)\mathbf{P}^j_{\tilde{X}}(k/k-1)\mathbf{C}^T(k)+\mathbf{R}(k)\right]^{-1} \tag{5}$$

filtering

$$\hat{\mathbf{X}}^j(k/k)=\hat{\mathbf{X}}^j(k/k-1)+\mathbf{K}_j(k)\left[\mathbf{Z}(k)-\mathbf{C}(k)\hat{\mathbf{X}}^j(k/k-1)\right] \tag{6}$$

A. Filtering error variance array

$$\mathbf{P}_{\tilde{X}}^{j}(k/k)=\left[\mathbf{I}-\mathbf{K}_{j}(k)\mathbf{C}(k)\right]\mathbf{P}_{\tilde{X}}^{j}(k/k-1) \tag{7}$$

B. Model probability update

$$\mu_j(k)=\frac{1}{c}\Lambda_j(k)\sum_{i=1}^{r}p_{ij}\mu_i(k-1)=\Lambda_j(k)\bar{c}_j/c \tag{8}$$

C. Output interaction

$$\hat{\mathbf{X}}(k/k)=\sum_{j=1}^{r}\hat{\mathbf{X}}^{j}(k/k)\mu_j(k) \tag{9}$$

$$\mathbf{P}(k/k)=\sum_{j=1}^{r}\mu_j(k)\left\{\mathbf{P}^{j}(k/k)+\left[\hat{\mathbf{X}}^{j}(k/k)-\hat{\mathbf{X}}(k/k)\right]\left[\hat{\mathbf{X}}^{j}(k/k)-\hat{\mathbf{X}}(k/k)\right]^{T}\right\} \tag{10}$$

D. Simulation experiment

As shown, 2 seconds of radar scan, x and y independent observations, standard deviation of measurement noise are 100 met. observation noises are 100 meters standard.

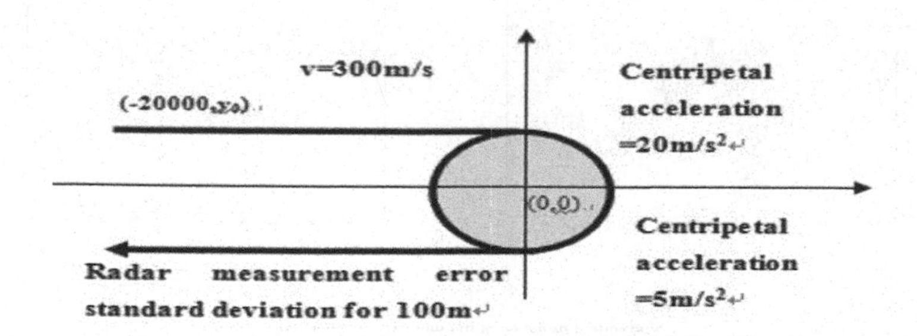

First, according to the nature of circular motion out the target of real trajectories:

$$\omega=\frac{v}{r}=\frac{300}{4500}=\frac{1}{15}\,rad/s \tag{11}$$

When centripetal acceleration (=20m/s2,

$$r = \frac{v^2}{a} = \frac{300^2}{20} = 4500\text{m} \tag{12}$$

When centripetal acceleration (=5m/s2,

$$r = \frac{v^2}{a} = \frac{300^2}{5} = 18000\text{m} \tag{13}$$

In two centripetal acceleration cases which use the above algorithm simulation experiment.

When the centripetal acceleration (=20m/s2, the following is the target trajectory curve, measured data curve, filtering the data curve and the location estimation error standard deviation curve.

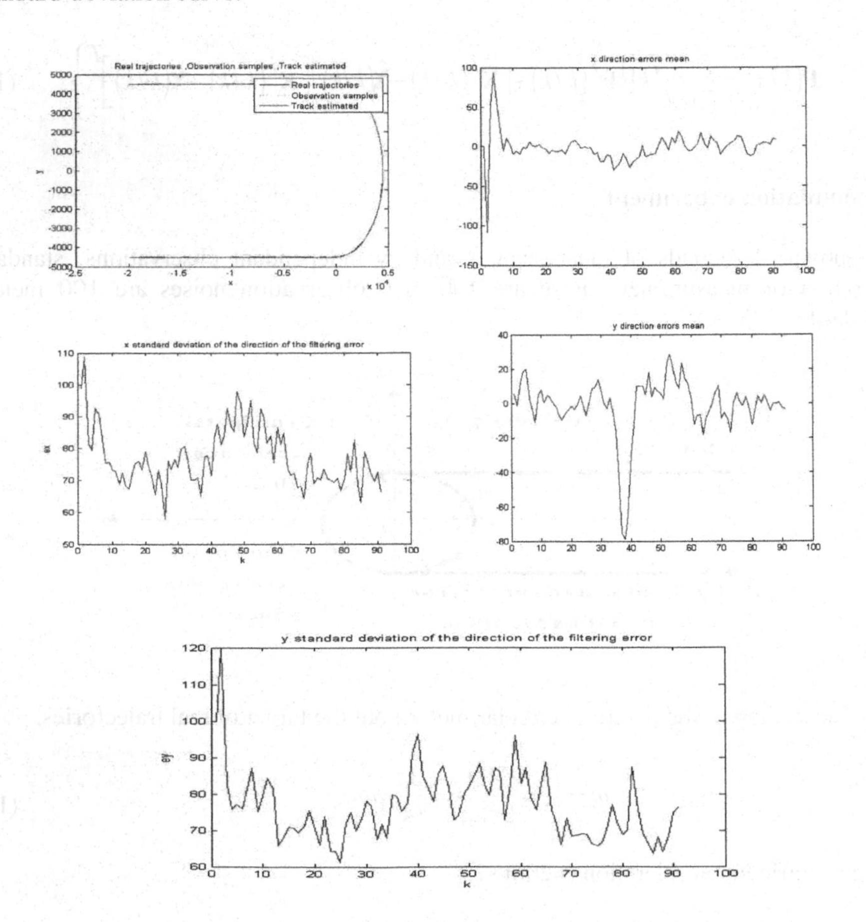

4 Conclusion

IMM algorithm given above target trajectory curve, curve measurements, filter data curve and the curve of the standard deviation of position estimation error, as can be seen from these figures, larger at the beginning of the filtering error, but time goes on, filtering error decreases rapidly, gradually approach the true trajectory estimates, when the conversion between models, will bring greater error. IMM used when maneuvering target tracking algorithm, filtering effect of controlled Markov chain model transformation of the transition probability matrix of the larger, with the initial set has a certain relationship. When the centripetal acceleration is larger, angular velocity of the larger movement, estimation error is relatively large at this time. The smaller the centripetal acceleration, smaller angular velocity of the movement, estimation error is relativeely small at this time. When the scanning time is longer, since fewer observantions, relatively low precision of the filter. When the scan time short-term changes, as more observations, filtering accuracy is also improved. Practice has proved , Interactive multiple model IMM algorithm is a good engineering practical value.

Acknowledgment. The research was supported in part by Liaoning Province Educational Foundation project under 2008501.

References

1. Hou, H.L., Zhou, D.Y.: An Optimization Algorithm for Improving Markedly the Tracking of a Maneuvering Target. Journal of Northwestern Poly Technical University 25(4), 562–565 (2007)
2. Li, R., Jilkov, V.P.: Survey of Maneuvering Target Tracking: Dynamic Models. IEEE Transaction on Aerospace and Electronic Systems 39(4) (2003)
3. Kalman, R.E.: A new approach to linear filtering and prediction problems. Trans. of the ASME Journal of Basic Engineering 82(Series D), 35–45 (1960)
4. Sehei, T.S.: Afinite-diffrence method for linearization in nonlinear estimation algoritllnls. Automatica 33(11), 2053–2058 (1997)
5. Julier, S.J., Uhlmann, J.K.: Redueed sigma point filters for the propagation of means and covariance thlough nonlinear transformations. In: Proeeedings of the IEEE Ameriean Control Conference, pp. 8–10. IEEE, Anehorage AK (2002)
6. Julier, S.J., Uhlmann, J.K., Durrant-Whyte, H.F.: A new approach for Filtering nonlinear system. In: American Control Conference Proeeedings, vol. 13, pp. 1628–1632 (1995)

The System Design of Intelligent Teaching Robot Based on Remote Control

Yi Gao[1], Xing Pan[2], and Yong Pan[2]

[1] College of Information Technical Science,
NanKai University,
TianJin, China
Gaoyi@mail.nankai.edu.cn
[2] Research and Development Department,
Qicheng Science and Technology Co., Ltd.,
TianJin, China

Abstract. This paper designed the teaching robot to be easy-assemble and module-building intelligent robot, which could automatically complete some basic functionality such as obstacle avoidance and path tracing, with the cooperation with the sensor. Take advantage of these basic functions, we could achieve a variety of seemingly complex tasks, such as obstacle avoidance, path tracing and target searching used by the fire-fighting robot. The robots are wheel driven, model founded and computer programmed. Its advantages are as follows. ARM is adopted as the core controller, and it has low consumption and outstanding performance. The embedded Linux is used as the operating system of the bottom controlling software, and its performance is stable and its real-performance is good.

Keywords: intelligent robot, remote control, ARM11, motor control, embedded system.

1 Introduction

As a frontier of information technology development, intelligent robot is a highly integrated, forward-looking, innovative practical disciplines, contains a very rich educational resources. Students could be more comprehensive and integrated understand modern industrial design, mechanical, electronics, sensors, computer hardware and software, bionic science, artificial intelligence and many other fields of advanced technology, personally experience modern technology, gain scientific knowledge, improve practical ability, develop team spirit by designing the intelligent robot as a common educational platform for robot teaching and participation in various scientific and technological innovation and competition activities, which is called as teaching robot[1][2].

Development and application of embedded systems, especially open source projects make installing and development of the operating system in the robot control system and programming the independent control software become possible, robot control program development easier and convenient to maintain, and the system more intelligent, more stable and more reliable. At present, the researches on intelligent

A. Xie & X. Huang (Eds.): Advances in Electrical Engineering and Automation, AISC 139, pp. 485–490.
springerlink.com

robots are mainly focused on the hardware platform design [3-6], the embedded software system [7、8] and relevant algorithms [9].

We designed the teaching robot to be easy-assemble and module-building intelligent robot, which could automatically complete some basic functionality such as obstacle avoidance and path tracing, with the cooperation with the sensor. The use of wheel-driven, model building and computer programming by the robot is as following:

1) Using ARM as the core controller, which makes the robot low power consuming and high performance.
2) Using embedded Linux as the operating system for the underlying control software makes the performance stable and the real-time better.
3) Modular expansion makes the secondary development easier.
4) Graphical programming interface makes the user's secondary development fast.

2 Hardware System Design

The control systems in this paper mainly include: micro-controller module, sensor module, motor drive module, wireless communication module and other parts. The frame diagram of the system structure is shown below as Figure 1.

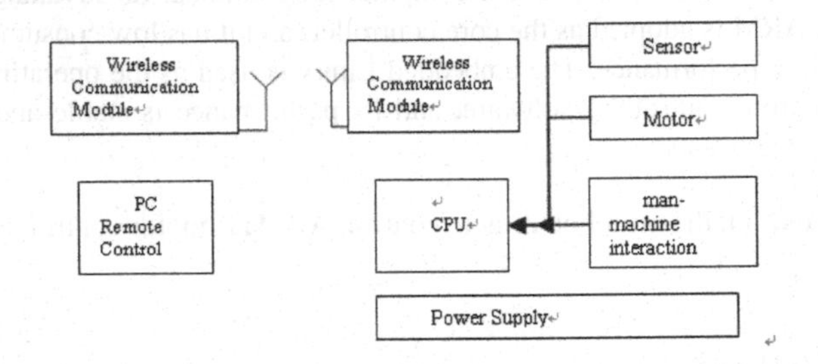

Fig. 1. Hardware System Design

As the kernel of the system, the micro-controller module mainly processes different kinds of information and date, coordinating all the modules in the system to complete their specific tasks. Sensor module is composed of ultrasonic ranging sensor, infrared sensor and other sensors. The function of sensor is detecting information of the robot's surrounding atmosphere, for example, detecting obstacles in the process of moving. Ultrasonic ranging sensor is used to detect remote obstacles while infrared sensor to detect close obstacles. The independent drive of the robot's left and right wheels is realized by the motor drive module, which is made up by motor conversion module and PWM unit installed in the micro-controller module. The differential control can be done by the motor drive module. And the wireless communication system is mainly responsible for the communication between the robot and the remote control terminal.

2.1 Micro-controller Module

The core processor of the embedded control system, S3C6410, is a low-cost, low-power, high performance microprocessor solution based on 16/32-bit RISC core. Besides, the CPU module also includes 128MB DDR RAM and 1GB NAND FLSH. DDR and RAM used are two K4X51163PE. We use K9K8G08 in NAND FLASH.

2.2 Lcd Display Module

Intelligent robot requires sophisticated human-machine interaction system. Adopted as the display module in this design, the 4.3 inches TFTLCD is used to display the image acquisition information and the condition of the robot. In the specific plan, an Innolux 4.3 inches AT043TN24 V1 LCD is used.

2.3 Electric Motor Drive Module

The L298 used in this robot can drive two motors simultaneously, control clockwise veer and anticlockwise veer of the robot, and modulate pulse width in the enable pin so that clockwise and anticlockwise turn of the wheel can be realized and the speed of the wheel can be controlled.

2.4 Sensor Module

Sensor module, an important part that enables the robot to detect information from the environment, involves many peripheral detecting types of equipment, for example, infrared photoelectric sensor, ultrasonic ranging sensor, tracking photoelectric sensor, image acquisition sensor, contact switch sensor and temperature sensor. This paper takes the temperature sensor and infrared encoder as an example.

Infrared photo electronic sensor installed on front side of the robot is a infrared sensor used to detect obstacles standing ahead which realize the function of evading obstacles automatically. The operating principle is that the infrared light transmitted by infrared light transmitting diode D1 which is controlled by I/O in the CPU is absorbed by the receiving module when there are obstacles before the robot and the information received is delivered to CPU to be processed, consequently, whether there is an obstacle standing before the robot can be easily detected. In order to reduce interference from the sun and to raise the performance and sensitivity of the receiving module, we can use infrared receiver module produced by the KODENSHI Company, whose receiving frequency range is 37.9kHz, which directly demodulates 37.9kHz modulating signal into baseband signal and delivers the signal to the CPU. The infrared photoelectric sensor can also be employed for the robot to track and to identify black and while lines of the identification. The design of the infrared photoelectric sensor module is shown in the diagram followed. The design is shown as Figure 2.

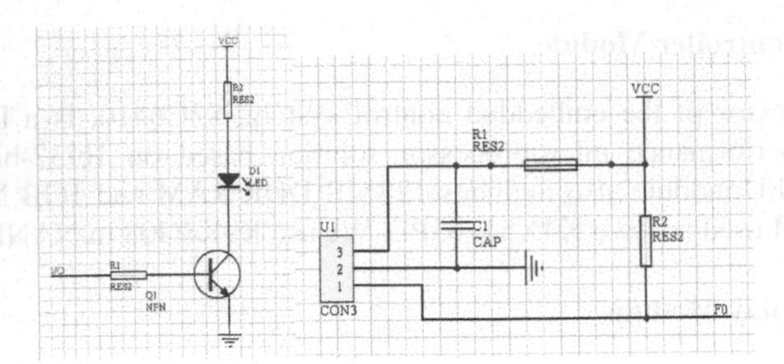

Fig. 2. Infrared photo electronic sensor

2.5 Wireless Communication Module

The principal part of the wireless communication module is nRF905, a single-chip radio frequency transceiver produced by Nordic VLSI in Norway, whose operating frequency is three ISM frequency ranges, 433/868/915 MHz3 with 11 working channels and channel switch time less than 650s. With high noise GFSK modulation and maximum data rate 100 Kbit/s, nRF905 is very convenient in configuration for it uses SPI interface to communicate with microcontroller. Its integration module comprises a fully integrated frequency modulator, a receiver with its demodulator, a power amplifier and a crystal oscillator and regulator, possessing low power consumption ShockBurst working mode, generating lead code and CRC(Cyclic Redundancy Check) automatically. The current consumption of the nRF905 is quite low. When the transmitting power is -10 dBm, the transmitting current will be 11mA and the receiving current will be 12.5 mA. The module has idle mode and off mode, which is energy efficient. The advantages of communication system nodes formed by this wireless communicating module are simple structure and low cost, which enables the application of large-scale intelligent blockade land mines system. The future of this module, low energy consumption, renders the blockade land mines used this technology applicable in wilderness over a long period of time. This module integrates reception and transmission, which further simplify the nude structure of the land mines network. The design of the wireless communicating module is shown below as Figure 3:

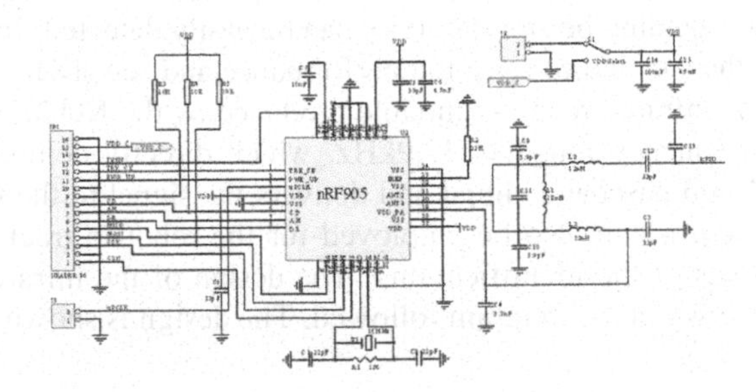

Fig. 3. Wireless Communication Module Design

3 Software System Design

At present, embedded system mainly includes: Windows CE, VxWorks, QNX, etc. The positive features of these software are good real-time, reliable system and randomness in task processing, etc. The embedded Linux, with its low price, can reduce cost substantially, thus becoming the first choice of the embedded operating system. However, applied in general operating system, it has certain technical barrier in real-time applications, as a result, the kernel of the Linux shall adjusted when the Linux is applied in embed system. Because the ability of the hard real-time interaction with the external environment is demanded by many embedded equipments, it is important to transform the Linux which primarily targets at the time-sharing system into a system that can support real-time operating systems of software.

The software system of the intelligent robot uses the embedded Linux which has went through real-time optimization as the operating system of the bottom-layer controlling software. It has stable performance as well as good real time quality. What's more, the modular structure can be expanded and further developed.

At present, design of the embedded application programmer in the embedded system is mainly the development of the real-time multi-task software based on the embedded real-time operating system. This paper is therefore based on RTOS and conforms to the process of developing a real-time multi-task software. Currently, the design tasks are: (1) seek tasks(photoelectric detection tasks)automatically (2) infrared detection tasks (3)ultrasonic detection tasks (4)image accumulation tasks (5)temperature detection tasks (6)key board scan tasks (7) liquid crystal display tasks (8) tasks of transmitting and receiving through ultrasonic (9) electric motor controlling tasks. The flow-process diagram is shown in Figure 4.

After powering up the system, we shall firstly initialize the liquid crystal, all kinds of sensors and the communicating modules. With the completion of initialization of the hardware, we shall initialize the operating system, followed by the establishment of the semaphore. In the end, we shall start multi-task management.

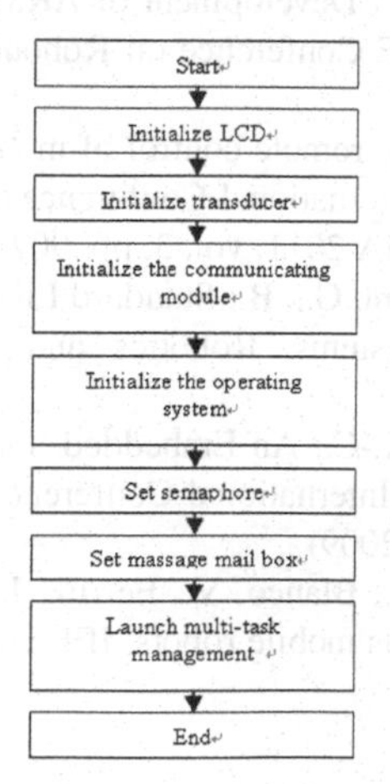

Fig. 4. The Initial flow-process diagram

4 Conclusion

The teaching robot based on ARM uses 32 bit high performances ARM CPU as the control kernel of the whole teaching robot. Taking advantage the features of the embedded processor including high performance, multiple access interface and convenience in implant and that of the embedded operating system including high real-time, high reliability and abundant resources, we use embed technology as the development platform of the teaching robot, changing the application of the technology concerning robots into re-development on the embedded platform, which shortens the time consuming in development and saves resources. The tests have proved this intelligent robot system can accurately accomplish various actions and it has good dynamic performance and controlling precision. Besides, it possess other advantages of modularity, easiness in enlargement, translatability, small hardware, low power consumption, high real-time and high reliability , etc.

References

1. Ferguson, M., Webb, N., Strzalkowski, T.N.: A low-cost social robot for research and education. In: SIGCSE 2011 - Proceedings of the 42nd ACM Technical Symposium on Computer Science Education, pp. 225–229 (2011)
2. Kang, S., Kim, J., Sohn, J., Cho, H.: Development of an experimental platform for child friendly emotional robot. In: International Conference on Control, Automation and Systems 2007, ICCAS 2007, pp. 1155–1158 (2007)
3. Rajaie, H., Zweigle, O., Häussermann, K., Käppeler, U.-P., Tamke, A., Levi, P.: Hardware design and distributed embedded control architecture of a mobile soccer robot. Mechatronics 21(2), 455–468 (2011)
4. Hu, L.J., Song, H., Zeng, X.H., Li, W.: The research of embedded techniques in control systems of service robots. Advanced Research on Information Science, Automation and Material System 219-220, 219–220 (2011)
5. Yu, S., Chen, W., Li, L., Qin, J.: Development of ARM-based Embedded System for Robot Applications. In: 2006 IEEE Conference on Robotics, Automation and Mechatronics, pp. 1–6 (December 2006)
6. Ding, C., Yan, B., Duan, P.: The remote control of mobile robot based on embedded technology. In: Proceedings - 3rd International Conference on Measuring Technology and Mechatronics Automation, ICMTMA 2011, vol. 3, pp. 907–910 (2011)
7. Bruzzone, G., Caccia, M., Ravera, G.,, B.: Standard Linux for embedded real-time robotics and manufacturing control systems. Robotics and Computer-Integrated Manufacturing 25(1), 178–190 (2009)
8. Koong, C.-S., Lai, H.-J., Lai, K.-C.: An Embedded Software Architecture for Robot with Variable Structures. In: Fourth International Conference on Frontier of Computer Science and Technology, pp. 478–484 (2009)
9. Baturone, I., Moreno-Velo, F.J., Blanco, V., Ferruz, J.: Design of embedded DSP-Based fuzzy controllers for autonomous mobile robots. IEEE Transactions on Industrial Electronics 55(2), 928–936 (2008)

Research of the Evaluation Index System of Learning City–Based on the Data of China's Four Municipalities

Jianyuan Yan, Hailing Guo, and Lei Zhang

Business School of Nankai University, Tianjin, 300071, China

Abstract. Based on the core elements of the learning city and in accordance with the design principles of science, hierarchy, operability, comparability and orientation, this paper builds a learning city index system which is made up of indicators from three levels and by using the method of combinational weight to determine weight factors of indicators, a comprehensive evaluation model of learning city has been established. The analysis of data of the four municipalities explains the practicality and operability of this evaluation index system.

Keywords: Learning city, Evaluation index system, Combinational weight method.

1 Introduction

As the new conception of city development, the construction of learning city has improved. Research on the index system of learning city has drawn more and more attention of intuitions and scholars. Through the evaluation of specific indicators, we can know the development trend and the shortage in the formation of the learning city, which is helpful for the construction of a learning city so as to be pointed.

The index system of learning city has been researched early aboard, which provides references and lessons to us. For example, the EU's TELS project developed an interactive questionnaire through investigating the performance of more than 80 cities from 18 countries in the process of constructing a learning city. Although the research went on a comprehensive study of the index system, it didn't give specified evaluation standard and evaluation method and there may be evaluation deviation of its specific utilization among different cities. In England, the evaluation index system of the learning city is built from the perspective of partnership, participation and performance, but it didn't give the evaluation criteria either. The related studies abroad are acted as guide for the construction of the index system in China, However, due to large population base, we need to improve the construction of the evaluation index system of learning city through trial based on our national situation.

Some domestic organizations and experts also give the evaluation index system of the learning city. For example, The Shanghai Ming De Institute of Learning Organization evaluates learning city construction from seven aspects respectively and it adopts 5 evaluation methods for rating criteria, but the index system is more suitable to the learning society. Research Team on Index System and Authentication of the Learning Organization of Nanjing Committee Propaganda Department evaluates the learning city by using the method of questionnaire to grade and has its measurement criteria, but the

A. Xie & X. Huang (Eds.): Advances in Electrical Engineering and Automation, AISC 139, pp. 491–497.
springerlink.com

index system includes some indexes that need a lot of social survey data which is strongly subjective. To sum up, we find that the operational research of learning city has been few studied in China, thus lacking comparability between cities. In view of this, based on the research at home and abroad, given the specific situation of China, a scientific and reasonable evaluation index system of learning city is established, which is very important to evaluate and guide the construction of learning city.

2 The Selecting Principles of Evaluation Indexes

The inadequateness of the previous studies of the index system is that the indexes are either too abstract to operationalize therefore causing data obtaining difficultly, or the coverage of the indexes is not complete thus leading to the lack of the comprehensiveness of the index system, or the index system is lack of generality and learning situations between cities can't be compared. This paper establishes evaluation index system by following such principles:

1. **Scientific.** The physical meaning of the selected indexes must be clear, from macro to micro with the indexes level by level in depth.
2. **Hierarchical.** In order to ensure the readability and understandability of the indexes, the index system should be a multi-level structure under a three-dimensional framework.
3. **Operational.** The index system is a measurement tool. Specifically speaking, in the premise of guaranteeing a real objective assessment result, the more easily to operate and enforce the more efficiency for our evaluation work.
4. **Comparative.** Each city in our country faces different situations, in order to make the index system versatile, the statistical coverage of the same evaluation index must be consistent. We use the standardized indexed as much as possible to study cities comparatively from vertical and horizontal dimensions.
5. **Oriented.** The evaluation system should play a leading role in guiding the government to pay attention to those easily overlooked areas in the process of constructing the learning city, and make full use of the advantage of the competitive aspects of the city.

3 The Establishment of Evaluation Index System

Learning city is a comprehensive concept. Our previous study has given a combined definition. Based on the definition, we extract eight central elements: lifelong learning, public participation, government driven, various forms, resource sharing, mutual cooperation, knowledge innovation and sustainable development. The central elements reflect the logical framework of the learning city's connotation from different perspectives. So we select the eight central elements as secondary indicator of cities' comprehensive evaluation.

On the basis of previous studies, viewed from learning system and learning mechanism, by reading the publications to refer to the academic theses openly distributed domestic and abroad, we collect the involved index of comprehensive evaluation about learning city, and then detail and crystallize the secondary indicators to adequately reflect the city's internal mechanism and its development. The framework of the evaluation index system of learning city is given in this study (see Table 1).

Table 1. Evaluation index system of the learning city

	Secondary indexes	Tertiary indexes	Source of indexes
Evaluation index of learning city	lifelong learning R_1	The proportion of illiterate population from aged 15 and above I_1	Li,M.R. (2006)[1]
		Average education years I_2	Li,M.R. (2006)[1]
	public participation R_2	The proportion of special education population in disabled population I_3	National development plan document
		The proportion of the total circulation times of public library in total population I_4	Wang, X.J. (2010)[2]
		Education expenditure per capita I_5	Wang, X.J. (2010)[2]
		The average number of students above high school per One hundred thousand I_6	Li,M.R. (2006)[1]
	government driven R_3	The ratio of education expenditure to GDP I_7	Li,M.R. (2006)[1]
		The ratio of science & technology expenditure to GDP I_8	Li,M.R. (2006)[1]
		The ratio of financial allocation on cultural institution to GDP I_9	Zhi, M. (2009)[3]
	various forms R_4	The proportion of the number of private vocational training institutions in total population I_{10}	Donovan, P. (2007)[4]
		The proportion of the audience of the public science lectures, exhibitions in total population I_{11}	Donovan, P. (2007)[4]
		The proportion of online education in total population I_{12}	National development plan document
	resource sharing R_5	the number of the theater per million people I_{13}	Wu, Y.H. (2007)[5]
		The amount of books of public library per capita I_{14}	Wu, Y.H. (2007)[5]
		The number of the museum per million people I_{15}	Zhi, M. (2009)[3]
		The number of cultural centers per million people I_{16}	Zhi, M. (2009)[3]
	mutual cooperation R_6	The ratio of scale industrial enterprises' R&D external expenditure to GDP I_{17}	Balézs Németh(2003)[6]
		The ratio of high schools' R&D external expenditure to GDP I_{18}	Balézs Németh(2003)[6]
		The ratio of R&D institution' R&D external expenditure to GDP I_{19}	Balézs Németh(2003)[6]
	knowledge innovation R_7	The number of domestic patent applications from authorization per million people I_{20}	Wang, X.J. (2010)[2]
		R&D personnel full-time equivalent per million people I_{21}	Zhi, M. (2009)[3]
		The ratio of the amount of contracts traded in technology market to GDP I_{22}	Wang, X.J. (2010)[2]
	sustainable development R_8	The average household disposable per urban person I_{23}	Report of CDI research
		GDP per capita I_{24}	Li,M.R. (2006)[1]
		The ratio of added value of cultural industry to GDP I_{25}	Report of CDI research
		The proportion of the high-tech product output value in total industrial output value I_{26}	Report of CDI research

4 The Model of Evaluation Index System of Learning City

In addition to establishing the index system, we need to design appropriate evaluation methods. Most of previous studies use subjective or objective evaluation separately as a measure of weight. So far, Delphi method and AHP methods are used more in subjective evaluation method, which just reflects the decision maker's preferences. As an objective evaluation method, entropy method is based on information characteristics of the sample data to determine the weight, so such subjective methods can avoid deviation caused by human factors, but it often ignores the importance of the indicators themselves, and sometimes the weights is inconsistency with the expectations.

Given that the subjective or objective evaluation method has its limit, this paper uses a combinational weight method which considers both subjective evaluation and objective evaluation. Namely, by using Delphi method to get a subjective weight and entropy method to get an objective weight, both of which constitute the combinational weight by certain organized way of formula.

4.1 Delphi Method

As an important method to determine the weight in subjective weighting methods, Delphi method takes advantage of the experience and knowledge of experts, which means each expert to make an independent judgment anonymously. Based on the integration of domestic and foreign literature, we point preliminarily out the framework of the evaluation index system of learning city. Furthermore, Delphi method is used for the further amending by sending email to a dozen experts and then we use statistical method to estimate the average weight of each index : $w_j^{'} = (w_1^{'}, w_2^{'}, \ldots, w_n^{'})$, with

$0 \le w_j^{'} \le 1, \sum_{j=1}^{n} w_j^{'} = 1.$

4.2 Entropy Method

Because entropy method reflects the utility value of information entropy, compared with Delphi and AHP methods, the weight calculated from entropy method has a higher reliability. Supposing that there are evaluation subjects of m, evaluation indexes of n and d_{ij} is the indicator value which adopts extreme method in dimensionless treatment. The decision-making information of each index can be demonstrated by entropy value:

$$e_j = -k \sum_{i=1}^{m} d_{ij} \ln d_{ij}, (j = 1,2,\ldots,n) \quad (1)$$

With $k = 1/\ln m$, k is a constant to make sure $0 \le e_j \le 1$.The weight factor of the j th index is:

$$w_j^{''} = \frac{1-e_j}{\sum_{j=1}^{n} 1-e_j} \tag{2}$$

4.3 Combinational Weight Method

In our analysis, we use a combination of both methods, aiming at avoiding the defects caused by a single method and improving the level and accuracy of the evaluation. The weight obtained by adopting the combinational weight methods is:

$$w_j = \alpha w_j^{'} + \beta w_j^{''} \tag{3}$$

Here $w_j^{'}$ is the objective weight, $w_j^{''}$ is the subjective weight, α and β express the importance of $w_j^{'}$ and $w_j^{''}$.According to the literature [7], we transfer the solution of the weight factor into the model of solving single-objective optimization problem.

$$\max\ Z = \sum_{i=1}^{m} d_i = \sum_{i=1}^{m}\sum_{j=1}^{n} d_{ij}(\alpha w_j^{'} + \beta w_j^{''}) \tag{4}$$

$$s.t.\ \alpha^2 + \beta^2 = 1 \tag{5}$$

$$\alpha\ ,\ \beta \geq 0 \tag{6}$$

The result of the optimal solution is that:

$$\overline{\alpha}^* = \sum_{i=1}^{m}\sum_{j=1}^{n} d_{ij} w_j^{'} \Big/ \sum_{i=1}^{m}\sum_{j=1}^{n} d_{ij}(w_j^{'} + w_j^{''}) \tag{7}$$

$$\overline{\beta}^* = \sum_{i=1}^{m}\sum_{j=1}^{n} d_{ij} w_j^{''} \Big/ \sum_{i=1}^{m}\sum_{j=1}^{n} d_{ij}(w_j^{'} + w_j^{''}) \tag{8}$$

Finally, the comprehensive evaluation value is:

$$d_i = \sum_{j=1}^{n} d_{ij}(\bar{\alpha}^* w_j' + \bar{\beta}^* w_j''), \quad i = 1,2,\ldots m \tag{9}$$

5 Example Analysis and Discussions

Due to space limitations, we select China's four municipalities to test the evaluation index in real data; meanwhile, we will compare their evaluation value. We obtain the data of indexes from available relative yearbooks, such as 2010 China statistical yearbook, 2010 China statistical yearbook on science and technology, 2010 China city statistical yearbook, etc.

Table 2. Weight of the tertiary indexes

Index Number	w_j'	w_j''	w_j	Index Number	w_j'	w_j''	w_j
I_1	0.0875	0.0041	0.0916	I_{14}	0.0428	0.0617	0.1044
I_2	0.0578	0.0002	0.0580	I_{15}	0.0277	0.0107	0.0384
I_3	0.0307	0.0070	0.0376	I_{16}	0.0295	0.0673	0.0967
I_4	0.0496	0.0132	0.0628	I_{17}	0.0330	0.0171	0.0501
I_5	0.0609	0.0019	0.0628	I_{18}	0.0350	0.1096	0.1445
I_6	0.0483	0.0149	0.0631	I_{19}	0.0308	0.1333	0.1641
I_7	0.0576	0.0120	0.0696	I_{20}	0.0431	0.0324	0.0754
I_8	0.0432	0.0451	0.0883	I_{21}	0.0361	0.0347	0.0698
I_9	0.0420	0.0094	0.0513	I_{22}	0.0362	0.0677	0.1038
I_{10}	0.0282	0.0014	0.0297	I_{23}	0.0323	0,0169	0.0492
I_{11}	0.0248	0.0215	0.0462	I_{24}	0.0231	0.0179	0.0409
I_{12}	0.0263	0.1574	0.1837	I_{25}	0.0260	0.0892	0.1152
I_{13}	0.0270	0.0341	0.0611	I_{26}	0.0216	0.0200	0.0416

By Eq.9, we get the comprehensive evaluation value of Beijing, Tianjin, Shanghai and Chongqing is (45.37, 14.50, 28.12, 12.00), from which we find that the development level of learning city of Beijing is much higher than Tianjin and Chongqing.

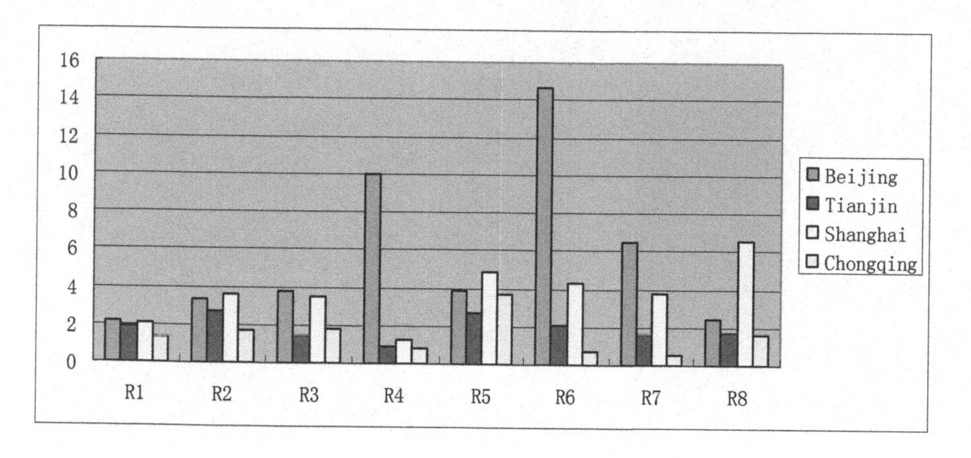

Fig. 1. The development of sub-indexes of learning city about municipalities in 2009

With further analysis comparing of the central elements (see Fig.1), the result shows that compared with other municipalities, Beijing has obvious advantages in mutual cooperation, various forms and knowledge innovation. Sustainable development of Shanghai develops well and its high level of economic development and health &sustainable development has been fully validated. As for Tianjin, in addition to slow development of these above aspects, the government driven and resource sharing are also the weak points, the two core elements of index value rank last in the four municipalities. From the analysis of the result, we should strengthen the weak links and make use of their advantage. At the same time, in the process of constructing learning city, we can learn from other cities of their advantages.

Acknowledgement. We thank the Social Science Foundation of Tianjin, China (No.TJGL10-862) for its financial support.

References

1. Li, M.R.: Evaluation of Economic Competitiveness about Province. Management World 5, 52–61 (2006)
2. Wang, X.J., Chen, W.: Regional Intellectual Capital and Regional Innovation Capability. Journal of Industrial Engineering & Engineering Management 3, 1–6 (2010)
3. Zhi, M.: The Study City Construction Research of Jinan. Shandong Normal University, Jinan (2009)
4. Donovan, P., Andrew, L.: The Learning City in a 'Planet of Slums'. Studies in Continuing Education 29, 37–50 (2007)
5. Wu, Y.H., Cai, B., Gu, X.: Research on Construction Performance Evaluation of the Western Learning City. Science and Technology Management Research 9, 112–114 (2007)
6. Németh, B.: CONFINTEA VI Follow-up and the Role of University Lifelong Learning: Some Issues for European Higher Education. International Review of Education 5 (2011)
7. Fan, Z.P., Zhao, X.: An Objective and Subjective Synthetic Approach to Determine Weights for Multiple Attribute Decision Making. Journal of Decision Making and Decision Support Systems 4, 87–91 (1997)

A Project Post-Occupancy Evaluation System and Knowledge Base Model Based on Knowledge Management

Cunbin Li[1], Si Wu[1], Ruhang Xu[1], and Fan Yang[2]

[1] School of Economics and Management, North China Electric Power University, 102206, Beijing, China
[2] School of Economics and Management, Yibin University, Yibin, China

Abstract. Knowledge management has been playing an increasingly significant role in modern management campaign in that it can efficiently manage to create, transfer, maintain and evaluate knowledge within organizations. By analyzing the problem of lacking support knowledge for decision-making in current project management, this paper put forward a thought of introducing knowledge management into project management. Initially, the significance of building timely updating decision knowledge system and then project post-occupancy evaluation is discussed. Next, on the basis of the special features of project management as well as the discussion above, this article solves the efficiency and time effectiveness of knowledge by establishing a novel post-occupancy evaluation system which is also an automatic self-fresh-circulation. Finally, a fussy model of information management system, namely a knowledge base application model, to make this whole theory real is proposed.

Keywords: post-occupancy evaluation, knowledge base, information management system, knowledge management, project management.

1 Introduction

Nowadays, the ever increasing application of computers and the Internet leads project managers into the overfull Information Age, in which numerous design and control support systems emerge now and then. These novel information systems are able to meet users' different demands at anytime and anywhere because of their amazing capacity to integrate and automatically deal with information, and thus dramatically accelerate the development of highly effective management systems such as rapid decision and project cluster management.

Nevertheless, several current information systems that are widely used in project management usually focus on providing support for explicit knowledge, such as network scheduling plan and network resource plan, while the decision support for specific project in accordance with its special features is in great need. Moreover, in order to build a decision base which concentrate on multi-goals within a specific realm and can constantly provide decision support, a timely updated knowledge base is a necessity. Fortunately, project post-occupancy evaluation is a potential and formal resource of knowledge in project management.

A. Xie & X. Huang (Eds.): Advances in Electrical Engineering and Automation, AISC 139, pp. 499–505.
springerlink.com

This paper put concentrations on how to exploit a set of knowledge creating system which meets the requirement of knowledge management by utilizing the practice process of realistic activities and incorporating the working mechanism of project management. Moreover, a comprehensive knowledge base and its implementation method of informationization are established on the foundation of above knowledge creating system.

2 Research Actuality

The study on Knowledge Management had an early start overseas. Many specialists in management, such as Peter Drucker and Paul Strassmann, have done great contributions in this field for their study put much emphasis on the rapidly increasing importance of information and tacit knowledge as precious enterprise resources[1],[2]. Peter Senge paid more attention on managing the cultural factors of knowledge. A Japanese scholar, named Ikujiro Nonaka, proposed a SECI model which included socialization, externalization, combination and internalization to make mutual conversion between tacit knowledge and explicit knowledge. Moreover, several relative models were successively proposed, such as KMMM for measuring the consistency and effectiveness of knowledge management and CKMM for enlarging manage performance and enhancing organization capacity. All in all, researches on Knowledge Management abroad usually focus on the application, transmission and saving of knowledge as well as on the design of its processes and models[3]. However, specific methods of knowledge creating is rare to see.

In China, the study on Knowledge Management started relatively late and has been in a primary stage which always focus on the design of Knowledge Management Information System. For instance, JiangCuiqing discussed the performance evaluation of Knowledge Management and put forward relative methods[4]. KeXianda considered the way to ensure the consistency of knowledge evolution and also proposed corresponding model[5]. All in all, researches on Knowledge Management In China are theoretical and lack real experience in project management field.

3 Significance of Building Timely Updating Decision Knowledge System and Project Post-Occupancy Evaluation System

The building of timely updating decision knowledge system exerts great influence on supporting decision goals. For one thing, managers usually make decisions in accordance with their own decision trees, which means they will lack shares knowledge. However, once the decision knowledge system is constructed, they can obtain useful decision knowledge from it efficiently in the process of project management. For another, the decision knowledge system must be timely updated. The reason why project management is often in shortage of knowledge is that the rapidly developing of domestic economy leads to increasing utilize of project management methods. Therefore, a sound decision knowledge system must keep abreast with current economic development.

The building of project post-occupancy evaluation system exerts great influence on timely updating decision knowledge system. On the one hand, theoretical methods of management usually generate from academic platform. However, knowledge emerging from managing practice find no efficient way to flow into this platform. Hence, an inner-open timely updating decision knowledge system is needed to connect practice experience with academic knowledge and construct a well-functioning knowledge flowing system. Its whole structure is shown below in Fig. 1.

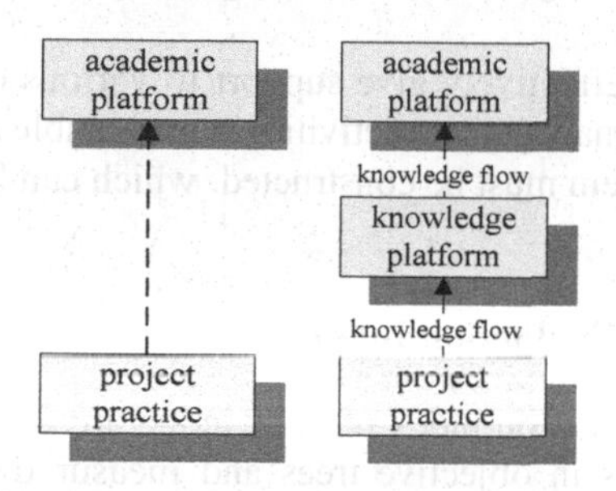

Fig. 1. The function of knowledge platform

On the other hand, suppose that the process of knowledge emerging with human activities is a process of knowledge creating, and that every individual branch of knowledge developing, namely developing from abstract and initial knowledge to specific and reasonable one, can be considered as PDCA process of creating knowledge. At the moment, the knowledge flowing is retarded in realistic managing activities, and thus knowledge cluster cannot enter the PDCA cycle. Meanwhile, an effective C(Check) stage is vacant, which also results in break-off of PDCA cycle. Consequently, a project post-occupancy evaluation system is needed to solve these problems.

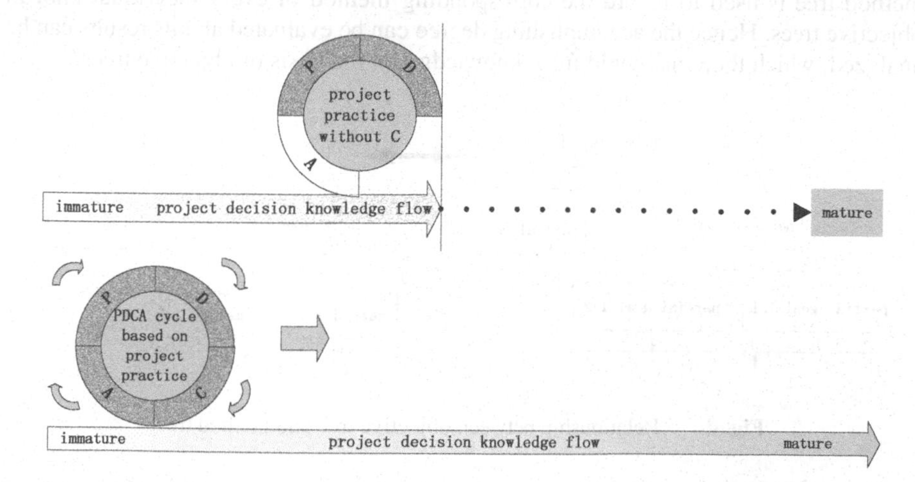

Fig. 2. Comparison between project practice without and with stage C

As presented in Fig. 2, when lacking stage C(Check), the PDCA cycle cannot function effectively. Nevertheless, once a project post-occupancy evaluation system is proposed to complement the cycle, the decision knowledge flowing system is correspondingly perfected. Moreover, this system is an automatic working system and can respond well to the knowledge demand of up-to-date project management practice.

4 The Design of Project Post-Occupancy Evaluation System

As discussed above, to effectively give support to various decision goals in different decision trees in project management activities, a reasonable and sound timely updating decision knowledge system must be constructed, which can be done through managing decision knowledge.

4.1 Post-Occupancy Evaluation Index

The post-occupancy evaluation index of knowledge are defined as the accomplishing degree of separate knots in objective trees and measured by quantitative index. By doing this, results of factor analysis of decision objectives and goals can be concluded, which then will yield useful knowledge back for decision objectives.

4.2 Post-Occupancy Evaluation Method

As stated before, factor analysis will be carried according to post-occupancy evaluation index. Therefore, to make it work successfully, this paper adopts an improved logical framework method which is also on the foundation of objective trees. The improvement is illustrated as below: On the one hand, the effect analysis of output and input is transferred into the effect analysis of accomplishing degree of separate knots in objective trees. Meanwhile, relative outcomes should be recorded. On the other hand, method tree is used to record the corresponding method of every individual knot in objective trees. Hence the accomplishing degree can be evaluated and its results can be analyzed, which then may yield fresh knowledge on the basis of objective trees.

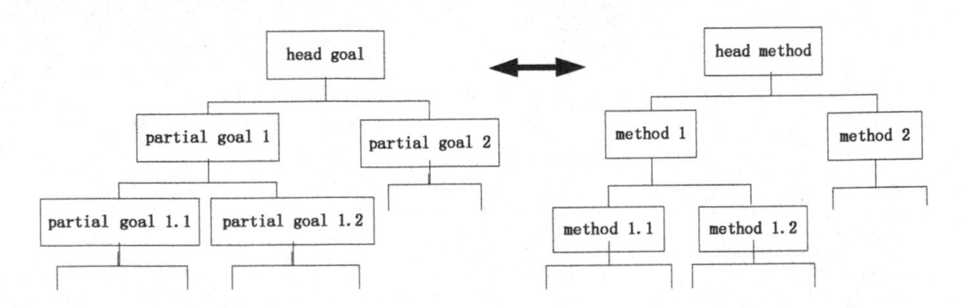

Fig. 4. Relationship between objective tree and method tree

As illustrated in Fig. 4, the dialectical relationship between objective tree and method tree is that method tree comes from human production practice under the guidance of objective tree, and conversely objective tree depends on method tree to reach its goals and may vary with method tree. Here, on the basis of objective tree and method tree, by utilizing factor analysis matrix, we can reach some outcomes.

Table 1. Factor analysis matrix

	goal-related index	Index verification	relative factors	factor analysis outcomes
head goal	corresponding quantitative index		relative factors that affect goals	necessary changes
partial goal 1	corresponding quantitative index	Verifying the accomplishing degree of goals	relative factors that affect goals	necessary changes
method 1		Verifying the accomplishing degree of goals	relative factors that affect goals	necessary changes

As shown in Table 1, a comprehensive analysis of objective tree can be carried in this matrix in that the accomplishing degree of every individual knot in objective trees can help evaluate the accomplishing degree of head goal and head method.

4.3 Post-Occupancy Evaluation Outcomes

As in Fig. 5, initially, the overall objective tree will be altered, which means that a new objective tree will emerge or the structure of the former objective tree will be different. Secondly, knots of objective trees will be altered, which means that a fresh knot will replace the old one or the content of the old knot will be shifted. Thirdly, the corresponding method of every knot will be altered, which means that novel knowledge will come into being in an old method or a new method will be created.

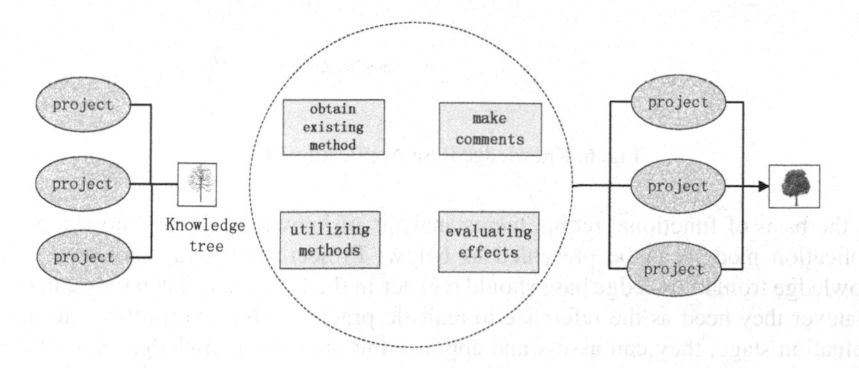

Fig. 5. Working process of post-occupancy evaluation system

5 The Establishing of Knowledge Base Based on Information Technology

5.1 Functional Requirements Analysis of Knowledge Base

First of all, the requirements analysis of managers who are in charge of project management includes the function of registration and administration, the function of saving and supervising decision tree and method tree, the function of supporting post-occupancy evaluation, and thc function of automatically creating trees. Secondly, the requirements analysis of administrations in specific realms includes the function of knowledge converge and recommendation. Finally, the requirements analysis of system performance includes timeliness of information and system security.

5.2 Application Model of Knowledge Base

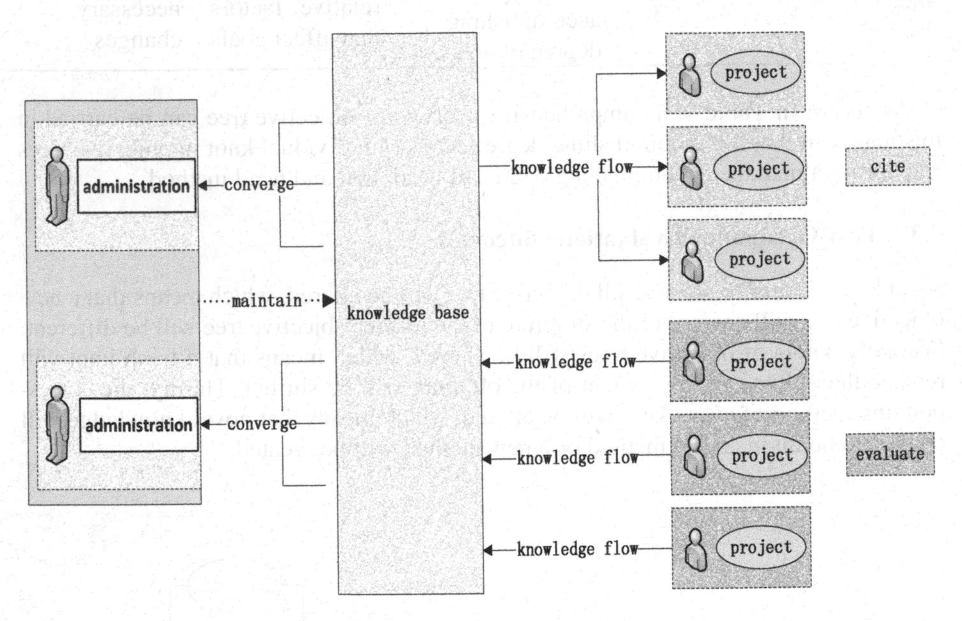

Fig. 6. Knowledge Base Application Model

On the basis of functional requirements analysis of knowledge base, knowledge base application model can be presented as below: Project managers who need to cite knowledge from knowledge base should register in the first place. Then they can utilize whatever they need as the reference to realistic practice. Moreover, after entering the evaluation stage, they can assess and appraise the obtained knowledge in accordance with their practice experience and results, and also send feedback to knowledge base. Next, administration of specific industry can acquire converged overall knowledge information from the knowledge base to get to know the developing tendency in this field. What's more, these administrations also play a significant role in maintaining the post-occupancy evaluation mechanism and guiding the developing trend. Finally, knowledge base system will analyze and deal with the knowledge that flows in, automatically consummate and perfect the corresponding knowledge tree structure, and output required knowledge when receiving requests. This process is presented in Fig. 6.

6 Conclusion

Under the situation of ever-changing methods of project management, obtaining the most current decision-support knowledge becomes increasingly crucial to managers. Only by adopting sound managing methods and avoiding low-efficiency mistakes while further processing the obtained knowledge, can novel and innovative methods have the opportunity to be put forward. As for these problems, this article establishes a creative post-occupancy evaluation system, which is an automatic self-fresh-circulation, to solve the efficiency and time effectiveness of knowledge. Moreover, a fussy model of information management system, namely a knowledge base on the basis of information technology, to make this whole theory real is proposed, and its functional requirements analysis, application model and overall conceptual structure are all illustrated.

Acknowledgements. This research is supported by the National Natural Science Foundation of China under Grant 71071054 and "211 Project" Foundation of North china Electric Power University. The authors also gratefully acknowledge the helpful comments and suggestions of the reviewers, which have improved the presentation.

References

1. Drucker, P.: Examining the theoretical inspirations of amanagement guru. Management Decision (1999)
2. Koenig, M.E.D.: Intellectual capital and how to leverage it. The Bottom Line: Managing Library Finances, P112–P118 (1997)
3. Loermans, J.: Synergizing the learning organization and knowledge management. Journal of Knowledge Management, P285–P294 (2002)
4. Jiang, C., Ye, C.: Performance evaluation on knowledge management based on knowledge circulation process. Journal of Harbin Institute of Technology (2009)
5. Ke, X., Wang, Y.: Framework for Ontology Evolution in Knowledge Management System. Journal of Computer Engineering (2009)

Management Risks and Solutions in New Energy Enterprise

Si Wu, Ruhang Xu, Cunbin Li, and Fan Yang

School of Economics and Management, North China Electric Power University, 102206, Beijing, China

Abstract. Nowadays, the new energy power generation industry has been in a status of rapid development. Every individual regional branch of power generation groups is always new project-oriented organization whose management exists various imperfections and risks. In this paper, current management situation of new energy power generation enterprises is analyzed at first, and then two major risks are pointed out. Finally, corresponding measure which includes two modules based on web technology, namely the task-flow management and workflow management is proposed to address the risks. The concept and methodology of these two management modules can well serve as reference to the information management of new energy power generation enterprises.

Keywords: management risk, new energy, power generation enterprise, information system, system architecture.

1 Introduction

The Chinese government attaches great importance to the research and development of renewable energy. At the moment, five power generation groups in China have all set up special organizations to develop new energy. Although they differ in the degree of development, their general tendency is promising. Generally, their developing pattern is as follow[1]: power generation groups first establish new energy sub-company, and then these sub-companies set up regional branches in different areas subsequently. The regional branches are project-oriented organizations, in which headquarters are responsible for preparatory work while construction and operation department are built to deal with the construction and operation of projects. Hence, it is sheer possible that management risks will occur.

2 Analysis of Management Risks

The objective characteristics of new energy project management result in two risks. On the one hand, management procedures are numerous and complicated. Due to a great number of projects, the matrix management mode is mainly adopted, which leads to individual's low and inefficient coordination[2]. It means that employees always serve as intersection of various management procedures and can be easily affected by interference of different procedures. Its specific behaviors are such as follow: Departments shirk their duties to each other to ensure their own responsibility in a relatively low

A. Xie & X. Huang (Eds.): Advances in Electrical Engineering and Automation, AISC 139, pp. 507–513.
springerlink.com

level, therefore to avoid criticisms from supervisors. Changes in policy are frequent, which is due to the uncertainty of situations, inaccuracy of leaders' decisions and drawbacks of matrix management mode, namely dual commands. These behaviors bring about two questions to deal with. The first is how to clearly define job roles and establish performance management system. The second is how to set up an effective task-flow management system.

On the other hand, management implementation is restricted by areas for the wide regional ranges of projects. For instance, if managers can only participate in local projects, procedures of other projects involved will be delayed or influenced. Thus, to enhance the management level and achieve cross-boundary management, modern information technologies can be well utilized and information transmission and management system can be established. However, these measures then bring about another two questions to handle. First of all, the security issue of information processing, transmission and storage is basic. For example, how to identify procedure participants, and how to guarantee the information not to be intercepted, tampered or deleted. Second, process control is of great importance, which requires effective automatic control system of workflow to automatically control and prompt the launch, transfer and settlement of various works.

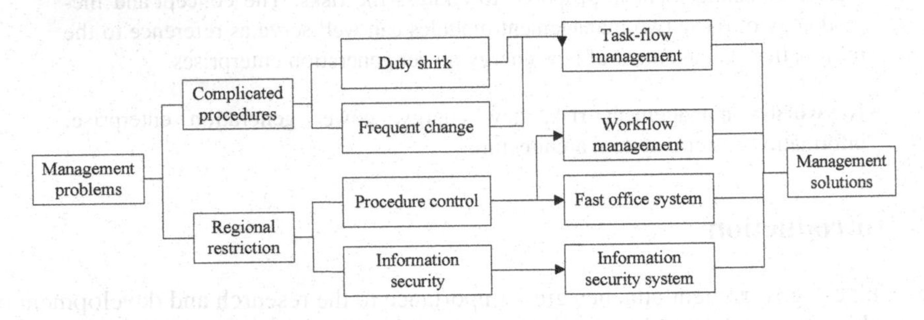

Fig. 1. Main structure of the solution

3 Analysis of Solutions

According to the risks mentioned above and integrating with real practice, information management solution based on web is proposed which includes two objectives and corresponding function system, as presented in Fig.1 below.

3.1 Objective of Task-Flow Management and Its System

The basic objective of task-flow management is to establish effective task-flow transmission system which can reduce task conflicts and loss and eliminate chaos in task management. In accordance with related management principles, there are four sub-goals. First, find out the core workflow which is on the basis of reasonable work range segmentation. Second, build the single task transmission mechanism from command source to receiver, by which to ensure the only upper command source for each receiver. Third, to realize closed-loop management and singularity of task, command source is in charge of task management, such as assignment, modification, revoke and review of tasks, while command receiver is responsible for implementation and feedback. Finally, achieve the automatic management of task launch, transmission and prompt.

3.2 Objective of Workflow Control and Its System

The fundamental objective of workflow management is to handle the flow and transfer of tasks automatically, which means encapsulating the complicated workflow network and presenting a succinct operation platform for users[3]. In fact, the essential contradiction is between the complexity of workflow and the efficiency of project management. According to reality, there are six sub-goals. First, privileged staff have right to launch relevant workflow. Second, individual can deal with the task and receive related prompt when tasks arrive. Third, launchers and stakeholders of the workflow are able to see the results and get corresponding prompts when the flow and transfer of tasks finish. Fourth, all participants in the workflow are capable of obtaining necessary data mining information. Fifth, launchers have right to manage the workflow. Finally, associated staff can respond to any alteration of the workflow.

3.3 Objective of Information Security and Its Assurance System [4]

The main objective of information security is to guarantee the authenticity, confidentiality and integrity of information. Based on real practice, the information security system consists of three parts, namely the hardware, software and safety regime. Hardware includes server, hardware firewall, and so on. Software can be divided into underlying protocols and applications. At last, safety regime includes safety training for users and administration, as well as safe operation rules.

3.4 Objective of Fast Office and SMS Office System

By fast office, it means that staff can express their approval comments in the easiest and fastest way[5]. Therefore, considering the reality, the most effective and practical method would be using the SMS to approve the workflow with low security requirement, while adopting the Internet and other identity authentication system to approve the workflow with high security requirement. Consequently, following sub-goals are demanded. First, build encoding system of workflow to allow SMS office system to discern the workflow in business office system. Second, establish the approval framework of SMS system to be capable of receiving and analyzing the response SMS, handling relevant tasks, and giving prompts to related staff. As for approval process with high security requirement, electronic signature with security technology and encrypted Internet transmission channel can be used to meet the demand of workflow information security.

4 Analysis of Design Patterns

4.1 Task-Flow System Pattern

The aim of task-flow system is to accomplish the target of task collaboration management. It is needed to work out the core task-flow of the company, and build core task tree map based on a full understanding of the company's present task-flow, as shown in Fig.2 . The aim of core task-flow is to restrict the company's regular managing pattern. It is set to guide the assignment of task between upper and lower in the organization based on the seriousness and mandatory of a management information system. Core task is made according to an effective human source position responsibility direction in a real working circumstance. It is possible that a position has more than one upper in specific circumstance, but there would be only one upper in a core task-flow system. On the basis of core task tree, the architecture of the management information system which is divided into function pattern and technical realization can be built.

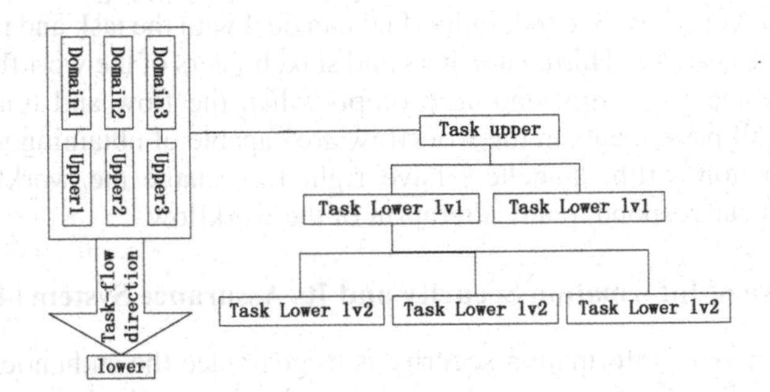

Fig. 2. Core task tree map

In a function pattern, there are two kind of operator, one is task upper and another is task lower. The relation of upper and lower is according to the abstraction of core task tree map. Uppers are responsible to launch a task, define the properties and manage his/her task. Launching task is defined as an action of registering a task in the system, and making a related assignment. Defining task properties is to specify the time interval, and content of the task. Management of one's own task contains the work of updating, revocation and review. Lowers are responsible to accept task from his/her upper. They are required to acquire the information and make response according to the time line. This function pattern has three advances: restriction of the task-flow pattern, and reduction of the interference from double command; reduction of changes in one task. Changes are limited to one task-flow; It is possible to fulfil close-loop management and promote efficiency through the task feedback mechanism.

When it comes to technical realization, the concept model of database is stated below in Table 1. The table of task record the life time of a task by recording some basic information, in which column taskState is used to control the flow of the task. The table of task_relation record a related person to a task in which taskSuper and taskLower record upper and lower of one task respectively, and taskId is used as a foreign key to make relation to table task. Based on the principle of object-oriented design, some classes were designed to fulfill the system mechanism. In task management, class task Business is designed to accomplish the logic work, including low and transmission of tasks and system analysis. Class taskDal is designed to provide data for the module of task by providing basic data for business layer through action.

Table 1. Concept model of database in task-flow

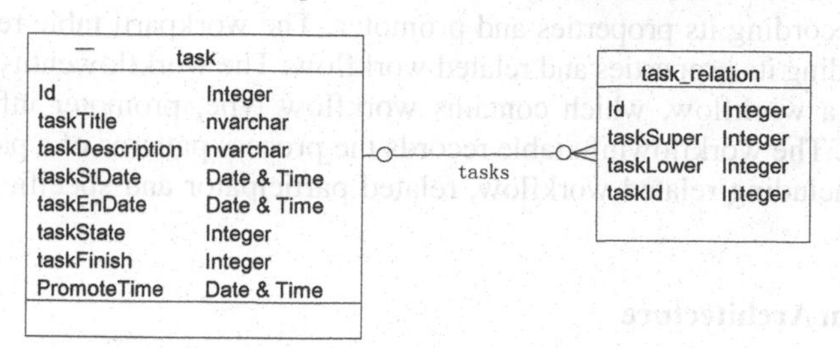

4.2 Workflow System Pattern

The aim of workflow system is to handle the complex work in a company to make an arranged system. This system seals its complex internal work logic and presents arranged work schedules to the users.

Table 2. Concept model of database in workflow

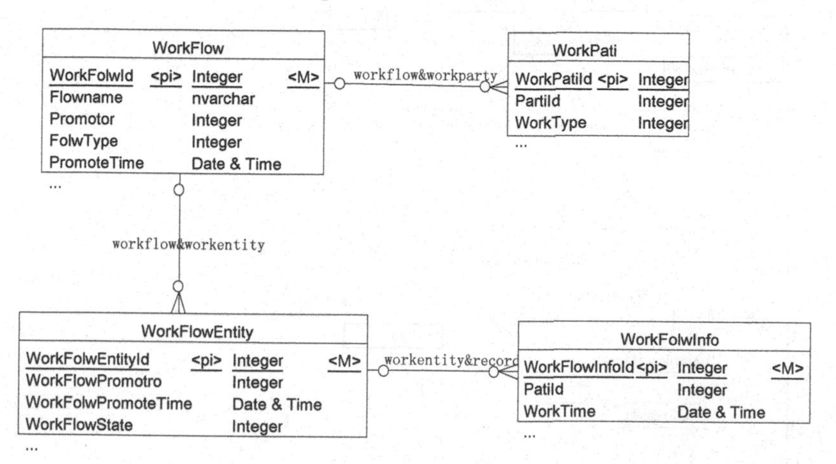

To realize this system, first, it is needed to optimize present work process. The aim of the optimization is to separate the most complex and repeatable from work system and leave the work to computers, and reorganize the work process. On the basis of work process optimization, it is needed to analysis the workflow properties, thus making an abstract model of workflow by defining the participants, directions and contents.

For function pattern, the design should base on the real circumstance of the company; the model is illustrated in Table 2. The participants of the system should be divided into process promoter and process participators. Process promoter promotes a workflow, and defines the properties of it. Then the system will transmit the work to an aim participator(s). Participators will handle the work and feedback some information. After this, the system will automatically do another transition. The system will automatically analysis the state of the workflow, and notice related participator(s). When the workflow terminate, the system will do a finishing job, and realize close-loop management. This system also provides custom work process definition function. It provides an user-friendly data collecting module through JavaScript.

When it comes to technical realization, the table of workflow can register a new workflow, recording its properties and promoter. The workparti table records participators, including its properties and related workflow. The workflowentity table records a launch of a workflow, which contains workflow type, promoter information and present state. The workflowinfo table records the process patterns of a participator to a workflow, including related workflow, related participator and specific propertied of the workflow.

4.3 System Architecture

Based on the principles of object, the whole architecture of the system using the software of power designer is proposed in Fig.3. The architecture generally considered the requirements of Weak coupling and Code reuse.

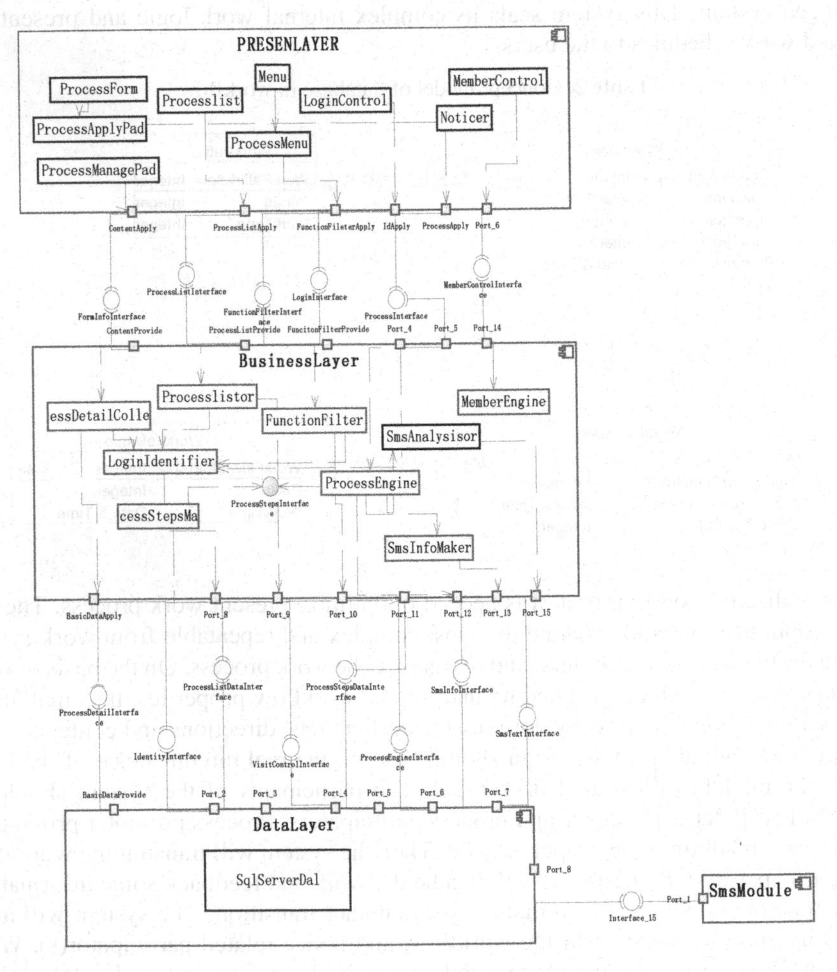

Fig. 3. Whole architecture of the system

5 Conclusion

On the basis of web technology, and integrating with the technology of information security and SMS, this paper propose a solution of developing task-flow management, workflow management, and corresponding SMS safe and fast office, which can, therefore, effectively handle the risks of complicated procedure and cross-boundary office.

Acknowledgements. This research is supported by the National Natural Science Foundation of China under Grant 71071054 and "211 Project" Foundation of North china Electric Power University. The authors also gratefully acknowledge the helpful comments and suggestions of the reviewers, which have improved the presentation.

References

1. Yang, B., Yan, K., Jiang, J.: The collaborative workflow model for web service oriented architecture. Computer Engineering and Design (3), 44–47 (2011)
2. Fy, D., Wang, Q.: The application of XML digital signature in workflow System. Computer Applications (3), 78–81 (2011)
3. Wang, H.: The design and implementation of grid workflow engine. Computer Engineering and Design (02), 65–68 (2011)
4. Fy, D., Wang, Q.: The application of XML digital signature in workflow System. Computer Applications (3), 78–81 (2011)
5. Wang, H.: The design and implementation of grid workflow engine. Computer Engineering and Design (02), 65–68 (2011)

Author Index